T0181883

Vibrations and Stability

Jon Juel Thomsen

Vibrations and Stability

Advanced Theory, Analysis, and Tools

Third Edition

With 193 Figures

 Springer

research. Necessarily incomplete in covering a very wide and active field, its purpose remains to try cover some of the gap between a typical elementary-level vibration course and interesting and relevant real vibration problems in engineering science and applications.

The Second Edition (2003) was a revised and expanded version of the first edition published by McGraw-Hill in 1997, reflecting the experience gathered during its six years in service as a classroom or self-study text for students and researchers. The second edition added a major new chapter (Chap. 7), three new appendices, many new exercise problems, more than 120 new and updated bibliographic references, and hundreds of minor updates, corrections, and clarifications. Also, the subtitle was changed to better reflect the scope and intended readership of the book. In deciding what to change or include, I have drew partly on a large number of inputs from students and other readers to improve it as a textbook and partly on a selfish wish to make it a self-contained handbook for use in my own work with vibration problems. Examples of the latter are the inclusion of lists of expressions of natural frequencies and mode shapes for basic structures such as rods, beams, plates, and membranes; tables of constants for engineering materials (elasticity, density, friction); and mathematical topics such as stability criteria for Mathieu's equation and the Routh–Hurwitz stability criterion.

Why Learn About Nonlinear Vibrations? Many do not bother with this question at all, once they caught by the delights and horrors of nonlinear phenomena. However, there are sound reasons for bothering engineers-to-be with this knowledge:

- Real systems are nonlinear. Nevertheless, we attempt linearizing whenever possible, because linear theory is well established and rather straightforward. However, when we linearize, we also lose essential information. Thus, students should be capable of recognizing nonlinear mechanisms and understand their possible significance.
- Nonlinearities may account for significant deviations between experimental observations and linear model predictions. Students should be trained to recognize significant nonlinear phenomena as these show up in experiments and computer simulations.
- Using nonlinear computer models without the analyst, possessing a firm theoretical background is meaningless, at best. One cannot, as with linear systems, obtain a general 'feel' for the dynamics of a nonlinear system simply by running the model with a few sets of parameters. There can be multiple solution branches, extreme sensitivity to initial conditions, discontinuities in response, special high-frequency effects, and several other non-trivial effects. Output from computers (and experiments) is always viewed in the light of the theoretical knowledge and expectations. Thus, students should know what to expect from nonlinear systems and be well prepared for the unexpected.
- Nonlinearities may *qualitatively* alter the response of a system. Students should know the universal nonlinear bifurcations.

- Nonlinearities may *quantitatively* alter the response of a system. Students should know the most common methods of perturbation analysis and acquire considerable working experience with at least one method.
- Modern mechanical structures are often flexible, lightweight, operate at a high speed or are dynamically controlled. Such structures are more easily forced into a nonlinear regime than are traditional stiff, heavy, and passive structures.
- Nonlinear problems can no longer be considered intractable, as was often the only reason for linearization in the past. Using mathematics, computers, and experience with commonly encountered types of nonlinearities—we are now in a much better position to analyze nonlinear systems.
- There is currently a significant interest in the subject, as browsing any scientific journal or conference program on dynamics and vibrations will show. To keep pace with the rapid evolution in vibration research, one has to know about nonlinear models, phenomena, and tools.

Why a New Book on Such an Old Topic? (Glimpse of Vibration History). Much research and teaching in mechanical vibrations concentrates on describing, understanding, predicting, measuring, and possibly controlling the free vibrations of mechanical system, or their response to resonant or near-resonant periodic excitation. In fact, this has been the case since Pythagoras (570–497 b.c.) quantified the theory of music and related it to his theory of numbers. It continued via Galileo's (1564–1642) work on pendulum oscillations (Galilei 1638) and Sir Isaac Newton's (1642–1727) subsequent formulation of his laws of motion (Newton 1696) to Robert Hooke's (1635–1703) formulation of the relationship between stress and strain in elastic bodies and the subsequent solution of the differential equation for a vibrating beam by Leonard Euler (1707–1783) and Daniel Bernoulli (1700–1782) (see Shabana 1996; Dimarogonas and Haddad 1992). A major culmination occurred in 1877 with Lord Rayleigh's masterpiece *The Theory of Sound* (Rayleigh 1877). Since Rayleigh, the character of problems facing the vibration theoretician or practitioner has hardly changed (which perhaps explains why Rayleigh's books appear surprisingly modern even today). But, the *tools* for solving vibration problems have improved immensely by the emergence of (1) computers and software that can be used to simulate the behavior of virtually any dynamic system; (2) new or improved mathematical tools for solving nonlinear differential equations approximately; and (3) computerized and miniaturized measurement equipment for experimentally based testing and improvement of theories. These corners of the 'mechanicians triangular toolbox' are centers of fields (numerical/theoretical/ experimental analysis) that are all expanding significantly in these years, seemingly in a mutually fertilizing manner. New or updated books are needed to reflect these developments and explain them to students.

Why *This* Book? There is, in fact, a plenitude of good books on nonlinear dynamics/vibrations. Many of these are on my shelf. I use them heavily, and had no intention of writing a competitor. Though, for my course in advanced vibration analysis, I found that none of them treated the relevant range of subjects in the way that I wanted to present them, that is: Treating classical as well as modern nonlinear analysis (excluding texts dealing with only one area), focusing almost exclusively on mechanical systems (precluding texts on general nonlinear systems), and balancing theory and applications in a way where the latter is the goal and the former is the tool (excluding excessively mathematical texts).

The book title reflects the intimate connection between a phenomenon (*vibrations* or vibration-related) and the presence of this phenomenon as something observable in reality (*stability*). Using vibration theory, we are mostly concerned with predicting equilibrium states, dynamical or statical, of mathematical models that are supposed to model selected aspects of real physical systems. But knowing such states is not enough: Typically, at least with nonlinear models, there are several states of equilibrium, some being stable to perturbations, and some being *un*stable (think of the down- and up-pointing equilibriums of a pendulum). In reality, and with computer models, we never observe systems in their unstable states for very long; they will either reside in a stable state or be on their way to one. So, stability analysis is an integral and crucial part of vibration analysis, in particular for nonlinear systems, and therefore also forms an important part of the book.

Notes for Teachers The book centers on the analysis of a limited number of simple generic models, which serve to illustrate qualitative behaviors for a wide variety of mechanical systems. The students are trained to analyze simple models and to recognize phenomena associated with them. Also, they are taught how to extract simple models out of complicated systems. When combined, this will allow them to appropriately benefit from computer simulations of large-scale systems, e.g., nonlinear finite element models.

A limited number of nonlinear *phenomena* are described rather than the rich variety of nonlinear *systems*. A few simple systems provide the backbone through several chapters, illustrating tools such as perturbation analysis and bifurcation analysis, as well as essential phenomena such as nonlinear frequency response, chaos, and stabilization by high-frequency fields.

I encourage the use of classroom demonstrations for supporting the examples given in the book. An instrumented clamped beam and a pair of magnets can be used for illustrating a whole range of subjects covered, e.g., resonance, mode shapes, statical buckling, hysteresis, bifurcations, and chaotic motion. Similarly, a pendulum on vibrating support (a hobby jigsaw, e.g.) can be used for illustrating parametrical resonance, softening frequency response, chaos, and stabilization (of the upside-down equilibrium) using high-frequency excitation; and a simple pendulum, hanging in a spring, illustrates internal resonance and modal interaction. Also, classroom computer simulations are helpful in illustrating concepts such as phase planes, Poincaré map and frequency spectra, and for showing animated models of physical systems while integrating numerically their equations of motion. Learning advanced vibration and stability analysis is hard work for most students, but, in my experience, students are significantly motivated by the first-hand encounter with some of the strange, amazing, and counter-intuitive phenomena which the world of nonlinear dynamics is so abundant of.

Acknowledgements I am deeply indebted to a great number of students, too many too mention by name, who are probably unaware of how much they have influenced the book. For the first edition—whose manuscript was written along with designing and giving the associated course in vibrations and stability—it was a challenging privilege to write for a specific group of critical, highly responsive, and diligent error-reporting students, rather than for a general, anonymous readership. Since then substantial heaps of paper slips and digital notes accumulated, listing comments and suggestions for improvements—most of them from students, or directly from classroom experience. It is with great relief and gratitude to these students that I now again discard such a pile, in the belief that the new edition of the book takes proper care of most of the comments. Also, I am thankful to colleagues, reviewers, and other readers around the world for fruitful discussions, useful comments, and suggestions for improvements. As for the two first editions of the book, special thanks go to Ilya I. Blekhman, Sergey Sorokin, Matthew P. Cartmell, and J. M. Hale. I am grateful to the professors, Ph.D. students, master students, and technical staff at MEK/Solid Mechanics at The Technical University of Denmark for providing a stimulating and obliging working environment. Special thanks here goes to Viggo Tvergaard, who initially suggested bringing the manuscript to a broader audience; Frithiof Niordson, who let me use his personal lecture notes as inspiration for Chap. 2; Jarl Jensen, who supplied the complete set of problems for Chap. 2; Pauli Pedersen, head of department during the first many years of the books life, for his steady support and encouragement over many years and projects; and last but definitely not the least to the other members of the 1990s-'Fast Vibration Club' (Alexander Fidlin, Morten H. Hansen, Jakob S. Jensen, and Dmitri M. Tcherniak), whom I owe more than can be listed in acceptable space. To finish, I need to thank my nearest and dearest, whose steady flow of encouragement, care, and interest (though not in vibrations) remains a fundamental source of energy and inspiration for me.

Kgs. Lyngby, Denmark   Jon Juel Thomsen
December 2020

Contents

About the Author

Jon Juel Thomsen is Ph.D. (1988) and dr.techn. (2003) in mechanical vibrations and dynamics and associate professor at the Technical University of Denmark, Department of Mechanical Engineering. He has authored 100+ peer-reviewed scientific publications, covering many aspects of theoretical and experimental vibration analysis, in particular nonlinear effects and analysis and has 30+ years of experience in university teaching in advanced and basic vibration theory, dynamics, computer modeling, and experimental mechanics. He has been engaged in fundamental and applied research with a special focus on the description, analysis and synthesis of nonlinear phenomena, e.g. vibration suppression using nonlinearity, object transport using vibrations, non-trivial effects of high-frequency excitation, deterministic chaos, wave propagation in periodic media, discontinuous processes (e.g., vibro-impact, friction, clearance), fluid flow affected vibrations, vibration analysis for non-destructive testing and structural health monitoring, and nonlinear modal interaction; see more at www.staff.dtu.dk/jjth.

Notations

x, X	Scalar variable or function (italics)
$\mathbf{x}$	Vector (lowercase, bold)
x_i	Vector element
$\mathbf{X}$	Matrix (upper case, bold)
X_{ij}	Matrix element
$\mathbf{x}^T, \mathbf{X}^T$	Vector or matrix transpose
$|x|$	Absolute value
$|\mathbf{X}|, \det(\mathbf{X})$	Matrix determinant
$\mathrm{tr}(\mathbf{X})$	Matrix trace
$\bar{x}$	Complex conjugate
$\dot{x}, \ddot{x}$	Time derivatives
x', x''	Spatial derivatives
$x^{(i)}$	ith spatial derivative
$x^{[i]}, x_{[i]}$	ith iterate
$O(x)$	Order of magnitude
∇/∇^2	Gradient/Laplacian operator
D_i^j	Multiple scales differential operator $\left(= \partial^j / \partial T_i^j \right)$
δ_{ij}	The Kronecker delta ($= 1$ for $i = j$; $= 0$ for $i \neq j$)
$\delta(x)$	Dirac's delta-function $\left(\int_{-\infty}^{+\infty} \delta(x)dx = 1 \right)$
i	Counter index or $\sqrt{-1}$ (as appears from context)
$\ln/\log$	Base-e / base-10 logarithm
$\equiv$	Assigning equality
cc	Complex conjugates of preceding terms
hh(n)	Harmonics of nth and higher order
$j = 1, n$	$j = 1, 2,..., n$

1 Vibration Basics

1.1 Introduction

This chapter surveys some fundamental concepts, methods, and phenomena associated with vibrations. Included only as a reference for subsequent chapters, it assumes the reader to be reasonably familiar with most of the topics described. The presentation will be brief, with few examples and no proofs.

For further reference regarding basic vibration theory, you may consult, e.g., Den Hartog (1985), Ginsberg (2001), Harris (1996), Inman (2001, 2014), Kelly (1993, 2007), Meirovitch (2001), Shabana (1996, 1997), Thomson and Dahleh (1998), or Timoshenko et al. (1974). Numerical tools are described in, e.g., Bathe and Wilson (1976), Cook et al. (1989), Morton and Mayers (1994), Press et al. (2002), and Zienkiewicz (1982) – and experimental methods and data analysis in, e.g., Brandt (2011), Broch (1984), Ewins (2000), Holman (1994), and Kobayashi (1993). You might even enjoy reading some of the great ancestors in this field, e.g., Lord Rayleigh's *The Theory of Sound* (Rayleigh 1877).

As should be well known, vibrations of physical objects are described in terms of *degrees of freedom* (DOFs). A model of a mechanical system has as many DOFs as are required for uniquely specifying its state of deformation with respect to some fixed reference configuration. The specific choice of DOFs depends on the problem to be solved. Thus, an aircraft wing may be adequately modeled as a *continuous structure* (having infinitely many DOFs), as a *multiple-DOF* system (having a finite number of DOFs), or even as a *single-DOF* system. Excitations may exist in the form of non-zero initial conditions only, giving rise to *free vibrations* of the object. Or there may be time-varying loads, in which case the object performs *forced vibrations*.

The survey is organized accordingly, that is: we consider systems having single, multiple, and infinitely many DOFs, and for each DOF-class we consider different types of excitation. Then we summarize two important energy methods for setting up equations of motions: Lagrange's equations and Hamilton's principle, both of

© The Author(s), under exclusive license to Springer Nature Switzerland AG 2021
J. J. Thomsen, *Vibrations and Stability*,
https://doi.org/10.1007/978-3-030-68045-9_1

which are to be employed in subsequent chapters. Finally some additional aids useful in vibration analysis are described: Nondimensionalizing equations of motion, relating various types and measures of damping, and deriving equations of motions using the flexibility and stiffness methods.

1.2   Single Degree of Freedom Systems

Fig. 1.1 shows a model of a linear single-DOF system, characterized by mass m, stiffness coefficient k, viscous damping coefficient c, time-varying excitation force $F(t)$, and off-equilibrium position $x(t)$. The equation of motion is:

$$m\ddot{x} + c\dot{x} + kx = F(t), \tag{1.1}$$

subjected to prescribed initial conditions $x(0) = x_0$ and $\dot{x}(0) = \dot{x}_0$, where $\dot{x} \equiv dx/dt$. Dividing the equation with m, an equivalent and often used form is obtained:

$$\ddot{x} + 2\zeta\omega\dot{x} + \omega^2 x = f(t), \tag{1.2}$$

where $f = F/m$, $\omega^2 = k/m$, and $2\zeta\omega = c/m$. Some special solutions are given below.

1.2.1   Undamped Free Vibrations

With $c = 0$ and $F(t) = 0$ the solution is a pure time-harmonic:

$$x(t) = A\sin(\omega t - \psi), \tag{1.3}$$

where the amplitude A and phase ψ are determined by the initial conditions, and ω is the *un-damped natural frequency*:

$$\omega = \sqrt{k/m}. \tag{1.4}$$

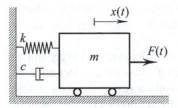

Fig. 1.1   Model of a single-DOF system

1.2.2 Damped Free Vibrations

With $F(t) = 0$ and $0 < c < 2\sqrt{km}$ the solution to (1.1)/(1.2) becomes:

$$x(t) = Ae^{-\zeta\omega t}\sin(\tilde{\omega}t + \psi), \qquad (1.5)$$

where ζ is the *damping ratio* (actual to critical damping):

$$\zeta = \frac{c}{c_{cr}} = \frac{c}{2\sqrt{km}}, \qquad (1.6)$$

and $\tilde{\omega}$ is the *damped natural frequency*:

$$\tilde{\omega} = \omega\sqrt{1 - \zeta^2}. \qquad (1.7)$$

The solution (1.5) describes an exponentially damped harmonic oscillation (see Fig. 1.2). This is the *under-damped* case ($\zeta < 1$). *Over-damped* ($\zeta > 1$) and *critically* damped ($\zeta = 1$) motions damp out without oscillating.

1.2.3 Harmonic Forcing

With $F(t) = F_0\sin(\Omega t)$ the solution to consists of two parts: the solution to the *homogeneous* equation (for $F(t) = 0$), and a *particular* solution. The homogeneous solution corresponds to damped harmonic oscillations at the damped natural frequency, as discussed in Sect. 1.2.2. The particular solution governs the motion which remains when the damped transients have decayed. It takes the form of a steady-state harmonic oscillation at the excitation frequency Ω, with $x(t)$ lagging $F(t)$ by a certain phase φ :

$$x(t) = H\sin(\Omega t - \varphi), \qquad (1.8)$$

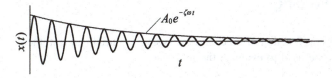

Fig. 1.2 Free oscillations of an under-damped single-DOF system. Initial conditions are $x(0) = 0$ and $\dot{x}(0) = A_0\tilde{\omega}$

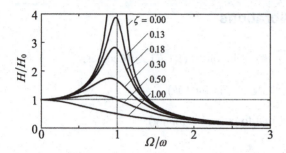

Fig. 1.3 Amplitude response of the harmonically excited single-DOF oscillator at different levels of damping $\zeta = c/c_{cr}$

where

$$H = \frac{F_0}{\sqrt{\left(k - m\Omega^2\right)^2 + \left(c\Omega\right)^2}} \; ; \quad \tan\varphi = \frac{c\Omega}{k - m\Omega^2} . \tag{1.9}$$

The *response curve* (Fig. 1.3) depicts the function $H(\Omega)$ in nondimensional form:

$$\frac{H}{H_0} = \frac{1}{\sqrt{\left(1 - (\Omega/\omega)^2\right)^2 + (2\zeta\Omega/\omega)^2}} \; ; \quad \left(\tan\varphi = \frac{2\zeta\Omega/\omega}{1 - (\Omega/\omega)^2}\right), \tag{1.10}$$

where $H_0 = F_0/k$ is the zero-frequency deflection, ω is defined by (1.4), and ζ by (1.6). When $\Omega \approx \omega$ large-amplitude *resonant* vibrations occur, and the natural frequency ω is also termed the *resonance frequency*[1]. At sharp resonance, $\Omega = \omega$, the response magnitude is $H/H_0 = 1/(2\zeta)$, which approaches infinity as the damping ratio vanishes. Actually the response is strongest for Ω slightly less than ω, namely at $\Omega = \omega\sqrt{1 - 2\zeta^2} \leq \tilde{\omega} \leq \omega$, where $H/H_0 = 1 \big/ \left(2\zeta\sqrt{1 - \zeta^2}\right)$. For $\Omega \ll \omega$ the deflection is in phase with the exciting force, $\varphi \to 0$, while at resonance $\varphi = \pi/2$, and as $\Omega \gg \omega$ then $\varphi \to \pi$ (i.e. deflection and force are in antiphase).

1.2.4   Arbitrary Forcing

If the excitation is a general function of time, and the system is initially at rest, then *Duhamel's Integral* can be used to evaluate the response:

---

[1]Sometimes resonance frequency is defined as the frequency of maximum response (e.g. Harris 1996). Then $\omega\sqrt{1 - 2\zeta^2}$ is the *(displacement) resonance frequency*, ω is the *velocity resonance frequency*, and $\omega \big/ \sqrt{1 - 2\zeta^2}$ is the *acceleration resonance frequency*.

$$x(t) = \int_0^t F(\tau)g(t-\tau)\,d\tau, \tag{1.11}$$

in which $g(t)$ is the *impulse response function*:

$$g(t) = \frac{1}{m\omega}\frac{e^{-\zeta\omega t}}{\sqrt{1-\zeta^2}}\sin(\tilde{\omega}t), \tag{1.12}$$

where ω, ζ and $\tilde{\omega}$ are defined by (1.4), (1.6), and (1.7), respectively.

If the system is not initially at rest, i.e. with non-zero initial conditions, the homogeneous (freely damped) solution should be added.

1.3 Multiple Degree of Freedom Systems

Fig. 1.4 shows an example model of a multiple-DOF system with mass m, moment of inertia I, stiffness coefficients k_1 and k_2, viscous damping coefficients c_1 and c_2, excitation force $F(t)$, and two degrees of freedom $x(t)$ and $\theta(t)$.

Some notions of single-DOF systems readily extend to the multiple-DOF case, some do not. In particular we are forced to consider *systems* of differential equations, generally coupled in the degrees of freedom.

1.3.1 Equations of Motion

The equations of motion for a general, linear n-DOF system may be written:

$$\sum_{j=1}^n \left(M_{ij}\ddot{x}_j + C_{ij}\dot{x}_j + K_{ij}x_j\right) = f_i(t), \quad i = 1, n, \tag{1.13}$$

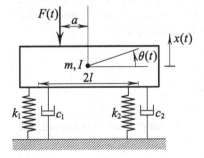

Fig. 1.4 Model system having two degrees of freedom, $x(t)$ and $\theta(t)$

with initial conditions $x_i(0) = x_{0i}$, $\dot{x}_i(0) = \dot{x}_{0i}$. Here, x_i denotes the i'th *generalized coordinate*, and f_i the associated *generalized force*. In matrix notation these equations take the form:

$$\mathbf{M}\ddot{\mathbf{x}} + \mathbf{C}\dot{\mathbf{x}} + \mathbf{K}\mathbf{x} = \mathbf{f}(t), \quad \mathbf{x}(t) \in R^n, \tag{1.14}$$

with initial conditions $\mathbf{x}(0) = \mathbf{x}_0$ and $\dot{\mathbf{x}}(0) = \dot{\mathbf{x}}_0$. Here, variables in uppercase bold denote matrices whereas variables in lowercase bold are vectors. Thus, $\mathbf{M}$, $\mathbf{C}$ and $\mathbf{K}$ are $n \times n$-matrices, whereas $\mathbf{x}$, $\dot{\mathbf{x}}$, $\ddot{\mathbf{x}}$ and $\mathbf{f}(t)$ are n-vectors. The vectors $\mathbf{x}$, $\dot{\mathbf{x}}$, and $\ddot{\mathbf{x}}$ denote generalized coordinates, velocities and accelerations, respectively, and $\mathbf{f}(t)$ holds the generalized time-dependent forces. $\mathbf{M}$ is the system *mass matrix*, $\mathbf{C}$ the *damping matrix* and $\mathbf{K}$ the *stiffness matrix*.

For setting up the equations of motion for an n-DOF system, one may employ Newton's second law, or the flexibility or stiffness method combined with the principle of d'Alembert (Sect. 1.8), or the principle of virtual work, Lagrange's equations (Sect. 1.5.1), Hamilton's principle (Sect. 1.5.2), or whatever convenient. Any of them yields – at least after linearization – a set of n second-order differential equations of the form (1.13) or (1.14).

As an example, employing Newton's second law for the 2-DOF model of Fig. 1.4 readily gives:

$$\begin{aligned} -F(t) - k_1(x - l\theta) - c_1(\dot{x} - l\dot{\theta}) - k_2(x + l\theta) - c_2(\dot{x} + l\dot{\theta}) &= m\ddot{x} \\ F(t)a + k_1(x - l\theta)l + c_1(\dot{x} - l\dot{\theta})l - k_2(x + l\theta)l - c_2(\dot{x} + l\dot{\theta})l &= I\ddot{\theta}, \end{aligned} \tag{1.15}$$

which has the form of (1.14), with:

$$\begin{aligned} \mathbf{x} &= \begin{Bmatrix} x \\ \theta \end{Bmatrix}, \quad \mathbf{f}(t) = \begin{Bmatrix} -F(t) \\ F(t)a \end{Bmatrix}, \quad \mathbf{M} = \begin{bmatrix} m & 0 \\ 0 & I \end{bmatrix}, \\ \mathbf{C} &= \begin{bmatrix} c_1 + c_2 & (c_2 - c_1)l \\ (c_2 - c_1)l & (c_1 + c_2)l^2 \end{bmatrix}, \quad \mathbf{K} = \begin{bmatrix} k_1 + k_2 & (k_2 - k_1)l \\ (k_2 - k_1)l & (k_1 + k_2)l^2 \end{bmatrix}. \end{aligned} \tag{1.16}$$

Observe that n-DOF models may also arise by purely mathematical discretizations of *continuous* systems. For example, the partial differential equations governing the vibrations of a continuous beam may be discretized in the space coordinate using finite differences, finite elements, mode shape expansion, the methods of Ritz or Galerkin, or other approximate techniques.

1.3.2   Undamped Free Vibrations

In the simplest case of n-DOF vibrations there is no damping and no forcing, and the equations of motion are:

$$\mathbf{M}\ddot{\mathbf{x}} + \mathbf{K}\mathbf{x} = \mathbf{0}. \tag{1.17}$$

To determine the time-harmonic solutions one inserts $\mathbf{x}(t) = \boldsymbol{\varphi}\cos(\omega t)$, and obtain an *eigenvalue problem* for the determination of ω and $\boldsymbol{\varphi}$:

$$(\mathbf{K} - \omega^2\mathbf{M})\boldsymbol{\varphi} = \mathbf{0}, \tag{1.18}$$

where ω^2 is an *eigenvalue* and $\boldsymbol{\varphi}$ the associated *eigenvector*. For nontrivial solutions $\boldsymbol{\varphi} \neq \mathbf{0}$ to exist, the determinant of the coefficient matrix must vanish:

$$|\mathbf{K} - \omega^2\mathbf{M}| = 0. \tag{1.19}$$

Expanding the determinant one obtains an n-degree polynomial in ω^2, the so-called *frequency-equation*. The n zeroes of this polynomial provides a set of eigenvalues ω_i^2, $i = 1, n$, and the corresponding values ω_i are the *undamped natural frequencies*. Substituting each ω_i into (1.18) one obtain the associated eigenvectors $\boldsymbol{\varphi}_i$, $i = 1, n$, also termed *(linear) normal modes* or *mode shapes*. If $\boldsymbol{\varphi}_i$ solves (1.18), then so does $\gamma\,\boldsymbol{\varphi}_i$, where γ is an arbitrary constant. Thus, the scale of the mode shapes remains undetermined. Mode shapes are typically *normalized*, such that a particular vector-component (or some other vectorial norm) takes on some prescribed value.

The most general motion of which an n-DOF system is capable is a linear combination of all possible modes superimposed:

$$\mathbf{x}(t) = \sum_{i=1}^{n} q_i\boldsymbol{\varphi}_i\cos(\omega_i t + \psi_i), \tag{1.20}$$

where the *mode participation factors* q_i and the *phase angles* ψ_i are determined by the initial conditions:

$$q_i^2 = \alpha_i^2 + (\beta_i/\omega_i)^2; \quad \tan\psi_i = -\beta_i/(\alpha_i\omega_i), \tag{1.21}$$

where the constants α_i and β_i, $i = 1, n$, solves the linear systems of equations:

$$\sum_{i=1}^{n}\boldsymbol{\varphi}_i\alpha_i = \mathbf{x}_0; \quad \sum_{i=1}^{n}\boldsymbol{\varphi}_i\beta_i = \dot{\mathbf{x}}_0, \tag{1.22}$$

where $\mathbf{x}_0 = \mathbf{x}(0)$ and $\dot{\mathbf{x}}_0 = \dot{\mathbf{x}}(0)$.

1.3.3   Orthogonality of Modes

If $\mathbf{M}$ and $\mathbf{K}$ are both symmetrical then $\boldsymbol{\varphi}_i^T\mathbf{M}\boldsymbol{\varphi}_j = 0$ for all $i \neq j$. This property of *orthogonality* of any two modes proves to be useful in many contexts. Assuming that each mode has been normalized such that $\boldsymbol{\varphi}_i^T \mathbf{M} \boldsymbol{\varphi}_i = 1$, we may state the following relations of *orthonormality*:

$$\boldsymbol{\varphi}_i^T \mathbf{M} \boldsymbol{\varphi}_j = \delta_{ij}, \quad \boldsymbol{\varphi}_i^T \mathbf{K} \boldsymbol{\varphi}_j = \omega_i^2 \delta_{ij}, \quad i, j = 1, \ldots, n, \tag{1.23}$$

where δ_{ij} is the Kronecker delta ($\delta_{ij} = 1$ for $i = j$ and is otherwise zero), and where the second relation follows from the first and (1.18).

1.3.4   Damped Free Vibrations

For this case $\mathbf{C} \neq \mathbf{0}$ and $\mathbf{f}(t) = \mathbf{0}$ in (1.14). Inserting an assumed solution $\mathbf{x}(t) = \boldsymbol{\varphi} e^{\lambda t}$ a linear system of equations is obtained:

$$\left(\lambda^2 \mathbf{M} + \lambda \mathbf{C} + \mathbf{K} \right) \boldsymbol{\varphi} = \mathbf{0}, \tag{1.24}$$

where the determinant of the coefficient matrix must vanish for nontrivial solutions $\boldsymbol{\varphi} \neq \mathbf{0}$ to exist. This requirement produces a polynomial of degree $2n$ in the eigenvalue λ. Roots λ_j of the polynomial will generally be complex-valued, as will the corresponding eigenvectors $\boldsymbol{\varphi}_j$ of (1.24). The complex roots occur in conjugate pairs $\lambda_j = \beta_j \pm i \omega_j$, which implies that the time-dependent part of the response has the form $e^{\beta_j t} \cos(\omega_j t + \psi_j)$. The total response becomes that of (1.20), with each modal contribution being multiplied by $e^{\beta_j t}$. That is, with $\beta_j < 0$ each mode performs damped oscillations at frequency ω_j.

1.3.5   Harmonically Forced Vibrations, No Damping

With $\mathbf{C} = \mathbf{0}$ and $\mathbf{f}(t) = \mathbf{f}_0 \cos(\Omega t + \psi)$ in (1.14) the n-DOF oscillator is given by:

$$\mathbf{M} \ddot{\mathbf{x}} + \mathbf{K} \mathbf{x} = \mathbf{f}_0 \cos(\Omega t + \psi). \tag{1.25}$$

Solutions are obtained by adding the solution of the homogeneous equation to an arbitrary particular solution. The homogeneous equation was discussed in Sect. 1.3.2 and the solution is given by (1.20). For positively damped systems this solution only contributes to the *transient response*. The *stationary response*, remaining when transients have decayed, is governed by the particular solution.

For the particular solution one may assume $\mathbf{x}(t) = \mathbf{a} \cos(\Omega t + \psi)$, substitute into (1.25), and solve the resulting algebraic equations for the vector $\mathbf{a}$:

$$(\mathbf{K} - \Omega^2\mathbf{M})\mathbf{a} = \mathbf{f}_0. \tag{1.26}$$

If $\Omega = \omega_i$ the coefficient matrix $\mathbf{K} - \Omega^2\mathbf{M}$ becomes singular (see Sect. 1.3.2) and the solution will be unbounded. Hence, exciting an undamped linear n-DOF system at any natural frequency ω_i, the amplitude of the excited mode will grow infinitely. For real systems, though, damping and/or nonlinearities will limit the response at some finite amplitude.

The particular solution can be more conveniently expressed in terms of the undamped mode shapes φ_i, that is, one assumes:

$$\mathbf{x}(t) = \sum_{i=1}^{n} q_i\boldsymbol{\varphi}_i \cos(\Omega t + \psi). \tag{1.27}$$

To find the constants q_i, insert (1.27) into (1.25), pre-multiply by φ_j^T and employ the orthonormality relations (1.23) for obtaining:

$$q_i = \frac{\varphi_i^T\mathbf{f}_0}{\omega_i^2 - \Omega^2}, \quad i = 1, n. \tag{1.28}$$

As appears, the modal amplitude q_i of the stationary response approaches infinity as the excitation frequency Ω approaches an undamped natural frequency ω_i.

1.3.6 Harmonically Forced Vibrations, Damping Included

With $\mathbf{C} \neq \mathbf{0}$ and $\mathbf{f}(t) = \mathbf{f}_0\cos(\Omega t + \psi)$ in (1.14) we consider the system:

$$\mathbf{M}\ddot{\mathbf{x}} + \mathbf{C}\dot{\mathbf{x}} + \mathbf{K}\mathbf{x} = \mathbf{f}_0 \cos(\Omega t + \psi). \tag{1.29}$$

The particular solution governing the stationary response will generally not be in phase with the excitation. One may write out the components of the matrix-Eq. (1.29), insert assumed solutions of the form $x_i(t) = a_i\cos(\Omega t + \psi_i)$, expand and separate trigonometric terms, and solve the resulting $2n$ algebraic equations for a_i and ψ_i, $i = 1, n$.

Things are considerably simplified if damping is assumed to be *mass-proportional* and/or *stiffness-proportional*, that is, if to a fair approximation:

$$\mathbf{C} = \alpha\mathbf{M} + \beta\mathbf{K}, \tag{1.30}$$

where the constants of mass- and stiffness-proportionality α and β are non-negative real numbers. It is then possible to *decouple* the equations of motion (1.29) in terms of the undamped mode shapes. For this, assume that

$$\mathbf{x}(t) = \sum_{i=1}^{n} q_i(t)\boldsymbol{\varphi}_i, \tag{1.31}$$

where $q_i(t)$ are unknown time-functions to be determined. Substitute (1.31) and (1.30) into (1.29), pre-multiply by $\boldsymbol{\varphi}_j^T$, and utilize the orthonormality-relations (1.23) for obtaining n decoupled equations in the unknown functions $q_i(t)$:

$$\ddot{q}_i + (\alpha + \beta\omega_i^2)\dot{q}_i + \omega_i^2 q_i = \boldsymbol{\varphi}_i^T \mathbf{f}_0 \cos(\Omega t + \psi), \quad i = 1, n. \tag{1.32}$$

Each equation here has the form of the harmonically forced single-DOF system discussed in Sect. 1.2.3. It can readily be solved, and each solution $q_i(t)$ in turn substituted back into (1.31) to yield the total response $\mathbf{x}(t)$. The frequency responses for q_i will exhibit large amplitudes, *resonance peaks*, for excitation frequencies approaching any damped natural frequency

$$\tilde{\omega}_i = \omega_i\sqrt{1 - \zeta_i^2}, \tag{1.33}$$

where ζ_i is the damping ratio of the i'th mode:

$$\zeta_i = \frac{1}{2}\left(\frac{\alpha}{\omega_i} + \beta\omega_i\right). \tag{1.34}$$

There may also be *antiresonances*, i.e. "inverted peaks" with no or very little response, to the particular excitation, of the corresponding mode at certain frequencies. By contrast to resonances, which are system properties (i.e. independent of the excitation parameters $\mathbf{f}_0$ and Ω), the antiresonances depend on the excitation, and will generally change between the frequency response for each modal coordinate q_i.

1.3.7 General Periodic Forcing

Let all components of the vector $\mathbf{f}(t)$ in (1.14) be T-periodic. Then $\mathbf{f}(t) = \mathbf{f}(t + T)$ at all times t, and one can expand $\mathbf{f}(t)$ in a Fourier series:

$$\mathbf{f}(t) = \frac{1}{2}\mathbf{a}_0 + \sum_{k=1}^{\infty}\left(\mathbf{a}_k \cos\left(\frac{2\pi kt}{T}\right) + \mathbf{b}_k \sin\left(\frac{2\pi kt}{T}\right)\right), \tag{1.35}$$

where

$$\mathbf{a}_k = \frac{2}{T}\int_{-T/2}^{T/2} \mathbf{f}(t)\cos\left(\frac{2\pi kt}{T}\right) dt; \quad \mathbf{b}_k = \frac{2}{T}\int_{-T/2}^{T/2} \mathbf{f}(t)\sin\left(\frac{2\pi kt}{T}\right) dt; \quad k = 0, 1, \ldots \tag{1.36}$$

Using the principle of superposition (valid for linear systems) we may compute the response to a sum of excitations as a sum of responses to single excitation terms. That is, one computes the response to each Fourier-term in (1.35) (using results from Sect. 1.3.6), and add up all individual responses to obtain the full solution. For many applications the infinite Fourier series is truncated at some finite value of k, typically quite low.

1.3.8 Arbitrary Forcing, Transients

For the system (1.14) with general excitation $\mathbf{f}(t)$ we may still decouple the equations in terms of the undamped mode shapes, at least if proportional damping as in (1.30) is assumed. Assuming again a solution of the form (1.31) one obtains, instead of (1.32):

$$\ddot{q}_i + (\alpha + \beta\omega_i^2)\dot{q}_i + \omega_i^2 q_i = \boldsymbol{\varphi}_i^T \mathbf{f}(t), \quad i = 1, n. \tag{1.37}$$

Each of the n equations has the form of the arbitrarily excited single-DOF oscillator discussed in Sect. 1.2.4. Duhamel's integral can be applied to each equation, and the full response obtained through back-substitution into (1.31).

1.4 Continuous Systems

Continuous systems are characterized by having infinitely many degrees of freedom. Or rather, it is not obvious how to describe their state of deformation by a finite set of numbers. Strings, rods, beams, plates and shells are examples of continuous systems. Vibrations of continuous systems are governed by *partial differential equations* – typically of time-order two and space-order two or four, in one, two or three space coordinates. We focus here on the *transverse vibrations of beams*, being illustrative of the most important aspects of continuous systems in general.

1.4.1 Equations of Motion

Consider as an example the beam of Fig. 1.5, having continuously varying bending stiffness $EI(x)$, mass distribution $\rho A(x)$, transverse load per unit length $q(x, t)$ and length l. The unknown state of deformation is characterized by $u(x, t)$. Assume that transverse and rotational vibration amplitudes are small, and that shear, longitudinal and torsional deformations are negligible, as are the rotary inertia of cross-sections.

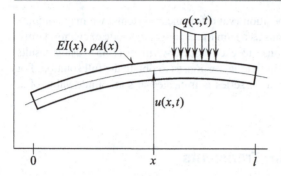

Fig. 1.5 Continuous beam

For setting up the equation of motion we apply the three conditions of dynamic equilibrium (Newton's 2nd law for forces and moment in the (x, u)-plane) to a differential element of the beam. Then take the x-derivative of the moment-equation, substitute the force-equations, substitute Hooke's law $M = EIu''$ for the bending moment, and obtain the partial differential equation:

$$\rho A(x)\ddot{u} + (EI(x)u'')'' - Nu'' + q(x,t) = 0, \qquad (1.38)$$

where $u = u(x, t)$, primes denote differentiation with respect to x, and N is the internal force in the x-direction (N is independent of x, since by ignoring axial inertia and Newton's 2nd law, $N' = 0$). Initial conditions are given by $u(x,0) = u_0(x)$ and $\dot{u}(x, 0) = \dot{u}_0(x)$.

To be properly defined, the fourth-order beam equation requires four *boundary conditions*. Some typical boundary conditions are: hinged ($u = u'' = 0$), clamped ($u = u' = 0$), guided ($u' = u''' = 0$), free ($u'' = u''' = 0$), linear spring support having stiffness κ_1 ($\kappa_1 u = (EIu'')' - Nu'$), and linear rotational spring having stiffness κ_2 ($\kappa_2 u' = -EIu''$). For example, a hinged-hinged beam obeys the boundary conditions $u(0, t) = u''(0, t) = u(l, t) = u''(l, t) = 0$.

1.4.2 Undamped Free Vibrations

For an unloaded beam of uniform cross-section and material one has $N = 0, q(x, t) = 0$ and constant values of EI and ρA. Equation (1.38) becomes:

$$\rho A\ddot{u} + EIu'''' = 0. \qquad (1.39)$$

We assume hinged-hinged boundary conditions, that is: $u(0,t) = u''(0,t) = u(l,t) = u''(l,t) = 0$. Substituting an assumed solution of the form

$$u(x,t) = \varphi(x)\sin(\omega t + \psi), \tag{1.40}$$

an *ordinary* differential equation is obtained:

$$EI\varphi'''' = \omega^2 \rho A \varphi, \tag{1.41}$$

with boundary conditions $\varphi(0) = \varphi''(0) = j(l) = \varphi''(l) = 0$. The general solution is:

$$\varphi(x) = C_1 \sin \lambda x + C_2 \cos \lambda x + C_3 \sinh \lambda x + C_4 \cosh \lambda x, \tag{1.42}$$

where $C_1, \ldots, C_4$ are arbitrary constants, and

$$\lambda^4 \equiv \rho A \omega^2 / EI. \tag{1.43}$$

Requiring (1.42) to fulfill $\varphi(0) = \varphi''(0) = 0$ one obtains $C_2 = C_4 = 0$. Requiring also the conditions $\varphi(l) = \varphi''(l) = 0$ to be fulfilled, a pair of homogeneous equations in C_1 and C_3 is obtained:

$$\begin{aligned} C_1 \sin \lambda l + C_3 \sinh \lambda l &= 0 \\ -C_1 \sin \lambda l + C_3 \sinh \lambda l &= 0. \end{aligned} \tag{1.44}$$

These yield nontrivial solutions if and only if $C_3 = 0$ and $\sin(\lambda l) = 0$, that is:

$$\lambda l = j\pi, \quad j = 0, 1, 2, \ldots \tag{1.45}$$

By (1.43) we then obtain an infinite series of *undamped natural frequencies* of the hinged-hinged uniform beam:

$$\omega_j = \left(\frac{j\pi}{l}\right)^2 \sqrt{\frac{EI}{\rho A}}, \quad j = 1, 2, \ldots \tag{1.46}$$

The corresponding *eigenfunctions* or *mode shapes* $\varphi_j(x)$ are found by substituting (1.45) and $C_2 = C_3 = C_4 = 0$ into (1.42), giving:

$$\varphi_j(x) = C_{1j} \sin(j\pi x/l), \quad j = 1, 2, \ldots \tag{1.47}$$

In the j'th mode the beam will oscillate according to (1.40):

$$u_j(x,t) = C_{1j} \sin(j\pi x/l) \sin(\omega_j t + \psi_j), \tag{1.48}$$

where ω_j is given by (1.46). The full solution of the free vibration problem (1.38) consists of all modes superimposed:

$$u(x,t) = \sum_{j=1}^{\infty} C_{1j} \sin(j\pi x/l) \sin(\omega_j t + \psi_j), \qquad (1.49)$$

where

$$C_{1j}^2 = A_j^2 + B_j^2; \quad \tan \psi_j = A_j/B_j, \quad j = 1, 2, \ldots, \qquad (1.50)$$

where A_j and B_j are determined by the initial conditions $u_0(x)$ and $\dot{u}_0(x)$, through:

$$A_j = 2l^{-1} \int_0^l u_0(x) \sin(j\pi x/l) \, dx; \quad B_j = 2(\omega_j l)^{-1} \int_0^l \dot{u}_0(x) \sin(j\pi x/l) \, dx. \quad (1.51)$$

Relations of orthogonality have been employed for obtaining (1.51). These will be presented next.

1.4.3 Orthogonality of Modes

For a variable section beam subjected to any combination of clamped, free, hinged or guided boundary conditions, the following relations of *orthogonality* between any two eigenfunctions $\varphi_i(x)$ and $\varphi_j(x)$ hold true:

$$\begin{aligned} \int_0^l \rho A(x) \varphi_i(x) \varphi_j(x) \, dx &= m_j \delta_{ij}, \\ \int_0^l EI(x) \varphi_i''(x) \varphi_j''(x) \, dx &= \omega_j^2 m_j \delta_{ij}, \quad i, j = 1, 2, \ldots, \end{aligned} \qquad (1.52)$$

where ω_j^2 is the eigenvalue corresponding to $\varphi_j(x)$, δ_{ij} is the Kronecker delta, and m_j is the *generalized mass* of the j'th mode:

$$m_j = \int_0^l \rho A(x) \left(\varphi_j(x)\right)^2 dx. \qquad (1.53)$$

As for the similar relations (1.23) for eigen*vectors* of finite-DOF systems, the orthogonality of continuous eigenfunctions proves to be useful in many ways.

1.4.4   Normal Coordinates

Assume the free mode shapes $\varphi_j(x)$ to constitute an infinite set of orthogonal functions (as in the above example). Any sufficiently smooth deflection field $u(x, t)$ satisfying the boundary conditions can then be represented as a weighted sum of mode shapes:

$$u(x, t) = \sum_{j=1}^{\infty} q_j(t)\varphi_j(x), \qquad (1.54)$$

where the time functions $q_j(t)$ are termed *normal coordinates, modal factors, modal coefficients, modal amplitudes,* or *mode participation factors.*. Many problems of vibrating structures simplify considerably when written in terms of normal coordinates. For example, the kinetic and potential energies of a beam subjected to any combination of clamped, free, hinged or guided ends are:

$$T = \frac{1}{2} \int_0^l \rho A(x)(\dot{u}(x, t))^2 \, dx; \quad V = \frac{1}{2} \int_0^l EI(x)(u''(x, t))^2 \, dx, \qquad (1.55)$$

which simplify to sums of squares of the normal coordinates q_j and $\dot{q}_j$:

$$T = \frac{1}{2} \sum_{j=1}^{\infty} m_j \dot{q}_j^2; \quad V = \frac{1}{2} \sum_{j=1}^{\infty} m_j \omega_j^2 q_j^2, \qquad (1.56)$$

where m_j is the generalized mass given by (1.53).

1.4.5   Forced Vibrations, No Damping

Consider the equation of motion governing transverse vibrations $u(x, t)$ of a transversely loaded beam with variable cross-section:

$$\rho A(x)\ddot{u} + [EI(x)u'']'' + q(x, t) = 0, \qquad (1.57)$$

with initial conditions $u(x,0) = u_0(x)$, $\dot{u}(x,0) = \dot{u}_0(x)$, and any combination of clamped, free, hinged or guided boundary conditions.

For obtaining a solution $u(x, t)$ in terms of normal coordinates one performs the following steps: Assume a solution of the form (1.54), insert this into (1.57), multiply by $\varphi_i(x)$, integrate over the length of the beam, utilize that $\varphi_j(x)$ satisfies (1.41), employ the orthogonality relations (1.52) and obtain:

$$\ddot{q}_j + \omega_j^2 q_j = Q_j(t)/m_j, \quad j = 1, 2, \ldots, \tag{1.58}$$

where $q_j = q_j(t)$ is the j'th normal coordinate, m_j is the generalized mass given by (1.53), and $Q_j(t)$ is the j'th *generalized force*:

$$Q_j(t) = -\int_0^l \varphi_j(x) q(x, t)\, dx, \quad j = 1, 2, \ldots. \tag{1.59}$$

By this procedure a partial differential equation has been transformed into an infinite set of ordinary differential equations. Each equation in (1.58) corresponds to a forced single-DOF oscillator, for which the methods of Sects. 1.2.3 or 1.2.4 can be applied.

For applications the infinite series (1.54) is typically *truncated*, that is, only a finite number N of modes is retained in the expansion:

$$u(x, t) = \sum_{j=1}^{N} q_j(t) \varphi_j(x). \tag{1.60}$$

To warrant truncation at $j = N$, the frequency content of the external load $q(x, t)$ must be limited well below ω_N. With non-smooth excitations (steps, impulses, random loads, e.g.) the frequency content will generally be broadband, and a large value of N may be required. However, for many *periodically* loaded structures of practical interest, by far the most vibrational energy is concentrated in a few of the lowest modes. For example, for studying vibrations of an aircraft wing subjected to periodic excitation from the engine, including only the lowest (say ten, five or even one) modes may be adequate.

1.4.6 Forced Vibrations, Damping Included

Assume the beam in Fig. 1.5 is subjected to distributed damping forces $c(x)\dot{u}(x, t)$ per unit length, with $c(x)$ being the viscous damping coefficient. The equation of motion (1.38) becomes (with $N = 0$):

$$\rho A(x)\ddot{u} + c(x)\dot{u} + (EI(x)u'')'' + q(x, t) = 0. \tag{1.61}$$

Expanding the solution in terms of normal coordinates, as in Sect. 1.4.5, we obtain instead of (1.58):

$$\ddot{q}_j + m_j^{-1} \sum_{i=1}^{\infty} c_{ij}\dot{q}_j + \omega_j^2 q_j = Q_j(t)/m_j, \quad j = 1, 2, \ldots, \tag{1.62}$$

where c_{ij} denote components of the *generalized damping matrix*:

$$c_{ij} = c_{ji} = \int_0^l c(x)\varphi_i(x)\varphi_j(x)\,dx, \quad i,j = 1,2,\dots \quad (1.63)$$

Equations (1.62) for the modal coordinates $q_j(t)$ are now *coupled* in the velocities $\dot{q}_i$, that is, the equations do not take the form of independent single-DOF oscillators. For some special cases of damping the coupling terms vanish. For example, if damping is assumed to be *mass-proportional*, $c(x) = \alpha\rho A(x)$, the orthogonality relations (1.52) imply that $c_{ij} = \alpha m_j \delta_{ij}$. Then (1.62) decouples into the following single-DOF equations:

$$\ddot{q}_j + \alpha\dot{q}_j + \omega_j^2 q_j = Q_j(t)/m_j, \quad j = 1,2,\dots \quad (1.64)$$

Observe that for beams having constant distribution of mass $\rho A(x)$ and constant coefficient of damping $c(x) = c$, the damping is indeed mass proportional.

With mass-proportional damping one may write down the full solution of (1.61), as a sum of the homogeneous solution (freely damped vibrations) and a particular solution (cf. Sects. 1.2.2 and 1.2.4):

$$u(x,t) = \sum_{j=1}^{\infty} q_j(t)\varphi_j(x), \quad (1.65)$$

where

$$\begin{aligned} q_j(t) = & A_j e^{-\alpha t/2}\sin(\tilde{\omega}_j t + \psi_j) \\ & + (m_j\tilde{\omega}_j)^{-1}\int_0^t Q_j(\tau)e^{-\alpha/2(t-\tau)}\sin\big(\tilde{\omega}_j(t-\tau)\big)\,d\tau, \end{aligned} \quad (1.66)$$

where

$$\tilde{\omega}_j^2 = \omega_j^2 - \tfrac{1}{4}\alpha^2, \quad A_j^2 = a_j^2 + b_j^2, \quad \tan\psi_j = a_j/b_j,$$

$$a_j = m_j^{-1}\int_0^l u(x,0)\rho A(x)\varphi_j(x)\,dx, \quad (1.67)$$

$$b_j = \tfrac{1}{2}\alpha a_j\tilde{\omega}_j^{-1} + (m_j\omega_j)^{-1}\int_0^l \dot{u}(x,0)\rho A(x)\varphi_j(x)\,dx.$$

Here $\tilde{\omega}_j$ defines the j'th *damped natural frequency*, and all other quantities are as previously defined.

The usefulness of mode shapes and natural frequencies is not restricted to theoretical vibration analysis. *Modal analysis* is a cornerstone also of *experimental vibration analysis* (e.g., Ewins 2000). For example, the lowest natural frequencies, mode shapes and damping ratios of a real beam may be determined experimentally by measuring the acceleration response to the impact of a light hammer tip at a few beam points. A frequency analyzer and a modal extraction procedure is then used to estimate ω_j, β and $\varphi_j(x)$. Knowing these modal characteristics, Eqs. (1.65)–(1.67) could be employed to predict the response of the beam to arbitrary loads $q(x, t)$.

1.4.7 Complex-Valued Eigenvalues and Mode Shapes

Typically, when solving unforced vibration problems with linear continuous systems, one inserts into the equation of motion a solution form $u(x, t) = \varphi(x)e^{\lambda t}$, where $(\varphi, \lambda) \in C^2$ is an unknown mode shape and eigenvalue to be determined. Generally, without assumptions on the character or presence of damping, both λ and φ will complex-valued. Having determined these, the question may then arise as how to come from (λ, φ) to the natural frequencies (damped and undamped), the damping ratios, and the actual motions of the system.

Using nothing but trigonometric identities and algebra, one can show that the system motions $\varphi(x)e^{\lambda t}$ can be written as $u(x,t) = A(x)e^{-\zeta\omega t}\cos(\tilde{\omega}t + \psi(x))$, where $\omega = |\lambda| = \sqrt{\text{Re}(\lambda)^2 + \text{Im}(\lambda)^2}$ is the *undamped* natural frequency, $\zeta = -\text{Re}(\lambda)/\omega$ is the *damping ratio*, $\tilde{\omega} = \text{Im}(\lambda) = \sqrt{1 - \zeta^2}\,\omega$ is the *damped* natural frequency, $A(x) = |\varphi(x)|$ is the *spatial amplitude function*, and $\psi(x) = \arctan(\text{Im}(\varphi(x)/\text{Re}(\varphi(x)))$ is the *spatial phase shift*.

With *mass- or stiffness-proportional damping* the mode shapes will be real. Then $\psi = 0$ for all x, implying motions of the form $u(x,t) = \varphi(x)e^{-\zeta\omega t}\cos(\tilde{\omega}t)$, i.e. with all points on the structure moving synchronously in either phase or antiphase (depending on the sign change of $\varphi(x)$); these are *standing wave* motions, with all *nodes* (points with no motion) and *antinodes* (largest motion) fixed in space, and all system points attaining maximum or minimum simultaneously.

Conversely, with *non-proportional damping*, the spatial phase ψ will generally depend on x, meaning that different system points attain maximum and minimum at different times, and that nodal points will move back and forth in time along the structure; such motions are not standing waves but *traveling waves*.

1.4.8 Rayleigh's Method

The *Rayleigh quotient* $R[\tilde{u}]$ yields an upper-bound estimate for the lowest natural frequency ω_1 of any conservative elastic system:

$$R[\tilde{u}] = \frac{V_{\max}(\tilde{u})}{\omega^{-2}T_{\max}(\tilde{u})} = \tilde{\omega}_1^2 \geq \omega_1^2, \tag{1.68}$$

where V_{max} and T_{max} are, respectively, the maximum potential and kinetic energy associated with an oscillating field $\tilde{u}(x)(\sin\omega t)$, and $\tilde{\omega}_1$ is the Rayleigh quotient estimate of ω_1. The function $\tilde{u}(x)$ is arbitrary, except that it must be C^{2m} ($2m$ is the order of the differential equation), and satisfy the *essential* boundary conditions (those having derivatives of order 0, 1, ..., $m-1$). We shall return to Rayleigh's quotient in Chap. 2, defining it there in terms of differential operators.

Considering as an example the problem of estimating the lowest natural frequency of the beam in Fig. 1.5, one finds:

$$R[\tilde{u}] = \frac{\frac{1}{2}\int_0^l EI(x)(\tilde{u}''(x))^2 \, dx}{\frac{1}{2}\int_0^l \rho A(x)(\tilde{u}(x))^2 \, dx} = \tilde{\omega}_1^2 \geq \omega_1^2, \qquad (1.69)$$

where $\tilde{u}(x)$ is any function satisfying the boundary conditions in u and u'. Choosing $\tilde{u}(x)$ as the first mode shape $\varphi_1(x)$ of the beam, the approximation to ω_1 becomes exact. Other functions produce ω_1-estimates that are higher than the true value.

The Rayleigh quotient is rather insensitive to the particular choice of $\tilde{u}(x)$, as long as $\tilde{u}(x)$ qualitatively resembles the true first mode shape φ_1: Generally relative errors of the order $\varepsilon \ll 1$ in $\tilde{u}(x)$ cause errors of order ε^2 in the estimate $\tilde{\omega}_1$ of ω_1. (cf. Sect. 2.8.7), so that even a rough guess on φ_1 may yield a reasonable estimate of ω_1. (e.g. with about 2% error in $\tilde{\omega}_1$ with a 20% error in the estimate of φ_1).

1.4.9 Ritz Method

As an extension of Rayleigh's method, one may choose $\tilde{u}(x)$ as an N-term series:

$$\tilde{u}(x) = \sum_{j=1}^{N} a_j \tilde{u}_j(x), \qquad (1.70)$$

where each function $\tilde{u}_j(x)$ satisfies the essential boundary conditions of the problem. The coefficients a_j are chosen so as to minimize $R[\tilde{u}]$. A necessary condition for $R(\tilde{u})$ to be minimal is that (cf. (1.68)):

$$\frac{\partial V_{max}}{\partial a_j} - R\frac{\partial (T_{max}/\omega^2)}{\partial a_j} = 0, \quad j = 1, N, \qquad (1.71)$$

where the energies V_{max} and T_{max} are evaluated as for Rayleigh's method. Equation (1.71) constitutes a set of N linear and homo-geneous equations in a_j. Equating to zero the determinant of the coefficient matrix, a frequency equation is obtained having n roots, $R_1, R_2, \ldots, R_N$. These are upper-bound approximations to the N lowest natural frequencies squared, that is: $R_1 \geq \omega_1^2$, $R_2 \geq \omega_2^2$, etc. Approximations for the corresponding mode shapes are obtained by substituting back a root R in (1.71), and then solve for $a_j, j = 1, N$. Generally, the accuracy of the method degenerates progressively for higher natural frequencies. The lower frequency approximations may be very accurate.

1.5 Energy Methods for Setting up Equations of Motion

Among numerous energy methods for setting up equations of motion, we present here only the two to be employed in subsequent chapters: *Lagrange's equations* and *Hamilton's principle*. For more complete treatments of these and other methods see, e.g., Arczewski et al. (1993), Chen (1966), El Naschie (1990a), Greenwood (2003), Lanczos (1962), or Tabarrok and Rimrott (1994).

Energy methods are occasionally preferred over force–balance methods (e.g., Newton's 2nd law). This is because scalar measures of energy are typically easier to calculate than forces, being vectorial quantities. For a particular system there will typically be common agreement upon the energies involved, whereas the decision of which forces are relevant, and how they relate to the system state, may take some discussion. In particular this holds true when rotating coordinate systems and/or nonlinearities are involved. Using energy or force methods, the efforts spent in setting up equations of motion may be the same. However, with energy methods the main efforts are spent on performing trivial mathematical operations, the correctness of which may be easily checked.

1.5.1 Lagrange's Equations

Lagrange's equations are applicable for systems having a finite number of degrees of freedom, those we call *multi-DOF*, *discrete* or *lumped* systems.

With N degrees of freedom, choose a set of *generalized coordinates* $q_i(t)$, $i = 1$, N, uniquely defining the state of the system. Express the potential and kinetic energy V and T of the system in terms of generalized coordinates q_i and generalized velocities $\dot{q}_i$. Consider p vectors of external forces $\mathbf{F}_k$ which are *not* derivable from potentials, and denote by $\mathbf{r}_k$ the *natural coordinates* (Cartesian, e.g.) of the associated system masses. Lagrange's equations then take the form:

$$\frac{d}{dt}\left(\frac{\partial L}{\partial \dot{q}_i}\right) - \frac{\partial L}{\partial q_i} = Q_i, \quad i = 1, N, \tag{1.72}$$

where

$$L \equiv T - V, \quad Q_i \equiv \sum_{k=1}^{p} \mathbf{F}_k \cdot \frac{\partial \mathbf{r}_k}{\partial q_i}. \tag{1.73}$$

The function L is termed the *Lagrangian*. Expressing the excess of kinetic energy over potential energy, the Lagrangian may be considered the most fundamental quantity in the mathematical analysis of mechanical problems.

For example, a simple pendulum of length l and mass m in gravity g has energies $T = 1/2m(l\dot\theta)^2$ and $V = mgl(1 - \cos\theta)$, where $\theta(t)$ is the off-gravity angle. Using (1.72) one obtains the familiar pendulum equation $\ddot\theta + (g/l)\sin\theta = 0$.

For conservative systems all forces are derivable from potentials, and hence $Q_i = 0$, $i = 1, N$ in (1.73). For *non*-conservative systems, instead of computing generalized loads Q_i, one can subtract from V the work done by non-potential forces, and then apply Lagrange's equations with $Q_j = 0$.

The *power dissipation function* is a convenient means for including the effect of certain types of damping into a Lagrangian formulation. A term $\partial D/\partial \dot q_i$ is then added to the left-hand side of (1.72), D being the dissipation function. For viscous damping one takes $D = 1/2\sum_{i,j}c_{ij}\dot q_i\dot q_j$, commonly referred to as the *Rayleigh dissipation function*. For the pendulum example we may assume an external torque $ml^2F(t)$, doing work $ml^2F(t)\theta(t)$, and viscous damping with dissipation function $D = 1/2cm(l\dot\theta)^2$. Then $T = 1/2m(l\dot\theta)^2$ and $V = mgl(1 - \cos\theta) - ml^2F(t)\theta(t)$. Adding $\partial D/\partial\dot\theta$ to the left-hand side of (1.72) one arrives at the equation of motion for the damped and harmonically forced pendulum, $\ddot\theta + c\dot\theta + (g/l)\sin\theta = F(t)$.

Algebraic constraints can be accounted for: Let $\mathbf{q} = \mathbf{q}(t)$ hold the generalized coordinates $q_i(t)$, subject to m holonomic (i.e. involving only position variables and time) constraints f_j in the form $\mathbf{f}(\mathbf{q}, t) = \mathbf{0}$. Using Hamilton's principle (see next section) one can show that the corresponding extended Lagrange's equation become:

$$\frac{d}{dt}\left(\frac{\partial L}{\partial \dot q_i}\right) - \frac{\partial L}{\partial q_i} + \sum_{j=1}^{m} \lambda_j \frac{\partial f_j}{\partial q_i} = Q_i, \quad i = 1, N, \tag{1.74}$$

which are called Lagrange's equation *of the first kind* (while the equations with a minimal set of generalized coordinates and no constraints are of the *second* kind). Here λ_j is the *Lagrange multiplier* for the constraint f_j; it can be shown to express the *constraint force*. Along with the m constraints, the N Lagrange-equations give $N + m$ equations for the $N + m$ unknown variables q_i and λ_j. The resulting equations are *differential algebraic equations* (DAE's). For the pendulum example one could take the Cartesian coordinate $(x, y) = (l\sin\theta, l\cos\theta)$ as the generalized coordinates, rather than the off-gravity angle θ. Then the supplemental constraint $x^2 + y^2 = l^2$ is needed (since x and y are not independent), and the constraint function is $f(x, y) = x^2 + y^2 - l^2$. Application of (1.74) then gives two differential equations and one algebraic equation for determining the three unknowns (x, y, λ).

1.5.2   Hamilton's Principle

Hamilton's principle holds for any mechanical systems subjected to *monogenic forces* and *holonomic constraints* (to be explained). By contrast to Lagrange's equations, it applies too for systems characterized by infinitely many degrees of freedom, that is, for *continuous* or *distributed* systems. By Hamilton's principle,

$$\delta H = 0, \quad \text{where} \quad H \equiv \int_{t_1}^{t_2} L dt, \tag{1.75}$$

where H defines the *action integral*, L is the Lagrangian, t_1 and t_2 are arbitrary instants of time, and δH is the *variation* of H.

The principle states that the motion of a mechanical system, from an initial configuration at time t_1 to a final configuration at time t_2, occurs in such a manner that the action integral attains a stationary value with respect to arbitrary admissible variations of system configurations – provided that the variations of displacements vanish at times t_1 and t_2.

At static equilibrium the kinetic energy T vanishes, and the potential energy V becomes independent of time; Hamilton's principle then reduces to the principle of minimum potential energy. Also, one can deduce Lagrange's equations from Hamilton's principle.

Some definitions may now be in order:

Monogenic forces are derivable from a scalar quantity, typically some work or energy function or potential. For example, the restoring force kx in a linear spring is monogenic, since one can derive it from the energy function $1/2 kx^2$ by differentiation. Forces that cannot be derived from a scalar quantity (dry friction, e.g.) are *polygenic*.

Holonomic constraints express relations between system coordinates $\mathbf{q} = \mathbf{q}(t) \in R^n$ in the form $\mathbf{f}(\mathbf{q}, t) = \mathbf{0}, \mathbf{f} \in R^m$, i.e. they depend only on position variables and possibly time, but not on higher time-derivatives such as velocities, and does not contain inequalities. For example, with transverse beam vibrations $u(x,t)$ the constraints $u(0, t) = 0$, $u''(0, t) = 0$, $u(0, t) = u(l, t)$, and $u(0, t) = u_0 \sin(\Omega t)$ are all holonomic, while $u(0, t) \geq 0$, $\dot{u}(0, t) = k \dot{u}(l, t)$, and $\dot{u}(0, t) = \Omega u_0 \cos(\Omega t)$ are *non-holonomic*.

A *variation* δu of a function u is a virtual infinitesimal change of all function values. This change, by contrast to the infinitesimal d-process of ordinary calculus, is not caused by an actual change of an independent variable, but is imposed on a set of dependent variables as a kind of 'mathematical experiment'. Imagine some deformation pattern $u = u(x)$. Then change all values of u by slight amounts $\delta u(x)$, and you have a variation in $u(x)$. An *admissible variation* make $u + \delta u$ satisfy the boundary conditions or initial conditions of the problem. Thus, with u specified at the boundaries, $\delta u = 0$ at these boundaries for $\delta u(x)$ to be an admissible variation.

Certain rules apply for the calculus of variations of *functionals* (functions of functions). For example, the (first) variation δF of a functional $F(x, u(x), u'(x), u''(x), \ldots, u^{(n)}(x))$ is given by a simple chain rule:

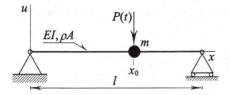

Fig. 1.6 Example system for applying Hamilton's principle

$$\delta F = \frac{\partial F}{\partial u}\,\delta u + \frac{\partial F}{\partial u'}\,\delta u' + \frac{\partial F}{\partial u''}\,\delta u'' + \cdots + \frac{\partial F}{\partial u^{(n)}}\,\delta u^{(n)}. \tag{1.76}$$

Also, variation and differentiation are interchangeable[2]: $\delta(du/dx) \equiv \delta u' = d(\delta u)/dx$, as are variation and integration: $\delta \int F dx = \int \delta F dx$. Variations of independent variables are not allowed, that is, for the functional $F(x, u(x), u'(x), \cdots)$ one has $\delta x \equiv 0$ by definition. Stationarity of the functional F requires $\delta F = 0$ for arbitrary admissible variations δu.

Expressions for *strain energy* are required when using Hamilton's principle for elastic structures. For one-dimensional bending of beams, the strain energy per unit beam length is $1/2 M^2/EI$, where M is the internal bending moment and EI the bending stiffness. For one-dimensional extension of rods the similar expression is $1/2 N^2/EA$, where N is the internal extensional force and EA the longitudinal stiffness (e.g., El Naschie 1990a).

With non-conservative systems, when calculating L one can subtract from V the work done by non-potential forces, just as with Lagrange's equations (Sect. 1.5.2).

Example Using Hamilton's principle to derive the equation of motion governing transverse vibrations $u(x, t)$ of the beam in Fig. 1.6. The beam is simply supported at $x = 0$ and $x = l$, has length l and bending stiffness EI, distributed mass ρA per unit length, a concentrated mass m at $x = x_0$, and is subjected to a point load $P(t)$ at $x = x_0$. The kinetic and potential energies, Lagrangian and action integral are, respectively:

$$T = \int_0^l \frac{1}{2}\rho A \dot{u}^2\, dx + \frac{1}{2}m(\dot{u}(x_0, t))^2 = \int_0^l \left(\frac{1}{2}\rho A + \frac{1}{2}m\tilde{\delta}(x - x_0)\right)\dot{u}^2 dx,$$

$$V = \int_0^l \frac{1}{2}\frac{M^2}{EI}\, dx - (-Pu(x_0, t)) = \int_0^l \left(\frac{1}{2}EI(u'')^2 + P(t)\tilde{\delta}(x - x_0)u\right)dx, \tag{1.77}$$

$$L = T - V = \int_0^l h(x, t, u, \dot{u}, u'')\, dx, \quad H = \int_{t_1}^{t_2} L\, dt,$$

---

[2]Here the assumption on holonomic constraints is important. With *non*-holonomic systems Hamilton's principle takes the form $\int_{t_1}^{t_2} \delta L dt = 0$ (Greenwood 2003), which is *not* a statement of stationarity, and generally not the same as (1.75).

where $\tilde{\delta}(x)$ is Dirac's delta function (see e.g. App. C), $M(x,\,t)$ is the bending moment of the beam, $M \approx EIu''$ for $(u')^2 \ll 1$, and

$$h \equiv \left(\tfrac{1}{2}\rho A + \tfrac{1}{2}m\tilde{\delta}(x - x_0)\right)\dot{u}^2 - \tfrac{1}{2}EI(u'')^2 - P(t)\tilde{\delta}(x - x_0)u. \qquad (1.78)$$

The variation of H becomes:

$$\delta H = \int_{t_1}^{t_2} \delta L\,dt = \int_{t_1}^{t_2} \delta \int_0^l h\,dx\,dt = \int_{t_1}^{t_2} \int_0^l \delta h\,dx\,dt$$

$$= \int_{t_1}^{t_2} \int_0^l \left(\frac{\partial h}{\partial u}\delta u + \frac{\partial h}{\partial \dot{u}}\delta \dot{u} + \frac{\partial h}{\partial u''}\delta u''\right) dx\,dt. \qquad (1.79)$$

Then employ integration by parts to express all variations of the last integral in terms of δu. For the integrand containing $\delta \dot{u}$ this yields:

$$\int_{t_1}^{t_2} \int_0^l \left(\frac{\partial h}{\partial \dot{u}}\delta \dot{u}\right) dx\,dt = \int_0^l \left(\int_{t_1}^{t_2} \frac{\partial h}{\partial \dot{u}}\delta \dot{u}\,dt\right) dx$$

$$= \int_0^l \left(\left[\frac{\partial h}{\partial \dot{u}}\delta u\right]_{t_1}^{t_2} - \int_{t_1}^{t_2} \frac{\partial}{\partial t}\left(\frac{\partial h}{\partial \dot{u}}\right)\delta u\,dt\right) dx$$

$$= \int_0^l \left(\left[\left(\rho A + m\tilde{\delta}(x - x_0)\right)\dot{u}\delta u\right]_{t_1}^{t_2}\right.$$

$$\left. - \int_{t_1}^{t_2} \left(\rho A + m\tilde{\delta}(x - x_0)\right)\ddot{u}\delta u\,dt\right) dx, \qquad (1.80)$$

and for the integrand containing $\delta u''$, similarly:

$$\int_{t_1}^{t_2} \int_0^l \left(\frac{\partial h}{\partial u''}\delta u''\right) dx\,dt = \int_{t_1}^{t_2} \left(\left[\frac{\partial h}{\partial u''}\delta u'\right]_0^l - \int_0^l \frac{\partial}{\partial x}\left(\frac{\partial h}{\partial u''}\right)\delta u'\,dx\right) dt$$

$$= \int_{t_1}^{t_2} \left(\left[\frac{\partial h}{\partial u''}\delta u'\right]_0^l - \left[\frac{\partial}{\partial x}\left(\frac{\partial h}{\partial u''}\right)\delta u\right]_0^l + \int_0^l \frac{\partial^2}{\partial x^2}\left(\frac{\partial h}{\partial u''}\right)\delta u\,dx\right) dt$$

$$= \int_{t_1}^{t_2} \left([-EIu''\delta u']_0^l - [-EIu'''\delta u]_0^l + \int_0^l -EIu''''\delta u\,dx\right) dt. \qquad (1.81)$$

The boundary term of (1.80) vanishes since by definition $\delta u = 0$ at $t = t_1, t_2$. The boundary terms of (1.81) vanishes because of the boundary conditions: $u'' = 0$ and $\delta u = 0$ at $x = 0$ and $x = l$. Thus, inserting (1.78) and (1.80)–(1.81) into (1.79):

$$\delta H = \int_{t_1}^{t_2} \int_0^l \left(-P(t)\tilde{\delta}(x - x_0) - \left(\rho A + m\tilde{\delta}(x - x_0) \right) \ddot{u} - EIu'''' \right) \delta u \, dx \, dt.$$

$$(1.82)$$

Finally, requiring $\delta H = 0$ for arbitrary variations δu, the integrand must vanish identically, giving the equation of motion for the beam:

$$\left(\rho A + m\tilde{\delta}(x - x_0) \right) \ddot{u} + EIu'''' + P(t)\tilde{\delta}(x - x_0) = 0. \qquad (1.83)$$

This particular equation could be obtained more easily by using Newton's 2nd law. But if the point-mass could slide along the beam, the power of Hamilton's principle would show up: The energies involved would still be quite easy to pose, and from there on the Hamiltonian approach follows a strict scheme.

1.5.3   From PDEs to ODEs: Mode Shape Expansion

With continuous systems the equations of motion typically come in the form of one or more *partial* differential equations (PDEs). These may be a result of applying Hamilton's principle or a force balance method. For solving PDEs one is likely to rely on approximate methods, in particular when nonlinearities are involved. Computer-based numerical methods (finite element or finite difference) can here be used for discretizing PDEs into a large number of approximate *ordinary* differential equations (ODEs). One then uses a computer for calculating particular solutions to these. Results obtained, of course, will be similarly particular to the parameter values and initial conditions chosen. Computer methods resemble laboratory experiments in this respect, both providing very specific answers to very specific questions. Especially when studying nonlinear systems and phenomena, the mere observation of output (numerical or experimental) is meaningless, at best.

To attain *understanding* – by contrast to *information* – we need a method for converting intractable PDEs into more manageable low-order sets of ODEs. This will require approximations, as will subsequent attempts to analytically solve the nonlinear ODEs. However, chances are that the essential behavior of the system can be revealed through a few ODEs. The approximations involved may then be subsequently checked using more accurate models, and perhaps laboratory experiments.

Mode shape expansion is a simple and workable approach for converting a vibration related PDE into a set of ODEs. In Sects. 1.4.4–1.4.6 we stated some results of applying this method. Here, to clear up the steps involved, we elaborate a little more on a simple example.

The steps are as follows: for a one-dimensional structure, let $u(x, t)$ denote the dependent variable of the PDE. Assume the expansion (1.60), i.e.

$$u(x,t) = \sum_{j=1}^{N} q_j(t)\varphi_j(x), \tag{1.84}$$

with $\varphi_j(x)$ being the free mode shapes (eigenfunctions) associated with the PDE, and q_j the unknown time functions (called normal coordinates, mode participation factors, or modal coefficients). Insert the expansion into the PDE, multiply by φ_i, integrate over the length of the structure, and obtain a set of ODEs in the variables q_j. The mode shapes φ_i typically constitute an orthogonal set of functions, and the expansion (1.84) yields exact results as $N \to \infty$.

Consider as an example the PDE (1.83) for the continuous beam of Fig. 1.6, with $m = 0$ and including a viscous damping-term:

$$\rho A \ddot{u} + c\rho A \dot{u} + E I u'''' + P(t)\tilde{\delta}(x - x_0) = 0. \tag{1.85}$$

The free mode shapes φ_j and natural frequencies ω_j of a simply supported beam are given by (cf. (1.46), (1.47)):

$$\varphi_j(x) = \sin(j\pi x/l), \quad \omega_j^2 = \left(\frac{j\pi}{l}\right)^4 \frac{EI}{\rho A}, \quad j = 1, N. \tag{1.86}$$

One can easily show that the mode shapes $\varphi_i(x)$ in (1.86) satisfy the following relations of orthogonality:

$$\int_0^l \varphi_i \varphi_j \, dx = \frac{1}{2} l \delta_{ij}, \quad \int_0^l \varphi_i \varphi_j'''' \, dx = \int_0^l \varphi_i'' \varphi_j'' \, dx = \frac{1}{2} l (j\pi/l)^4 \delta_{ij}, \tag{1.87}$$

where δ_{ij} denote the Kronecker delta and $i, j = 1, N$. Now, insert the expansion (1.84) into (1.85), multiply by φ_i and integrate over $x \in [0, l]$ to obtain:

$$\rho A \int_0^l \varphi_i \sum_j \ddot{q}_j \varphi_j \, dx + c\rho A \int_0^l \varphi_i \sum_j \dot{q}_j \varphi_j \, dx$$
$$+ EI \int_0^l \varphi_i \sum_j q_j \varphi_j'''' \, dx + P(t) \int_0^l \varphi_i \tilde{\delta}(x - x_0) \, dx = 0, \tag{1.88}$$

or, by interchanging integration and summation and using (1.86)–(1.87):

$$\rho A \sum_j \int_0^l \varphi_i \varphi_j dx \ddot{q}_j + c\rho A \sum_j \int_0^l \varphi_i \varphi_j dx \dot{q}_j$$

$$+ EI \sum_j \int_0^l \varphi_i \varphi_j'''' dx \, q_j + P(t)\varphi_i(x_0) = 0$$

$$\Rightarrow \rho A \sum_j \frac{1}{2} l \delta_{ij} \ddot{q}_j + c\rho A \sum_j \frac{1}{2} l \delta_{ij} \dot{q}_j \qquad (1.89)$$

$$+ EI \sum_j \frac{1}{2} l (j\pi/l)^4 \delta_{ij} q_j + P(t)\varphi_i(x_0) = 0$$

$$\Rightarrow \rho A \frac{1}{2} l \ddot{q}_i + c\rho A \frac{1}{2} l \dot{q}_i + EI \frac{1}{2} l (i\pi/l)^4 q_i + P(t)\varphi_i(x_0) = 0$$

$$\Rightarrow \quad \ddot{q}_i + c\dot{q}_i + \omega_i^2 q_i = -2(\rho A l)^{-1} P(t) \sin(i\pi x_0/l), \quad i = 1, N,$$

Thus, the PDE has been transformed into a set of N uncoupled ODEs, each constituting a single-DOF oscillator, which can be readily solved by the methods of Sect. 1.2.3 (if $P(t)$ is harmonic) or Sect. 1.2.4 (for general forcing). The solutions for $q_i(t)$ are then substituted into (1.84) for obtaining $u(x, t)$.

This approach works equally well for nonlinear problems, though in general the ODEs will be nonlinearly coupled. Also, it readily extends to higher dimensions, e.g., to plate and shell problems (Leissa 1993ab).

Mode shape expansion goes under several other names, e.g., the *normal mode method* or the *eigenfunction expansion method*. In the *Galerkin Method*, also called *Galerkin discretization*, the functions φ_j are not required to be mode shapes of the corresponding eigenvalue problem, but only to be *test* or *comparison* functions. Test and comparison functions satisfy all boundary conditions of the problem, and are differentiable to the order of the differential equation. Naturally, for a given number of functions, the accuracy is then less than for mode shape expansion. The requirements on the functions φ_j can be further relaxed, e.g. to fulfill only the geometrical boundary conditions, or to be any set of functions one think will be able to suitably represent the basic motions of the system of concern; the discretization process is then called the *assumed-modes method* (a bit misleading, since the functions are usually not mode shapes). Note that the procedure for calculating the unknown coefficients q_j are the same for all methods – only the requirements on the functions φ_j differs, and thus the accuracy for a given number of functions. Then why not use eigenfunctions for best accuracy all the time? Because these may be very difficult or impossible to calculate for a given system, while test functions and 'assumed modes' are often quite easy to suggest. Usually these methods work well, but as with most approximate methods there are some pitfalls and possible errors to be aware of (Lacarbonara 1999; Nayfeh 1998).

1.5.4   Using Lagrange's Equations with Continuous Systems

For continuous systems we may sometimes want to bypass setting up partial differential equations, heading on directly for an approximating set of ordinary differential equations. This can be accomplished by combining the mode shape expansion technique described above with Lagrange's equations, as illustrated by the following example.

Consider again the continuous beam of Fig. 1.6, but for the purpose of illustration without the point-mass ($m = 0$). For setting up the equations of motion one could employ Newton's second law or Hamilton's principle for obtaining the PDE, and then turn the PDE into a set of approximating ODEs by using mode shape expansion; this approach was used in Sect. 1.5.3. Here we choose instead to employ mode shape expansion already at the stage of defining the energies. The expansion (1.84) approximates the continuous variable $u(x, t)$ in terms of a finite set of discrete variables $q_j(t), j = 1, N$. So, by expressing energies in terms of discrete variables q_j instead of a continuous variable u, the system transforms into a multi-DOF system. Of course, since N needs to be finite, this system only approximates the original continuous system. This approximation, however, would have to be introduced anyway, we merely incorporate it at an earlier stage. The potential and kinetic energies and the dissipation function become:

$$T = \int_0^l \frac{1}{2}\rho A \dot{u}^2 \, dx = \int_0^l \frac{1}{2}\rho A \left(\sum_j \dot{q}_j \varphi_j\right)^2 dx,$$

$$V = \int_0^l \frac{1}{2}EI(u'')^2 dx - (-P(t)u(x_0, t))$$

$$= \int_0^l \frac{1}{2}EI\left(\sum_j q_j \varphi_j''\right)^2 dx + P(t)\sum_j q_j \varphi_j(x_0), \tag{1.90}$$

$$D = \int_0^l \frac{1}{2}c\rho A \dot{u}^2 dx = cT.$$

Inserting into Lagrange's equations:

$$\frac{d}{dt}\left(\frac{\partial L}{\partial \dot{q}_i}\right) - \frac{\partial L}{\partial q_i} + \frac{\partial D}{\partial \dot{q}_i} = 0; \quad L \equiv T - V, \quad i = 1, N, \tag{1.91}$$

one obtains:

$$\int_0^l \rho A\left(\varphi_i \sum_j \ddot{q}_j \varphi_j\right) dx + \int_0^l EI\left(\varphi_i'' \sum_j q_j \varphi_j''\right) dx$$

$$+ P(t)\varphi_i(x_0) + c \int_0^l \rho A\left(\varphi_i \sum_j \dot{q}_j \varphi_j\right) dx = 0 . \tag{1.92}$$

Using the orthogonality relations (1.87) and inserting (1.86) this reduces to:

$$\ddot{q}_i + \omega_i^2 q_i + 2P(t)(\rho Al)^{-1} \sin(i\pi x_0/l) + c\dot{q}_i = 0, \quad i = 1, N. \tag{1.93}$$

which is identical to the final result in (1.89). Hence, for the above continuous system we have arrived at a set of approximating ordinary differential equations without considering the PDE.

This approach is simple and workable, even for nonlinear problems. Some information is lost however, since one never gets a chance of inspecting the underlying PDE.

1.6 Nondimensionalized Equations of Motion

Often the equations of motion for a mechanical system are *nondimensionalized* before the analysis starts, i.e. new variables and parameters are introduced and substituted, that express ratios to other quantities having the same physical units. This will typically reduce the number of parameters to those that are necessary and sufficient for analyzing the system, make the equations of motion appear simpler, and considerably ease interpretation of results.

In textbooks and scientific papers, it is common practice just to state the final nondimensionalized equations of motion, along with the definitions of nondimensional variables and parameters. This may leave newcomers rather puzzled as to how and why just these variables and parameters were chosen for the nondimensionalization, and they might conclude that special insights or smart ideas are required. However, the only thing required is a systematic procedure, and knowledge of what the different parameters and equation terms means physically. Here we illustrate the idea in terms of a simple 1-DOF system (a more involved 2-DOF case is exercised in Problem 4.5b).

Consider a typical equation of motion for a displacement variable $u = u(t)$,

$$m\ddot{u} + c\dot{u} + ku = Q\sin(\Omega t) \quad [\text{N}], \tag{1.94}$$

With all parameters and variables having physical units, and brackets in this section indicating the physical unit of the preceding equation, variable, or parameter. To nondimensionalize (1.94) We first introduce a *dimensionless independent variable* τ, by scaling t:

$$\tau = \omega t \quad [1], \tag{1.95}$$

where for now $\omega[\text{s}^{-1}]$ is a free constant. We can also introduce a *dimensionless dependent variable*, by scaling u:

$$y = u/\ell \quad [1],\tag{1.96}$$

where ℓ [m] is yet a free constant? Then we have a *dimensionless time* and a *dimensionless displacement* y, and are free to choose the constants ω and ℓ. They can be chosen as some *characteristic frequency* ω and some *characteristic length* ℓ for the system. However, any choice is allowed, and for now, the decision is just postponed, until we can see which choice makes the resulting dimensionless equation appear as simple as possible.

From (1.95) It follows that differentiation wrt. physical time t turns into differentiation wrt. dimensionless time τ as follows:

$$
\begin{aligned}
\dot{u} &= \frac{du}{dt} = \frac{du}{d\tau}\frac{d\tau}{dt} = \omega\frac{du}{d\tau}, \\
\ddot{u} &= \frac{d\dot{u}}{dt} = \frac{d\dot{u}}{d\tau}\frac{d\tau}{dt} = \omega\frac{d}{d\tau}\left(\omega\frac{du}{d\tau}\right) = \omega^2\frac{d^2u}{d\tau^2}.
\end{aligned}
\tag{1.97}
$$

Insert (1.95)–(1.97) into (1.94) and obtain:

$$m\omega^2\frac{d^2(y\ell)}{d\tau} + c\frac{d(y\ell)}{d\tau}\omega + k(y\ell) = Q\sin(\Omega\tau/\omega).\tag{1.98}$$

Then redefine overdots to mean differentiation wrt. τ, divide by $m\omega^2\ell$ to free the highest-order derivative term (here $d^2y/d\tau^2$), and find:

$$\ddot{y} + \frac{c}{\omega m}\dot{y} + \frac{k}{m\omega^2}y = \frac{Q}{m\omega^2\ell}\sin\left(\frac{\Omega}{\omega}\tau\right) \quad [1].\tag{1.99}$$

Now one can choose ω and ℓ to give a simple equation, with interpretable parameters. For example, we can choose:

$$\frac{k}{m\omega^2} = 1 \quad\Rightarrow\quad \omega = \sqrt{\frac{k}{m}},\tag{1.100}$$

$$\frac{Q}{m\omega^2\ell} = 1 \quad\Rightarrow\quad \ell = \frac{Q}{m\omega^2} = \frac{Q}{mk/m} = \frac{Q}{k},\tag{1.101}$$

$$\frac{c}{\omega m} = 2\zeta \quad\Rightarrow\quad \zeta = \frac{c}{2\sqrt{km}},\tag{1.102}$$

$$\tilde{\Omega} = \frac{\Omega}{\omega}.\tag{1.103}$$

Note that all of the new parameters introduced by (1.100)–(1.103) have a simple physical interpretation: ω is the undamped natural frequency, ℓ the static deformation of the elastic element having stiffness k when the load is Q, ζ is the damping ratio, and $\tilde{\Omega}$ the ratio of excitation frequency to natural frequency. Substituting (1.100)–(1.103) into (1.99) we find:

$$\ddot{y} + 2\zeta\dot{y} + y = \sin(\tilde{\Omega}\tau) \quad [1]. \tag{1.104}$$

This form of the equation of motion (1.94) is defined in terms of just two parameters: The excitation frequency ratio $\tilde{\Omega}$, and the damping ratio ζ. One could then describe the whole range of possible behaviors, e.g. by plotting frequency responses (stationary amplitudes of $y(\tau)$ versus $\tilde{\Omega}$) for a few representative values of ζ; The *parameter space* $(\tilde{\Omega}, \zeta)$ is just 2-dimensional, and thus not difficult to explore. This contrasts to the original system (1.94), where a 5-dimensional parameter-space (m, c, k, Q, Ω) makes a thorough exploration of parameter dependencies very time-consuming to calculate, and virtually impossible to communicate in a few graphs.

Some may think that with nondimensionalization the connection to the "real" physical system is somehow lost, and the results therefore not "practically" applicable. This is incorrect. The nondimensional equation of motion (1.104) is completely equivalent to the original Eq. (1.94), but just much more convenient to analyze, since there are fewer parameters. It can be used for conveniently illustrating different kinds of solutions, knowing that all inessential parameters have been stripped of. And for calculating solutions for a specific set of real physical parameters (m, c, k, Q, Ω), one can always compute the corresponding nondimensional parameters $(\tilde{\Omega}, \zeta)$ using (1.100)–(1.103), then solve (1.104) for $y(\tau)$, and finally rescale back to the original variables $u(t)$ using (1.95)–(1.96).

1.7 Damping: Types, Measures, Parameter Relations

This section summarizes some useful facts on damping in vibration analysis. Much of it can be derived from the fundamental relations in Sects. 1.2-4. For more detailed treatments see e.g. Adhikari (2013, 2014), Inman (2014), Lazan (1968), Crandall (1970), Woodhouse (1998), Bert (1973), Dahl (1976), Liang and Feeny (1998).

1.7.1 Damping in Equations of Motion

Damping and friction often constitutes the most troublesome part of equations of motions for mechanical systems. The equations of motion can be seen as composed as four main parts: *Inertia forces, elastic restoring forces, damping/friction/ dissipative* (lossy) *forces*, and *excitation* (input) *forces*. Of these the inertia and elastic forces constitutes the energy-conserving part, composed of kinetic and potential energy, respectively. The damping/friction forces and the excitation forces make up the *non*-conservative part, corresponding to dissipated and induced energy, respectively. For the case of linear viscous damping the balances of energy or power can be illustrated by multiplying the SDOF system standard equation of motion (1.1) by the velocity $\dot{x}$, so as to obtain a *power* (rather than force) balance equation, and then integrate wrt. time to obtain an *energy* balance equation of motion:

$$\frac{1}{2}m\dot{x}^2 + \int c\dot{x}^2\,dt + \frac{1}{2}kx^2 == \int F(t)\dot{x}dt, \qquad (1.105)$$

or

$$E = T + V = E_{\text{in}} - E_{\text{out}}, \qquad (1.106)$$

where E is the instantaneous total mechanical energy, $T = \frac{1}{2}m\dot{x}^2$ the kinetic energy, $V = \frac{1}{2}kx^2$ the potential energy, $E_{\text{in}} = \int F(t)\dot{x}dt$ the externally induced energy, and $E_{\text{out}} = \int c\dot{x}^2\,dt$ the energy dissipated via damping and friction.

Of these the dissipated energy E_{out} is by far the most difficult to model, due to the highly complex nature of the numerous mechanisms leading to loss of energy, – mostly *micro*mechanical, as a consequence of the interaction of a large number of individual entities (surface asperities, air or fluid molecules, etc.). Luckily, however, when vibrations are of concern, $E_{\text{out}} \ll E$ is the typical case; otherwise vibrations would quickly fade or be insignificantly small. This means that we can often get away with using even very crude *macro*mechanical models of the energy dissipating, to obtain reasonable agreement of predicted output with laboratory experiments, as long as the dissipated energy is small compared to the total energy.

The best known and most often used example of such crude modeling is linear viscous damping, i.e. assuming the damping forces are simply proportional to relative velocity. This is not based on any deeper principles or believe that real damping forces behave like this, but rather on two facts: (1) If linear modeling is aimed at, then the damping forces *have* to be simply proportional to velocity, and (2) good agreement is typically obtained between theory and experiments with such a simple damping model, as long as damping forces are small compared to other forces involved.

1.7.2   Damping Models

There are many different ways of modelling the energy dissipation of vibrating systems, e.g. see Adhikari (2013, 2014), Lazan (1968), Crandall (1970), Bert (1973). Inman (2014) lists the more common of these as in Table 1.1. Except for the linear viscous damping model, all of them are nonlinear. Nonlinear models usually implies much more complicated vibration analysis (as will be exemplified throughout this book). Sometimes nonlinear analysis cannot be warranted, e.g. with very low levels of vibration and damping. Then a workable option may be to just postulate linear viscous damping anyway, with a damping constant c_{eq}, which is

Table 1.1 Damping force models, in terms of instantaneous displacement $x = x(t)$ and velocity $\dot{x}$, and equivalent linear viscous damping constants c_{eq} at vibration frequency ω and amplitude X

Model name	Damping force	c_{eq}	Source
Linear viscous damping	$c\dot{x}$	c	Slow fluid flow
Air or quadratic damping	$c_1 \text{sgn}(\dot{x})\dot{x}^2 = c_1\dot{x}\lvert\dot{x}\rvert$	$\frac{8c_1\omega X}{3\pi}$	Fast fluid flow
Coulomb damping/friction	$\mu\,\text{sgn}(\dot{x}) = \mu\dot{x}/\lvert\dot{x}\rvert$	$\frac{4\mu}{\pi\omega X}$	Sliding friction
Displacement-squared damping	$c_2\text{sgn}(\dot{x})x^2$	$\frac{4c_2 X}{3\pi\omega}$	Material damping
Solid or structural damping	$c_3\text{sgn}(\dot{x})\lvert x\rvert$	$\frac{2c_3}{\pi\omega}$	Internal damping

calculated by equating the work done by damping forces through one full vibration cycle for, respectively, linear viscous damping the nonlinear damping law in question. Table 1.1 lists the resulting *equivalent linear viscous damping coefficients* c_{eq}; as appears these are not really system constants, but depend on vibration frequency ω and amplitude X.

1.7.3 Damping Measures and Their Relations

Table 1.2 summarizes some key damping parameters and their interrelations. Most of these follows directly from the definitions and relations in Sects. 1.2-4, others can be found in, e.g., Inman (2014), Crandall (1970), Lazan (1968), Adhikari (2013).

Table 1.2 Damping measures and interrelations for a linear SDOF-system with displacement coordinate $x(t)$, concentrated mass m, linear stiffness k, undamped natural frequency $\omega_0 = \sqrt{k/m}$, oscillation period $T = 2\pi/\omega_0$, (displacement/force) frequency response function $H(\omega)$, energy loss during a single oscillation cycle ΔE, maximum potential energy U_{max}, and number of oscillation cycles n

Symbol	Quantity	Interrelations
c	Viscous damping coefficient	$c = \zeta c_{cr} = 2\zeta\sqrt{km} = 2\zeta m\omega_0$
c_{cr}	Critical viscous damping coeff.	$c_{cr} = \frac{c}{\zeta} = 2\sqrt{km} = 2m\omega_0$
ζ	Damping ratio	$\zeta = \frac{c}{c_{cr}} = \frac{\Delta\omega_{3db}}{2\omega_0} = \frac{c}{2\sqrt{km}} = \frac{c}{2m\omega_0} = \frac{\delta}{\sqrt{4\pi^2 + \delta^2}}\ \left(\approx \frac{\delta}{2\pi}\right)$
δ	Logarithmic decrement	$\delta = \ln\left(\frac{x(t)}{x(t+T)}\right) = \frac{1}{n}\ln\frac{x(t)}{x(t+nT)} = \zeta\omega_0 T \approx 2\pi\zeta$
$\Delta\omega_{3dB}$	3 dB bandwidth	$\Delta\omega_{3dB}$ = Width of resonance peak in $H(\omega)$ where the peak response $H(\omega_0)$ has dropped a factor $\sqrt{2}$ (=3 dB)
η	Loss factor/ coefficient	$\eta = \frac{\Delta E}{2\pi U_{max}}$ (=2ζ at resonance for linear visc. damp.)
Q	Quality factor	$Q = \frac{1}{\eta}$

Table 1.3 Freely damped oscillation cycles to reach half amplitude ($n_{1/2}$) or to decimate amplitude ($n_{1/10}$) in dependency of damping ratio ζ

ζ (%)	$n_{1/2}$	$n_{1/10}$
10	1	3
1	11	37
0.1	110	365
0.01	1103	3664

1.7.4 Damping Influence on Free Vibration Decay

With free vibrations the level and character of the damping forces determine how fast, and with which time envelope, vibrations decay. So if this is of concern, careful consideration to damping is necessary. With linear viscous damping the displacement (and thus also velocity and acceleration) amplitude will decay exponentially with time, as illustrated by Fig. 1.2 in Sect. 1.2.2, and quantified further below. With dry friction the decay envelope is different, e.g. pure Coulomb friction/damping gives an acceleration amplitude that decays linearly with time (Lorenz 1924; Lian 2005).

For *linear viscous damping* with damping ratio ζ one can show, using (1.5), that the number $n_{1/r}$ of free oscillations taken to reduce the initial maximum displacement amplitude by a factor of r is:

$$n_{1/r} = \frac{\ln r}{2\pi} \sqrt{\frac{1}{\zeta^2} - 1} \quad \rightarrow \quad \frac{\ln r}{2\pi\zeta} \text{ for } \zeta \rightarrow 0. \tag{1.107}$$

Table 1.3 quantifies examples of using this relation to calculate $n_{1/2}$ and $n_{1/10}$, i.e. the number of oscillation cycles required to, respectively, halve and decimate the initial displacement amplitude. As a simple memo it takes about 10 oscillations to half the amplitude at 1% damping; then just scale up and down in proportion: about 100 oscillations to half the amplitude at 0.1% damping, etc.

1.7.5 Damping Influence on Resonance Buildup

With forced vibrations the level of damping also determines how quickly the response settles into stationary vibrations, i.e. how fast the effect if initial conditions (i.e. the free, unforced part of the vibrations) disappears. Or in mathematical terms: The total response consists of the homogeneous solution plus the particular solution, and damping determines how fast the homogeneous part decays towards zero. With very low damping it may take a very long time for stationary vibrations to settle. This holds with real systems, and also with numerical simulation. So, if damping is ignored in numerical simulation, the homogeneous part of the solution

Table 1.4 Number of forcing cycles N_R for stationary vibrations to reach the fraction R of full steady-state resonant amplitude at damping ratio ζ

ζ (%)	$R_{90\%}$	$R_{99\%}$	$R_{99.9\%}$
10	3.66	7.33	11
1	36.6	73.3	110
0.1	366	733	1100
0.01	3660	7330	11,000
0.001	36,600	73,300	110,000

will never decay, and the predicted output will be a mix of free and forced vibrations at all times – with no resemblance to reality, where the free vibrations will always decay. Here we quantify the relation between damping and the time-to-settle into stationary forced vibrations, i.e. the transient period.

For linear viscous damping one can show that with zero initial conditions (i.e. starting from rest) the *response envelope* with resonant harmonic excitation of a linear single-DOF oscillator is $X(1 - e^{-\zeta\omega_0 t})$, where ζ is the damping ratio, ω_0 the linear natural frequency, $\omega_d = \sqrt{1 - \zeta^2}\omega_0$ the damped natural frequency, and $X = p_0/(2\zeta\omega_0)$ the stationary amplitude of forced harmonic vibrations with forcing amplitude p_0 and frequency $\Omega = \omega_d$. Thus the response is:

$$x(t) = X\left(1 - e^{-\zeta\omega_0 t}\right)\sin(\omega_d t), \tag{1.108}$$

with an amplitude that in the beginning increases from zero almost linearly with time $(1 - e^{-\zeta\omega_0 t} \to \zeta\omega_0 t$ for $t \to 0)$, but later turns into a saturated response with stationary amplitude X (since $1 - e^{-\zeta\omega_0 t} \to 1$ for $t \to \infty)$.

The time taken to reach a certain fraction $R \in [0;1[$ of the full resonant stationary amplitude X is:

$$t_R = \frac{-\ln(1 - R)}{\zeta\omega_0}, \tag{1.109}$$

with a corresponding number N_R of oscillation periods $T(= 2\pi/\Omega = 2\pi/\omega_d)$:

$$N_R = \frac{t_R}{T} \approx \frac{-\ln(1 - R)}{2\pi\zeta}. \tag{1.110}$$

Table 1.4 illustrates how N_R increases as $R \to 1$ and decreases with ζ. For example, with a damping ratio of $\zeta = 0.1\%$ (typical of many joint-less steel structures), the number of oscillations required to reach 99% of the stationary resonant amplitude is about 733 full oscillations. For a wind turbine wing with a lowest natural frequency of 1 Hz this would be about 12 min time, while for a vibrating beam sensor element with a natural frequency of 50 kHz it would be about 15 ms. Such time scales dictates e.g. how long one has to simulate numerically to determine (or wait before experimentally measuring) a stationary response after changing initial conditions, or how fast vibration based sensors can react to changing load.

1.7.6   Estimating Mass/Stiffness-Proportional Damping Constants

The two constants α and β in the stiffness/mass-proportional model of linear viscous damping (cf. Sect. 1.3.6) can be estimated by fitting the model-predicted damping ratios (cf. (1.34)) to experimentally measured quantities. Let $(\omega_j, \zeta_j), j = 1, n, n \geq 2$ be a set of n measured natural frequencies and damping ratios, and define:

$$r_1 = \frac{1}{n}\sum_{j=1}^{n} \frac{1}{\omega_j^2}, \quad r_2 = \frac{1}{n}\sum_{j=1}^{n} \omega_j^2,$$

$$p_1 = \frac{1}{n}\sum_{j=1}^{n} \frac{\zeta_j}{\omega_j}; \quad p_2 = \frac{1}{n}\sum_{j=1}^{n} \zeta_j\omega_j. \tag{1.111}$$

One can then show that the estimate of α and I minimizing the mean square error between measured and model-predicted (by (1.34)) modal damping ratios is:

$$\alpha = \frac{2(r_2 p_1 - p_2)}{r_1 r_2 - 1}, \quad \beta = \frac{2(r_1 p_2 - p_1)}{r_1 r_2 - 1}. \tag{1.112}$$

1.7.7   Mass/Stiffness Damping Proportionality Constants for Beams

Consider the equation of motion for transverse vibrations of a beam with both internal (cf. Sect. 1.4.6) and external linear viscous damping:

$$\rho A \ddot{u} + EI u'''' + c_1 \dot{u} + c_2 \dot{u}'''' + q(x, t) = 0, \tag{1.113}$$

where c_1 is the coefficients of *external damping* (proportional to velocity, expressing e.g. external air resistance), while c_2 is the coefficient of *internal damping* (proportional to rate of change of curvature or bending moment, expressing energy loss associated with beam material deformation). Performing a mode shape expansion as in Sect. 1.5.3, one arrives at a system in the form (1.29), with a damping matrix of the form (1.30), where the mass- and stiffness-proportionality constants are:

$$\alpha = \frac{c_1}{\rho A}, \quad \beta = \frac{c_2}{EI}. \tag{1.114}$$

Thus there is a simple and direct relation between the external damping constant c_1 and the mass-proportional damping constant α, and also between the internal damping constant c_2 and the stiffness-proportional damping constant β.

1.8 The Stiffness and Flexibility Methods for Deriving Equations of Motion

Newton's 2nd and 3rd third law can always be used for setting up equations of motion. For systems with several degrees of freedom, direct use of Newton's laws can be quite elaborate, and we may resort to energy methods like Lagrange's equations or Hamilton's principle (Sect. 1.5). But several other methods can be derived from Newton's laws. Here we summarize the *flexibility method* and the *stiffness method*, which work for lumped-mass type structures.

1.8.1 Common Basis

Assume n masses m_j, $j = 1, \ldots, n$ being attached to a flexible, massless structure, and the problem is to find the equations governing their time-dependent motions $\mathbf{x} = \{x_1 \ x_2 \ \cdots \ x_n\}^T$. First separate the masses from the structure, replacing their actions with unknown forces, that is, make free-body diagrams for each mass. Then, supposing a linear model is aimed at, a massless structure will remain, where the forces $\mathbf{f} = \{f_1 \ f_2 \ \cdots \ f_n\}^T$ and displacements $\mathbf{x}$ are related by

$$\mathbf{f} = \mathbf{Kx}, \tag{1.115}$$

where $\mathbf{K}$ is the *stiffness matrix*. The lack of acceleration terms does not mean there are no accelerations, but is a consequence of the lack of mass, so that the mass $\times$ acceleration terms in Newton's 2nd law vanish. Note that $\mathbf{x}$ should be understood as *generalized deformations* (translations, rotations, etc.), and $\mathbf{f}$ as *generalized forces* (linear forces, twisting moments, bending moments, etc.).

Multiplying (1.115) by the *flexibility matrix* $\mathbf{A} = \mathbf{K}^{-1}$, the force-deformation relation can also be expressed in the inverse form:

$$\mathbf{x} = \mathbf{Af}. \tag{1.116}$$

The forces $\mathbf{f}$ arising from detached masses can be calculated from the free-body diagrams. Using Newton's 2nd law for mass j gives[3] $-f_j = m_j\ddot{x}_j$, where the negative sign on f_j reflects we are dealing with a reaction forces to those contained in $\mathbf{f}$ in (1.115) and (1.116). Expressing this in matrix form as $\mathbf{f} = -\mathbf{M}\ddot{x}$, where $\mathbf{M}$ = diag $(m_1 \, m_2 \, \cdots \, m_n)$, one obtains by insertion into (1.115) and (1.116), respectively:

$$\mathbf{M}\ddot{x} + \mathbf{K}x = \mathbf{0}, \tag{1.117}$$

$$\mathbf{A}\mathbf{M}\ddot{x} + x = \mathbf{0}, \tag{1.118}$$

which are two equivalent forms of equations of motion for the system. The stiffness- and flexibility methods, respectively, results in these two forms of equations of motion. With the stiffness method one has to set up the stiffness matrix $\mathbf{K}$, while with the flexibility method the flexibility matrix $\mathbf{A}$ is required. Typically, for a given problem, one of these matrices is easier to obtain than the other. Next we outline and exemplify how each method is used.

1.8.2   The Flexibility Method

The flexibility method will produce at system in the form (1.118) with $\mathbf{M}$ = diag(m_1 $m_2 \cdots m_n$). The elements a_{ij} of the flexibility matrix $\mathbf{A}$ are called *flexibility influence coefficients* or just *flexibility coefficients*. They are determined this way:

a_{ij} is the value of x_i when f_j=1 and all other f-variables are zero.

To see why this is so it will suffice to write out the components of relation (1.116) for a system with three degrees of freedom ($n = 3$):

$$\begin{Bmatrix} x_1 \\ x_2 \\ x_3 \end{Bmatrix} = \begin{bmatrix} a_{11} & a_{12} & a_{13} \\ a_{21} & a_{22} & a_{23} \\ a_{31} & a_{32} & a_{33} \end{bmatrix} \begin{Bmatrix} f_1 \\ f_2 \\ f_3 \end{Bmatrix} = \begin{Bmatrix} a_{11}f_1 + a_{12}f_2 + a_{13}f_3 \\ a_{21}f_1 + a_{22}f_2 + a_{23}f_3 \\ a_{31}f_1 + a_{32}f_2 + a_{33}f_3 \end{Bmatrix}. \tag{1.119}$$

As appears one can compute a coefficient such as a_{23} as the value of x_2 obtained when letting $f_3 = 1$ and $f_1 = f_2 = 0$, since then the equation for x_2 becomes $x_2 = a_{23}$; this is just what the rule dictates.

The computation of flexibility coefficients is often particularly convenient for beam-type problems where the required displacements due, to a single imposed force, can be determined from standard lookup tables.

---

[3]Sometimes this is expressed instead as $f_j = -m\ddot{x}_j$ and then termed *d'Alembert's Principle*, but actually it follows directly from Newton's 2nd and 3rd law.

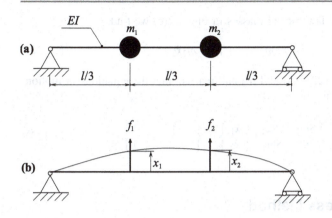

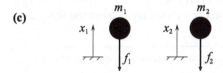

Fig. 1.7 Stiffness method example system

Example The massless beam with two point-masses in Fig. 1.7(a) has two degrees of freedom. Fig. 1.7(b) shows the system stripped from its masses, whose influence are replaced by forces f_1 and f_2. These same forces appear as oppositely directed reaction forces in Fig. 1.7(c), showing the free-body diagrams for the masses. Using the flexibility method we start by writing the components of the linear force relation $\mathbf{x} = \mathbf{A}\mathbf{f}$:

$$\left\{ \begin{array}{c} x_1 \\ x_2 \end{array} \right\} = \left\{ \begin{array}{c} a_{11}f_1 + a_{12}f_2 \\ a_{21}f_1 + a_{22}f_2 \end{array} \right\}. \tag{1.120}$$

Using the above boxed rule and Fig. 1.7(b) we find, using a lookup table (e.g. Gere and Timoshenko 1997; Gross et al. 2011) for a simply supported beam:

$$
\begin{aligned}
a_{11} &= (\text{ value of } x_1 \text{ when } f_1 = 1 \text{ and } f_2 = 0) = \frac{4l^3}{243EI}, \\
a_{12} &= (\text{ value of } x_1 \text{ when } f_2 = 1 \text{ and } f_1 = 0) = \frac{7l^3}{486EI}, \\
a_{21} &= (\text{ value of } x_2 \text{ when } f_1 = 1 \text{ and } f_2 = 0) = \frac{7l^3}{486EI}, \\
a_{22} &= (\text{ value of } x_2 \text{ when } f_2 = 1 \text{ and } f_1 = 0) = \frac{4l^3}{243EI}.
\end{aligned}
\tag{1.121}
$$

Using Newton's 2nd law for the masses in Fig. 1.7(c) we find:

$$-f_1 = m_1\ddot{x}_1; \quad -f_2 = m_2\ddot{x}_2. \tag{1.122}$$

Inserting (1.121)–(1.122) into (1.120) and rearranging, the equations of motion take the form of (1.118):

$$\frac{l^3}{486EI}\begin{bmatrix} 8m_1 & 7m_2 \\ 7m_1 & 8m_2 \end{bmatrix}\begin{Bmatrix} \ddot{x}_1 \\ \ddot{x}_2 \end{Bmatrix} + \begin{Bmatrix} x_1 \\ x_2 \end{Bmatrix} = \begin{Bmatrix} 0 \\ 0 \end{Bmatrix}. \tag{1.123}$$

1.8.3 The Stiffness Method

The stiffness method will produce at system in the form (1.117) with $\mathbf{M} = \text{diag}(m_1\ m_2 \cdots m_n)$. The elements k_{ij} of the stiffness matrix $\mathbf{K}$ are called *stiffness influence coefficients* or just stiffness coefficients. The coefficients are determined this way:

k_{ij} is the value of the force f_i required to keep the system in a position where x_j = 1 while all the other x-values are zero.

To see this we write out the components of relation (1.115) for a system with three degrees of freedom:

$$\begin{Bmatrix} f_1 \\ f_2 \\ f_3 \end{Bmatrix} = \begin{bmatrix} k_{11} & k_{12} & k_{13} \\ k_{21} & k_{22} & k_{23} \\ k_{31} & k_{32} & k_{33} \end{bmatrix}\begin{Bmatrix} x_1 \\ x_2 \\ x_3 \end{Bmatrix} = \begin{Bmatrix} k_{11}x_1 + k_{12}x_2 + k_{13}x_3 \\ k_{21}x_1 + k_{22}x_2 + k_{23}x_3 \\ k_{31}x_1 + k_{32}x_2 + k_{33}x_3 \end{Bmatrix}. \tag{1.124}$$

As appears, a coefficient such as k_{23} can be computed as the value of f_2 obtained when letting $x_1 = x_2 = 0$ and $x_3 = 1$, because in that case the equation for f_2 becomes $f_2 = k_{23}$, as the rule says.

The computation of stiffness coefficients is often convenient in cases where flexible elements are connected in series, as in the below example.

Example The system in Fig. 1.8(a) has two degrees of freedom. Figure 1.8(b) shows the system stripped from its masses, whose influence are replaced by forces f_1 and f_2. These same forces appear as reaction forces in Fig. 1.8(c), showing the free-body diagrams of the masses. Using the stiffness method we start by writing the linear force relation $\mathbf{f} = \mathbf{K}x$ component-wise:

$$\begin{Bmatrix} f_1 \\ f_2 \end{Bmatrix} = \begin{Bmatrix} k_{11}x_1 + k_{12}x_2 \\ k_{21}x_1 + k_{22}x_2 \end{Bmatrix}. \tag{1.125}$$

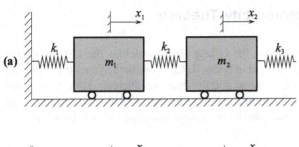

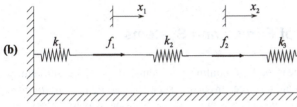

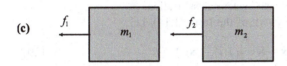

Fig. 1.8 Flexibility method example system

Using the above boxed rule and Fig. 1.8(b) we find:

$$
\begin{aligned}
k_{11} &= (\text{ value of } f_1 \text{ when } x_1 = 1 \text{ and } x_2 = 0 \text{ fixed }) = k_1 + k_2, \\
k_{12} &= (\text{ value of } f_1 \text{ when } x_2 = 1 \text{ and } x_1 = 0 \text{ fixed }) = -k_2, \\
k_{21} &= (\text{ value of } f_2 \text{ when } x_1 = 1 \text{ and } x_2 = 0 \text{ fixed }) = -k_2, \\
k_{22} &= (\text{ value of } f_2 \text{ when } x_2 = 1 \text{ and } x_1 = 0 \text{ fixed }) = k_2 + k_3 ,
\end{aligned}
\tag{1.126}
$$

while using Newton's 2nd law for the masses in Fig. 1.8(c) gives:

$$
-f_1 = m_1 \ddot{x}_2; \quad -f_2 = m_2 \ddot{x}_2.
\tag{1.127}
$$

Inserting (1.126)–(1.127) into (1.125) and rearranging, the equations of motion take the form of (1.117):

$$
\begin{bmatrix} m_1 & 0 \\ 0 & m_2 \end{bmatrix}
\begin{Bmatrix} \ddot{x}_1 \\ \ddot{x}_2 \end{Bmatrix}
+
\begin{bmatrix} k_1 + k_2 & -k_2 \\ -k_2 & k_2 + k_3 \end{bmatrix}
\begin{Bmatrix} x_1 \\ x_2 \end{Bmatrix}
=
\begin{Bmatrix} 0 \\ 0 \end{Bmatrix}.
\tag{1.128}
$$

1.8.4   Maxwell's Reciprocity Theorem

The reciprocity theorem states that, for a linear elastic structure, the stiffness matrix and the flexibility matrix (as defined in Sects. 1.8.2-3) are both symmetric. This can ease the burden of computing stiffness- and flexibility-coefficients, since $k_{ij} = k_{ji}$ and $a_{ij} = a_{ji}$ for $(i, j) = 1, n$. The theorem can be proved by calculating and equating the work done by two forces f_i and f_j in different sequence (first j after i, then i after j).

1.9   Classification of Forces and Systems

Linear systems (or the linear part of nonlinear systems, if any) are sometimes classified according to the kind of forces involved, such as *instationary, dissipative, (non-)circulatory*, and *gyroscopic*. Here we just present a simple way to perform the classification, referring to the Ziegler (1968) for a thorough treatment, and to problems 4.7–4.10 that involves also system classification.

Consider a linear/linearized system of the form (1.14), i.e.:

$$\mathbf{M}\ddot{\mathbf{x}} + \mathbf{C}\dot{\mathbf{x}} + \mathbf{K}\mathbf{x} = \mathbf{f}(t), \quad \mathbf{x}(t) \in R^n, \tag{1.129}$$

which could model a multiple-DOF discrete mechanical system, or arise from using mode shape expansion (Sect. 1.5.3) for a continuous system (then $\mathbf{x}$ would be the modal amplitudes). Here the mass matrix $\mathbf{M}$ is assumed to be symmetric and positive definite; this will normally be the case when the system is properly restricted in space, and when symmetry has not been destroyed by manipulating individual equations (e.g. multiplying any of the n equations in (1.129) by a constant would preserve model validity but destroy the symmetry of $\mathbf{M}$ and $\mathbf{K}$). For this rather general linear system we next describe the classification of *forces*, which is then subsequently used as a basis for classifying *systems*, which in turn provides a basis for *stability considerations*.

1.9.1   Force Classification

Any square matrix $\mathbf{B}$ can be split into symmetric and antisymmetric parts as follows:

$$\mathbf{B} = \mathbf{B}_S + \mathbf{B}_A, \tag{1.130}$$

where $\mathbf{B}$'s symmetric part $\mathbf{B}_S = \mathbf{B}_S^T$ and antisymmetric part $\mathbf{B}_A = -\mathbf{B}_A^T$ can be calculated as:

$$\mathbf{B_S} = \frac{1}{2}(\mathbf{B} + \mathbf{B}^T),$$
$$\mathbf{B_A} = \frac{1}{2}(\mathbf{B} - \mathbf{B}^T). \tag{1.131}$$

Based on just the time-dependency and symmetry properties of $\mathbf{M}$, $\mathbf{C}$, $\mathbf{K}$, and $\mathbf{f}$, the forces involved can be classified as follows:

1. *Instationary forces* are present if $\mathbf{M}$, $\mathbf{C}$, $\mathbf{K}$, or $\mathbf{f}$ depend on time t. Examples are externally applied time-varying transverse loads $\mathbf{f}$ from e.g. vibration exciters, connected machinery, or natural sources like traffic or wind, or axially applied time-varying loads leading to time-varying $\mathbf{K}$, or loading from propulsion engines leading to time-varying $\mathbf{M}$.
2. *Circulatory forces* are present if the stiffness matrix has an antisymmetric part, i.e. if $\mathbf{K_A} = -\mathbf{K_A}^T = \frac{1}{2}(\mathbf{K} - \mathbf{K}^T) \neq \mathbf{0}$. Examples include follower-type forces where the load direction follows the structural deformation, as with water expelling from a hose and aeroelastic loads.
3. *Non-circulatory forces* are present if the stiffness matrix has a symmetric part, i.e. $\mathbf{K_S} = \mathbf{K_S}^T = \frac{1}{2}(\mathbf{K} + \mathbf{K}^T) \neq \mathbf{0}$. Examples include potential forces like gravity and linear restoring forces, i.e. some of the most common forces in mechanics.
4. *Gyroscopic forces* are present if the damping matrix has an antisymmetric part, i.e. $\mathbf{C_A} = -\mathbf{C_A}^T = \frac{1}{2}(\mathbf{C} - \mathbf{C}^T) \neq \mathbf{0}$. Examples are centrifugal and Coriolis (fictitious/pseudo) forces.
5. *Dissipative forces* are present if the damping matrix has a symmetric part, i.e. $\mathbf{C_S} = \mathbf{C_S}^T = \frac{1}{2}(\mathbf{C} + \mathbf{C}^T) \neq \mathbf{0}$. The damping forces are *completely* dissipative if $\mathbf{C_S}$ is positive definite (implying that every possible system movement $\mathbf{x}$ with finite velocity will be accompanied by energy dissipation). Dissipative forces, though the linear form is an idealization, are present in any real mechanical system, accompanying relative motion with, e.g., air resistance, sliding surfaces, and material deformation.

Among these forces only the non-circulatory and the gyroscopic forces are conservative (doing or extracting no net work on/from the system), while instationary, circulatory, and dissipative forces are nonconservative.

Note that the classification only involves the symmetry properties and time-dependency (presence or absence) of the system matrices and vectors; the actual values of $\mathbf{M}$, $\mathbf{C}$, $\mathbf{K}$, and $\mathbf{f}$ need not to be known. For example, a system of the form (1.129) could arise from mode shape expansion of the partial differential equation for a continuous elastic beam, with elements of the damping matric $\mathbf{C}$ being $c_{ij} = \int_0^l c(x)\varphi_i(x)\varphi_j(x)\,dx$ for an axial damping distribution $c(x)$ (cf. Sect. 1.4.6). Thus here $c_{ij} = c_{ji}$ for all $i, j = 1, n$, independent of the n mode shapes φ_j, which do not need to be known or assumed to conclude that $\mathbf{C}$ is symmetric, so that there are dissipative forces but no gyroscopic forces involved. A mode shape expansion utilized this way, with no actual mode shapes known or assumed, is called a *formal mode shape expansion*.

1.9.2 System Classification

Systems of the type (1.129) can be classified according to the forces involved:

1. *Conservative system*: All forces are conservative

   1.1. *Gyroscopic conservative system*: Conservative system with gyroscopic forces
   1.2. *Non-gyroscopic conservative system*: Conservative system with*out* gyroscopic forces

2. *Non-conservative system*:

   2.1. *[Purely] Dissipative system*: [Only conservative and] Dissipative forces
   2.2. *[Purely] Circulatory system*: [Only conservative and] Circulatory forces
   2.3. *[Purely] Instationary system*: [Only conservative and] Instationary forces

Some of the descriptors may combine, so that a system can be classified as, for example, 'purely and completely dissipative', or 'circulatory and dissipative', or 'gyroscopic conservative with non-circulatory forces'.

System classification may point to certain types of dynamic behavior to expect, in particular to the types of instabilities that may occur, and – as described next – to possible ways of examining this.

1.9.3 Stability Assessment

Without external forcing ($\mathbf{f} = \mathbf{0}$) Eq. (1.129) has a static equilibrium solution $\mathbf{x} = \dot{\mathbf{x}} = \mathbf{0}$. This equilibrium can be *stable* or *unstable*, or *marginally stable* (i.e. at the border separating stable from unstable). Image the system starting at the equilibrium at rest, i.e. $\mathbf{x}(0) = \dot{\mathbf{x}}(0) = \mathbf{0}$, and then subjected to a small disturbance, e.g. $\dot{\mathbf{x}}(0_+) \neq \mathbf{0}$. As time t increases the state $\mathbf{x}(t)$ may reapproach the equilibrium asymptotically so that $\mathbf{x} \to 0$ for t→∞; in that case the equilibrium $\mathbf{x} = \mathbf{0}$ is *stable*. Or $\mathbf{x}$ may diverge exponentially away from $\mathbf{0}$ so that $|\mathbf{x}| \to \infty$, in which case $\mathbf{x} = \mathbf{0}$ is *unstable*. Or the state may stay close to the equilibrium, and $\mathbf{x} = \mathbf{0}$ is *marginally unstable*.

Note that it is the (equilibrium) *state* of the system that can be stable or unstable, not 'the system' itself. Think of the unforced mathematical pendulum, having two equilibria – a down-pointing/stable, and an up-pointing/unstable (in the absence of damping the down-pointing equilibrium is only marginally stable). It makes no sense to address stability of 'the pendulum', but the pendulum *state* has two static equilibria, one stable (down) and the other unstable (up).

There are several ways to calculate equilibrium stability (more on this in Chaps. 3–5.). Ziegler (1968) provides a number of theorems for assessing how stability of the zero solution of (1.129) (with $\mathbf{f} = \mathbf{0}$) can be assessed in a possibly simple manner, dependent on the system class. It is useful to be able to identify such cases, which can save efforts as compared to the more elaborate methods in Chaps. 3–5.

Most important in this respect is *Lagrange's theorem*, from which follows that for *purely dissipative systems* the equilibrium $\mathbf{x} = \mathbf{0}$ of (1.129) (with $\mathbf{f} = \mathbf{0}$) is stable if $\mathbf{K}$ is positive definite. Since $\mathbf{K}$ involves only the static part of the system, a full dynamic analysis is then not needed to assess the stability of $\mathbf{x} = \mathbf{0}$.

When the system is not only purely but also *completely* dissipative (i.e. $\mathbf{C_S}$ is positive definite) a stronger statement can be made (Ziegler 1968, Theorem 8): The equilibrium $\mathbf{x} = \mathbf{0}$ is stable if $\mathbf{K}$ is positive definite, and unstable if $\mathbf{K}$ is *not* positive definite. Again, dynamic analysis is not needed.

Another important theorem says that the same holds for *non-gyroscopic conservative systems*, i.e. $\mathbf{x} = \mathbf{0}$ is (un)stable if $\mathbf{K}$ is (not) positive definite (Ziegler 1968, Theorem 1). Also, it can be shown that gyroscopic forces cannot destabilize $\mathbf{x} = \mathbf{0}$ for a conservative system, implying that stability can be checked with gyroscopic forces ignored (Ziegler 1968, Theorem 5).

For most other system classes the stability of $\mathbf{x} = \mathbf{0}$ cannot be assessed by considering just the static part (i.e. $\mathbf{K}$). Then a full dynamic analysis of stability is required, involving all terms in (1.129), as described in Chaps. 3-5 of this book.

1.10 Problems

Problem 1.1 Solve $\ddot{u} + 2\zeta\omega\,\dot{u} + \omega^2 u = 0$ for $u(t)$ when $u(0) = u_0$, $\dot{u}(0) = 0$, and $0 < \zeta < 1$, and sketch the solution.

Problem 1.2 Solve $\ddot{u} + \zeta\omega\,\dot{u} + \omega^2 u = p\cos(\Omega t)$ for $u(t)$, and sketch the stationary amplitude of $u(t)$ as a function of Ω when, respectively, $\zeta = 0$ and $0 < \zeta \ll 1$.

Problem 1.3 Solve $\ddot{u} + \zeta\omega\,\dot{u} + \omega^2 u = p\delta(t - t_0)$ and sketch the solution for $\zeta \ll 1$.

Problem 1.4 For the model system shown in Fig. 1.4 (Sect. 1.3.1), Calculate the undamped natural frequencies and associated mode shapes when, respectively, $k_1 = k_2 = k$ and $k_1 = k_2/2 = k$.

Problem 1.5 For the model system shown in Fig. 1.4 (Sect. 1.3.1), Calculate the forced response when $F(t) = F_0\sin(\Omega t)$, $a = l/2$, $k_1 = k_2 = k$ and $c_1 = c_2 = c$.

Problem 1.6 The mass m in Fig. P1.6 moves in a plane (x, y), restricted by springs with linear stiffness k.

a) Use Lagrange's equations for setting up the equations of motion.
b) Expand nonlinear terms to order three to yield polynomial nonlinearities.
c) Linearize the equations of motion for small oscillations near $(x, y) = (0, 0)$.
d) Compute the undamped natural frequencies and mode shapes.

Problem 1.7 The beam in Fig. P1.7 is simply supported, has length l, flexural stiffness EI and mass per unit length ρA. It is subjected to a time-varying load $P(t)$ at $x = x_0$, where a linear spring of relatively small stiffness k restricts the motion.

a) Use Hamilton's principle to set up the equation of motion governing small-amplitude transverse vibrations $w(x, t)$.
b) Employ a mode shape expansion of $w(x, t)$, and obtain a set of ordinary differential equations governing the modal amplitudes.
c) Discuss the validity of the expansion when, respectively, $k \to 0$, $k \to \infty$, $x_0 \to 0$ (or $x_0 \to l$) and $x_0 \to l/2$.

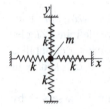

Fig. P1.6   **Fig. P1.7**

Problem 1.8 Assume the mass m in Fig. P1.6 has a rotary inertia J, which is too large to be ignored.

a) Use Hamilton's principle to set up the equation of motion governing small-amplitude transverse vibrations $u(x, t)$.
b) Employ a mode shape expansion of $u(x, t)$, and obtain a set of ordinary differential equations governing the modal amplitudes.

Problem 1.9 The figure shows *Stutts' sliding bar experiment*[4], where a rigid bar slides on a pair of *inward* (Case 1) or *outward* (Case 2) grooved discs driven by a crossed V-belt. The bar has mass m in gravity g, and slides with a kinetic coefficient of friction μ_k. The center-to-center distance between the discs is $2L$, and the instantaneous horizontal displacement of the center of mass is $x(t)$.

a) Use simple physical reasoning to predict the nature of the motion of the bar in the x-direction.
b) Show that the equation for the x-motions is

$$\ddot{x} \pm \frac{\mu_k g}{L} x = 0, \tag{1.132}$$

where the plus sign is for case 1, and the minus sign for case 2.

c) Inspecting the equation of motion, without actually solving it – what does it tell about the nature of the motions?

---

[4]Prof. Daniel S. Stutts, Missouri Univ. of Science and Technology.

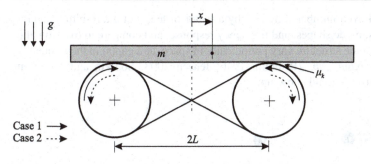

Fig. P1.9

Problem 1.10 Consider the linear flexible structure in Fig. P1.10, with n masses and equation of motion:

$$\mathbf{M}\ddot{\mathbf{x}} + \mathbf{K}\mathbf{x} = \mathbf{F}\sin(\Omega t), \tag{1.133}$$

where $\mathbf{M} = \text{diag}\{ m_1 \quad m_2 \quad \cdots \quad m_n \}$, $\mathbf{F} = \{f_1 \quad 0 \quad \cdots \quad 0\}^T$, and:

$$\mathbf{K} = \begin{bmatrix} k_1 & -k_1 & 0 & 0 & \cdots & 0 \\ -k_1 & k_1+k_2 & -k_2 & 0 & \cdots & 0 \\ 0 & -k_2 & k_2+k_3 & -k_3 & \cdots & 0 \\ \vdots & \vdots & \vdots & \ddots & \vdots & \vdots \\ 0 & 0 & 0 & -k_{n-2} & k_{n-2}+k_{n-1} & -k_{n-1} \\ 0 & 0 & 0 & 0 & -k_{n-1} & k_{n-1} \end{bmatrix}. \tag{1.134}$$

a) Show that the stationary output at mass n is:

$$x_n = a_n \sin(\Omega t), \tag{1.135}$$

where $a_n = \{\mathbf{a}\}_n$ is the amplitude of the n'th mass, and

$$\mathbf{a} = (\mathbf{K} - \Omega^2 \mathbf{M})^{-1}\mathbf{F}. \tag{1.136}$$

b) Let $f_1 = 1$, $m_i = k_i = 1$, $i = 1, n$, $n = 10$, and use a numerical tool to compute and plot the natural frequencies, the mode shapes, and the frequency response $|H(\Omega)| = |a_n(\Omega)|/f_1$. This corresponds to a uniform elastic structure, so $|H(\Omega)|$ should show distributed resonance spikes.

c) Increase all even-numbered masses by a factor of ten, plot the resulting natural frequencies, mode shapes, and frequency response, and compare to (b). Now the properties of the structure vary periodically in space, and gap(s) in the frequency response – *band gaps* (Brillouin 1946; Jensen 2003; Sorokin and Thomsen 2015a, 2016) – should show up.

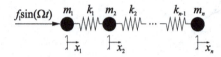

Fig. P1.10

2 Eigenvalue Problems of Vibrations and Stability

2.1 Introduction

Solving problems of vibrations and stability almost inevitably involves solving *eigenvalue problems* (EVPs). In vibration analysis one is faced with an EVP when free responses are to be determined, or when forced responses are to be expanded in terms of free mode shapes. Also, EVPs occur when critical buckling loads and modes are to be determined in problems of elastic stability. EVPs occur repeatedly in subsequent chapters of this book.

There are two types of EVPs: *algebraic* (or *matrix*) EVPs, associated with discrete finite-DOF systems, and *differential* EVPs, associated with systems described by ordinary or partial differential equations. Algebraic and differential EVPs share many concepts and features, though posed in forms mathematically quite distinct. However, whereas all algebraic EVPs are alike and rather simple to solve, differential EVPs come in various forms and the mathematical analysis is more involved.

This chapter focuses on differential EVPs for ordinary differential equations. To some extend we shall abstract from the details of specific EVPs. Instead, using the notion of *operators*, we present a *formal* analysis of a rather broad class of EVPs to which many problems of vibration and stability belongs.

The theory of EVPs is well established, with a formal set of theorems and proofs. We state those theorems that are important for applications, and a few proofs illustrating techniques that one should master when coping with EVPs.

Certainly, the complete and true story on differential EVPs cannot be told in a single chapter of acceptable size. Collatz (1963) devoted an entire textbook to the subject, the main reference of this chapter. Other references are Flügge (1962), Inman (2014), Leipholz (1977), and Meirovitch (1967, 2001).

© The Author(s), under exclusive license to Springer Nature Switzerland AG 2021
J. J. Thomsen, *Vibrations and Stability*,
https://doi.org/10.1007/978-3-030-68045-9_2

2.2   The Algebraic EVP

Some familiarity with *algebraic* EVPs is required when dealing with *differential* EVPs. For example, many methods for obtaining approximate solutions to differential EVPs require one to solve algebraic EVPs. The two types of EVPs share many concepts, though the meaning of these may be different.

2.2.1   Mathematical Form

Algebraic EVPs take the form:

$$\mathbf{K}\boldsymbol{\varphi} = \lambda \mathbf{L}\boldsymbol{\varphi}, \tag{2.1}$$

where φ is an n-vector of state variables, $\mathbf{K}$ and $\mathbf{L}$ are $n \times n$ matrices with components depending on the data of the system, and λ is a scalar. All quantities can be complex-valued.

Equation (2.1) is solved by $\varphi = \mathbf{0}$, $\lambda \in R$. However, solving the algebraic EVP means to find those values of λ for which *nontrivial solutions* $\varphi \neq \mathbf{0}$ exist. The values of λ allowing this are called *eigenvalues*, and the associated vectors φ are called *eigen-vectors*. In general there are n eigenvalues and eigenvectors of the EVP. The pairs (λ_j, φ_j), $j = 1, n$ are called *eigenpairs*.

Example 2.1. In locating undamped natural frequencies and mode shapes for multi-DOF systems, one is lead to solve the system of homogeneous equations $(\mathbf{K} - \omega^2 \mathbf{M})\varphi = \mathbf{0}$, where $\mathbf{K}$ is the stiffness matrix, $\mathbf{M}$ the mass matrix, ω^2 the squared natural frequency, and φ the mode shape (cf. (1.18)). This is an algebraic EVP of the form (2.1), with ω^2 being the eigenvalue.

2.2.2   Properties of Eigenvalues and Eigenvectors

For assessing questions of stability, it is often essential to know whether the eigenvalues of a given EVP are real-valued. Here is a sufficient condition:

Theorem 2.1 : Real-valueness of eigenvalues. If the EVP (2.1) is *Hermetian*, and if $\mathbf{K}$ is *definite* or $\mathbf{K}$ and $\mathbf{L}$ *commute*, then all eigenvalues λ are real.

Some definitions may be in order to appreciate this theorem: The EVP (2.1) is *Hermetian* if $\mathbf{K}$ and $\mathbf{L}$ are both Hermetian matrices. A matrix $\mathbf{K}$ is Hermetian or *self-adjoint* if $\mathbf{K} = \bar{\mathbf{K}}^T$, where the overbar denotes *complex conjugation* and $\mathbf{K}^T$ is the *matrix transpose* of $\mathbf{K}$. A *definite matrix* is positive or negative definite. A matrix $\mathbf{K}$ is *positive definite* if $\mathbf{z}^T\mathbf{Kz} > 0$ and *negative definite* if $\mathbf{z}^T\mathbf{Kz} < 0$ for any *test vector* $\mathbf{z}$, where a *test vector* is any vector containing at least one non-zero element.

Example 2.2. Consider a vibration-related EVP (as in Example 2.1) having real-valued, symmetric stiffness and mass matrix, and positive definite stiffness matrix. By Theorem 2.1, the eigenvalues are all real-valued.

Many algebraic EVPs of applied mechanics possess eigenvectors that are characterized by *orthogonality*. This property can be used for decoupling otherwise coupled equations of motions, as illustrated in Sect. 1.3.6. We have:

Theorem 2.2 Orthogonality of eigenvectors. If the matrices $\mathbf{K}$ and $\mathbf{L}$ of the EVP (2.1) are *symmetric*, then any two eigenvectors $\boldsymbol{\varphi}_i$ and $\boldsymbol{\varphi}_j$ are *orthogonal* in the sense that $\boldsymbol{\varphi}_i^T\mathbf{K}\boldsymbol{\varphi}_j = \boldsymbol{\varphi}_i^T\mathbf{L}\boldsymbol{\varphi}_j = 0$ for any $i \neq j$; $i, j = 1, n$.

2.2.3 Methods of Solution

There are numerous methods for calculating eigenvalues and eigenvectors of algebraic EVPs. The most simple relies on a *direct calculation of the characteristic polynomial*, whereby roots λ_j (generally complex-valued) of the polynomial determinant equation $|\mathbf{K} - \lambda\mathbf{L}| = 0$ are calculated, and in turn substituted back into the EVP to yield the associated eigenvectors $\boldsymbol{\varphi}_j$. This method is inefficient if $\mathbf{K}$ and $\mathbf{L}$ are large matrices, and if more eigenvalues than the lowest few are in need. Then various iterative methods become relevant, among these *Jacobi's method, inverse iteration* and (in particular powerful) *subspace iteration* (e.g., Bathe and Wilson 1976).

We illustrate here the method of inverse iteration. By this technique vector-iterates $\mathbf{u}_{[k+1]}$ are generated by solving the linear system of equations $\mathbf{Ku}_{[k+1]} = \mathbf{Lu}_{[k]}$ for $k = 0, 1, \ldots$, starting with an arbitrary vector $\mathbf{u}_{[0]} \neq \mathbf{0}$. The sequence $\mathbf{u}_{[1]}, \mathbf{u}_{[2]}, \ldots$ then converges towards the first eigenvector of the system. For computational convenience a slightly modified algorithm is often used:

Inverse Iteration. For the EVP $\mathbf{K}\boldsymbol{\varphi} = \lambda\mathbf{L}\boldsymbol{\varphi}$:

1. Let $\mathbf{u}_{[0]} = \{1 \ 1 \ \cdots \ 1\}^T$, $\lambda_{[0]} = 1$ and $\mathbf{z}_{[0]} = \mathbf{Lu}_{[0]}$
2. Iterate for $k = 0, 1, \ldots$, until $|\lambda_{[k+1]} - \lambda_{[k]}|/\lambda_{[k]} < \varepsilon \ll 1$:

   (a) Solve $\mathbf{K}\tilde{\mathbf{u}}_{[k+1]} = \mathbf{z}_{[k]}$ to find $\tilde{\mathbf{u}}_{[k+1]}$

   (b) Compute $\tilde{\mathbf{z}}_{[k+1]} = \mathbf{L}\tilde{\mathbf{u}}_{[k+1]}$ and $r_{[k+1]}^2 = \tilde{\mathbf{u}}_{[k+1]}^T\tilde{\mathbf{z}}_{[k+1]}$

   (c) Compute $\lambda_{[k+1]} = \tilde{\mathbf{u}}_{[k+1]}^T\mathbf{z}_{[k]}/r_{[k+1]}^2$ and $\mathbf{z}_{[k+1]} = \tilde{\mathbf{z}}_{[k+1]}/r_{[k+1]}$

3. If iterations has converged at step $k = p$, then $\lambda_1 \approx \lambda_{[p+1]}$ and $\boldsymbol{\varphi}_1 \approx \tilde{\mathbf{u}}_{[p+1]}/r_{[p+1]}$

The method of inverse iterations can be extended to yield also higher eigen-values and eigen-vectors (λ_j, φ_j), $j = 2, 3, \ldots$. However, for that purpose the method of subspace iteration (Bathe and Wilson 1976) is more efficient and stable.

2.3 The Differential EVP

Algebraic EVPs, as described above, are defined in terms of matrices and vectors. *Differential* EVPs, to be considered here, are characterized by differential operators, continuous functions, and boundary conditions. Eigenvalues appear in either type of EVP. Though differential EVPs generally have infinitely many eigenvalues, whereas algebraic EVPs have only a finite number.

2.3.1 Mathematical Form

Most differential EVPs of vibrations and stability have the form:

$$
\begin{aligned}
K\varphi(\mathbf{x}) &= \lambda L\varphi(\mathbf{x}) &&\text{for } \mathbf{x} \in \Omega \\
B_\mu\varphi(\mathbf{x}) &= 0 &&\text{for } \mathbf{x} \in \partial\Omega_\mu, \ \mu = 1, k.
\end{aligned}
\tag{2.2}
$$

The first line describes a differential equation, and the second a set of boundary conditions, both in the formal clothing of operators. The function $\varphi(\mathbf{x})$ is some measure of deflection, $\mathbf{x}$ a vector of spatial variables (generally three-dimensional), λ a scalar variable, and Ω a bounded region with boundaries $\partial\Omega_\mu$.

The quantities K, L and B_μ are *linear differential operators* of the spatial variables. An *operator* is a rule of transformation: it assigns to each function $\varphi(\mathbf{x})$ belonging to a certain class another function, perhaps belonging to a different class (think of the Laplace or the Fourier transform/operator). A *linear* operator obeys the rules of linear algebra: If K is a linear operator and (α_1, α_2) are constants, then K $(\alpha_1\varphi_1 + \alpha_2\varphi_2) = \alpha_1 K\varphi_1 + \alpha_2 K\varphi_2$. A *differential* operator acts through functional derivatives, e.g., $K = \partial^2/\partial x\partial y$ is a differential operator. The *unitary* operator I leaves any function unaltered.

Equation (2.2) merely defines a differential equation and a set of boundary conditions. To make up an EVP, some unprescribed parameter λ must appear in the differential equation. The function $\varphi(\mathbf{x}) = 0$ always solves the differential equation. However, to solve the EVP means to compute those values of λ for which nontrivial solutions $\varphi(\mathbf{x}) \neq 0$ exist, satisfying all boundary conditions. Such special values of λ are called *eigenvalues*, and the associated functions $\varphi(\mathbf{x})$ are called *eigen-functions*. There are generally infinitely many eigenvalues and eigenfunctions of a differential EVP. The pairs $(\lambda_j, \varphi_j(\mathbf{x}))$, $j = 1, \infty$ are called *eigenpairs*.

Example 2.3. For computing the undamped natural frequencies and mode shapes of a simply supported beam of variable cross-section, one arrives at the differential EVP (cf. Sect. 1.4.2):

$$(EI(x)\varphi''(x))'' = \omega^2 \rho A(x)\varphi(x) , \quad x \in [0; l] ,$$
$$\varphi(0) = \varphi''(0) = \varphi(l) = \varphi''(l) = 0. \tag{2.3}$$

This EVP has the form (2.2), with $\mathbf{x} = x$, $\varphi(\mathbf{x}) = \varphi(x)$, $\lambda = \omega^2$, $k = 4$, $\Omega = [0; l]$, $L = \rho A(x)$, $K = d^2/dx^2[EI(x)d^2/dx^2]$, $\partial\Omega_1 = \partial\Omega_2 = \{0\}$, $\partial\Omega_3 = \partial\Omega_4 = \{l\}$, $B_1 = B_3 = I$, and $B_2 = B_4 = d^2/dx^2$.

There is a striking similarity between the differential EVP (2.2), and the algebraic EVP (2.1). The operators K and L of (2.2) correspond to matrices $\mathbf{K}$ and $\mathbf{L}$ of (2.1), the eigen*function* $\varphi(\mathbf{x})$ to the eigen*vector* $\boldsymbol{\varphi}$, and there is an eigenvalue λ which is defined so as to yield nontrivial solutions. However, algebraic EVPs have no explicitly stated boundary conditions (they are incorporated into the matrices), and have only a finite number of solutions.

Leaving for a while the formal EVP, we proceed now to a number of examples illustrating how differential EVPs may arise, and how they can be solved.

2.4   Stability-Related EVPs

In EVPs originating from problems of elastic stability the eigenvalue λ typically represents a critical load parameter. Beyond this critical value, the undeformed position of static equilibrium becomes unstable in favor of a *buckled* shape of deformation. The eigen*function* then represents the functional form of a buckled shape.

One basic method for arriving at the mathematical formulation of a buckling problem is to assume that the structure *is* buckled, and then derive the equations necessary to make the buckled mode shape compatible with static equilibrium, material properties and support conditions. Some examples will illustrate this.

2.4.1   The Clamped-Hinged Euler Column

Consider the clamped-hinged column of Fig. 2.1(a), which has length l and bending stiffness EI, and is centrally loaded by a compressive force P. Requiring an infinitesimal element (Fig. 2.1(b)) of the buckled column to be in static equilibrium, we obtain for the forces $N(x)$ and $T(x)$ and the moment $M(x)$:

$$(N + dN) - N = 0 \quad \Rightarrow \quad N' = 0$$
$$(T + dT) - T = 0 \quad \Rightarrow \quad T' = 0 \tag{2.4}$$
$$(M + dM) - M + T\,dx - N\,dy = 0 \quad \Rightarrow \quad M' + T - Ny' = 0,$$

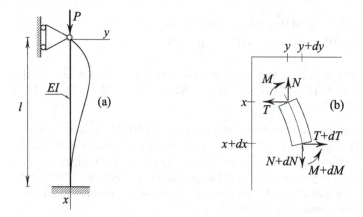

Fig. 2.1 **a** Clamped-hinged Euler column, **b** Differential column element

where the implications follows from divisions by dx, and $(\ )' \equiv d/dx$. Differentiating the third equation with respect to x we obtain, upon inserting the first and the second equation, that:

$$M'' - Ny'' = 0. \tag{2.5}$$

Since $N' = 0$ and $N(0) = -P$, it holds that $N(x) = -P$. Further, according to Hooke's law, $M(x) = EIy''$. Inserting this into (2.5) one then obtains the differential equation governing the shape $y(x)$ of the column:

$$y'''' = -\lambda y'', \quad \lambda \equiv P/EI. \tag{2.6}$$

As for the boundary conditions, the three are purely geometric, while the fourth is obtained from applying Hooke's law to the static condition $M(0)=0$, thus:

$$y(0) = y(l) = y'(l) = y''(0) = 0. \tag{2.7}$$

The differential Eq. (2.6) with boundary conditions (2.7) defines a differential EVP of the general form (2.2).

Now, obviously $y = 0$ is a solution of (2.6) satisfying (2.7) for any value of λ. However, nontrivial solutions $y \neq 0$ might exist. To check this we need to solve the EVP, which precisely means to calculate those values of λ for which (2.6) with (2.7) has nontrivial solutions $y \neq 0$. To solve the EVP we consider the general solution of (2.6), which is:

$$y(x) = c_1 \sin(\gamma x) + c_2 \cos(\gamma x) + c_3 x + c_4, \quad \gamma^2 \equiv \lambda, \tag{2.8}$$

where $c_1, \ldots, c_4$ are arbitrary constants of integration. Requiring this solution to satisfy the boundary conditions (2.7), we obtain from $y(0) = y''(0) = 0$ that $c_2 = c_4 = 0$, and from $y(l) = y'(l) = 0$ that:

$$c_1 \sin(\gamma l) + c_3 l = 0$$
$$c_1 \gamma \cos(\gamma l) + c_3 = 0. \tag{2.9}$$

The trivial solution $c_1 = c_3 = 0$ corresponds to the straight equilibrium $y = 0$. For nontrivial solutions to exist the determinant of coefficients must vanish, that is, $\sin(\gamma l) - \gamma l \cos(\gamma l) = 0$, or:

$$\tan \alpha = \alpha, \quad \alpha \equiv \gamma l. \tag{2.10}$$

This transcendent equation has an infinity of solutions, as appears from Fig. 2.2. Using, e.g., Newton-Raphson iteration, the lowest three solutions turns out to be $\alpha_1 = 4.4934095\cdots$, $\alpha_2 = 7.7252518\cdots$ and $\alpha_3 = 10.904121\cdots$. The eigenvalues λ_i are then given by:

$$\lambda_i = \gamma_i^2 = (\alpha_i/l)^2, \quad i = 1, \infty. \tag{2.11}$$

With $c_2 = c_4 = 0$ and $c_3 = -c_1 \gamma \cos(\gamma\, l)$ inserted into (2.8) the solution $y(x)$ becomes

$$y(x) = c_1 (\sin(\gamma x) - \gamma x \cos(\gamma l)). \tag{2.12}$$

Hence, for each eigenvalue λ_i an associated eigenfunction (*buckling mode*) exists (see Fig. 2.3) :

$$y_i(x) = c_{1i} \left(\sin(\sqrt{\lambda_i} x) - \sqrt{\lambda_i} x \cos(\sqrt{\lambda_i} l) \right). \tag{2.13}$$

For each eigenfunction $y_i(x)$ one constant c_{1i} remains undetermined. This reflects the incapability of linear models to capture any *post-buckling* behavior. Linear theory predicts buckling amplitudes to grow unbounded when $\lambda \geq \lambda_1$. As amplitudes grow, however, nonlinearities come into play that will inevitably limit

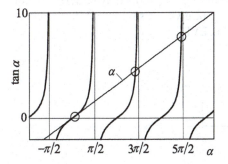

Fig. 2.2 Locating solutions of $\tan\alpha = \alpha$

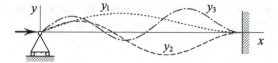

Fig. 2.3 Buckling modes of the clamped-hinged column

the growth to some finite value. To capture post-buckling behavior one needs to drop the assumption of small rotations, introduced through formulating Hooke's law as $M(x) = EIy''$. The correct expression is $M = EI\kappa$ with κ being the curvature, $\kappa = y''/(1 + (y')^2)^{3/2}$. Thus, the linear approximation $\kappa = y''$ is only valid for small rotations, $(y')^2 \ll 1$. Still, linear theory is sufficient to predict the *onset* of buckling and the *initial* post-buckling behavior.

We observe from the example that:

- The constituents of the EVP (differential equation + boundary conditions) are both *linear* and *homogeneous*. Many, but not all, EVPs possess this property.
- For *every* value of λ the EVP has the trivial solution $\varphi(x)=0$. All EVPs possess this property.
- For *some* values of λ the EVP has additional nontrivial solutions. This is *not* a property of all EVPs (cf. the following example).

2.4.2 The Paradox of Follower-Loading

The clamped-free column in Fig. 2.4 is loaded by a special type of force, known as a *follower-load*. It acts along tangents to the free tip of the deformed column.

The differential Eq. (2.6) for the previous example still applies. However, at $x = 0$ the geometric boundary condition $y(0) = 0$ in (2.7) is replaced by the static condition $M' = 0$, that is (by Hooke's law) $y'''(0) = 0$.

The general solution (2.8) applies here too, since the differential equation of the EVP is unchanged. Requiring this solution to satisfy $y''(0) = y'''(0) = 0$ one obtains that either $c_1 = c_2 = 0$, or that $\gamma = 0$. Substituting the remaining conditions $y(l) = y'(l) = 0$, one obtain for the case $c_1 = c_2 = 0$ that $c_3 = c_4 = 0$, implying $y(x) = 0$. For the case $\gamma = 0$ one finds that $c_3 = c_2 + c_4 = 0$, giving also $y(x) = 0$. So, in either case there are no nontrivial solutions to this EVP.

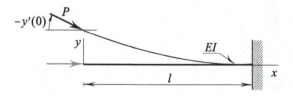

Fig. 2.4 Follower-loaded column

We may conclude, seemingly, that the follower-loaded column is elastically stable in its straight position, however large the value of P. Of course it is not – or rather: the conclusion is right but the premises fail. The paradox puzzled scientists for many years, until in the early 1950s it became clear that a follower-load requires some sort of *dynamic* device for its realization. Consequently, the problem calls for a dynamic analysis. In fact, including mass and acceleration into the formulation, it is revealed that large values of P cause the column to lose stability through the so-called *flutter*-mechanism (e.g. Bolotin 1963; Pedersen 1986; Sugiyama et al. 1999; Ziegler 1968; Sugiyama and Langthjem 2007; the survey by Langthjem and Sugiyama 2000; the critical review by Elishakoff 2005; and the book by Sugiyama et al. 2019). During flutter a structure oscillates at growing amplitudes, until failure occurs or nonlinearities limit the response.

2.4.3   Buckling by Gravity

The columns of the preceding two examples were assumed to have constant cross-sectional properties. The differential equations of the EVPs turned out to have constant coefficients. However, differential equations with non-constant coefficients may appear, even for uniform beams.

To see this, consider the clamped-free column of Fig. 2.5(a), loaded by its own weight in a gravity field g in addition to a central compressive force P. The bending stiffness EI and mass per unit length ρA is constant along the beam. Static equilibrium of the infinitesimal element in Fig. 2.5(b) requires:

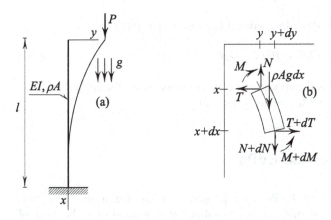

Fig. 2.5 **a** Clamped-free Euler column; **b** Differential column element

$$(N + dN) + \rho A g dx - N = 0 \quad \Rightarrow \quad N' = -\rho A g$$
$$(T + dT) - T = 0 \quad \Rightarrow \quad T' = 0 \tag{2.14}$$
$$(M + dM) - M + T dx - N dy = 0 \quad \Rightarrow \quad M' + T - N y' = 0.$$

Solving the first equation with the boundary condition $N(0) = -P$, one finds that $N(x) = -\rho A g x - P$. Solving the second equation with the boundary condition $T(0) = 0$ yields $T(x) = 0$. Inserting into the third equation, and using Hooke's law ($M = E I y''$), one obtains the differential equation for the EVP:

$$E I y''' + (\rho A g x + P) y' = 0. \tag{2.15}$$

The three boundary conditions required for this third-order equation are:

$$y''(0) = y(l) = y'(l) = 0. \tag{2.16}$$

On introducing the column rotation (assumed small) as a new variable,

$$\theta(x) = y'(x), \tag{2.17}$$

the differential equation is reduced to order two:

$$E I \theta'' + (\rho A g x + P) \theta = 0, \quad \theta'(0) = \theta(l) = 0. \tag{2.18}$$

This equation allows a variety of EVPs to be posed. For example, one may define the eigenvalue so as to describe a critical load P:

$$E I \theta'' + \rho A g x \theta = -\lambda \theta, \quad \lambda \equiv P, \tag{2.19}$$

or the eigenvalue could describe a critical column weight $\rho A g l$:

$$E I l \theta'' + P l \theta = -\lambda x \theta, \quad \lambda \equiv \rho A g l. \tag{2.20}$$

What is to be noted is the appearance of a non-constant coefficient ($\rho A g x$ or λx, respectively), even though the beam is uniform.

2.5 Vibration-Related EVPs

In problems of vibrations EVPs arise primarily during the analysis of free, undamped vibrations. The study of free vibrations in turn constitutes the basis of many exact and approximate methods for calculating damped and/or forced responses for linear as well as nonlinear problems. We consider two examples.

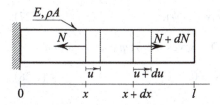

Fig. 2.6 Axially vibrating uniform rod

2.5.1 Axial Vibrations of Straight Rods

The straight rod in Fig. 2.6 has constant cross-sectional area A, Young's modulus E, mass density ρ and length l. Dynamic equilibrium of an infinitesimal element dx of the rod requires:

$$(N + dN) - N = \rho A\, dx\, \ddot{u} \quad \Rightarrow \quad N' = \rho A \ddot{u}, \tag{2.21}$$

where $u = u(x, t)$ measures the axial deformation, and $N = N(x, t)$ is the internal axial force. Using Hooke's law with stress σ and strain ε:

$$N(x,t) = A\sigma(x,t) = AE\varepsilon(x,t) = AEu'(x,t), \tag{2.22}$$

the following differential equation of motion is obtained:

$$c^2 u'' = \ddot{u}, \quad c^2 \equiv E/\rho, \tag{2.23}$$

where the constant c is known as the *wave speed*. Assuming the solution to be a time harmonic with frequency ω:

$$u(x,t) = \varphi(x)\sin(\omega t), \tag{2.24}$$

an ordinary differential equation is obtained for $\varphi(x)$:

$$c^2 \varphi'' = -\omega^2 \varphi. \tag{2.25}$$

The boundary conditions for a fixed-free rod is $u(0, t) = N(l, t) = 0$, or, using Hooke's law and (2.24):

$$\varphi(0) = \varphi'(l) = 0. \tag{2.26}$$

Equation (2.25) with boundary conditions (2.26) constitutes an EVP with $\lambda = \omega^2$ as the eigenvalue and $\varphi(x)$ as the eigenfunction. For solving it we require the general solution of Eq. (2.25), which is

$$\varphi(x) = B\sin(\omega x/c + \psi), \tag{2.27}$$

to satisfy the boundary conditions (2.26). This gives the *frequency equation*:

$$\cos(\omega l/c) = 0, \tag{2.28}$$

which is readily solved to yield:

$$\omega_j = (2j-1)\pi c/(2l), \quad j = 1, \infty. \tag{2.29}$$

The corresponding eigenfunctions $\varphi_j(x)$ are then given by (2.27):

$$\varphi_j(x) = B_j \sin(\omega_j x/c), \quad j = 1, \infty. \tag{2.30}$$

Hence, besides the trivial solution $u(x, t) = 0$ there are infinitely many nontrivial solutions. These are oscillatory, with frequencies and mode shapes given by (2.29) and (2.30) respectively.

2.5.2 Flexural Vibrations of Beams

Consider the clamped-hinged beam of Fig. 2.7(a), having length l, bending stiffness EI and mass per unit length ρA. Requiring an infinitesimal element of the beam (Fig. 2.7(b)) to be in dynamic equilibrium, one obtains, on assuming small transverse deflections $u(x, t)$:

$$\begin{aligned}
(N+dN) - N &= 0 &\Rightarrow\quad N' &= 0 \\
(T+dT) - T &= \rho A\,dx\,\ddot{u} &\Rightarrow\quad T' &= \rho A\ddot{u} \\
(M+dM) - M + T\,dx - N\,du &= 0 &\Rightarrow\quad M' + T - Nu' &= 0.
\end{aligned} \tag{2.31}$$

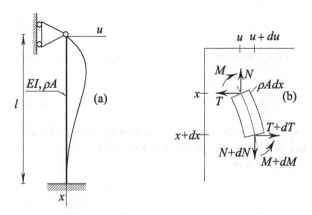

Fig. 2.7 a Flexural vibrations of a slender beam; **b** Beam element

The first equation and the boundary condition $N(0, t) = 0$ implies that $N(x, t) = 0$. Then differentiate the third equation with respect to x, insert the first and the second equation, insert $N = 0$ and $M = EIu''$, and obtain:

$$EIu'''' + \rho A\ddot{u} = 0. \tag{2.32}$$

Assuming a time harmonic solution:

$$u(x,t) = \varphi(x) \sin(\omega t + \psi), \tag{2.33}$$

an ordinary differential equation results:

$$\varphi'''' = \lambda^4 \varphi, \quad \lambda^4 \equiv \rho A\omega^2/EI. \tag{2.34}$$

The boundary conditions are $u(0, t) = u''(0, t) = u(l, t) = u'(l, t) = 0$, or, expressed in terms of $\varphi(x)$:

$$\varphi(0) = \varphi''(0) = \varphi(l) = \varphi'(l) = 0 \tag{2.35}$$

Equations (2.34)–(2.35) constitute an EVP. The general solution of (2.34):

$$\varphi(x) = c_1 \sin(\lambda x) + c_2 \cos(\lambda x) + c_3 \sinh(\lambda x) + c_4 \cosh(\lambda x), \tag{2.36}$$

must satisfy the boundary conditions (2.35). This gives $c_2 = c_4 = 0$, and:

$$\begin{aligned} c_1 \sin(\lambda l) + c_3 \sinh(\lambda l) &= 0 \\ c_1 \cos(\lambda l) + c_3 \cosh(\lambda l) &= 0. \end{aligned} \tag{2.37}$$

For nontrivial solutions (c_1, c_3) to exist the determinant of coefficients must vanish; This gives the frequency equation:

$$\tan \alpha = \tanh \alpha, \quad \alpha \equiv \lambda l. \tag{2.38}$$

Solutions to this transcendent equation are to be found numerically. The lowest three are $\alpha_1 = 3.9266023\cdots$, $\alpha_2 = 7.0685716\cdots$ and $\alpha_3 = 10.210164$. The eigenvalues λ_j are given by $\lambda_j = \alpha_j/l$, and the *natural frequencies* ω_j of the beam are then given by the definition of λ in (2.34):

$$\omega_j = (\alpha_j/l)^2 \sqrt{EI/\rho A}, \quad j = 1, \infty. \tag{2.39}$$

The corresponding *eigenfunctions* – with vibration problems better known as *mode shapes* or *normal modes* – are obtained by inserting $c_2 = c_4 = 0$, and one of the equations in (2.37) into (2.36), giving:

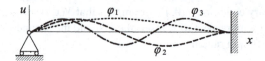

Fig. 2.8 Mode shapes $\varphi_j(x)$, $j = 1, 3$

$$\varphi_j(x) = c_{1j}\big(\sin(\alpha_j x/l) - (\sin \alpha_j/\sinh \alpha_j)\sinh(\alpha_j x/l)\big), \quad j = 1, \infty. \qquad (2.40)$$

Figure 2.8 depicts the lowest three mode shapes, clearly resembling the static buckling modes of Fig. 2.3 for the same column.

2.6 Concepts of Differential EVPs

With the above examples in mind we now return to more general aspects of differential EVPs. Restricting attention to the most important subclass of the general EVP (2.2), in which the differential equation is *ordinary*, the EVP is given by:

$$\begin{aligned} K\varphi &= \lambda L\varphi, \quad \varphi = \varphi(x), \quad x \in [a; b] \\ B_\mu \varphi &= 0, \quad x \in \{a, b\}, \end{aligned} \qquad (2.41)$$

where the operators K, L and B_μ take the forms:

$$K = \sum_{v=0}^{m}(-1)^v \frac{d^v}{dx^v}\left(f_v(x)\frac{d^v}{dx^v}\right), \quad f_v \in C^m[a;b], \quad f_m(x) \neq 0$$

$$L = \sum_{v=0}^{n}(-1)^v \frac{d^v}{dx^v}\left(g_v(x)\frac{d^v}{dx^v}\right), \quad g_v \in C^n[a;b], \quad g_n(x) \neq 0 \qquad (2.42)$$

$$B_\mu = \sum_{v=0}^{2m-1}\left(\alpha_{\mu v}\frac{d^v}{dx^v}\bigg|_{x=a} + \beta_{\mu v}\frac{d^v}{dx^v}\bigg|_{x=b}\right), \quad \mu = 1, 2m .$$

Here $2m$ is the order of the differential equation, and $m > n \geq 0$. The $2m$ homogeneous boundary conditions are assumed to be linearly independent, with real-valued coefficients $\alpha_{\mu v}$ and $\beta_{\mu v}$. This subclass includes numerous EVPs originating from applied mechanics.

Example 2.4. The EVP of a gravity-loaded column has differential equation $EI\varphi'' + \rho Agx\varphi = -\lambda\varphi$ and boundary conditions $\varphi'(0) = \varphi(l) = 0$ (cf. Sect. 2.4.3). The EVP is of the form (2.41)–(2.42) with $a = 0$, $b = l$, $m = 1$, $n = 0$, $f_0(x) = \rho Agx$, $f_1(x) = -EI$, $g_0(x) = -1$, $\alpha_{11} = 1$, $\beta_{20} = 1$, and all other quantities zero.

The purpose of writing EVPs in general form is to provide theorems and reach conclusions of similar generality. Before turning to theorems, some more concepts and definitions are in order.

2.6.1   Multiplicity

If, for an eigenvalue λ, there are $p > 1$ linearly independent eigenfunctions $\varphi(x)$, then λ is a *p-fold* eigenvalue – or we say that λ has *multiplicity p* or is a *repeated eigenvalue*. Most eigenvalues are *simple*, however; they occur only once, with a single eigenfunction. Two or more simple eigenvalues are called *distinct*.

2.6.2   Boundary Conditions: Essential or Natural/ Suppressible

The $2m$ boundary conditions $B_\mu \varphi = 0$ in (2.41)–(2.42) generally contain derivatives of the orders 0, 1, ..., $2m - 1$. Assume that derivatives of order m and higher have been eliminated as far as possible – e.g., by forming linear combinations of the original conditions. Then boundary conditions containing derivatives of order 0, 1, ..., $m - 1$ are called *essential*, whereas those containing derivatives of order m, $m + 1$, ..., $2m-1$ are called *natural* (or *suppressible*).

Example 2.5.   Consider an EVP with a second order differential equation and boundary conditions $\varphi(0) = c_1 \varphi'(0)$, $\varphi(l) = c_2 \varphi'(0)$. In this case $m = 1$, and we should first eliminate, as far as possible, derivatives of order one. We can do no better than eliminate $\varphi'(0)$ from one of the conditions. Substituting $\varphi'(0) = \varphi(l)/c_2$ from the second into the first condition yields the new set: $c_2 \varphi(0) - c_1 \varphi(l) = 0$ and $\varphi(l) - c_2 \varphi'(0) = 0$. The first condition is seen to be essential whereas the second is natural.

Every boundary condition is either essential or natural/suppressible. However, it is not unusual to encounter EVPs totally lacking one or the other class of conditions.

Example 2.6.   The free flexural vibrations of a uniform beam are governed by an EVP with differential equation $EI\varphi'''' = \omega^2 \rho A \varphi$ (cf. Sect. 2.5.2). Since $m = 2$, the essential boundary conditions are those containing derivatives of order zero and one; all others are natural. Thus, if the beam is clamped-free there are two essential boundary conditions: $\varphi(0) = \varphi'(0) = 0$, and two natural: $\varphi''(l) = \varphi'''(l) = 0$. With clamped-clamped supports: $\varphi(0) = \varphi'(0) = \varphi(l) = \varphi'(l) = 0$, all boundary conditions are essential, whereas with free-free supports: $\varphi''(0) = \varphi'''(0) = \varphi''(l) = \varphi'''(l) = 0$ they are all natural. Note that the geometric boundary conditions are essential, and that the static conditions are natural.

2.6.3   Function Classes: Eigen-, Test-, and Admissible Functions

For every EVP on the form (2.41)–(2.42), we define this hierarchy of functions:

1. *Eigenfunction*: Any non-zero function $u(x) \in C^{2m}[a;b]$, satisfying the differential equation as well as all boundary conditions of the EVP.
2. *Test function*: Any non-zero function $u(x) \in C^{2m}[a;b]$ satisfying all boundary conditions of the EVP. (Also called a *comparison function*.)
3. *Admissible function*: Any non-zero function $u(x) \in C^{m}[a;b]$ satisfying all essential boundary conditions of the EVP.

We shall employ the shorthand notation $u \in u_{EF}$, $u \in u_{TF}$ and $u \in u_{AF}$, respectively, to indicate the class of a function $u(x)$. Note that eigenfunctions are also test- and admissible functions, and that test functions are also admissible functions.

Example 2.7. Consider the differential equation $c^2 \varphi'' = -\omega^2 \varphi$ of a vibrating straight rod with fixed-free boundary conditions $\varphi(0) = \varphi'(l) = 0$ (cf. Sect. 2.5.1). Then, $\varphi(x) = \sin(\omega\, x/c)$ with $\omega = \pi c/(2l)$ is an eigen-function of the EVP, whereas $\varphi(x) = x(x - 2l)$ is a test function, and $\varphi(x) = x$ is an admissible function.

2.6.4   Adjointness

The EVP (2.41) is termed *self-adjoint* if, for any two test functions $u(x)$ and $v(x)$:

$$\int_a^b (uKv - vKu)\,dx = 0 \quad \text{and} \quad \int_a^b (uLv - vLu)\,dx = 0, \quad u, v \in u_{TF}. \quad (2.43)$$

Self-adjointness of differential operators parallels symmetry of matrices.

To test for self-adjointness one inserts the actual operators K and L and integrate by parts, attempting to achieve integrands that vanish due to symmetry in u and v, and limit-terms that vanish due to boundary conditions (which are satisfied by any test function u, v).

Example 2.8. The EVP of the fixed-free vibrating rod (cf. Sect. 2.5.1) has differential equation $c^2 \varphi'' = -\omega^2 \varphi$ and boundary conditions $\varphi(0) = \varphi'(l) = 0$. Then, with $K = c^2 d^2/dx^2$ and $L = -1$:

$$\int_a^b (uKv - vKu)\,dx = \int_0^l \left(u(c^2 v'') - v(c^2 u'') \right) dx$$

$$= \left[uc^2 v' \right]_0^l - \left[vc^2 u' \right]_0^l - \int_0^l (u' c^2 v' - v' c^2 u')\,dx = 0 \quad (2.44)$$

$$\int_a^b (uLv - vLu)\,dx = \int_0^l (u(-v) - v(-u))\,dx = 0,$$

where the integrands vanish due to symmetry in u and v, and the bracketed terms vanish due to the boundary conditions; Hence, the EVP is self-adjoint.

Any EVP associated with a linear conservative system may be posed in self-adjoint form. Non-conservative systems generally cause non-self-adjoint EVPs:

Example 2.9. A follower-loaded clamped-free column obeys the differential equation $y'''' = -\lambda y''$ and boundary conditions $y''(0) = y'''(0) = y(l) = y'(l) = 0$ (cf. Sect. 2.4.2). Follower-loads are inherently non-conservative, so we do not expect this EVP to be self-adjoint. To show this we consider the L-integral:

$$\int_a^b [uLv - vLu]\,dx = \int_0^l (u(-v'') - v(-u''))\,dx$$

$$= -[uv']_0^l + [vu']_0^l + \int_0^l (u'v' - v'u')\,dx \quad (2.45)$$

$$= u(0)v'(0) - v(0)u'(0),$$

where the last expression is not guaranteed to be zero for all $u, v \in u_{TF}$; (e.g. for $u(x) = (x - l)^2$ and $v(x) = (x - l)^3$ one has $u(0)v'(0) - v(0)u'(0) = l^4 \neq 0$); Hence the EVP is not self-adjoint.

2.6.5 Definiteness

The EVP (2.41) is said to be *completely definite* if, for any test function $u(x)$:

$$\int_a^b uKu\,dx > 0 \quad \text{and} \quad \int_a^b uLu\,dx > 0 \quad \text{or if}$$

$$\int_a^b uKu\,dx < 0 \quad \text{and} \quad \int_a^b uLu\,dx < 0, \quad u \in u_{TF}. \quad (2.46)$$

An EVP satisfying only the L-condition (i.e. $\int_a^b uLu\,dx \neq 0$ for all $u \in u_{TF}$) is *semidefinite*.

Example 2.10. The EVP of the fixed-free vibrating rod (Sect. 2.5.1) has differ-
ential equation $c^2\varphi'' = -\omega^2\varphi$ and boundary conditions $\varphi(0) = \varphi'(l) = 0$. The EVP
is completely definite for any $c \neq 0$, since:

$$\int_a^b (uKu)\,dx = \int_0^l u(c^2u'')\,dx = \left[uc^2u'\right]_0^l - \int_0^l c^2(u')^2\,dx$$

$$= -\int_0^l c^2(u')^2\,dx < 0 \quad \text{for all } u \in u_{TF}, \qquad (2.47)$$

$$\int_a^b (uLu)\,dx = \int_0^l u(-u)\,dx = -\int_0^l u^2\,dx < 0 \quad \text{for all } u \in u_{TF}.$$

The attribute of complete definiteness is associated with the *operators* of an
EVP. Other notions of definiteness are tied to the *sign of the eigenvalues*. Thus, an
EVP with eigenvalues λ_j, $j = 1, \infty$, is said to be *positive definite* if $\lambda_j > 0$ (all
j implied) and *negative definite* if $\lambda_j < 0$. It is *positive semidefinite* if $\lambda_j \geq 0$ and
negative semidefinite if $\lambda_j \leq 0$. A *definite* EVP is either positive or negative
definite, and a *semidefinite* EVP is either positive or negative semidefinite.

2.6.6 Orthogonality

Two real-valued functions $u(x)$ and $v(x)$, integrable on the interval $[a;b]$, are said to
be *orthogonal* if:

$$\int_a^b uv\,dx = 0 \qquad (2.48)$$

Similarly, the functions are called *G-orthogonal* if, for a given linear operator G:

$$\int_a^b uGv\,dx = 0 \qquad (2.49)$$

where the letter 'G' should be replaced by the symbol of the operator in each case.
This kind of orthogonality is sometimes referred to as *general orthogonality*.

Example 2.11. The functions $u = \sin(x)$ and $v = \cos(x)$ are orthogonal on $x \in$
$[0;\pi]$, satisfying (2.48). Also, the functions are K-orthogonal for the operator $K = -$
d^2/dx^2 on $x \in [0;\pi]$, satisfying (2.49) with $G = K$.

2.6.7   Three Classes of EVPs

For the EVP (2.41) we define three sub-classes, dependent on the operator L:

1. The *general EVP*: L contains at least one derivative, $n > 0$.
2. The *one-term EVP*: L is single-termed and $n \geq 0$.
3. The *special EVP*: L is single-termed and free of derivatives, $n = 0$.

2.6.8   The Rayleigh Quotient

For any test function $u(x)$ of a semidefinite and self-adjoint EVP, we define the associated *Rayleigh quotient R[u]* by:

$$R[u] = \frac{\int_a^b uKu\,dx}{\int_a^b uLu\,dx}, \quad u \in u_{TF}. \tag{2.50}$$

The Rayleigh quotient is a *functional*, that is, a 'function of a function': It's argument is a function (here $u(x)$) – not an independent variable as for a usual function.

2.7   Properties of Eigenvalues and Eigenfunctions

The above definitions may now be put into use in a series of theorems concerning mathematical properties of eigenvalues and eigenfunctions.

2.7.1   Real-Valueness of Eigenvalues

It is often essential to ascertain whether the eigenvalues of a given EVP are real-valued; Here is a sufficient condition:

Theorem 2.3.   The eigenvalues of a semidefinite and self-adjoint EVP with real coefficients are all real.

Proof.   Assume the EVP $K\varphi = \lambda L\varphi$ possesses a complex-valued eigenpair (λ, φ). The complex conjugated pair $(\bar{\lambda}, \bar{\varphi})$ is then also an eigenpair. The functions φ and $\bar{\varphi}$ are both test functions, since every eigenfunction is also a test function. If the EVP is self-adjoint we know that

$$\int_a^b (\bar{\varphi} K\varphi - \varphi K\bar{\varphi})\, dx = 0 \quad \text{and} \quad \int_a^b (\bar{\varphi} L\varphi - \varphi L\bar{\varphi})\, dx = 0. \tag{2.51}$$

Since φ and $\bar{\varphi}$ are eigenfunctions of the EVP, it holds that $K\varphi = \lambda L\varphi$ and, if the EVP has real coefficients, also that $K\bar{\varphi} = \bar{\lambda} L\bar{\varphi}$. Substituting this into the first integral in (2.51) we obtain:

$$(\lambda - \bar{\lambda}) \int_a^b \bar{\varphi} L\varphi\, dx = 0. \tag{2.52}$$

If the EVP is *semidefinite* we know that $\int uLu\, dx \neq 0$ for any $u \in u_{TF}$, so that the integral in (2.52) is non-zero for every eigenfunction φ. Consequently $(\lambda - \bar{\lambda})$ must be zero, which is the case only when λ is real-valued.

2.7.2   Sign of the Eigenvalues

> **Theorem 2.4.**   A self-adjoint and completely definite EVP is also positive definite, with all eigenvalues being real and positive.

Proof That the eigenvalues are real-valued follows from Theorem 2.3. For showing that they are also positive, we assume that (λ_j, φ_j) is an eigenpair of an EVP, that is, $K\varphi_j = \lambda_j L\varphi_j$. Pre-multiply this expression by φ_j, integrate over $x \in [a; b]$ and rearrange to obtain:

$$\lambda_j = \int_a^b \varphi_j K\varphi_j\, dx \Big/ \int_a^b \varphi_j L\varphi_j\, dx. \tag{2.53}$$

If the EVP is completely definite the two integrals are both negative, or both positive. In either case $\lambda_j > 0$ for all j, that is, the EVP is positive definite.

Example 2.12. According to Examples 2.8 and 2.10, the EVP of the vibrating rod ($c^2\varphi'' = -\omega^2\varphi$, $\varphi(0) = \varphi'(l) = 0$) is self-adjoint and completely definite. Hence, all eigenvalues ω^2 are real and positive. This implies that also ω must be real, and thus that free vibrations of the rod are indeed possible – as assumed by the term $\sin(\omega t + \psi)$ in (2.24).

2.7.3 Orthogonality of Eigenfunctions

Many differential EVPs from applied mechanics possess eigenfunctions that are characterized by *orthogonality*:

Theorem 2.5. For a self-adjoint EVP, $K\varphi = \lambda L\varphi$, any two eigenfunctions φ_i and φ_j corresponding to distinct eigenvalues λ_i and λ_j are *K-orthogonal* and *L-orthogonal*, that is:

$$\int_a^b \varphi_i K\varphi_j\, dx = \int_a^b \varphi_i L\varphi_j\, dx = 0 \quad \text{for } \lambda_i \neq \lambda_j. \tag{2.54}$$

Proof. Since (λ_i, φ_i) and (λ_j, φ_j) are eigenpairs, we know that $K\varphi_i = \lambda_i L\varphi_i$ and $K\varphi_j = \lambda_j L\varphi_j$. Multiply the first equation by φ_j and the second by φ_i, subtract the two equations from each other, and integrate over $x \in [a;b]$ to obtain:

$$\int_a^b \left(\varphi_j K\varphi_i - \varphi_i K\varphi_j \right) dx = \int_a^b \left(\lambda_i \varphi_j L\varphi_i - \lambda_j \varphi_i L\varphi_j \right) dx, \tag{2.55}$$

or, since the EVP is assumed to be self-adjoint:

$$0 = (\lambda_i - \lambda_j) \int_a^b \varphi_j L\varphi_i\, dx. \tag{2.56}$$

With distinct eigenvalues $\lambda_i \neq \lambda_j$ the integral must vanish identically, that is, φ_i and φ_j must be *L-orthogonal*. Multiplying $K\varphi_i = \lambda_i L\varphi_i$ by φ_j and integrating, it is seen that φ_i and φ_j are also *K-orthogonal*.

Eigenfunctions are often *normalized* so that $\int \varphi_j L\varphi_j\, dx = 1$, which implies that $\int \varphi_j K\varphi_j\, dx = l_j$. The eigenfunctions are then said to be *orthonormal*, and to satisfy the relations of *orthonormality*:

$$\int_a^b \varphi_i K\varphi_j\, dx = \lambda_j \delta_{ij}; \quad \int_a^b \varphi_i L\varphi_j\, dx = \delta_{ij}, \quad i,j = 1, \infty. \tag{2.57}$$

Example 2.13. For the EVP of the fixed-free vibrating rod with differential equation $c^2\varphi'' = -\omega^2\varphi$ and boundary conditions $\varphi(0) = \varphi'(l) = 0$, the natural frequencies are given by $\omega_j = (2j - 1)\pi c/2l$, $j = 1, \infty$, and the mode shapes by $\varphi_j(x) = B_j \sin(\omega_j x/c)$, where the B_js are undetermined constants (cf. Sect. 2.5.1). From Example 2.8 we know that the EVP is self-adjoint and thus, by Theorem 2.5, that any two mode shapes φ_i and φ_j are orthogonal. To *normalize* these we require that $\int \varphi_j L \varphi_j dx = 1$, which yields $B_j^2 l/2 = 1$. Thus $\varphi_j(x) = \sqrt{2/l}\sin(\omega_j x/c)$ are orthonormal eigenfunctions (mode shapes) for the EVP.

2.7.4 Minimum Properties of the Eigenvalues

> **Theorem 2.6.** For a self-adjoint and completely definite EVP, the first eigen-function $\varphi_1(x)$, among all test functions, makes the Rayleigh quotient a *minimum* with corresponding minimum value λ_1, that is:
>
> $$\min_{u \in u_{TF}} R[u] = R[\varphi_1] = \lambda_1 \quad \text{and} \quad R[u] \geq \lambda_1. \tag{2.58}$$

Proof. A lengthy proof of the first equality is omitted (see Collatz 1963). The second equality follows directly from the EVP and the definition of Rayleigh's quotient:

$$R[\varphi_1] = \int_a^b \varphi_1 K \varphi_1 \, dx \bigg/ \int_a^b \varphi_1 L \varphi_1 \, dx = \int_a^b \varphi_1 (\lambda_1 L \varphi_1) \, dx \bigg/ \int_a^b \varphi_1 L \varphi_1 \, dx = \lambda_1. \tag{2.59}$$

Finally, since φ yields a minimal value λ_1 of R, any other function must yield values at least as large as λ_1, that is, $R[u] \geq \lambda_1$.

Knowing that $R[u] \geq \lambda_1$ may be useful for estimating the lowest eigenvalue of a completely definite and self-adjoint EVP:

Example 2.14. Consider the EVP of a vibrating rod with differential equation $c^2\varphi'' = -\omega^2\varphi$ and boundary conditions $\varphi(0) = \varphi'(l) = 0$. The function $u(x) = x(x-2l)$ satisfy the boundary conditions, hence $u \in u_{TF}$. The EVP is self-adjoint and completely definite (cf. Examples 2.8 and 2.10). Theorem 2.6 then yields:

$$R[u] = \frac{\int_0^l u(c^2 u'') \, dx}{\int_0^l u(-u) \, dx} = \frac{2c^2 \int_0^l x(x-2l) \, dx}{-\int_0^l x^2(x-2l)^2 \, dx} = \frac{5}{2}\frac{c^2}{l^2} \geq \omega_1^2. \tag{2.60}$$

Thus, $\omega_1 \approx \sqrt{5/2}\, c/l$ is an upper bound estimate for the lowest natural frequency of the rod – being only 0.7% higher than the exact value, $\omega_1 = (\pi/2)c/l$.

There is also a theorem for higher eigenvalues (proof omitted):

> **Theorem 2.7.**   For a self-adjoint and completely definite EVP, the q'th eigen-function $\varphi_q(x)$, among all test functions that are L-orthogonal to the first $q - 1$ eigenfunctions, makes the Rayleigh quotient a *minimum* with corresponding minimum value λ_q. That is, if $\int uL\varphi_j dx = 0$ for $j = 1, q{-}1$ and $u \in u_{TF}$, then:
>
> $$\min R[u] = R[\varphi_q] = \lambda_q \quad \text{and} \quad R[u] \geq \lambda_q. \tag{2.61}$$

Example 2.15.   For the vibrating rod with differential equation $c^2\varphi'' = -\omega^2\varphi$ and boundary conditions $\varphi(0) = \varphi'(l) = 0$, the first natural frequency is $\omega_1 = \pi c/2l$ with corresponding mode shape $\varphi_1 = \sin(\pi x/2l)$. For estimating the second natural frequency we suggest the test function $u = \sin(3\pi x/2l)$, which has one more zero-crossing (nodal point) than φ_1. This function is L-orthogonal to φ_1. The Rayleigh quotient of u becomes:

$$R[u] = \int_0^l u(c^2 u'')\, dx \Big/ \int_0^l u(-u)\, dx = \frac{9}{4}\frac{\pi^2 c^2}{l^2} \geq \omega_2^2. \tag{2.62}$$

The value $\omega_2 \approx 3\pi c/2l$ is then an upper-bound approximation to the second natural frequency of the rod. (In fact, it *is* the exact value, since the test function employed also happens to be the exact second eigenfunction.)

For applying Theorem 2.7 one needs to set up test functions that are L-orthogonal to lower eigenfunctions of a given EVP. *Schmidt-orthogonalization* may be useful for this: Assume that a test function $u(x)$ is required which is L-orthogonal to some known eigenfunction $\varphi_k(x)$. Start by suggesting a test function $f(x)$. Then any function of the form $u(x) = f(x) - \kappa \varphi_k(x)$, where κ is a free constant, will too be a test function. Choosing $\kappa = \int fL\varphi_k dx \int \varphi_k L\varphi_k dx$ one finds $\int uLj_k dx = 0$, that is, u is orthogonal to φ_k. This technique can be extended to form test functions that are orthogonal to several eigenfunctions, such as required when estimating a third or higher eigenvalue.

2.7.5   The Comparison Theorem

Sometimes simple EVPs with known eigenvalues may be used for placing bounds on the eigen-values of more complicated EVPs:

> **Theorem 2.8 (Comparison)**   If for two EVPs, $K\varphi = \lambda L\varphi$ and $K\varphi = \lambda^* L^* \varphi$, that are self-adjoint with identical K-operators and boundary conditions, it holds that:

$$\int_a^b uKu\,dx > 0 \quad \text{and} \quad \int_a^b uLu\,dx \ge \int_a^b uL^*u\,dx > 0 \quad \text{for all } u \in u_{TF},$$

$$\text{then } \lambda_j \le \lambda_j^* \quad \text{for } j = 1, \infty. \tag{2.63}$$

The theorem may yield upper as well as lower bounds on eigenvalues:

Example 2.16. Consider a hinged-hinged elastic beam having length $l = 1$, constant bending stiffness EI, and non-constant mass per unit length $\rho A = \rho A_0(1 + x/l)$. The EVP associated with transverse harmonic vibrations at frequency ω becomes $\varphi'''' = \lambda(1 + x)\varphi$ with $\varphi(0) = \varphi''(0) = \varphi(l) = \varphi''(l) = 0$, $\lambda \equiv \rho A_0 \omega^2 / EI$. This EVP is self-adjoint, and the K-operator satisfies the K-inequality in (2.63):

$$\int_a^b uKu\,dx = \int_0^1 uu''''\,dx = \int_0^1 (u'')^2\,dx > 0 \quad \text{for all } u \in u_{TF} \tag{2.64}$$

We then employ Theorem 2.8 for estimating natural frequencies of the beam. Two comparison EVPs are considered: 1) $\varphi'''' = \lambda^* \varphi$, with known eigenvalues $\lambda_j^* = (j\pi)^4$, and 2) $\varphi'''' = 2\lambda^{**}\varphi$ with eigenvalues $\lambda_j^{**} = (j\pi)^4/2$, $j = 1, \infty$. Using the first EVP we obtain:

$$\int_a^b uLu\,dx = \int_0^1 (1+x)u^2\,dx \ge \int_a^b uL^*u\,dx = \int_0^1 u^2\,dx \quad \text{for all } u \in u_{TF}, \tag{2.65}$$

so that $\lambda_j \le \lambda_j^*$ for $j = 1, \infty$. Using the second EVP we obtain:

$$\int_a^b uL^{**}u\,dx = \int_0^1 2u^2\,dx \ge \int_a^b uLu\,dx = \int_0^1 (1+x)u^2\,dx \quad \text{for all } u \in u_{TF}, \tag{2.66}$$

so that $\lambda_j^{**} \le \lambda_j$ for $j = 1, \infty$. Hence $\lambda_j^{**} \le \lambda_j \le \lambda_j^*$, and the natural frequencies ω_j of the beam are bounded by $(j\pi)^4/2 \le \rho A_0 \omega_j^2 / EI \le (j\pi)^4$, $j = 1, \infty$.

2.7.6 The Inclusion Theorem for One-Term EVPs

One-term EVPs have the form (cf. Sect. 2.6.7):

$$K\varphi = \lambda L\varphi = (-1)^n \lambda \left(g_n(x)\varphi^{(n)}\right)^{(n)},$$

$$B_\mu \varphi = 0, \quad n \ge 0, \quad x \in [a; b]. \tag{2.67}$$

This theorem locates, for certain one-term EVPs, a finite interval containing at least one eigenvalue:

Theorem 2.9 (Inclusion).   For an EVP of the form (2.67) let the following conditions apply: The EVP is self-adjoint, $\int_a^b uKu\,dx > 0$ for $u \in u_{TF}$, g_n has fixed sign on $x \in [a;b]$, and

$$(-1)^n \int_a^b u\left[g_n v^{(n)}\right]^{(n)} dx = \int_a^b g_n u^{(n)} v^{(n)}\, dx \quad \text{for } u, v \in u_{TF}. \qquad (2.68)$$

Then define two functions $u_0(x) \in C^{2n}[a;b]$ and $u_1(x) \in u_{TF}$ so that $Ku_1 = Lu_0$, and a third function $\Phi(x)$ through:

$$\Phi(x) = u_0^{(n)} \big/ u_1^{(n)} . \qquad (2.69)$$

If $\Phi(x)$ remains between finite and positive limits $\Phi_{\min}$ and $\Phi_{\max}$ for $x \in [a; b]$, then the interval $[\Phi_{\min}; \Phi_{\max}]$ contains at least one eigenvalue.

The theorem is applied in three steps: 1) Choose a test function $u_1(x)$; 2) calculate $u_0(x)$ as the solution of $Ku_1 = Lu_0$; 3) determine the minimum and maximum of $\Phi(x)$ on $[a;b]$.

The function u_0 is not required to be a test function (satisfying all boundary conditions). However, a closer interval ($\Phi_{\max} - \Phi_{\min}$) is obtained if u_0 satisfies some or all boundary conditions, since u_0 then resembles more of an eigenfunction. Thus, it is advisable to leave some free constants in u_1 that allows u_0 to be matched to the boundary conditions.

Example 2.17 A vibrating rod having length $l = 1$ obeys the differential equation $-c^2 \varphi'' = \omega^2 \varphi$ and boundary conditions $\varphi(0) = \varphi'(1) = 0$. The EVP is self-adjoint (cf. Example 2.8). Further, $\int_a^b uKu\,dx = \int_0^1 u(-c^2 u'')dx = c^2 \int_0^1 (u')^2 dx > 0$ for $u \in u_{TF}$, and $g_0(x) = 1$ has fixed sign on $[0;1]$. Thus, the inclusion theorem is applicable.

For applying the theorem we choose a polynomial $u_1(x) = a_0 + a_1 x + a_2 x^2 + a_3 x^3 + a_4 x^4$, and solve $Ku_1 = Lu_0$ for u_0 to yield $u_0(x) = -c^2 u_1'' = -c^2(2a_2 + 6a_3 x + 12x^2)$. To be a test function u_1 must satisfy the boundary conditions, giving $a_0 = 0$ and $a_1 + 2a_2 + 3a_3 = -4$. Though not strictly necessary, we may use the remaining free constants to force also u_0 to satisfy the boundary conditions, giving $a_2 = 0$ and $a_3 = -4$. Consequently $u_0(x) = 12c^2 x(2 - x)$ and $u_1(x) = x(8 - 4x^2 + x^3)$, so that the function $\Phi(x)$ becomes:

$$\Phi(x) = \frac{u_0(x)}{u_1(x)} = \frac{12c^2x(2-x)}{x(8-4x^2+x^3)} = \frac{12c^2}{5-(1-x)^2}. \tag{2.70}$$

For the interval $x \in [0;1]$ one finds $\Phi_{min} = \Phi(1) = 12c^2/5$ and $\Phi_{max} = \Phi(0) = 3c^2$. According to the inclusion theorem the interval $[\Phi_{min};\Phi_{max}]$ contains at least one eigenvalue. Hence, $12/5 \leq (\omega_j/c)^2 \leq 3$ for at least one j. For this case, the lower bound of the inclusion interval (12/5) is close to the exact value of the first natural frequency of the rod, $(\omega_1/c)^2 = \pi^2/4$.

2.8 Methods of Solution

Closed-form solutions exist for certain differential EVPs. Often, however, one must rely on various numerical methods of approximation, some of which are presented in this section.

Some methods produce *bounds* only, on the true eigenvalues, such as upper, lower or two-sided bounds. Occasionally a successive refinement of such bounds may yield close approximations to the true eigenvalues. Locating eigenvalue bounds may be important, even if these are one-sided and perhaps far from a true eigen-value of the system. For an aircraft component or a power plant rotor, say, it may suffice to know only whether the lowest natural frequency is below or above some excitation frequency of the system. Similarly, for a structure to be assured against static buckling, it may suffice to know that the lowest buckling load is larger than some specified value.

Several of the theorems given in Sect. 2.7 may directly yield approximate solutions: The *minimum property of the Rayleigh quotient* (Sect. 2.7.4) produces up-per-bound estimates for the lowest eigenvalue. The *inclusion theorem* (Sect. 2.7.6) provides two-sided eigenvalue bounds. Occasionally the *comparison theorem* (Sect. 2.7.5) may also yield two-sided bounds.

Among the methods presented in this section the most important are:

1. The *method of eigenfunction iteration*. Provides upper bounds for a lowest eigenvalue. Two-sided bounds can be formed for self-adjoint and completely definite EVPs. May be extended to also yield higher eigenvalues.
2. The *Rayleigh–Ritz method*. For self-adjoint and completely definite EVPs. Provides upper-bound estimates for a number of lowest eigenvalues. The quality of estimates depends on the quality of a set of pre-selected test functions. An algebraic EVP has to be solved.
3. The *finite difference method*. Produces estimates for a number of lowest eigen-values and corresponding eigenfunctions. Works for almost any EVP, including those associated with nonlinear and/or partial differential equations. Requires an algebraic EVP to be solved.

2.8.1   Closed-Form Solutions

When the differential equation of the EVP (2.41) has constant coefficients, it is often possible to express its general solution in closed form in terms of $2m$ arbitrary constants α_v. Requiring the $2m$ boundary conditions to be satisfied, a set of $2m$ linear and homogeneous equations in the constants α_v results. For non-trivial solutions to exist, the coefficient-determinant of the equations must vanish. This gives a single algebraic equation, typically transcendent, for determining the eigenvalues λ. The corresponding eigenfunctions φ are then calculated by substituting back each eigenvalue into the system of equations, solving for the constants α_v, and in turn substituting these constants into the general solution of the differential equation. For each eigen-function one of the constants remain undetermined, free to be scaled for normalization. When applicable this technique is straightforward, as illustrated by the examples in Sects. 2.5.1 and 2.5.2.

2.8.2   The Method of Eigenfunction Iteration

This technique parallels the method of inverse iterations for algebraic EVPs:

Eigenfunction Iteration. For the differential EVP $K\varphi = \lambda L\varphi$, with boundary conditions $B_\mu \varphi = 0$ that are independent of λ:

1. Choose an arbitrary initial function $u_{[0]}(x)$ and let $\lambda_{[0]} = 1$
2. Iterate for $k = 0, 1, \ldots$, until $|\lambda_{[k+1]} - \lambda_{[k]}|/\lambda_{[k]} < \varepsilon \ll 1$:

   (a) Solve $Ku_{[k+1]} = \lambda L u_{[k]}$ with $B_\mu u_{[k+1]} = 0$ for the eigenfunction estimate $u_{[k+1]}$
   (b) Compute the associated eigenvalue estimate as the Rayleigh quotient of $u_{[k+1]}$:

$$\lambda_{[k+1]} = R[u_{[k+1]}] = \int_a^b u_{[k+1]} L u_{[k]} dx \Big/ \int_a^b u_{[k+1]} L u_{[k+1]} dx$$
$$(\geq \lambda_1)$$

$$(2.71)$$

If convergence has been achieved at step $k = p$, then let $\lambda_1 \approx \lambda_{[p+1]}$ and $\varphi_1(x) \approx u_{[p+1]}(x)$.

Example 2.18. The EVP associated with vibrations of an elastic rod of length $l = 1$ may be written $-\varphi'' = \lambda\varphi$ with $\varphi(0) = \varphi'(1) = 0$, where $\lambda \equiv (\omega/c)^2$. The scheme for eigenfunction iterations becomes:

Table 2.1 Eigenfunction iterations for Example 2.18

Iteration number k	Normalized eigenfunction $\tilde{u}_{[k]}(x) = u_{[k]}(x)/u_{[k]}(1)$	Eigenfunction RMS-error $\left(\int_0^1 \left(u_{[k]} - u_{\text{Exact}}\right)^2 dx\right)^{1/2}$	Eigenvalue $\lambda_{[k]}$	Eigenvalue error $|\lambda_{[k]} - \lambda_1|/\lambda_1$
0	1	4.8×10^{-1}	1.0	5.9×10^{-1}
1	$-x^2 + 2x$	3.6×10^{-2}	2.5	1.3×10^{-2}
2	$x(x^3 - 4x^2 + 8)/5$	4.0×10^{-3}	2.4677	1.2×10^{-4}
3	$x(-x^5 + 6x^4 - 40x^2 + 96)/61$	4.5×10^{-4}	2.467405	1.6×10^{-6}
Exact	$\sin(\pi/2)$	0	2.467401... $(=\pi^2/4)$	0

$$- u''_{[k+1]} = u_{[k]}, \quad u_{[k+1]}(0) = u'_{[k+1]}(1) = 0,$$

$$\lambda_{[k+1]} = \frac{\int_0^1 u_{[k+1]} u_{[k]}\, dx}{\int_0^1 u^2_{[k+1]}\, dx}, \quad k = 0, 1, \dots. \tag{2.72}$$

Starting with $u_{[0]}(x) = 1$ and $\lambda_{[0]} = 1$ we obtain, by integration of $-u''_{[1]} = u_{[0]}$ and insertion of boundary conditions, that $u_{[1]}(x) = -x^2/2 + x$. This in turn implies that $\lambda_{[1]} = \int_0^1 u_{[1]} u_{[0]} dx / \int_0^1 u^2_{[1]} dx = 5/2$. Compared to the exact solution $\lambda_1 = \pi^2/4$ the error is only 1.3%. Results for the first three iterations are given in Table 2.1, indicating rapid convergence towards the exact eigenvalue and eigenfunction. Already the first (normalized) function-iterate $\tilde{u}_{[1]}$ is close to the exact eigenfunction, with an integrated RMS-error of only 3.6%. The next iterate $\tilde{u}_{[2]}(x)$ (with RMS-error 0.4%) is virtually indistinguishable from the exact eigenfunction, when plotted.

The eigenvalue iterates $\lambda_{[k+1]}$ are *upper bound* estimates, $\lambda_{[k+1]} \geq \lambda_1$. On certain conditions one may establish two-sided bounds in terms of the so-called *Schwartz's quotients* (Flügge 1962). The method of eigenfunction iterations can be extended to yield also higher eigenvalues and eigenfunctions. However, the Rayleigh–Ritz and finite difference methods are more convenient for this purpose.

2.8.3 The Rayleigh–Ritz Method

The *Rayleigh–Ritz Method* transforms a differential EVP into a corresponding approximate algebraic EVP. Exploiting the extremum property of the Rayleigh quotient, the eigenvalues of the algebraic EVP become upper-bound approximations to the lowest eigenvalues of the differential EVP.

For an EVP with differential equation $K\varphi = \lambda L\varphi$, we assume as a test function for computing the Rayleigh quotient:

$$u(x) = \sum_{j=1}^{q} a_j v_j(x), \tag{2.73}$$

where $v_j(x)$, $j = 1, q$, is a set of chosen test functions, and the constants a_j are so determined as to minimize the Rayleigh quotient associated with $u(x)$. The Rayleigh quotient of u is:

$$R[u] = \frac{\int_a^b uKu\,dx}{\int_a^b uLu\,dx} \equiv \frac{Q_1}{Q_2}. \tag{2.74}$$

If the EVP is self-adjoint and completely definite, then any extremum Λ of $R[u]$ is an upper bound for an eigenvalue λ of the EVP (cf. Theorems 2.6–2.7), i.e.:

$$\min_{u \in u_{TF}} R[u] = \Lambda \geq \lambda. \tag{2.75}$$

For $R[u]$ to be an extremum with respect to the constants a_i it is required that:

$$\frac{\partial R}{\partial a_i} = 0, \quad i = 1, q, \tag{2.76}$$

that is, inserting (2.74):

$$\left(Q_2 \frac{\partial Q_1}{\partial a_i} - Q_1 \frac{\partial Q_2}{\partial a_i} \right) Q_2^{-2} = 0, \quad i = 1, q. \tag{2.77}$$

Dividing by Q_2 – which is non-zero for completely definite EVPs – and substituting the extremizing value Λ for Q_1/Q_2, we obtain:

$$\frac{\partial Q_1}{\partial a_i} - \Lambda \frac{\partial Q_2}{\partial a_i} = 0, \quad i = 1, q. \tag{2.78}$$

With the EVP being self-adjoint, and K a linear operator, we may elaborate Q_1, $\partial Q_1/\partial a_i$, and $\partial Q_2/\partial a_i$ as follows:

$$Q_1 = \int_a^b uKu\,dx = \int_a^b \left(\sum_{j=1}^{q} a_j v_j \right) K \left(\sum_{j=1}^{q} a_j v_j \right) dx, \tag{2.79}$$

$$\frac{\partial Q_1}{\partial a_i} = \int_a^b v_i K(\sum_{j=1}^q a_j v_j)\,dx + \int_a^b (\sum_{j=1}^q a_j v_j) K v_i\,dx = 2\int_a^b v_i K(\sum_{j=1}^q a_j v_j)\,dx$$

$$= 2\sum_{j=1}^q \int_a^b v_i K v_j\,dx\, a_j = 2\sum_{j=1}^q k_{ij} a_j,\quad i=1,q$$

$$\frac{\partial Q_2}{\partial a_i} = \cdots = 2\sum_{j=1}^q l_{ij} a_j,\quad i=1,q,$$

$$(2.80)$$

where the coefficients k_{ij} and l_{ij} are defined by:

$$k_{ij} \equiv \int_a^b v_i K v_j\,dx;\quad l_{ij} \equiv \int_a^b v_i L v_j\,dx,\quad i,j=1,q. \tag{2.81}$$

Substituting (2.79)–(2.80) into (2.78), a set of q linear and homogeneous equations in the coefficients a_i is obtained, the so-called *Galerkin equations*:

$$\sum_{j=1}^q \left(k_{ij} - \Lambda l_{ij}\right) a_j = 0,\quad i=1,q. \tag{2.82}$$

The Galerkin equations are easily recognized as an *algebraic* EVP:

$$\mathbf{Ka} = \Lambda \mathbf{La}, \tag{2.83}$$

where $\mathbf{K}$ and $\mathbf{L}$ are q by q symmetrical matrices with components k_{ij} and l_{ij}, respectively, Λ is an eigenvalue and $\mathbf{a}$ is an eigenvector with components a_i. Solving the algebraic EVP (e.g., by locating zeroes of $|\mathbf{K} - \Lambda\mathbf{L}|$) one obtains a set of eigenvalues $\Lambda_j, j = 1, q$, which can be arranged in ascending order:

$$\Lambda_1 \le \Lambda_2 \le \cdots \le \Lambda_q. \tag{2.84}$$

These values are regarded as approximations to the correspondingly ordered eigenvalues $\lambda_1, \lambda_2, \ldots, \lambda_q$ of the original differential EVP. For self-adjoint and completely definite EVPs, all Λ_j are upper bounds for the true eigenvalues:

$$\Lambda_j \ge \lambda_j,\quad j=1,q. \tag{2.85}$$

Summarizing we have:

Rayleigh–Ritz Method. For a differential EVP $K\varphi = \lambda L\varphi$:

1. Choose a set of test functions $v_j(x) \in u_{TF}$, $j = 1, q$
2. Compute matrices $\mathbf{K}$ and $\mathbf{L}$, with components k_{ij} and l_{ij} as given by (2.81)
3. Solve the algebraic EVP $\mathbf{Ka} = \Lambda \mathbf{La}$ for the eigenvalues Λ_j, $j = 1, q$

If $K\varphi = \lambda L\varphi$ is self-adjoint and completely definite, then $\lambda_j \leq \Lambda_j$ for $j = 1, q$.

The quality of eigenvalue estimates – in particular the higher ones – strongly depends on the number and quality of selected test functions. Most test functions will yield a fair approximation to at least the lowest eigenvalue. However, it is generally recommended to select test functions that qualitatively resemble the assumed true eigenfunctions. If this is impossible, one should choose $q \gg n$, where n is the number of eigenvalues to be estimated.

The Rayleigh–Ritz method is incapable of providing *lower*-bound estimates; *Temple quotients* may be useful for this purpose (Collatz 1963).

Example 2.19. The EVP associated with vibrations of an elastic rod of length $l = 1$ is $-c^2\varphi'' = \omega^2\varphi$ with $\varphi(0) = \varphi'(1) = 0$. The EVP is self-adjoint and completely definite (cf. Examples 2.8 and 2.10). We here employ Rayleigh–Ritz's method for estimating the lowest two eigen-values, using these test functions:

$$v_1(x) = x(x - 2); \quad v_2(x) = x\left(x - \frac{2}{3}\right)(x - 2)\left(x - 2 + \frac{2}{3}\right). \tag{2.86}$$

For the components of matrices $\mathbf{K}$ and $\mathbf{L}$ we have, using (2.81):

$$k_{ij} = \int_a^b v_i K v_j \, dx = \int_0^1 v_i(-c^2 v_j'') \, dx = c^2 \int_0^1 v_i' v_j' \, dx$$

$$l_{ij} = \int_a^b v_i L v_j \, dx = \int_0^1 v_i v_j \, dx, \quad i, j = 1, 2. \tag{2.87}$$

Inserting v_1 and v_2 and integrating it is found that:

$$\mathbf{K} = c^2 \begin{bmatrix} \frac{4}{3} & \frac{16}{135} \\ \frac{16}{135} & \frac{640}{1701} \end{bmatrix}; \quad \mathbf{L} = \begin{bmatrix} \frac{8}{15} & \frac{16}{945} \\ \frac{16}{945} & \frac{128}{8505} \end{bmatrix}. \tag{2.88}$$

The characteristic polynomial $|\mathbf{K} - \Lambda \mathbf{L}|$ becomes $\Lambda 2 - 28c^2\Lambda + 63c^4$ with roots $\Lambda_1 = (14 - \sqrt{133})c^2$ and $\Lambda_2 = \sqrt{7}(2\sqrt{7} + \sqrt{19})c^2$. So for the two lowest natural frequencies:

$$\omega_1 \leq \sqrt{\Lambda_1} \approx 1.57080 \; c, \quad \omega_2 \leq \sqrt{\Lambda_2} \approx 5.05297 \; c. \tag{2.89}$$

Comparing to the exact values $\omega_1 = \pi c/2 \approx 1.57079c$ and $\omega_2 = 3\pi c/2 \approx 4.71238c$, the first estimate is very close to the true first eigenvalue ω_1, whereas the second is 7% in error. Using more test functions improves the quality of all eigenvalue estimates. Thus if only v_1 is employed, the estimate of ω_1 changes to $\omega_1 \leq \sqrt{(5/2)}c \approx 1.581$, clearly less accurate than the estimate obtained using both v_1 and v_2.

2.8.4 The Finite Difference Method

Finite difference methods rely on approximating infinitely small *differentials* by finite-magnitude *differences*. Differential equations will then transform into difference equations. Finite difference methods are applicable to any type of boundary value and/or initial value problem, including those associated with partial and/or nonlinear differential equations. The methods are easily automated, and thus especially well suited to computer applications.

Here we illustrate a simple central-difference scheme for solving differential EVPs; Extensions and variants of the method are described in, e.g. Collatz (1963), Flügge (1962), and Press et al. (2002).

For an EVP with a given ordinary differential equation, one divides the x-interval $[a;b]$ into n equidistant parts of length $h = (b - a)/n$, such that only $n + 1$ discrete x-values x_k are considered (see Fig. 2.9):

$$\begin{aligned} x_k &= x_0 + kh, \quad k = 0, n \\ x_0 &= a, \quad x_n = b. \end{aligned} \tag{2.90}$$

At each point x_k we define φ_k as the pointwise approximation to the value $\varphi(x_k)$ of the true eigenfunction. Any derivative $\varphi^{(v)}(x_k)$ occurring in the EVP is then replaced by a corresponding finite difference approximation $\varphi_k^{(v)}$. As approximating expressions we may use central, backward, forward or other difference formulas. For example, the first four so-called *simple central difference* expressions are given by:

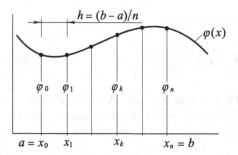

Fig. 2.9 Representing a continuous eigenfunction $\varphi(x)$ by a set of discrete points $\varphi_k = \varphi(x_k)$, $k = 0, n$

$$\varphi'_k = \frac{\varphi_{k+1} - \varphi_{k-1}}{2h}$$

$$\varphi''_k = \frac{1}{h}\left(\frac{\varphi_{k+1} - \varphi_k}{h} - \frac{\varphi_k - \varphi_{k-1}}{h}\right) = \frac{\varphi_{k+1} - 2\varphi_k + \varphi_{k-1}}{h^2}$$

$$\varphi'''_k = \frac{\varphi_{k+2} - 2\varphi_{k+1} + 2\varphi_{k-1} - \varphi_{k-2}}{h^3} \tag{2.91}$$

$$\varphi''''_k = \frac{\varphi_{k+2} - 4\varphi_{k+1} + 6\varphi_k - 4\varphi_{k-1} + \varphi_{k-2}}{h^4}.$$

In general, a simple central difference approximation of a $\varphi_k^{(v)}$ may be written:

$$\varphi_k^{(v)} = \begin{cases} \left(\Delta^v \varphi_{k-(v+1)/2} + \Delta^v \varphi_{k-(v-1)/2}\right) \big/ h^v, & v \text{ odd} \\ \left(\Delta^v \varphi_{k-v/2}\right) \big/ h^v, & v \text{ even}, \end{cases} \tag{2.92}$$

where Δ denotes the *difference operator*: $\Delta\varphi(x) = \varphi(x + h) - \varphi(x)$, or:

$$\Delta\varphi_k = \varphi_{k+1} - \varphi_k, \tag{2.93}$$

so that, for example:

$$\Delta^2\varphi_k = \Delta(\Delta\varphi_k) = \Delta(\varphi_{k+1} - \varphi_k)$$
$$= (\varphi_{k+2} - \varphi_{k+1}) - (\varphi_{k+1} - \varphi_k) = \varphi_{k+2} - 2\varphi_{k+1} + \varphi_k$$
$$\Delta^3\varphi_k = \Delta(\Delta^2\varphi_k) = \varphi_{k+3} - 3\varphi_{k+2} + 3\varphi_{k+1} - \varphi_k \tag{2.94}$$
$$\Delta^4\varphi_k = \Delta^2(\Delta^2\varphi_k) = \varphi_{k+4} - 4\varphi_{k+3} + 6\varphi_{k+2} - 4\varphi_{k+1} + \varphi_k.$$

Derivatives are approximated in the differential equation *and* boundary conditions of the EVP. A differential equation of order $2m$ on $x \in [a;b]$ thus becomes a finite *difference* equation on the mesh x_k, $k = 0, n$.

There will be $2m$ boundary conditions, of order up to r_a and r_b at, respectively, the left and right boundary, $r_a, r_b \leq 2m-1$. If $r_a > 0$ we need to extend the mesh at the left end to $x_A \leq x_0$ (i.e. $A \leq 0$), so as to express the corresponding boundary condition in difference form. (Example: With a boundary condition $\varphi'(x_a) = '_0 = 0$ so that $r_a = 1$, application of (2.91) for $k = 0$ gives $\varphi_1 = \varphi_{-1}$, and thus $A = -1$, i.e. the mesh needs extension to x_{-1}, left of the physical boundary at x_0.) Similarly for the other end – if $r_b > 0$ we need to extend the mesh at the *right* end, to x_B where $B \geq n$.

Thus the extended mesh holds $N = B - A + 1 \geq n + 1$ points $x_A, \dots, x_B$. To solve for the N corresponding discretized function values $\varphi_A, \dots, \varphi_B$ we need N independent algebraic equations. Of these, $2m$ equations are provided by the boundary conditions, while the remaining $N - 2m$ are provided by iterating the difference equation approximating the differential equation – starting with an index value of k that involves the leftmost mesh point A (and no lower), and ending with an index value that involves the rightmost mesh point B (and no higher).

For linear differential equations, the resulting N equations will form a homogenous set of linear algebraic equations. For non-trivial solutions $\varphi_A,\ldots, \varphi_B$ to exist, the determinant of the coefficient matrix must vanish. This requirement results in a single *characteristic equation*, the solutions of which contain approximations for a number of the lowest eigenvalues. For each eigenvalue thus determined, the corresponding approximate discretized eigenfunction is obtained by solving the system of equations for $\varphi_j, j = A, B$, with the corresponding eigenvalue substituted. The discrete vector with elements $\varphi_j, j = 0, n$ (i.e. disregarding the possibly extended parts of mesh) can then be presented as an approximation to the corresponding continuous eigen*functions*.

For large values of n the setup and subsequent solution of a characteristic equation becomes impractical. The set of algebraic equations may then be rewritten as an *algebraic* EVP, with block-diagonal matrices, for which highly efficient methods exist (e.g. Press et al. 2002).

In choosing an adequate number of subdivisions n, one needs to compromise between inconsistent requirements of small truncation and rounding errors, acceptable computation time and numerical stability. A coarse resolution (small value of n) may produce unacceptable truncation errors. As n is increased, more eigenvalues may be determined with better accuracy, until at some large value of n the accuracy falls off due to rounding errors.

Eigenfunctions $\varphi(x)$ for higher eigenvalues may change rapidly with x. Thus, a higher resolution (larger n) is required for approximating higher eigenfunctions. Lower eigenvalues and eigenfunctions are more accurately approximated. In applications one may increase n (e.g. in powers of 2), until the highest eigenvalue estimate in demand does not change within some acceptable number of significant digits.

Example 2.20. Free vibrations of an elastic rod of length l is governed by the EVP $-c^2\varphi'' = \omega^2\varphi$ with $\varphi(0) = \varphi'(l) = 0$. Using (2.91) to replace derivatives with simple central differences we obtain, for the differential equation and the boundary conditions:

$$\varphi_{k+1} + \gamma\varphi_k + \varphi_{k-1} = 0, \quad k = 1, n,$$
$$\varphi_0 = 0, \quad \varphi_{n+1} = \varphi_{n-1}, \tag{2.95}$$

where the lower index value $k = 1$ has been chosen so as to involve the left boundary condition on φ_0 as the leftmost point in the first difference equation, the upper value $k = n$ is chosen so as to involve the right boundary condition with φ_{n+1} as the rightmost point in the last of the n difference equations, and the constant γ incorporates the unknown eigenvalue ω^2 through:

$$\gamma \equiv (\omega h/c)^2 - 2 = (\omega l/nc)^2 - 2. \tag{2.96}$$

The difference equation in (2.95), when iterated, produces n algebraic equations in the $n + 2$ discrete eigenfunction components $\varphi_0, \ldots, \varphi_{n+1}$. Appending the two boundary equations for φ_0 and φ_{n+1}, a set of $n + 2$ algebraic equations is obtained:

$$
\begin{bmatrix}
1 & 0 & 0 & 0 & \cdots & 0 & 0 & 0 & 0 \\
1 & \gamma & 1 & 0 & \cdots & 0 & 0 & 0 & 0 \\
0 & 1 & \gamma & 1 & \cdots & 0 & 0 & 0 & 0 \\
0 & 0 & 1 & \gamma & \cdots & 0 & 0 & 0 & 0 \\
\vdots & \vdots & \vdots & \vdots & \vdots & \vdots & \vdots & \vdots & \vdots \\
0 & 0 & 0 & 0 & \cdots & \gamma & 1 & 0 & 0 \\
0 & 0 & 0 & 0 & \cdots & 1 & \gamma & 1 & 0 \\
0 & 0 & 0 & 0 & \cdots & 0 & 1 & \gamma & 1 \\
0 & 0 & 0 & 0 & \cdots & 0 & -1 & 0 & 1
\end{bmatrix}
\begin{Bmatrix}
\varphi_0 \\
\varphi_1 \\
\varphi_2 \\
\varphi_3 \\
\vdots \\
\varphi_{n-2} \\
\varphi_{n-1} \\
\varphi_n \\
\varphi_{n+1}
\end{Bmatrix}
=
\begin{Bmatrix}
0 \\
0 \\
0 \\
0 \\
\vdots \\
0 \\
0 \\
0 \\
0
\end{Bmatrix}. \tag{2.97}
$$

With $n = 2$ (an absurdly coarse subdivision used just for illustration) a system of four equations in $\varphi_0, \ldots, \varphi_3$ results. Requiring the coefficient determinant of the system to vanish, the characteristic equation becomes $\gamma^2 - 2 = 0$ with solutions $\gamma = \pm\sqrt{2}$. The corresponding approximate natural frequencies $\omega_{1,2}$ become, by (2.96):

$$
\omega_{1,2} l/c = 2\sqrt{2 \pm \sqrt{2}} \approx \begin{cases} 1.53 \\ 3.70 \end{cases} \tag{2.98}
$$

Compared to the exact values of $\pi/2$ and $3\pi/2$, these estimates are 2.6% and 21% in error, respectively. Using instead $n = 6$, one finds the characteristic equation becomes $(\gamma^2 - 2)(\gamma^4 - 4\gamma^2 + 1) = 0$. Of the six solutions, those corresponding to the lowest two natural frequencies $\omega_{1,2}$ are $\gamma_1 = -\sqrt{(2 + \sqrt{3})}$ and $\gamma_2 = \sqrt{2}$. This gives $\omega_1 l/c \approx 1.566$ and $\omega_2 l/c \approx 4.592$, and the errors drop to 0.3% and 2.5%, respectively.

The choice of approximating finite difference formulas is not unique. More accurate formulas, the so-called *higher order finite expressions*, can be formed through Taylor expansions. For example, we may choose to approximate a first-order derivative φ'_k as a linear combination of values of φ_k at the five points $x_{k-2}, \ldots, x_{k+2}$, that is:

$$
\varphi'_k \approx \sum_{j=-2}^{2} c_j \varphi_{k+j}. \tag{2.99}
$$

To determine the constants c_j we first consider the Taylor expansion for φ_{k+j} :

$$
\varphi_{k+j} = \varphi_k + \varphi'_k jh + \frac{1}{2!}\varphi''_k (jh)^2 + \frac{1}{3!}\varphi'''_k (jh)^3 + \frac{1}{4!}\varphi''''_k (jh)^4 + O(h^5). \tag{2.100}
$$

Substituting into (2.99) we obtain:

$$\varphi'_k \approx \varphi_k \sum_{j=-2}^{2} c_j + \varphi'_k h \sum_{j=-2}^{2} jc_j + \frac{1}{2!}\varphi''_k h^2 \sum_{j=-2}^{2} j^2 c_j$$

$$+\frac{1}{3!}\varphi'''_k h^3 \sum_{j=-2}^{2} j^3 c_j + \frac{1}{4!}\varphi''''_k h^4 \sum_{j=-2}^{2} j^4 c_j + O(h^5) . \tag{2.101}$$

Requiring the approximation to be exact to order h^4, it is seen that the five constants c_j must satisfy the five linear equations:

$$\sum_{j=-2}^{2} jc_j = \frac{1}{h}, \quad \sum_{j=-2}^{2} c_j = \sum_{j=-2}^{2} j^2 c_j = \sum_{j=-2}^{2} j^3 c_j = \sum_{j=-2}^{2} j^4 c_j = 0, \tag{2.102}$$

with solution $c_{-2} = -c_2 = 1/(12h)$, $c_1 = -c_{-1} = 2/(3h)$, $c_0 = 0$. Thus, for the derivative φ'_k the fourth order finite expression becomes:

$$\varphi'_k \approx \frac{\varphi_{k-2} - 8\varphi_{k-1} + 8\varphi_{k+1} - \varphi_{k+2}}{12h}. \tag{2.103}$$

The error of this expression is at most $(h^4/18)|\varphi^5|_{max}$ in the interval $[x_{k-2};x_{k+2}]$. By contrast, for the simple forward difference expression $\varphi'_k \approx (\varphi_{k+1} - \varphi_k)/h$ the truncation error can be as large as $(h/2)|\varphi''|_{max}$ in $[x_k;x_{k+1}]$. Hence, for a given level of accuracy, higher order expressions require fewer mesh-points than do simple expressions. With fewer mesh-points the total size of the coefficient matrix decrease, while its *bandwidth* (width of the diagonal band) will *increase*. Thus, the extra efforts associated with higher order expressions are not guaranteed to pay off when considering total computation time for attaining a given accuracy.

2.8.5 Collocation

A somewhat primitive method, though occasionally surprisingly accurate, collocation is simple to apply, and can be used for establishing a first rough estimate of lowest eigenvalues and eigenfunctions.

As with the Rayleigh–Ritz method, an unknown eigenfunction $\varphi(x)$ of the EVP $K\varphi = \lambda L\varphi$ is approximated as a sum of test functions $v_j(x)$:

$$\varphi \approx \tilde{\varphi} = \sum_{j=1}^{q} a_j v_j(x), \quad v_j \in u_{TF}, \tag{2.104}$$

where $a_j, j = 1, q$ are constants to be determined. Upon substituting $\tilde{\varphi}$ for φ in the EVP, an error $\varepsilon(x)$ will arise:

$$\varepsilon(x) = K\tilde{\varphi} - \tilde{\lambda}L\tilde{\varphi} \qquad (2.105)$$

where $\tilde{\lambda}$ denotes the approximate eigenvalue corresponding to $\tilde{\varphi}$. One then selects q points x_k, $k = 1, q$, uniformly distributed on $[a;b]$, at which the error is required to vanish, that is, $\varepsilon(x_k) = 0$ for $k = 1, q$. By (2.104)–(2.105) this requirement expands to q linear and homogeneous algebraic equations for the coefficients a_j:

$$\sum_{j=1}^{q} \left(Kv_j(x_k) - \tilde{\lambda}Lv_j(x_k) \right) a_j = 0, \quad k = 1, q. \qquad (2.106)$$

Thus, a differential EVP has been transformed into an approximating algebraic EVP. Equating to zero the determinant of coefficients, one obtains approximations $\tilde{\lambda}$ to the q lowest eigenvalues. Each $\tilde{\lambda}$ allows (2.106) to be solved for a_j, $j = 1, q$. The corresponding approximating eigenfunction $\tilde{\varphi}$ is then obtained by inserting the values of a_i into (2.104).

2.8.6 Composite EVPs: Dunkerley's and Southwell's Formulas

Sometimes eigenvalue bounds may be established by partitioning an EVP into sub-problems with known eigenvalues. The two formulas below apply to EVPs characterized by composite L-operator and K-operator, respectively.

Dunkerley's Formula. If the EVP $K\varphi = \lambda L\varphi$ with boundary conditions $B_\mu = 0$ has composite L-operator, i.e. the EVP has the form:

$$K\varphi = \lambda \sum_{p=1}^{r} L_p \varphi, \qquad (2.107)$$

and if each partial problem $K\varphi = \lambda^{(p)} L_p \varphi$ with $B_\mu = 0$ is self-adjoint and completely definite with smallest eigenvalue $\lambda_1^{(p)}$. Then for the smallest eigenvalue λ_1 of the composite EVP it holds that:

$$\lambda_1 \geq \left(\sum_{p=1}^{r} \frac{1}{\lambda_1^{(p)}} \right)^{-1} \qquad (2.108)$$

Southwell's Formula. If the EVP $K\varphi = \lambda L\varphi$ with boundary conditions $B_\mu = 0$ has composite K-operator, that is:

$$\sum_{p=1}^{r} K_p \varphi = \lambda L \varphi \qquad (2.109)$$

and if each partial problem $K_p \varphi = \lambda^{(p)} L \varphi$ with $B_\mu = 0$ is self-adjoint and completely definite with smallest eigenvalue $\lambda_1^{(p)}$. Then for the smallest eigenvalue λ_1 of the composite EVP it holds that:

$$\lambda_1 \geq \left(\sum_{p=1}^{r} \lambda_1^{(p)} \right) \qquad (2.110)$$

2.8.7 The Rayleigh Quotient Estimate and Its Accuracy

The Rayleigh quotient (Sects. 2.6.8 and 2.7.4) provides estimates of the lowest eigenvalues which are typically much more accurate than the trial functions used. For the general differential EVP (2.2) this is difficult to prove (Temple and Bickley 1933), but for the algebraic EVP (2.1) it is rather simple. Since differential EVPs can be approximated by algebraic EVPs (e.g. using Rayleigh-Ritz, finite difference, or finite element methods), the calculation of the accuracy of Rayleigh quotient estimates for standard linear algebraic EVPs is instructive.

For the algebraic EVP (2.1) the Rayleigh quotient becomes, similarly to (2.50):

$$R(\mathbf{u}) = \frac{\mathbf{u}^T \mathbf{K} \mathbf{u}}{\mathbf{u}^T \mathbf{L} \mathbf{u}} = \tilde{\lambda} \geq \lambda_1, \qquad (2.111)$$

where $\mathbf{u}$ is the guess or trial vector for the lowest eigenvector $\boldsymbol{\varphi}_1$, $\tilde{\lambda}$ is the Rayleigh quotient estimate of the true eigenvalue λ_1 corresponding to $\mathbf{u}$, and by (2.1) $R(\boldsymbol{\varphi}_1) = \lambda_1$. To estimate the error $\tilde{\lambda}_1 - \lambda_1$, we consider the vector variable $\mathbf{u} = \boldsymbol{\varphi}_1 + \boldsymbol{\varepsilon}$ as composed of the true lowest eigenvector $\boldsymbol{\varphi}_1$ and an error vector $\boldsymbol{\varepsilon}$, and calculate the corresponding change in the Rayleigh quotient by Taylor-expanding for $|\boldsymbol{\varepsilon}| = \varepsilon \ll 1$:

$$R(\mathbf{u}) = R(\boldsymbol{\varphi}_1 + \boldsymbol{\varepsilon}) = \tilde{\lambda}_1 = R(\boldsymbol{\varphi}_1) + (\nabla R(\boldsymbol{\varphi}_1))^T \boldsymbol{\varepsilon} + O(\varepsilon^2), \qquad (2.112)$$

where $O(\varepsilon^2)$ denotes small terms of order of magnitude ε^2. Calculating the gradient:

$$\nabla R(\mathbf{u}) = \frac{\partial R}{\partial \mathbf{u}} = \frac{\partial}{\partial \mathbf{u}} \left(\frac{\mathbf{u}^T \mathbf{K} \mathbf{u}}{\mathbf{u}^T \mathbf{L} \mathbf{u}} \right) = \frac{\frac{\partial}{\partial \mathbf{u}} (\mathbf{u}^T \mathbf{K} \mathbf{u}) \mathbf{u}^T \mathbf{L} \mathbf{u} - \mathbf{u}^T \mathbf{K} \mathbf{u} \frac{\partial}{\partial \mathbf{u}} (\mathbf{u}^T \mathbf{L} \mathbf{u})}{(\mathbf{u}^T \mathbf{L} \mathbf{u})^2}$$

$$= 2 \frac{(\mathbf{K} \mathbf{u}) \mathbf{u}^T \mathbf{L} \mathbf{u} - \mathbf{u}^T \mathbf{K} \mathbf{u} (\mathbf{L} \mathbf{u})}{(\mathbf{u}^T \mathbf{L} \mathbf{u})^2} = 2 \frac{\mathbf{K} \mathbf{u} - \frac{\mathbf{u}^T \mathbf{K} \mathbf{u}}{\mathbf{u}^T \mathbf{L} \mathbf{u}} \mathbf{L} \mathbf{u}}{\mathbf{u}^T \mathbf{L} \mathbf{u}}, \qquad (2.113)$$

For $\mathbf{u} = \varphi$ one finds, when inserting also $R(\varphi_1) = \lambda_1$ and (2.1) (i.e. $\mathbf{K}\varphi_1 = \lambda_1 \mathbf{L}\varphi_1$) that:

$$\nabla R(\varphi_1) = 2 \frac{\mathbf{K}\varphi_1 - \frac{\varphi_1^T \mathbf{K}\varphi_1}{\varphi_1^T \mathbf{L}\varphi_1} \mathbf{L}\varphi_1}{\varphi_1^T \mathbf{L}\varphi_1} = 2 \frac{\mathbf{K}\varphi_1 - \lambda_1 \mathbf{L}\varphi_1}{\varphi_1^T \mathbf{L}\varphi_1} = \mathbf{0}. \tag{2.114}$$

Thus, when $\mathbf{u}$ equals the eigenvector φ_1 the gradient vanishes, i.e. the Rayleigh quotient attains *stationarity* wrt. infinitesimal variations in $\mathbf{u}$. Inserting (2.114) into (2.112), again with $R(\varphi_1) = \lambda_1$, it is found that:

$$R(\varphi_1 + \boldsymbol{\varepsilon}) = \tilde{\lambda}_1 = \lambda_1 + O(\varepsilon^2), \tag{2.115}$$

so that the error $\tilde{\lambda}_1 - \lambda_1 = O(\varepsilon^2)$. Consequently, the Rayleigh quotient for a function $\mathbf{u}$, with some small deviation of order of magnitude ε wrt. to the true eigenvector φ_1, gives a corresponding eigenvalue which deviates with even smaller order ε^2 wrt. the true eigenvalue λ_1. In other words: For a reasonably close estimate of the lowest eigenvector, the Rayleigh quotient provides a *quadratically accurate* estimate of the corresponding eigenvalue.

In vibration analysis eigenvalues often represent squared eigenvalues, $\lambda = \omega^2$. If $\tilde{\lambda}_1$ is the Rayleigh quotient estimate of λ_1, the corresponding natural frequency estimate of ω_1 is $\tilde{\omega}_1 = \sqrt{\tilde{\lambda}_1} = \sqrt{\lambda_1 + O(\varepsilon^2)} = \sqrt{\lambda_1} + O(\varepsilon^2)$, i.e. *also* quadratically accurate (and not just $O(\varepsilon)$-accurate, as one might think.)

2.8.8 Other Methods

Other methods for differential EVPs exist, e.g., the methods of Grammel and Trefftz, the method of moments, the minimum mean-square method and various methods based on perturbation analysis (Collatz 1963; Flügge 1962). However, the methods already presented are well suited for most problems occurring with engineering structural components with reasonably simple geometry an material composition. For structures with complex geometry and distribution of materials the finite element method (FEM) is well suited (e.g., Zienkiewicz 1982; Bathe and Wilson 1976; Cook et al. 1989); essentially it resembles the Ritz /Galerkin /mode shape expansion methods, by expressing unknown deformation fields in terms of elementary spatial functions, and by producing eigenvalues and eigenfunctions as solutions of approximating algebraic eigenvalue problems.

2.9 Problems

Problem 2.1 The clamped-hinged elastic column in Fig. P2.1 has bending stiffness EI and length l, and is centrally loaded by a compressive force P.

a) Set up an EVP for the determination of buckling loads and associated buckling modes.
b) Examine whether the EVP is self-adjoint and completely definite.

Problem 2.2 A mass M hangs in a chain having length l and mass per unit length m. The chain is vertically suspended in a gravity field of strength g.

a) Set up an EVP for the determination of natural frequencies and mode shapes for small transverse oscillation of the chain.
b) Examine whether the EVP is self-adjoint and completely definite.

Problem 2.3 The beam in Fig. P2.3 has bending stiffness EI, and is clamped in one end and supported at the other end by springs having linear stiffness c.

a) Set up an EVP for the determination of a critical buckling load P.
b) Examine whether the EVP is self-adjoint and completely definite.

Problem 2.4 The clamped-free torsional rod in Fig. P2.4 has axially varying moment of inertia $I(x)$, density $\rho(x)$ and torsional stiffness $GK(x)$.

a) Set up an EVP for the determination of natural frequencies associated with small torsional vibrations of the rod.
b) Examine whether the EVP is self-adjoint and completely definite.

Problem 2.5 The clamped-free beam in Fig. P2.5 has mass per unit length $\rho A(x)$ and bending stiffness $EI(x)$, and is oriented vertically in a gravity field g.

a) Set up an EVP for the determination of natural frequencies associated with small amplitude bending vibrations of the beam.
b) Examine whether the EVP is self-adjoint and completely definite.

Problem 2.6 The elastic rope in Fig. P2.6 has constant density ρ and varying cross-sectional area $A(x) = A_0(1 + x^2)$, and is pre-tensioned by a force P between two rigid walls at $x = \pm 1$.

a) Set up an EVP for the determination of natural frequencies associated with small transverse vibrations of the rope.
b) Examine whether the EVP is self-adjoint and completely definite.

Problem 2.7 The stiff cylindrical tube in Fig. P2.7 has inner radius R, and rotates about its axis with angular speed Ω. A flexible column of circular cross-section has one end rigidly clamped at the inner rim of the rotating tube and the other end free. The column has length l, bending stiffness EI and mass per unit length ρA. The undeformed axis of the column coincides with a radius of the tube.

a) Set up an EVP for assessing whether the column will buckle in the plane of the tube cross-section.
b) Determine the values of $l \in]0;2R[$ for which the EVP is self-adjoint and completely definite.

Problem 2.8 For the elastic rope of Problem 2.6:

a) Calculate a pair of approximations for the lowest eigenvalue, using Rayleigh's quotient with test functions $u = 1 - x^2$ and $u = \cos(\pi x/2)$, respectively.
b) Calculate all eigenvalues approximately using the comparison theorem.
c) Calculate an approximate first eigenvalue, using the inclusion theorem with the test function $u_1 = x^4 + ax^3 + bx^2 + cx + d$.

Problem 2.9 Fig. P2.9 shows a column having circular cylindrical cross-section and bending stiffness EI. A twisting moment M and a compressive force P act along the undeformed axis of the column.

a) Set up an EVP for the critical buckling loads in terms of M and P.
b) Examine whether the EVP is self-adjoint.

Problem 2.10 The clamped-free elastic rod in Fig. P2.10 has length $l = 1$, Young's modulus $E(x)$, cross-sectional area $A(x)$ and density $\rho (x)$.

a) Set up an EVP for the determination of natural frequencies and mode shapes associated with small axial vibrations of the rod.
b) Examine whether the EVP is self-adjoint and completely definite.
c) Assuming $E(x) = \rho (x) = 1$ and $A(x) = 1 + x^2$, employ the inclusion theorem with test functions $x^4 + c_1 x^3 + c_2 x^2 + c_3 x + c_4$ and $A(x)^p \sin[q(x - 1)]$ for estimating upper and lower bounds for the lowest natural frequency.

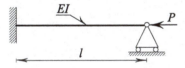

Fig. P2.1

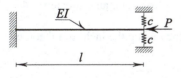

Fig. P2.3

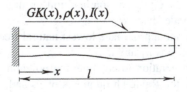

Fig. P2.4

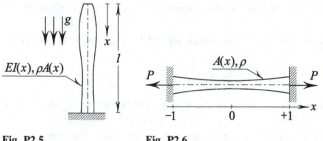

Fig. P2.5   **Fig. P2.6**

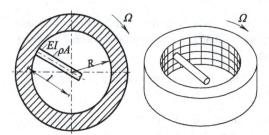

Fig. P2.7

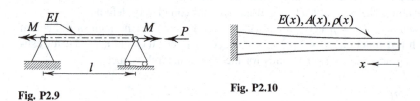

Fig. P2.9   **Fig. P2.10**

Problem 2.11 For the elastic rope of Problems 2.6 and 2.8:

a) Compute the Rayleigh quotient of u_1 as obtained in Problem 2.8c.
b) Perform one eigenfunction iteration, from u_1 to u_2, to obtain an estimate for the lowest mode shape φ_1.
c) Calculate an approximate lowest natural frequency using the inclusion theorem with functions u_1 and u_2.
d) Apply an iteration to compute the polynomial u_3.
e) Compute the Rayleigh quotients of u_2 and u_3, respectively.
f) Supply bounds for the lowest natural frequency using the Rayleigh quotients of u_1, u_2 and u_3.

Problem 2.12 Fig. P2.12 shows a column of length l, mass per unit length ρA (x) and bending stiffness $EI(x)$. The column is simply supported at one end, and further supported along its length by elastic springs of linear stiffness $c(x)$ per unit length. The springs act along $45°$ angles, and are pre-loaded by forces $q(x)$ per unit length (tensioned when $q > 0$). A force P, acts centrally at the tip of the column (compressive when $P > 0$).

a) Set up an EVP for the determination of natural frequencies and mode shapes corresponding to small transverse vibrations of the column.
b) Locate the sub-ranges of $(q, P) \in R^2$ for which the EVP is self-adjoint and completely definite.

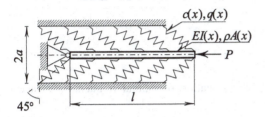

Fig. P2.12

Problem 2.13 For determining natural frequencies of a clamped-hinged beam one must solve the EVP $\varphi'''' = \lambda^4 \varphi$ with $\varphi(0) = \varphi''(0) = \varphi(l) = \varphi'(l) = 0$, where $\lambda^4 \equiv \rho A \omega^2 / EI$ (cf. Sect. 2.5.2). Here, ω is a natural frequency of a beam with bending stiffness EI, mass per unit length ρA and length l.

Select a pair of simple test functions for the EVP, and use Rayleigh-Ritz's method for obtaining approximations to ω_1 and ω_2. Compare results to the exact values given in Sect. 2.5.2.

Problem 2.14 For the EVP of Problem 2.13:

a) Set up a finite difference scheme for approximately solving the EVP.
b) Test the scheme by hand-calculating the two lowest beam natural frequencies ω_1 and ω_2, comparing results to the exact values given in Sect. 2.5.2.
c) Implement the finite difference scheme in a small computer program, and use it for obtaining approximations to ω_1 and ω_2 with, respectively, $n = 2, 4, 8, \ldots$ subdivisions of the range $x \in [0;l]$. (For solving the algebraic eigenvalue you may prefer using library software). List for each value of n the relative errors on ω_1 and ω_2 as compared to the exact values given in Sect. 2.5.2.

Problem 2.15 The figure shows a beam with two embedded piezo-ceramic elements, wired to control and measure its actual state (e.g. Høgsberg and Krenk 2012). The beam could model, e.g., a "smart" ski or part of an actively controlled helicopter rotor. The beam can work in *actuator mode*, with current supplied to the piezo elements, or in *sensor mode*, with current being generated by the piezo elements according to deformations of the beam. By appropriate wiring and poling of the voltage applied to the piezo-ceramic elements, extensional or flexural vibrations can be induced, or measured.

The beam has total length l, of which the piezo elements occupy a minor middle region of length εl, $\varepsilon \ll 1$. Outside this region the beam has uniform transverse bending stiffness EI_0, and mass per unit length ρA_0. At the region with piezo-elements the stiffness and mass per unit length is different, with effective ("homogenized") values αEI_0 and $\beta \rho A_0$, respectively.

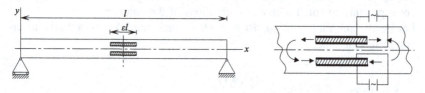

Fig. P2.15

The eigenvalue problem for the determination of natural frequencies ω and corresponding mode shapes $y(x)$ is:

$$(EIy'')''= \omega^2 \rho A y,$$
$$y(0) = y''(0) = y(l) = y''(l) = 0, \tag{2.116}$$

where $EI = EI(x)$ and $\rho A = \rho A(x)$ are as described above.

a) Use Rayleigh's quotient, $R[y] \geq \omega_1^2$, with a suitable test function y, to derive an approximate expression for the fundamental natural frequency ω_1 of the beam.
b) With $\varepsilon \ll 1$ it makes sense to Taylor-expand $R[y]$ for small values of ε, retaining only terms up to order ε^1. Do this, and derive the corresponding approximate expression for ω_1.
c) Using the result from b), calculate and discuss the relative change η in squared fundamental natural frequency caused by the presence of the piezo elements, $\eta = (\omega_1^2 - \omega_{10}^2)/\omega_{10}^2$, where ω_{10} is the fundamental natural frequency for the uniform beam without piezo elements.

Problem 2.16 (Requires MATLAB or similar numerical software). For a hinged-hinged uniform beam, the EVP for the determination of natural frequencies ω and mode shapes $\varphi(x)$ can be written (cf. Sect. 1.4.2):

$$\varphi'''' = \lambda \varphi, \quad \lambda = \rho A \omega^2 / EI,$$
$$\varphi(0) = \varphi''(0) = \varphi(l) = \varphi''(l) = 0. \tag{2.117}$$

a) Show that a (central) finite difference scheme leads to the following algebraic EVP for approximately calculating the natural frequencies and mode shapes:

$$\mathbf{K}\boldsymbol{\varphi} = \lambda\boldsymbol{\varphi},\qquad(2.118)$$

where $\boldsymbol{\varphi}^{\mathrm{T}} = \{\, \varphi_1 \quad \varphi_2 \quad \cdots \quad \varphi_{n-1} \,\}$, $\varphi_0 = 0$, $\varphi_n = 0$, n is the number of x-intervals, and:

$$\mathbf{K} = h^{-4}\begin{bmatrix}
5 & -4 & 1 & 0 & 0 & 0 & 0 & \cdots & 0 & 0 & 0 & 0 & 0 \\
-4 & 6 & -4 & 1 & 0 & 0 & 0 & \cdots & 0 & 0 & 0 & 0 & 0 \\
1 & -4 & 6 & -4 & 1 & 0 & 0 & \cdots & 0 & 0 & 0 & 0 & 0 \\
0 & 1 & -4 & 6 & -4 & 1 & 0 & \cdots & 0 & 0 & 0 & 0 & 0 \\
\vdots & \vdots & \vdots & \vdots & \vdots & \vdots & \vdots & \ddots & \vdots & \vdots & \vdots & \vdots & \vdots \\
0 & 0 & 0 & 0 & 0 & 0 & 0 & \cdots & 1 & -4 & 6 & -4 & 1 \\
0 & 0 & 0 & 0 & 0 & 0 & 0 & \cdots & 0 & 1 & -4 & 6 & -4 \\
0 & 0 & 0 & 0 & 0 & 0 & 0 & \cdots & 0 & 0 & 1 & -4 & 5
\end{bmatrix}$$

$$(2.119)$$

where $h = l/n$ is the interval width ("stepsize").

b) For $l = 1$, calculate the lowest four eigenvalues for the algebraic EVP using the MATLAB function EIGS, and plot the corresponding eigenfunctions (mode shapes). Calculate also the corresponding normalized natural frequencies $\tilde{\omega} = \omega/\sqrt{EI/\rho A} = \sqrt{\lambda}$, and compare these to the exact analytical beam frequencies $\tilde{\omega}_j = (j\pi/l)^2, j = 1, \ldots$ (cf. Eq. (1.43)). How large are the errors when the number of intervals is, respectively, $n = 8$ and $n = 25$?

c) Now assume that the middle part $\varepsilon l(\varepsilon < 1)$ of the beam has a differing stiffness EI and mass per unit length A, where α and β are positive constants, so that the beam is only *piecewise* uniform. The differential EVP for the natural frequencies then changes into[1]:

$$\varphi'''' = \lambda f(x)\varphi, \qquad f(x) = \begin{cases} \beta/\alpha, & x \in \Omega = [\tfrac{1}{2}l(1-\varepsilon); \tfrac{1}{2}l(1+\varepsilon)] \\ 1, & x \in [0; l]\backslash\Omega \end{cases}\qquad(2.120)$$

Employing a finite difference scheme, show that the corresponding algebraic EVP becomes $\mathbf{K} = \mathbf{L}$, where $\mathbf{L} = \mathrm{diag}\{f_1 f_2 \ldots f_{n-1}\}, f_j = f(x_j) = f(jh)$. Then use MATLAB to compute the change in the lowest two natural frequencies (in Hz) of a 1 m long Plexiglas beam, having a uniform 1 cm diameter circular cross section, due to a 10 cm Aluminium insert in the middle of the beam. (For plexiglas®/"PMMA" $E = 3.4$ GPa, $\rho = 1200$ kg/m³; for aluminum $E = 70$ GPa, $\rho = 2700$ kg/m³.)

---

[1]Actually $(EI\varphi'')'' = \omega^2\rho A\varphi$ leads to $EI\varphi'''' + 2EI'\varphi''' + EI''\varphi'' = \omega^2\rho A\varphi$, while (2.119) ignores (to keep the example simple) the 2nd and 3rd term of the left-hand-side, which are zero everywhere except at the discontinuous jumps in EI, i.e. at $x = \tfrac{1}{2}l(1 \pm \varepsilon)$. The consequence of this for the accuracy in results is not obvious. You could compare to exact results, e.g. by Krishnan (1998), and also see Jang and Bert (1989).

3 Nonlinear Vibrations: Classical Local Theory

3.1 Introduction

A *linear* system represents a mathematical abstraction – a useful and productive invention of the human mind. However, the physical systems found in nature and manmade devices are inherently *nonlinear*. If in doubt, try imagining any physical system subjected to a relevant excitation of increasing magnitude: pump energy into the system and it will respond in some way. Initially, the state of the system may change in proportion to the applied load, but at some point the change will inevitably be disproportional. Ultimately, given unlimited power supplies, the system may break apart into two or more subsystems, or reorganize itself into a qualitatively different type of system. Rarely does nature recognize linearity and straight lines.

So, linear models of physical systems are mathematical abstractions, and you may start wondering why engineering and other scientific educations are then so heavily dominated by linear systems and theories. One reason is that many real-life systems indeed operate within a limited range, for which a linear *approximation* is sufficient to understand, predict or control their behavior. Another reason is that all linear models mathematically are more or less alike, and are well understood within a framework of almost fully developed theories. Nonlinear systems, on the other hand, are highly diversified – like 'non-Europeans', 'non-metallics' and other non-specific descriptors. Few are well understood, and nonlinear theory is still, more than a hundred years after the pioneering work of Poincaré, by many aspects in its infancy. For nonlinear systems one has to cope with an impressive number of competing analytical and numerical tools, some of which are only partly developed and still subjected to controversies.

Sometimes nonlinear effects cannot be neglected, their inclusion in a model aids in understanding a problem, or a nonlinear phenomenon is to be suppressed or utilized in a specific device. We may then turn to the computer, and simulate the system numerically. Computers, however, provide highly specific answers to highly

© The Author(s), under exclusive license to Springer Nature Switzerland AG 2021
J. J. Thomsen, *Vibrations and Stability*,
https://doi.org/10.1007/978-3-030-68045-9_3

specific questions. One obtains the response of a system only for particular sets of parameters, and for particular intervals of time. For nonlinear systems one cannot just interpolate or extrapolate from discrete data. So, in particular for nonlinear systems, some theoretical insight is required for turning the huge amount of computer-generated *information* into useful *understanding*.

This chapter provides a set of classical tools for the local analysis of nonlinear vibrations. *Local* here means 'close-to-equilibrium'. A nonlinear system typically has several states of static and dynamic equilibrium. Local theories concern solutions in the immediate vicinity of such states. *Global* solutions, that may involve multiple equilibriums and far-from-equilibrium behavior, are usually inaccessible to the classical methods presented in this chapter. The term 'classical' serves to distinguish the methods described in this chapter from those described in Chap. 6 on chaotic and global dynamics. Classical methods, though constantly being developed and refined, are based on classical mathematical paradigms, whereas the study of global dynamics and chaos requires new mathematical concepts and tools. Classical local analysis is perfectly adequate for the study of numerous nonlinear problems, and is an important prerequisite for studying global behavior.

The diversity of nonlinear models and tools precludes a general presentation of the subject. Instead, after describing potential sources of nonlinearity, we go right into a nonlinear analysis of a simple and illustrative example. This will allow essential concepts, methods and phenomena to be presented where appropriate. Subsequent chapters will treat more complicated systems and phenomena.

The main references of this chapter are Blaquiére (1966), Bogoliubov and Mitropolskii (1961), Bolotin (1964), Cartmell (1990), Guckenheimer and Holmes (1983), Jackson (1991), Mitropolskii and Nguyen 1997, Nayfeh (1973), Nayfeh and Balachandran (1995), Nayfeh and Mook (1979), Sanders and Verhulst (1985), Schmidt and Tondl (2009), Stoker (1950), and Verhulst (1996a).

3.2 Sources of Nonlinearity

Nonlinearities may enter a model in many ways. Their origin may be *geometrical* or *material*, or associated with *nonlinear forces* or *physical configuration*. Whatever their origin, nonlinearities may enter the model equations in similar ways. So, it is rarely possible to deduce the physical origin of a nonlinearity from its mathematical representation. Any component of the equations of motion may be nonlinearly affected: The *inertial terms*, the terms describing elastic or inelastic *restoring forces*, the *dissipative terms*, terms describing *external excitation*, and the *boundary conditions*. Nonlinear terms are recognized by being nonlinear functions of the *dependent* variables of the equations of motion. For example, if $u(t)$ describes the motion of a system, then the terms u^3, $\ddot{u}$, $\sin u$ and $|u|$ are all nonlinear, whereas $t^2 u$, $u \sin t$ and $e^{-t}\ddot{u}$ are linear terms.

3.2.1 Geometrical Nonlinearities

Geometrical nonlinearities typically arise from large deflections or rotations, or other purely kinematic characteristics. For example, the dynamics of the pendulum in Fig. 3.1(a) is governed by

$$\ddot{\theta} + \omega^2 \sin\theta = 0. \tag{3.1}$$

The nonlinear sine-term expands to $\sin\theta = \theta - \frac{1}{6}\theta^3 + \cdots$, showing that the familiar linear approximation $\ddot{\theta} + \omega^2\theta = 0$ is valid only for small rotations θ.

The beam in Fig. 3.1(b) is axially coupled with a linear spring. Here the longitudinal displacement w of the moving end is nonlinearly related to the transverse deflection u. A single-mode, third-order approximation to the equations of dynamic motion takes the form

$$\ddot{a} + \omega_0^2(1 - P_0/P_c)a + \gamma^2 a^3 = 0, \tag{3.2}$$

where a is the mid-span deflection, γ^2 is a positive constant depending on the spring stiffness, P_0 is the spring pre-load and P_c the critical buckling load. Note that for post-critical loads one has $(1 - P_0/P_c) < 0$, so that there are three possible equilibrium positions: $a = 0$ and $a = \pm(\omega_0/\gamma)(P_0/P_c - 1)^{1/2}$. Nonlinear systems often have multiple states of equilibrium. Conversely, a system possessing more than one equilibrium is certainly nonlinear.

For the clamped Euler column in Fig. 3.1(c) the internal moment is given by $M = EI\kappa$, where κ is the curvature:

$$\kappa = \frac{u''(s)}{\sqrt{\left(1 - (u'(s))^2\right)}} = u''(s)\left(1 + \frac{1}{2}(u'(s))^2 + \cdots\right). \tag{3.3}$$

Thus, for rotations $u'(s)$ that are finite (but not very large), a third order nonlinearity will appear in the equation of motion. A single-mode approximation to the equation of motion takes the form of (3.2), with a positive linear stiffness coefficient. For small rotations, $u'(s) \ll 1$, one has $s \approx x$ and $\kappa \approx u''(s)$, and thus the well-known linear expression for the bending moment, $M = EIu''(x)$.

Fig. 3.1(d) shows a system causing nonlinear *inertial* terms to appear in the equations of motion. Transverse oscillations of the beam are accompanied by small horizontal displacements w of the moveable end-mass m. hence the mass exerts an axial force $-m\ddot{w}$) on the beam. For finite rotations, w is nonlinearly related to the transverse deflection u. A single-mode, third-order approximation to the equation describing transverse motions takes the form.

$$\ddot{a} + \omega_0^2 a + \eta(a\ddot{a} + \dot{a}^2)a = 0, \tag{3.4}$$

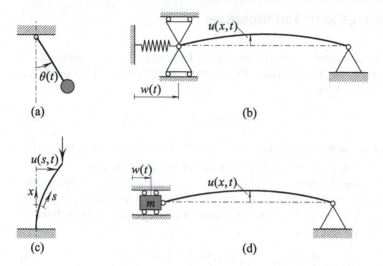

Fig. 3.1. Geometrical nonlinearities due to **(a,c)** large rotations, and **(b,d)** coupling between transverse and longitudinal displacements

where a is the mid-span deflection, and the constant η depends on the ratio of m to the beam mass. Again, the leading nonlinearity is of order three.

3.2.2 Material Nonlinearities

For the beam examples above, the structural material was assumed to be linearly elastic. However, all real materials obey a nonlinear relationship between stress and strain (and thus between force and deformation), which must be accounted for when strain variations are large.

In the system of Fig. 3.2(a), the spring is assumed to represent the stiffness of a nonlinear material. The equation of motion may take the form.

$$\ddot{x} + \omega^2 x + \gamma x^3 = 0 \tag{3.5}$$

If $\gamma > 0$ the stiffness increases with increased deformation, and we consider the nonlinearity (and the spring) to be *hardening*. The equation of motion is identical to (3.2) with $P_0 < P_c$, describing sub-critical loading of the beam in Fig. 3.1(b). If $\gamma < 0$ the stiffness decreases with increased deformation, and the nonlinearity (and spring) is said to be *softening*. Mathematically this case is similar to that of Fig. 3.1 (b) and Eq. (3.2) with $P_0 > P_c$, and to the pendulum case of Fig. 3.1(a) and Eq. (3.1) for $\sin\theta \approx \theta - \frac{1}{6}\theta^3$.

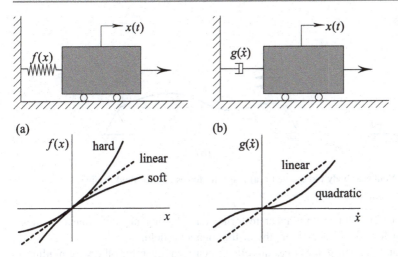

Fig. 3.2. Material nonlinearities due to **(a)** restoring force, and **(b)** damping

Most real materials obey relationships between stress σ and strain ε satisfying the expression $d^2\sigma / d\varepsilon^2 \leq 0$, i.e., they are softening. Some materials (rubbers, e.g.) are softening for low strains and hardening for larger strains.

Nonlinear *damping* may cause nonlinear dissipative terms to appear in the equations of motion, as may certain purely geometrical nonlinearities (Lazan 1968; Amabili 2018). The system in Fig. 3.2(b) exhibits *quadratic damping*. Quadratic damping forces take the form $g(\dot{x}) = \mu\dot{x}|\dot{x}|$, approximating the resistance experienced by a body that moves through a fluid at high Reynolds numbers. Forces due to dry friction are typically described by *Coulomb friction*, $g(\dot{x}) = \mu\dot{x}/|\dot{x}|$. The damping of certain alloys and composite materials is sometimes described by a power law, $g(\dot{x}) = c\dot{x}^\nu(\dot{x}/|\dot{x}|)$.

3.2.3 Nonlinear Body Forces

Certain kinds of body forces may vary nonlinearly with the state of a system. For a beam in a magnetic field (Fig. 3.3(a)), the total potential energy includes a magnetic potential V_m. The potential may be approximated by the first terms of a Taylor expansion:

$$V_m = \frac{1}{2}\gamma_1 a^2 + \frac{1}{4}\gamma_2 a^4, \tag{3.6}$$

where $a = a(t)$ is the displacement of the free beam end. The presence of terms higher than quadratic in the potential causes nonlinearities to appear in the equation of motion. For the beam, a single-mode approximation takes a form similar to

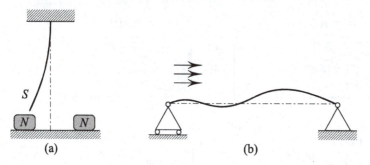

Fig. 3.3. Nonlinear body forces due to (**a**) magnetic forces and (**b**) fluid loading

Equation (3.2), with the nonlinear restoring term having negative linear stiffness coefficient for post-critical strengths of the magnetic field.

Fluid and *aerodynamic forces* are often linearized, even for otherwise nonlinear problems. However, some problems may require proper consideration to the non-linearity of such forces, e.g., problems involving structures subjected to supersonic flows (Fig. 3.3(b)).

In problems of *structural control*, the applied control forces can be any function of the state variables of the system to be controlled. Bang-bang control, various adaptive control schemes, piecewise linear control and other nonlinear control laws all correspond to nonlinear body forces.

Non-ideal power sources also correspond to nonlinear body forces. Forces delivered by non-ideal source are *limited*, that is, their magnitude depends of the state of the driven system in a nonlinear fashion.

3.2.4 Physical Configuration Nonlinearities

Even when the individual components of a system are linear, or operate in a linear range, specific physical configurations of these components may cause the com-bined system to behave nonlinearly. Most systems with plays, stops and other discontinuous couplings display this type of nonlinearity (Babitsky 1998; Burton 1968; Kobrinskii 1969).

For the systems in Fig. 3.4(a)–(b), the combined action of two linear springs corresponds to a nonlinear restoring force.

Fig. 3.4(c) shows a restricted pendulum, sometimes used for damping out vibrations of a connected structure. If the free play is small the pendulum behaves linearly in the periods of free flight, whereas the sudden resistance to motion occurring at the moments of collisions with the restricting wall corresponds to a discontinuous, and thus nonlinear, change in elastic stiffness.

Piece-wise linear behavior also characterizes the continuous beam of Fig. 3.4(d). Here the equation of motion is a linear fourth-order partial differential equation,

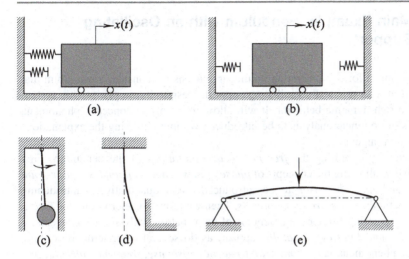

Fig. 3.4. Physical configuration nonlinearities due to **(a)** bi-linear spring, **(b)** dead-band spring, **(c)** play, **(d)** stop, and **(e)** midplane stretching

subjected to linear boundary conditions at the clamped end. Nonlinear boundary conditions apply to the end, which is alternately free and restricted.

The beam in Fig. 3.4(e) is subjected to *nonlinear stretching*: Due to the immovable ends, any transverse beam deflection is accompanied by longitudinal stretching, and thus by axial forces that are nonlinearly related to transverse deformations. The equation of motion takes the form

$$\rho A \ddot{u} + EI u'''' - \frac{EA}{2l} \left[\int_0^l (u')^2 \, dx \right] u'' = 0 \qquad (3.7)$$

where $u = u(x, t)$ is the transverse deflection, ρA is the mass per unit length, EI the bending stiffness, EA the longitudinal stiffness, and the integral expresses the axial force. A single-mode approximation to this equation has the form (3.2), with a positive linear stiffness coefficient. The nonlinear term disappears if one of the beam-ends is allowed to move freely in the longitudinal direction. Thus, even though this nonlinearity has a geometrical origin, it manifests itself only for certain physical configurations. Nonlinear stretching, also known as *midplane stretching*, provides a source of nonlinearity for many curved structures, such as arches and shells – and for plane structures, such as beams and plates, that are somehow restricted in performing plane displacements.

3.3 Main Example: Pendulum with an Oscillating Support

As a first object for demonstrating nonlinear analysis we consider a pendulum that is hinged at a harmonically oscillating support. Despite its simplicity, this system exhibits a rich dynamic behavior. It will allow many of the concepts, phenomena and tools of nonlinear analysis to be introduced without cluttering the explanations in inessential algebra.

We start by exploring the *free nonlinear oscillations* of the pendulum. First qualitatively, allowing the concepts of *phase planes, singular points, singular-point stability* and *local behavior* to be introduced. Next quantitatively, introducing various methods of *perturbation analysis*. Then we turn into determining the *forced response*, exploring further the potentials of perturbation analysis. The concepts of *limit cycles* and *limit cycle stability* appear, as do several other terms describing nonlinear phenomena, e.g., *nonlinear frequency response, frequency locking, amplitude jumps, subharmonics* and *superharmonics*.

The pendulum with an oscillating support is known to exhibit chaotic behavior for certain ranges of physical parameters; However, since chaos is generally a global phenomenon it is not revealed by the local methods described in this chapter. Only ordered, non-chaotic motions will be considered, whereas the study of chaotic motions will have to wait until Chap. 6.

3.3.1 Equation of Motion

Fig. 3.5 shows the pendulum, characterized by mass m, length l, and instantaneous angle of rotation $\theta(t)$. The pendulum is subjected to a gravity field g, and to a viscous damping moment $-cl^2\dot{\theta}$. The position $u(t)$ of the hinged support oscillates harmonically at a prescribed amplitude ql and frequency Ω.

The equation of motion is conveniently set up using Lagrange's equation for single-DOF non-conservative systems (*cf.* Sect. 1.5.1):

$$\frac{d}{dt}\left(\frac{\partial L}{\partial \dot{\theta}}\right) - \frac{\partial L}{\partial \theta} = Q, \quad L = T - V, \tag{3.8}$$

The kinetic energy T, potential energy V, and generalized non-conservative force Q become, respectively, with $x = l\cos\theta - u$ and $y = l\sin\theta$:

$$
\begin{aligned}
T &= \tfrac{1}{2}m\dot{x}^2 + \tfrac{1}{2}m\dot{y}^2 = \tfrac{1}{2}m\left(l^2\dot{\theta}^2 + \dot{u}^2 + 2l\dot{\theta}\dot{u}\sin\theta\right), \\
V &= -mgx = -mg(l\cos\theta - u), \\
Q &= -cl^2\dot{\theta}.
\end{aligned}
\tag{3.9}
$$

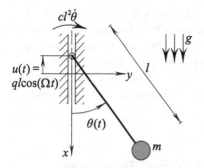

Fig. 3.5. Pendulum with an oscillating support

Inserting into (3.8) one obtains the equation of motion:

$$\ddot{\theta} + \frac{c}{m}\dot{\theta} + l^{-1}(g + \ddot{u})\sin\theta = 0 \tag{3.10}$$

Introducing $\omega_0^2 = g/l$ and $\beta = c/(2m\omega_0)$, and substituting $u(t) = ql\cos(\Omega t)$, the equation of motion becomes:

$$\ddot{\theta} + 2\beta\omega_0\dot{\theta} + (\omega_0^2 - q\Omega^2\cos\Omega t)\sin\theta = 0 \tag{3.11}$$

with prescribed initial conditions:

$$\theta(0) = \theta_0, \quad \dot{\theta}(0) = \dot{\theta}_0 \tag{3.12}$$

Note that ω_0 is the linear (small rotation) *natural frequency*, Ω is the *excitation frequency*, q measures the *support displacement* as a fraction of the pendulum length, and β is the *damping ratio* (actual to critical).

The pendulum equation is nonlinear due to the term $\sin\theta$. For finite (but not very large) rotations θ, we can approximate the nonlinearity by the first two terms of a Taylor expansion, $\sin\theta \approx \theta - \frac{1}{6}\theta^3$. The term $\omega_0^2\sin\theta$ of the pendulum equation is then recognized as a nonlinear restoring force of the *softening* type, since the coefficient of the cubic nonlinearity is negative (*cf.* Section 3.2.2).

We also note that the pendulum is *parametrically* excited, i.e. the external excitation acts through a parameter of the system, in this case the stiffness parameter. This implies that, for certain ranges of excitation-frequencies Ω, even small levels of excitation magnitude $q\Omega^2$ may cause large oscillations of the pendulum[1]. For the linearized system ($\sin\theta \approx \theta$) we may calculate those ranges of

---

[1]To demonstrate this: hold one end of a pendulum-like object so that it can swing freely. Shake it vertically at a frequency roughly twice the frequency of free oscillations. Even a weak shaking will then generate large oscillations of the object.

(q, Ω) for which the solution $\theta(t) = 0$ becomes unstable and oscillations start growing. This analysis will predict pendulum rotations $\theta (t)$ that approach infinity at exponential rate. As amplitudes grow, however, the linear model becomes increasingly inadequate, and one needs to drop the assumption $\sin\theta \approx \theta$. Thus, linear analysis will predict the state $\theta (t) = 0$ to be unstable for certain values of (q,Ω), while a nonlinear analysis is required for predicting the new (post-critical) state replacing $\theta (t) = 0$.

3.4   Qualitative Analysis of the Unforced Response

3.4.1   Recasting the Equations into First-Order Form

If the pendulum support is fixed ($q = 0$) we are left with the equation of motion of the ordinary, unforced pendulum:

$$\ddot{\theta} + 2\beta\omega_0\dot{\theta} + \omega_0^2 \sin \theta = 0, \quad \theta = \theta(t),$$
$$\theta(0) = \theta_0, \quad \dot{\theta}(0) = \dot{\theta}_0. \tag{3.13}$$

As a first step towards a nonlinear analysis of the nonlinear response, we recast the second-order equation into two first-order equations:

$$\dot{\theta} = v,$$
$$\dot{v} = -2\beta\omega_0 v - \omega_0^2 \sin \theta, \tag{3.14}$$

with initial conditions

$$\theta(0) = \theta_0, \quad v(0) = v_0 = \dot{\theta}(0), \tag{3.15}$$

where a new variable $v \equiv \dot{\theta}$ has been introduced. To allow for a more general discussion, we may as well write the set of autonomous first-order equations with initial conditions in the form

$$\dot{\mathbf{x}} = \mathbf{f}(\mathbf{x}), \quad \mathbf{x}(0) = \mathbf{x}_0 \tag{3.16}$$

where $\mathbf{x} = \mathbf{x}(t)$ is a vector of *state variables*, spanning the *state space*, and $\mathbf{f}(\mathbf{x})$ is a vector of generally nonlinear functions of the state variables. For the pendulum case one has $\mathbf{x} = \{\theta, v\}^T$ and $\mathbf{f} = \{v, -2\beta\omega_0 v - \omega_0{}^2\sin\theta\}^T$.

3.4.2   The Phase Plane

Motions of nonlinear systems are often presented in a *phase plane*. A phase plane is spanned by two arbitrary state-variables. Thus we may describe motions of the pendulum in a (θ,v)-plane, rather than in the familiar (t,θ) or $(t,\dot{\theta})$ plane. In a phase plane time is *implicit*: Though you may imagine the time axis running out of and behind the paper, we only consider motions projected on the (θ, v) plane. With the passage of time, the points $(\theta(t),v(t))$ describe a curve in the phase plane. Such a curve is called an *orbit*, a *trajectory*, or an *integral curve*. Sample pendulum orbits are shown in Fig. 3.6 for the undamped case $(\beta = 0)$, and in Fig. 3.7 for the case of subcritical damping $(0 < \beta < 1)$.

When there is no damping, we have one of the rare cases for which nonlinear phase plane orbits can be analytically determined. To see this, let $\beta = 0$ in the equations of motion (3.14), and divide the second equation by the first:

$$\frac{\dot{v}}{\dot{\theta}} \equiv \frac{dv/dt}{d\theta/dt} = \frac{dv}{d\theta} = \frac{-\omega_0^2 \sin\theta}{v} \tag{3.17}$$

Separating the variables of the last equality and integrating readily yields

$$\frac{1}{2}v^2 = \omega_0^2 \cos\theta + C \tag{3.18}$$

where the constant C is determined by the initial conditions (θ_0, v_0):

$$C = \frac{1}{2}v_0^2 - \omega_0^2 \cos\theta_0, \quad C \geq -\omega_0^2. \tag{3.19}$$

When $\theta \ll 1$ one has $\cos\theta \approx 1 - \theta^2/2$, so that the orbits (v,θ) corresponding to small pendulum rotations are ellipses centered at $(0,0)$, as determined by:

$$\frac{1}{2}v^2 + \frac{1}{2}\omega_0^2\theta^2 = C + \omega_0^2 \tag{3.20}$$

These orbits correspond to a small-amplitude linear solution $\theta(t) = A\cos(\omega_0 t + \psi)$, $v(t) = -A\omega_0\sin(\omega_0 t + \psi)$. At larger rotations the ellipses become nonlinearly distorted, according to (3.18):

$$v = \pm\sqrt{2(\omega_0^2 \cos\theta + C)} \tag{3.21}$$

If $C > \omega_0^2$, the orbits can never intersect the axis $v = 0$. This occurs whenever initial conditions are such that $v_0^2/2 > \omega_0^2(1 + \cos\theta_0)$, so that the pendulum performs full rotations through 2π, rather than oscillations around $\theta = 0$.

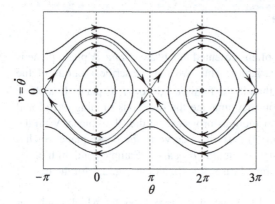

Fig. 3.6. Phase plane and orbits for the unforced pendulum: undamped case ($\beta = 0$)

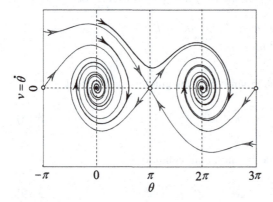

Fig. 3.7. Phase plane and orbits for the unforced pendulum: damped case ($0 < \beta < 1$)

If damping is present (Fig. 3.7), the dissipation of energy causes orbits to spiral towards the equilibriums at $\theta = 0$, 2π, ... With initial energies sufficiently large, several full rotations may precede the spiralling.

The arrows in Fig. 3.6 and Fig. 3.7 indicate directions of motion for increasing time t. This direction is easily established, by considering that θ must *increase* in the upper half-plane, where $v > 0$, and *decrease* in the lower half-plane, $v < 0$.

3.4.3 Singular Points

Certain points of a phase plane may correspond to states of static equilibrium. Such points are termed *singular points, fixed points, equilibriums* or *zeroes*. All other points of the phase plane are *regular*. For static equilibrium to occur we must have $\dot{\mathbf{x}} = \mathbf{0}$ in (3.16). Thus, singular points are located by solving the algebraic set of equations $\mathbf{f}(\mathbf{x}) = \mathbf{0}$. We denote a singular point by $\tilde{\mathbf{x}}$, so that by definition $\mathbf{f}(\tilde{\mathbf{x}}) = \mathbf{0}$.

The singular points of the pendulum equation (3.14) are obtained by solving the system $\dot{\theta} = \dot{v} = 0$. This gives $v = 0$ and $\sin\theta = 0$, that is:

$$(\tilde{\theta}, \tilde{v}) = (k\pi, 0), \quad k = \ldots -1, 0, 1, \ldots \tag{3.22}$$

In Fig. 3.6 and Fig. 3.7 the singular points are marked by filled or unfilled circles. Note that in the undamped case orbits encircle the singular points at $\theta = 0$, 2π, …, whereas in the damped case these points attract nearby orbits.

For deterministic systems only a single orbit $\mathbf{x}(t)$ can pass through each point of state space. Otherwise orbits could intersect themselves, or other orbits, and the state of the system would not be uniquely determined by the initial conditions. This property also holds for singular points: only a single orbit may pass through.

Reconsidering Fig. 3.6 and Fig. 3.7, it may seem as if several orbits pass through the singular points of the pendulum system. However, the only orbit passing through a singular point is the *singular orbit*. You may imagine the singular orbit as a straight line extending through a singular point and behind the paper for $t \rightarrow +\infty$, and out of the paper for $t \rightarrow -\infty$. Other orbits may at most *approach* this orbit, that is, it will take them infinitely long to actually reach it. Non-singular orbits are called *regular*.

One should be careful here to distinguish between the notions of state space and phase plane. The above observations on intersections of orbits apply to orbits in the *state space*, which is spanned by the state-variables of a set of autonomous, first-order differential equations. Phase planes are spanned by any two state-variables. For the unforced pendulum, since there are only two autonomous equations, the state space happens to be also a phase plane. Thus, for two-dimensional systems, orbits do not intersect in the phase plane. Now, considering instead the *forced* pendulum, we would have to write *three* autonomous equations[2]. Then orbits in the three-dimensional state space would not intersect, while orbits projected on a phase plane probably would; they are only *projections* of the state space orbits to the plane.

Orbits that connect distinct singular points are called *heteroclinic*. In Fig. 3.6 the orbits connecting the singular points at $\theta = -\pi, \pi$, are heteroclinic. Other orbits (though none for the pendulum) are *homoclinic*; they loop from a singular point and back to that same point.

3.4.4   Stability of Singular Points

A *stable* singular point *attracts* nearby orbits, an *unstable* singular point *repels* orbits, whereas a *marginally/neutrally stable* singular point acts as a *center*, neither repelling nor attracting orbits. Certain singular points have a mixed kind of stability,

---

[2] $\dot{\theta} = v, \dot{v} = 2\beta\omega_0 v - \omega_0^2 \sin\theta + q\Omega^2 \cos z$, and $\dot{z} = \Omega$.

attracting orbits from some directions while repelling orbits from other directions (e.g., $(\pi,0)$ in Fig. 3.7); these are called unstable.

We consider here only the *local* stability of singular points. The question to be answered is, whether orbits in the immediate vicinity of a singular point stay near or escape. For analyzing local stability, a study of the linearized system will suffice. In regular cases the stability properties of singular points for the linearized system hold too for the nonlinear system. That is, if a singular point of a linearized system is (un)stable, then so is the same singular point of the original nonlinear system.

Consider a general nonlinear system, written as a set of n first-order autonomous differential equations:

$$\dot{\mathbf{x}} = \mathbf{f}(\mathbf{x}), \quad \mathbf{x} = \mathbf{x}(t) \in R^n, \tag{3.23}$$

for which we aim at determining the local stability of a given singular point $\mathbf{x} = \tilde{\mathbf{x}}$. Taylor-expanding the right-hand side near $\mathbf{x} = \tilde{\mathbf{x}}$ one obtains:

$$\dot{\mathbf{x}} = \mathbf{f}(\tilde{\mathbf{x}}) + \left.\frac{\partial \mathbf{f}}{\partial \mathbf{x}}\right|_{\mathbf{x}=\tilde{\mathbf{x}}} (\mathbf{x} - \tilde{\mathbf{x}}) + O\big((\mathbf{x} - \tilde{\mathbf{x}})^T (\mathbf{x} - \tilde{\mathbf{x}})\big) \tag{3.24}$$

where the last term represents quadratic and higher-order terms. The first term of the expansion vanishes, since by definition $f(\tilde{\mathbf{x}}) = \mathbf{0}$. To indicate the nearness of an orbit $\mathbf{x}(t)$ to the singular point $\tilde{\mathbf{x}}$, we introduce a new dependent variable $\varepsilon(t) = \mathbf{x}(t) - \tilde{\mathbf{x}}$. For orbits near the singular point $|\varepsilon| \ll 1$, so that higher-order terms of the Taylor expansion can be dropped. Performing the variable shift from $\mathbf{x}$ to ε in (3.24) and neglecting higher-order terms, we find that small distances between orbits and the singular point $\tilde{\mathbf{x}}$ are governed by the linear set of equations:

$$\dot{\varepsilon} = \mathbf{J}(\tilde{\mathbf{x}})\varepsilon, \tag{3.25}$$

where $\mathbf{J}(\tilde{\mathbf{x}})$ denotes the *Jacobian* of the nonlinear system, evaluated at the singular point. The Jacobian $\mathbf{J}(\mathbf{x})$ of the system (3.23) is defined by:

$$\mathbf{J}(\mathbf{x}) \equiv \frac{\partial \mathbf{f}}{\partial \mathbf{x}} = \begin{bmatrix} \frac{\partial f_1}{\partial x_1} & \frac{\partial f_1}{\partial x_2} & \cdots & \frac{\partial f_1}{\partial x_n} \\ \frac{\partial f_2}{\partial x_1} & \frac{\partial f_2}{\partial x_2} & \cdots & \frac{\partial f_2}{\partial x_n} \\ \vdots & \vdots & \ddots & \vdots \\ \frac{\partial f_n}{\partial x_1} & \frac{\partial f_n}{\partial x_2} & \cdots & \frac{\partial f_n}{\partial x_n} \end{bmatrix} \tag{3.26}$$

For the approximation (3.25) to be valid, $\tilde{\mathbf{x}}$ must be an *isolated* singular point, i.e. $\mathbf{J}(\tilde{\mathbf{x}}) \neq 0$. Now, the solution of the linearized system (3.25) has the form:

$$\epsilon = \mathbf{a}e^{\lambda t}, \tag{3.27}$$

which upon substitution into (3.25) yields an algebraic eigenvalue problem:

$$(\mathbf{J}(\tilde{\mathbf{x}}) - \lambda \mathbf{I})\mathbf{a} = \mathbf{0}, \tag{3.28}$$

for the determination of a set of eigenvalues λ_j and corresponding eigenvectors $\mathbf{a}_j$, $j = 1, n$. Inspecting (3.27), it appears that any eigenvalue λ having a positive real part causes orbits of the linearized system (3.25) to escape from the singular point. An orbit of the corresponding nonlinear system (3.23) will escape as well, since near the singular point the nonlinear system is closely approximated by the linearized one.

Thus, the stability of a singular point of a nonlinear system is determined by examining eigenvalues of the Jacobian matrix evaluated at that point. The following rules apply:

- $\mathrm{Re}(\lambda_j) < 0$ for all $j = 1, n$: $\tilde{\mathbf{x}}$ is *stable*
- $\mathrm{Re}(\lambda_j) > 0$ for at least one $j = 1, n$: $\tilde{\mathbf{x}}$ is *unstable*
- $\max[\mathrm{Re}(\lambda_j)] = 0$ for $j = 1, n$: $\tilde{\mathbf{x}}$ may be stable or unstable

In the third case there is at least one eigenvalue with vanishing real part, but no eigenvalues with positive real parts. This is a *critical* case, for which the stability of the singular point cannot be deduced from the linearized system; higher-order nonlinear terms may render the point stable or unstable.

The unforced pendulum is governed by the form (3.23) with $\mathbf{x} = \{\theta, v\}^T$ and $\mathbf{f} = \{v, -2\beta\omega_0 v - \omega_0^2\sin\theta\}^T$ (*cf.* Sect. 3.4.1). The singular points are

$$\begin{Bmatrix} \tilde{\theta} \\ \tilde{v} \end{Bmatrix} = \begin{Bmatrix} k\pi \\ 0 \end{Bmatrix}, \quad k = \ldots, -1, 0, 1, \ldots, \tag{3.29}$$

and the Jacobian of the system is

$$\mathbf{J}\left(\begin{Bmatrix} \theta \\ v \end{Bmatrix}\right) = \begin{bmatrix} \frac{\partial f_1}{\partial \theta} & \frac{\partial f_1}{\partial v} \\ \frac{\partial f_2}{\partial \theta} & \frac{\partial f_2}{\partial v} \end{bmatrix} = \begin{bmatrix} 0 & 1 \\ -\omega_0^2 \cos\theta & -2\beta\omega_0 \end{bmatrix} \tag{3.30}$$

At the singular points, the Jacobian becomes

$$\mathbf{J}\left(\begin{Bmatrix} \tilde{\theta} \\ \tilde{v} \end{Bmatrix}\right) = \begin{bmatrix} 0 & 1 \\ -\omega_0^2(-1)^k & -2\beta\omega_0 \end{bmatrix} \tag{3.31}$$

with eigenvalues:

$$\lambda_{1,2} = \left(-\beta \pm \sqrt{\beta^2 - (-1)^k}\right)\omega_0 \tag{3.32}$$

For the case of *no damping* ($\beta = 0$) the eigenvalues are

$$\lambda_{1,2} = \begin{cases} \pm i\omega_0 & \text{for } k \text{ even,} \\ \pm\omega_0 & \text{for } k \text{ odd.} \end{cases} \tag{3.33}$$

For singular points of *even k* ($\theta = \dots, -2\pi, 0, 2\pi, \dots$) both eigenvalues are imaginary. This is the critical case described above ($\max[\text{Re}(\lambda_j)] = 0$). Comparing with Fig. 3.6 we see that these singular points are neither stable nor unstable: Nearby orbits stay near, but are neither attracted to, nor repelled from the singular points corresponding to the down-pointing position of the pendulum. Singular points with *odd k* ($\theta = \dots, -\pi, \pi, 3\pi, \dots$) are unstable, since one eigenvalue has a positive real part ($\lambda = +\omega_0$). This also appears from the orbits of Fig. 3.6.

For *subcritical damping* ($0 < \beta < 1$) we may write the Jacobian eigenvalues (3.32) in the form

$$\lambda_{1,2} = \begin{cases} \left(-\beta \pm i\sqrt{1-\beta^2}\right)\omega_0 & \text{for } k \text{ even,} \\ \left(-\beta \pm \sqrt{1+\beta^2}\right)\omega_0 & \text{for } k \text{ odd.} \end{cases} \tag{3.34}$$

Thus singular points of *even k* are stable, since both eigenvalues have negative real parts. Singular points of *odd k* are unstable, since $(1 + \beta^2)^{1/2} > \beta$ when $0 < \beta < 1$, implying that one of the eigenvalues has a positive real part. These results agree with the flow of orbits in Fig. 3.7.

3.4.5   On the Behavior of Orbits Near Singular Points

The above analysis of stability is based on Jacobian eigenvalues. These eigenvalues merely reveal whether a singular point attracts or repels nearby orbits. To obtain a more detailed picture on *how* orbits are disturbed in the vicinity of a singular point, one needs to consider also the associated eigen*vectors*. The eigenvectors of the linearized system will provide local information on orbits for the nonlinear system as well, since near singular points the orbits of the nonlinear system are well approximated by those of the linearized system.

Be careful to note that only the *local* behavior near singular points can be assessed this way. To obtain a global orbital picture is kind of an art, requiring all the local pictures to be tied together in the right manner; often this is not needed, though.

Inspecting (3.27) it appears that the character of the linearized solutions depends on whether the eigenvalues λ are real or complex, and whether the eigenvalues are distinct. Real eigenvalues cause the motion to grow or shrink exponentially in time, whereas complex eigenvalues introduce oscillatory components in the motion (since, with $\gamma, \omega \in R$: $\lambda = \gamma + i\omega \Rightarrow e^{\lambda t} = e^{\gamma t}(\cos(\omega t) + i\sin(\omega t))$). Non-distinct eigenvalues imply solutions of the form $\varepsilon = \mathbf{a}_0 e^{\lambda t} + \mathbf{a}_1 t e^{\lambda t} + \; + \mathbf{a}_m t^m e^{\lambda t}$, where m is the multiplicity of the non-distinct eigenvalue.

To keep the presentation on a manageable level, we restrict attention to the case $n = 2$, that is, to systems described by two ordinary, autonomous differential equations. For a discussion of the topology of higher-dimensional singular points see, e.g., Blaquiére (1966) or Guckenheimer and Holmes (1983).

Jacobian Eigenvalues for n = 2 With two first-order equations of motion, there are two eigenvalues of the Jacobian to be examined. One can easily show that these can be written in terms of the trace and the determinant of the Jacobian, thus:

$$\lambda_{1,2} = \frac{1}{2}\left(p \pm \sqrt{p^2 - 4q}\right), \quad p = tr(\mathbf{J}(\tilde{\mathbf{x}})), \quad q = det(\mathbf{J}(\tilde{\mathbf{x}})). \tag{3.35}$$

Hence, if $p^2 > 4q$ the two eigenvalues are *real and distinct*, if $p^2 = 4q$ they are *real and equal*, and if $p^2 < 4q$ they are *complex conjugates*.

Obtaining Qualitative Results: The Similarity Transform To picture the flow of orbits near singular points one needs to consider the Jacobian eigenvector $\mathbf{a}$ of (3.28). We aim here at obtaining only the *topology* of orbits, the qualitative features. So, to facilitate the interpretation of results we first perform a *similarity transform* of the linearized system (3.25). This is accomplished by introducing a new state vector $\mathbf{u}$, linearly related to the original state vector ε by:

$$\varepsilon = \mathbf{P}\mathbf{u}, \tag{3.36}$$

where $\mathbf{P}$ is a constant, non-singular matrix. This linear transformation preserves all topological features of the original system, that is: The origin maps onto itself, straight lines map into straight lines, and parallel lines into parallel lines. Thus, substituting (3.36) into (3.25) one arrives at a topological identical system:

$$\dot{\mathbf{u}} = \hat{\mathbf{J}}(\tilde{\mathbf{x}})\mathbf{u}, \tag{3.37}$$

where

$$\hat{\mathbf{J}}(\tilde{\mathbf{x}}) = \mathbf{P}^{-1}\mathbf{J}(\tilde{\mathbf{x}})\mathbf{P}. \tag{3.38}$$

The matrices $\mathbf{J}$ and $\hat{\mathbf{J}}$ are said to be *similarity matrices*; they have identical eigenvalues, whatever the particular choice of the (non-singular) transformation matrix $\mathbf{P}$. Thus, we are free to choose $\mathbf{P}$ so as to make $\hat{\mathbf{J}}$ have the simplest possible form, preferably a diagonal one.

Jordan Canonical Forms The 'simplest possible form' of the similarity matrix $\hat{\mathbf{J}}$ is termed a *Jordan Canonical form*. To obtain a Jordan form one may choose a transformation matrix $\mathbf{P}$ that is made up by the eigenvectors of (3.28), that is:

$$\mathbf{P} = [\,\mathbf{a}_1 \quad \mathbf{a}_2\,].$$ (3.39)

Then, due to the orthogonality of eigenvectors $\hat{\mathbf{J}}$ will be as close as possible to diagonal. To see this we note that the eigenvectors of (3.28) will satisfy $\mathbf{J}(\tilde{\mathbf{x}})$ $\mathbf{a}_1 = \lambda_1 \mathbf{a}_1$ and $\mathbf{J}(\tilde{\mathbf{x}})\mathbf{a}_2 = \lambda_2 \mathbf{a}_2$, or, written as matrix equations:

$$\mathbf{J}(\tilde{\mathbf{x}})[\,\mathbf{a}_1 \quad \mathbf{a}_2\,] = [\,\mathbf{a}_1 \quad \mathbf{a}_2\,]\begin{bmatrix} \lambda_1 & 0 \\ 0 & \lambda_2 \end{bmatrix}$$ (3.40)

Hence:

$$\hat{\mathbf{J}}(\tilde{\mathbf{x}}) = \mathbf{P}^{-1}\mathbf{J}(\tilde{\mathbf{x}})\mathbf{P} = [\,\mathbf{a}_1 \quad \mathbf{a}_2\,]^{-1}\mathbf{J}(\tilde{\mathbf{x}})[\,\mathbf{a}_1 \quad \mathbf{a}_2\,]$$
$$= [\,\mathbf{a}_1 \quad \mathbf{a}_2\,]^{-1}[\,\mathbf{a}_1 \quad \mathbf{a}_2\,]\begin{bmatrix} \lambda_1 & 0 \\ 0 & \lambda_2 \end{bmatrix} = \begin{bmatrix} \lambda_1 & 0 \\ 0 & \lambda_2 \end{bmatrix}; \quad \|[\,\mathbf{a}_1 \quad \mathbf{a}_2\,]\| \neq 0.$$ (3.41)

Thus, for $n = 2$ and $\|[\,\mathbf{a}_1 \ \mathbf{a}_2\,]\| \neq 0$ the Jordan form is diagonal with elements made up by the Jacobian eigenvalues of the system. This applies too when the eigenvalues are *equal* with *linearly independent eigenvectors* (e.g., the unit-matrix, has this property), since even in this case $\|[\,\mathbf{a}_1 \ \mathbf{a}_2\,]\| \neq 0$.

However, in case the eigenvalues are *equal* with *linearly dependent eigenvectors*, then $\|[\,\mathbf{a}_1 \ \mathbf{a}_2\,]\| = 0$ and $\mathbf{J}$ cannot be fully diagonalized. For this case one can choose $\mathbf{P}$ so as to partly diagonalize $\mathbf{J}$, for example by choosing:

$$\mathbf{P} = \begin{bmatrix} \mathbf{a}_1 & \begin{bmatrix} 1 & 0 \\ J_{12}^{-1} & 1 \end{bmatrix}\mathbf{a}_1 \end{bmatrix}$$ (3.42)

where $\mathbf{a}_1$ is the (one and only) eigenvector of $\mathbf{J}$, and J_{12} denotes the upper-right element of $\mathbf{J}$. The Jordan form then becomes:

$$\hat{\mathbf{J}}(\tilde{\mathbf{x}}) = \mathbf{P}^{-1}\mathbf{J}(\tilde{\mathbf{x}})\mathbf{P}$$
$$= \begin{bmatrix} \mathbf{a}_1 & \begin{bmatrix} 1 & 0 \\ J_{12}^{-1} & 1 \end{bmatrix}\mathbf{a}_1 \end{bmatrix}^{-1}\mathbf{J}(\tilde{\mathbf{x}})\begin{bmatrix} \mathbf{a}_1 & \begin{bmatrix} 1 & 0 \\ J_{12}^{-1} & 1 \end{bmatrix}\mathbf{a}_1 \end{bmatrix}$$
$$= \cdots \text{(some algebra)} \cdots$$
$$= \begin{bmatrix} \lambda_1 & 1 \\ 0 & \lambda_1 \end{bmatrix}, \quad \text{for } \lambda_1 = \lambda_2 \quad \text{and} \quad \|[\,\mathbf{a}_1 \quad \mathbf{a}_2\,]\| = 0.$$ (3.43)

Orbits for Diagonal Jordan Forms Due to the similarity transform employed, orbits of the Jordan forms will qualitatively resemble those of the linearized system (3.25), which in turn approximates the original nonlinear system (3.23) near the singular point $\tilde{\mathbf{x}}$. Thus, the simple Jordan forms provide information on the flow of orbits near singular points for a possibly highly complicated nonlinear system.

The diagonal Jordan form (3.41) applies when $|[a_1 \ a_2]| \neq 0$. The system (3.37) then becomes, writing out the components:

$$\begin{aligned} \dot{u}_1 &= \lambda_1 u_1, \\ \dot{u}_2 &= \lambda_2 u_2, \end{aligned} \tag{3.44}$$

with solution

$$\begin{aligned} u_1(t) &= u_{10} e^{\lambda_1 t}, \\ u_2(t) &= u_{20} e^{\lambda_2 t}, \end{aligned} \tag{3.45}$$

or, upon eliminating the time-variable t:

$$u_2 = u_{20}(u_1/u_{10})^{\lambda_2/\lambda_1} \tag{3.46}$$

where $u_{10} = u_1(0)$ and $u_{20} = u_2(0)$ define the initial conditions.

If the eigenvalues $\lambda_{1,2}$ are *real with equal sign*, the arrangement of orbits $(u_1(t), u_2(t))$ is as depicted in Fig. 3.8(a) for $\lambda_1 \neq \lambda_2$, and in Fig. 3.8(c) for $\lambda_1 = \lambda_2$. The singular point $(0,0)$ is in this case a *node*; in fact a *stable node* for $\lambda_{1,2} < 0$ and an *unstable node* for $\lambda_1 > 0$ or $\lambda_2 > 0$. The orbits in Fig. 3.8(a) are tangent to the u_1-axis (except the one on the u_2-axis). This occurs for $\lambda_2 > \lambda_1$, whereas for $\lambda_2 < \lambda_1$ the u_2-axis becomes the tangent.

If the eigenvalues are *real with different sign*, then $\lambda_2/\lambda_1 < 0$, and the orbits are as depicted in Fig. 3.8(b). The singular point is a *saddle*. A saddle point is always unstable, though occasionally we will refer to the stable and unstable *directions* or *branches* of a saddle (in Fig. 3.8(b), the u_2- and u_1-axis, respectively).

If the eigenvalues are *complex* they will be complex conjugates, that is, if $\lambda_1 = \lambda$ then $\lambda_2 = \bar{\lambda}$. The solution (3.46) remains valid, though u_1 and u_2 now turn complex with $u_2 = \bar{u}_1$. Introducing polar coordinates:

$$u_1 = re^{i\varphi}, \quad u_2 = \bar{u}_1 = re^{-i\varphi}; \quad r = r(t), \quad \varphi = \varphi(t), \tag{3.47}$$

equation (3.46) becomes

$$re^{-i\varphi} = r_0 e^{-i\varphi_0} \left(\frac{re^{i\varphi}}{r_0 e^{i\varphi_0}} \right)^{\bar{\lambda}/\lambda}, \tag{3.48}$$

which can be reduced to

$$r = r_0 \exp\left(\frac{\mathrm{Re}(\lambda)}{\mathrm{Im}(\lambda)} (\varphi - \varphi_0) \right). \tag{3.49}$$

Eigenvalues λ_1, λ_2	Jordan form $\hat{\mathbf{J}}(\tilde{\mathbf{x}})$	Phase plane orbits near $\mathbf{u} = \mathbf{0}$ for $\dot{\mathbf{u}} = \hat{\mathbf{J}}(\tilde{\mathbf{x}})\mathbf{u}$

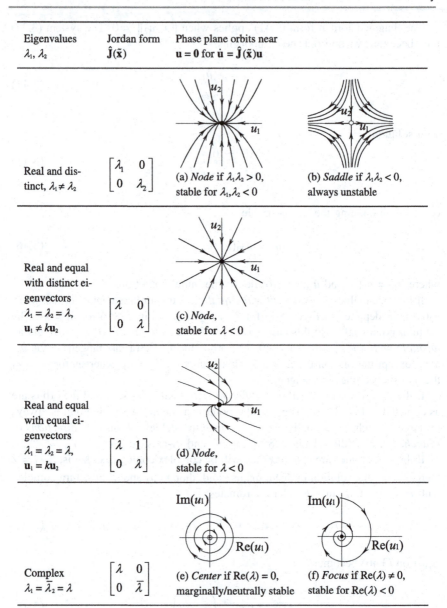

Real and distinct, $\lambda_1 \neq \lambda_2$ — $\begin{bmatrix} \lambda_1 & 0 \\ 0 & \lambda_2 \end{bmatrix}$

(a) *Node* if $\lambda_1\lambda_2 > 0$, stable for $\lambda_1, \lambda_2 < 0$

(b) *Saddle* if $\lambda_1\lambda_2 < 0$, always unstable

Real and equal with distinct eigenvectors $\lambda_1 = \lambda_2 = \lambda$, $\mathbf{u}_1 \neq k\mathbf{u}_2$ — $\begin{bmatrix} \lambda & 0 \\ 0 & \lambda \end{bmatrix}$

(c) *Node*, stable for $\lambda < 0$

Real and equal with equal eigenvectors $\lambda_1 = \lambda_2 = \lambda$, $\mathbf{u}_1 = k\mathbf{u}_2$ — $\begin{bmatrix} \lambda & 1 \\ 0 & \lambda \end{bmatrix}$

(d) *Node*, stable for $\lambda < 0$

Complex $\lambda_1 = \bar{\lambda}_2 = \lambda$ — $\begin{bmatrix} \lambda & 0 \\ 0 & \bar{\lambda} \end{bmatrix}$

(e) *Center* if $\mathrm{Re}(\lambda) = 0$, marginally/neutrally stable

(f) *Focus* if $\mathrm{Re}(\lambda) \neq 0$, stable for $\mathrm{Re}(\lambda) < 0$

Fig. 3.8. Qualitative flow of orbits near a singular point of dimension $n = 2$. Eigenvalues $\lambda_{1,2}$ are for the Jacobian $\mathbf{J}(\mathbf{x})$ of the system (3.16), evaluated at the singular point $\tilde{\mathbf{x}}$

Fig. 3.8(e)–(f) shows the corresponding orbits in a $(\text{Re}(u_1), \text{Im}(u_1))$ plane (those for u_2 are similar). If $\text{Re}(\lambda) = 0$, i.e. the eigenvalues are purely imaginary, the orbits are circles of radius r_0, and the singular point is called a *center* (Fig. 3.8(e)).

A center may be considered stable or unstable, according to various definitions of stability. For the moment we bypass this discussion by calling a center *marginally* or *neutrally stable*. Centers and other singular points with $\text{Re}(\lambda) = 0$, are significant for the study of *bifurcations* – the qualitative shifts in system behavior that we shall consider further in Chap. 5.

If $\text{Re}(\lambda) \neq 0$ the orbits spiral around the singular point, which is then called a *focus* (Fig. 3.8(f)). For $\text{Re}(\lambda) < 0$ the orbits *approach* the singular point, and the focus is *stable*. This completes our discussion of possible types of singular points for two-dimensional systems with diagonal Jordan form.

Orbits for Non-diagonal Jordan Forms The non-diagonal Jordan form (3.43) applies when $|[\mathbf{a}_1 \ \mathbf{a}_2]| = 0$, that is, when the eigenvalues of $\mathbf{J}(\tilde{\mathbf{x}})$ are real and equal and the eigenvectors are linearly dependent. The system (3.37) becomes:

$$\dot{u}_1 = \lambda u_1 + u_2, \quad \dot{u}_2 = \lambda u_2, \tag{3.50}$$

with solution

$$u_1(t) = (u_{10} + u_{20}t)e^{\lambda t}, \quad u_2(t) = u_{20}e^{\lambda t}, \tag{3.51}$$

or, by eliminating t:

$$u_1 = \left(\frac{u_{10}}{u_{20}} + \frac{1}{\lambda} \ln \left| \frac{u_2}{u_{20}} \right| \right) u_2 \tag{3.52}$$

The orbits $(u_1(t), u_2(t))$ are shown in Fig. 3.8(d). It appears from (3.44) that two half-orbits $(u_2(t) = 0, t \neq 0)$ coincide with the u_1-axis, whereas there are no orbits along the u_2-axis. The singular point is a *node*, being stable for $\lambda < 0$.

Orbital Topology for the Pendulum Case We now return then to the singular points $(\theta, v) = (k\pi, 0)$ of the unforced pendulum.

In the *undamped case*, according to Eq. (3.33), the two eigenvalues are purely imaginary for even k, and real and distinct with different signs for odd k. Hence, according to Fig. 3.8, the singular points corresponding to even k are *centers*, whereas those corresponding to odd k are *saddles*. Indeed, this knowledge is sufficient for sketching the phase plane orbits shown in Fig. 3.6. First we would draw the local arrangements of orbits near the individual centers and saddles. Next, observing that orbits must be smooth and cannot intersect, we could connect the stable and unstable saddle branches into heteroclinic orbits, and finally sketch the outermost running orbits, running outside the heteroclinic loops.

In the *damped case*, according to Eq. (3.34), the eigenvalues are complex with negative real parts for even k, and real and distinct with different signs for odd k. Hence, according to Fig. 3.8, singular points corresponding to even k are *stable foci*, whereas those corresponding to odd k are *saddles*. Again, one could sketch the phase plane orbits of Fig. 3.7 by using this information (and perhaps a bit of imagination and experience).

In Summary, Singular points for $n = 2$ The singular points of two-dimensional systems divide themselves into *nodes*, *saddles*, *centers* and *foci*. To classify a particular singular point, one computes the Jacobian of the system, inserting the singular point. The eigenvalues of this Jacobian determine the type of the singular point, as summarized in Fig. 3.8. If the eigenvalues are real and equal, it may be necessary to compute also the Jacobian eigen*vectors*, or at least to determine whether these are linearly dependent or not.

Linear systems have at most one singular point. Classifying the type and stability of this point, the qualitative flow of orbits is revealed for the entire phase plane.

Nonlinear systems may have any number of singular points. These points can be classified through linearization, that is, by evaluating eigenvalues of the Jacobian (if non-singular) at each singular point in turn. This will reveal the *local* arrangement of orbits in the immediate vicinity of the singular points.

For systems of arbitrary dimension, linear or nonlinear, the *stability* of a singular point can be determined by examining the Jacobian eigenvalue having the largest real part. If the largest real part is positive (negative), then the point is unstable (stable). For nonlinear systems, this method yields a *local* measure of stability, describing the consequences of small perturbations to orbits near a singular point.

3.5 Quantitative Analysis

3.5.1 Approximate Methods

This section presents three commonly applied methods for approximately solving nonlinear systems: the *method of multiple scales*, the *method of averaging*, and the *method of harmonic balance*. The two first of these are *perturbation* methods, that is, they work by applying small nonlinear perturbations to linearized solutions. Their application is restricted to *weakly nonlinear* systems, so the nonlinear terms should be small compared to linear terms[3]. Usually this is the case when motions are finite, but not very large. The correctness of perturbation solutions typically decreases for growing amplitudes of motion.

Other perturbation methods exist than those described here. To some extent, the choice of perturbation method for a particular problem is a matter of personal taste and habit. None of the existing perturbation methods appear to be superior in general. The methods to be presented seem to be those most widely used; they will

---

[3]Sometimes this assumption can be relaxed, e.g. Lakrad and Belhaq (2002) shows how to employ the multiple scales perturbation method for strongly nonlinear systems, by expressing the solutions in terms of Jacobian elliptic functions; see also Burton and Rahman (1986), Thomsen (2008b).

prove effective to most problems most of the time. Nayfeh (1973) offers a far more complete discussion of perturbation methods than can be given here.

Little is lost by presenting the methods in terms of the pendulum example, so we shall take this illustrative approach. For our purposes there is no reason to bother about damping and forcing yet, so at this first stage we consider the equation of the undamped and unforced pendulum (*cf.* (3.11)):

$$\ddot{\theta} + \omega_0^2 \sin\theta = 0, \quad \theta = \theta(t), \tag{3.53}$$

subjected to prescribed initial conditions $\theta(0)$ and $\dot{\theta}(0)$. Introducing dimensionless time $\tau = \omega_0 t$, and assuming pendulum rotations to be finite but not very large $(\sin\theta \approx \theta - \frac{1}{6}\theta^3)$, the equation to be solved becomes:

$$\ddot{\theta} + \theta = \frac{1}{6}\varepsilon\theta^3, \quad \theta = \theta(\tau), \tag{3.54}$$

with initial conditions $\theta(0)$ and $\dot{\theta}(0)$, where now overdots denote differentiation with respect to the dimensionless time τ. A small parameter $\varepsilon \ll 1$ has been introduced, merely as a book-keeping device, to indicate that the nonlinear term is assumed to be weak or 'small' compared to the linear terms. The equation has been arranged so that the left side constitutes a linear undamped system.

Next we illustrate three methods for obtaining approximate solutions $\theta(\tau)$ to (3.54), But first a little more on the small parameter ε, generally.

3.5.2 On the "Small Parameter" in Perturbation Analysis

The widespread characterization, in particular with multiple scales analysis, of ε as both a "small parameter" (compared to unity) and a "bookkeeping device" (which can be set to unity in the end) often causes confusion. What is really meant, but typically left just understood, is that it is the whole *term* with the ε in front that is assumed "small". And that "small" is as compared to the terms that does *not* have an ε in front. So what is meant by the ε in (3.54) is just that the nonlinear term on the right-hand side is assumed to be small compared to the (linear) left-hand side. This will hold true when $|\theta| \ll 1$, because in that case $\theta^2 \ll 1$ so that $|\theta^3| \ll |\theta|$, and thus also $|\frac{1}{6}\theta^3| \ll |\theta|$. Then ε is used just to keep track of the magnitude order of different terms during the analysis, also knowing that when the assumptions are fulfilled (here $|\theta| \ll 1$), then terms occurring during the analysis with ε^2 infront are of smaller magnitude than those with just ε^1 – which in turn are smaller than those of order $\varepsilon^0 = 1$. Still, ε has no physical interpretation in itself, and can freely be omitted in the final results (i.e. put to unity) – or could be kept, so as to indicate the order of approximation involved.

3.5.3 The Straightforward Expansion

This simplest and most obvious approach, unfortunately, does not produce useful solutions. It is included here merely to illustrate the fundamental principle of applying perturbations, and to show why more workable perturbation methods are necessarily more involved.

We seek a solution $\theta(\tau)$ to the nonlinear system (3.54) that is valid for small but finite amplitudes. By a straightforward expansion we assume the solution to be expandable in a series as follows, in terms of the small parameter ε:

$$\theta(\tau; \varepsilon) = \theta_0(\tau) + \varepsilon\theta_1(\tau) + \varepsilon^2\theta_2(\tau) + \cdots, \tag{3.55}$$

where $\theta_j(\tau), j = 1, 2, \ldots$ are unknown functions to be determined, and the terms form a sequence decreasing in magnitude, i.e. $|\varepsilon^j\theta_j(\tau)| > |\varepsilon^{j+1}\theta_{j+1}(\tau)|$ for $j = 0, 1, \ldots$. The expansion should be *uniformly valid*, that is, it should hold for all $\tau > 0$.

Substituting the expansion into (3.54) we can arrange the result as a polynomial in ε. Since the functions θ_j are independent of ε, the coefficient to each power of ε is required to vanish identically. This yields, in treating initial conditions by the same procedure, that to order ε^0:

$$\ddot{\theta}_0 + \theta_0 = 0$$
$$\theta_0(0) = \theta(0), \quad \dot{\theta}_0(0) = \dot{\theta}(0), \tag{3.56}$$

to order ε^1:

$$\ddot{\theta}_1 + \theta_1 = \frac{1}{6}\theta_0^3$$
$$\theta_1(0) = 0, \quad \dot{\theta}_1(0) = 0, \tag{3.57}$$

and to order ε^2:

$$\ddot{\theta}_2 + \theta_2 = \frac{1}{2}\theta_0^2\theta_1,$$
$$\theta_2(0) = 0, \quad \dot{\theta}_2(0) = 0. \tag{3.58}$$

The solution of (3.56) is

$$\theta_0 = a_0 \cos(\tau + \varphi_0), \tag{3.59}$$

where the constants a_0 and φ_0 are determined by the initial conditions:

$$a_0 = \left(\theta(0)^2 + \dot{\theta}(0)^2\right)^{1/2}, \quad \tan \varphi_0 = -\dot{\theta}(0)/\theta(0) \tag{3.60}$$

This is the zero-order solution, the solution to the linearized problem. Substituting then (3.59) into the first order-problem (3.57), one obtains an equation for the first-order solution θ_1:

$$\begin{aligned} \ddot{\theta}_1 + \theta_1 &= \tfrac{1}{6}a_0^3 \cos^3(\tau + \varphi_0) \\ &= \tfrac{1}{8}a_0^3 \cos(\tau + \varphi_0) + \tfrac{1}{24}a_0^3 \cos(3(\tau + \varphi_0)), \end{aligned} \tag{3.61}$$

where trigonometric identities have been used to expand $\cos^3(\tau + \varphi_0)$. The equation for θ_1 describes a linear, undamped oscillator with two harmonic forcing terms. It is readily solved to yield (*cf.* Section 1.2.3):

$$\theta_1 = a_1 \cos(\tau + \varphi_1) + \tfrac{1}{16}a_0^3\tau \sin(\tau + \varphi_0) - \tfrac{1}{192}a_0^3 \cos(3(\tau + \varphi_0)), \tag{3.62}$$

where the first term is the homogeneous solution (a_1, φ_1 being constants of integration), whereas the two last terms are particular solutions corresponding to each of the two forcing terms in (3.61).

There is no need to proceed further to see that this approach is doomed to fail. First, one should expect a free pendulum to oscillate periodically, but the term $\tau \sin(\tau + \varphi_0)$ in (3.62) implies the solution to be non-periodic. Second, this same term grows unbounded with time, which is incompatible with the assumption that the expansion (3.55) should be uniformly valid. And thirdly, as the terms grows in magnitude it will quickly become larger than θ_0, thus violating the assumption that $|\theta_1| < |\theta_0|$. Proceeding to calculate the higher-order terms θ_2, θ_3, … will not cure these problems, since then terms containing τ^2, τ^3 and so forth will enter the solution.

Solution terms such as $\tau \sin(\tau + \varphi_0)$ are called *secular*. Secular terms arise whenever an undamped oscillator is resonantly excited. This is the case with (3.61), where the forcing term $\tfrac{1}{8}a_0^3\cos(\tau + \varphi_0)$ is a *resonant term*, exciting the undamped oscillator exactly at the natural frequency. The straightforward expansion inevitably produces secular terms, and is thus inadequate for obtaining proper solutions. The methods described next will remedy this.

3.5.4 The Method of Multiple Scales

For linear systems the amplitudes and frequencies of oscillations are independent quantities. For example, the oscillation frequency of a freely swinging pendulum is independent of the oscillation amplitude, as long as the amplitudes are very small. Nonlinear systems, by contrast, typically display dependency between amplitude and frequency, For example, large-amplitude oscillations of a free pendulum occur at a lower frequency than do small oscillations.

The straightforward expansion technique described above fails to correctly represent a proper relation between amplitude and frequency. By the *method of multiple scales* this deficiency is overcome by permitting the solution to be a function of multiple independent time-variables, or -scales. For example, a fast scale can be used for capturing motions at frequencies comparable to the linear natural frequency of the system, while a slow scale accounts for slow modulations of amplitudes and phases. This will allow for a proper amplitude-frequency relation to be represented, which is free of secular terms.

By the method of multiple scales one generally assumes a uniformly valid expansion of the form:

$$\theta(\tau, \varepsilon) = \theta_0(T_0, T_1, T_2, \ldots) + \varepsilon\theta_1(T_0, T_1, T_2, \ldots) + \varepsilon^2\theta_2(T_0, T_1, T_2, \ldots) + \cdots$$

$$(3.63)$$

where θ_j, $j = 0$, 1, $\ldots$ are functions to be determined, and T_j, $j = 0$, 1, $\ldots$ are independent time-scales:

$$T_j = \varepsilon^j \tau, \quad j = 0, 1, \ldots, \quad \varepsilon \ll 1 \qquad (3.64)$$

As many independent time-scales are needed as there are terms in the expansion (3.63). That is, if the expansion is carried out to order $O(\varepsilon^n)$ one needs the scales T_j, $j = 0$, n. It will appear shortly, that the presence of several independent time-scales allows one to impose conditions that will eliminate secular terms of the solution.

As with the straightforward expansion, the assumed expansion (now (3.63)) is substituted into the equations of motion, and coefficients to like powers of ε are zeroed to yield a set of perturbation equations. Resonant terms will appear, causing secular terms in the solution. Due to the independent time-scales, however, the perturbation equations do not uniquely define the set of expansion functions θ_j. It is like fitting a parabola through two data points only; there is no unique way to do this, unless an additional restriction is imposed. For uniquely defining the expansion functions θ_j, one is thus free to impose the condition that the solution should contain no secular terms. There is no cheating or magic about this; the condition merely ensures the expansion to be uniformly valid, as assumed. That is, *if* a solution of the form (3.63) exists, then it must be free of secular terms because otherwise the solution would not be uniformly valid.

To illustrate the method for the problem (3.54) of the unforced and undamped pendulum, we seek a solution in the form of a uniformly valid expansion:

$$\theta(\tau, \varepsilon) = \theta_0(T_0, T_1) + \varepsilon\theta_1(T_0, T_1) + O(\varepsilon^2), \tag{3.65}$$

where T_0 is the *fast time* and T_1 the *slow time*:

$$T_0 = \tau, \quad T_1 = \varepsilon\tau. \tag{3.66}$$

Upon substituting (3.65) into (3.54) one needs to calculate derivatives with respect to τ. These become partial derivatives with respect to T_j, according to the chain rule:

$$\frac{d}{d\tau} = \frac{dT_0}{d\tau}\frac{\partial}{\partial T_0} + \frac{dT_1}{d\tau}\frac{\partial}{\partial T_1} = \frac{\partial}{\partial T_0} + \varepsilon\frac{\partial}{\partial T_1} = D_0 + \varepsilon D_1,$$
$$\frac{d^2}{d\tau^2} = \frac{\partial(D_0 + \varepsilon D_1)}{\partial T_0} + \varepsilon\frac{\partial(D_0 + \varepsilon D_1)}{\partial T_1} = D_0^2 + 2\varepsilon D_0 D_1 + O(\varepsilon^2), \tag{3.67}$$

where a partial differential operator $D_i^{\,j}$ has been defined through

$$D_i^j \equiv \frac{\partial^j}{\partial T_i^j}. \tag{3.68}$$

Substituting (3.65) into (3.54) we obtain:

$$\left(D_0^2 + 2\varepsilon D_0 D_1 + O(\varepsilon^2)\right)\left(\theta_0 + \varepsilon\theta_1 + O(\varepsilon^2)\right) + \left(\theta_0 + \varepsilon\theta_1 + O(\varepsilon^2)\right)$$
$$= \frac{1}{6}\varepsilon\left(\theta_0 + \varepsilon\theta_1 + O(\varepsilon^2)\right)^3. \tag{3.69}$$

Equating to zero the coefficients to like powers of ε yields, to order ε^0:

$$D_0^2\theta_0 + \theta_0 = 0, \tag{3.70}$$

and to order ε^1:

$$D_0^2\theta_1 + \theta_1 = \frac{1}{6}\theta_0^3 - 2D_0 D_1\theta_0. \tag{3.71}$$

The zero order problem (3.70) has the form of an undamped and unforced linear oscillator. If only a single time-scale $T_0 = \tau$ was present, the term $D_0^2\theta_0$ would be written $\ddot{\theta}_0$ and the solution would have the familiar form $\theta_0 = a\cos(\tau + \varphi)$. With *two* independent variables T_0 and T_1, and $D_0^2\theta_0 = \partial^2\theta_0/\partial T_0^2$, the same solution applies, but with the difference that a and φ can be functions of T_1. Thus, the solution to (3.70) has the form $\theta_0 = a(T_1)\cos(T_0 + \varphi(T_1))$. With the method of multiple scales it is convenient to employ complex exponentials, so we write this solution in the form (*cf.* App. C):

$$\theta_0 = A(T_1)e^{iT_0} + \overline{A}(T_1)e^{-iT_0}, \quad A(T_1) \in C, \tag{3.72}$$

where A is an unknown complex-valued function of the slow scale T_1, and is the complex conjugate of A. Substituting (3.72) into (3.71) yields:

$$D_0^2\theta_1 + \theta_1 = \frac{1}{6}\left(A^3e^{i3T_0} + 3A^2\overline{A}e^{iT_0}\right) - 2\left(iA'e^{iT_0}\right) + cc, \tag{3.73}$$

where $A' \equiv dA/dT_1$ and cc denotes complex conjugates of preceding terms.

Terms containing the factor e^{iT_0} are resonant terms; this is because $e^{iT_0} = \cos T_0 + i\sin T_0$ corresponds to harmonic excitation at the natural frequency of the linear system (3.73). Resonant excitation terms will produce secular terms of the form $T_0e^{iT_0}$ in the solution for θ_1. However, we are free to choose the function A so as to cancel out these. This is accomplished by requiring the coefficient of e^{iT_0} to vanish identically, that is, $A(T_1)$ should satisfy the following relation:

$$\frac{1}{2}A^2\overline{A} - i2A' = 0, \tag{3.74}$$

which is called the *solvability condition* for the multiple scales analysis. With the solvability condition fulfilled, (3.73) constitutes a linear oscillator with a single harmonic forcing term $\frac{1}{6}A^3e^{i3T_0} + cc$. Any particular solution to this system must be periodic with a frequency equal to that of the excitation. Inserting the assumed form $\theta_1 = B(T_1)e^{i3T_0}$ one finds that $B(T_1) = -\frac{1}{48}A(T_1)^3$, and thus that:

$$\theta_1 = -\frac{1}{48}A^3e^{i3T_0} + cc. \tag{3.75}$$

As a standard trick of the method one ignores the homogeneous part of the solution to (3.73), postponing the handling of initial conditions to the final step.

We now turn to the determination of the function $A(T_1)$ from the solvability condition (3.74). When solving equations of this type it is convenient to use polar notation, so we express A in the form:

$$A = \frac{1}{2}ae^{i\varphi}, \quad a(T_1), \varphi(T_1) \in R, \tag{3.76}$$

where a and φ are real-valued functions of T_1. Substitution into the solvability condition (3.74) yields, upon separating real and imaginary components:

$$a' = 0,$$
$$\frac{1}{16}a^3 + a\varphi' = 0.$$

(3.77)

Keeping in mind that $(\ )' = d/dT_1$, this pair of first-order, ordinary differential equations are readily solved to yield:

$$a = a_0,$$
$$\varphi = \varphi_0 - \frac{1}{16}a_0^2 T_1,$$

(3.78)

where a_0 and φ_0 are arbitrary constants of integration.

We are then able to write down the first-order approximate solution for the problem of the unforced and undamped pendulum. By (3.65), (3.72), (3.75), (3.76), (3.78) and (3.66) the solution becomes:

$$\begin{aligned}
\theta(\tau, \varepsilon) &= \theta_0 + \varepsilon\theta_1 + O(\varepsilon^2) \\
&= Ae^{iT_0} - \frac{1}{48}\varepsilon A^3 e^{i3T_0} + cc + O(\varepsilon^2) \\
&= \frac{1}{2}ae^{i\varphi}e^{iT_0} - \frac{1}{48}\varepsilon\frac{1}{8}a^3 e^{i3\varphi}e^{i3T_0} + cc + O(\varepsilon^2) \\
&= a\cos(T_0 + \varphi) - \frac{1}{192}\varepsilon a^3 \cos(3(T_0 + \varphi)) + O(\varepsilon^2) \\
&= a_0 \cos\left(\tau + \varphi_0 - \frac{1}{16}a_0^2\varepsilon\tau\right) \\
&\quad - \frac{1}{192}\varepsilon a_0^3 \cos\left(3(\tau + \varphi_0 - \frac{1}{16}a_0^2\varepsilon\tau)\right) + O(\varepsilon^2) ,
\end{aligned}$$

(3.79)

where a_0 and φ_0 are determined by the initial conditions, and where ε now merely serves to indicate the assumed magnitude of terms. Letting $\varepsilon = 1$ and returning to the original time-variable $t = \tau/\omega_0$, the first-order approximate solution takes the form:

$$\theta(t) = a_0 \cos(\hat{\omega}t + \varphi_0) - \frac{1}{192}a_0^3 \cos(3(\hat{\omega}t + \varphi_0)),$$

(3.80)

where

$$\hat{\omega} \equiv \left(1 - \frac{1}{16}a_o^2\right)\omega_0.$$

(3.81)

This solution is periodic, as expected. As appears the fundamental frequency of oscillations $\hat{\omega}$ depends on the oscillation amplitude a_0. This is a typical nonlinear phenomenon which is readily observed with an experimental pendulum. Only when $a_0^2 \ll 1$, as assumed by a linearized pendulum model, the frequency becomes virtually independent of the amplitude (i.e. $\hat{\omega} \approx \omega_0$).

Note that a *higher harmonic* $3\hat{\omega}$ appears in the response. This too is a nonlinear feature. It gradually vanishes as the amplitude $a_0 \to 0$ and the pendulum behaves increasingly linearly with purely harmonic (i.e. single-frequency sinusoidal) oscillations. Including more independent time-scales (T_2, T_3, ...) in the analysis, the procedure could be continued to yield approximations of even higher order; these would contain even higher harmonics.

Starting the pendulum from rest with $\dot{\theta}(0) = 0$ and $\theta(0) = \tilde{\theta}_0$, we obtain from (3.80) that $\varphi_0 = 0$ and $a_0 \approx \tilde{\theta}_0$, and hence that:

$$\theta(t) \approx \tilde{\theta}_0 \cos \hat{\omega}t - \frac{1}{192}\tilde{\theta}_0^3 \cos 3\hat{\omega}t , \quad \hat{\omega} = \left(1 - \frac{1}{16}\tilde{\theta}_o^2\right)\omega_0. \qquad (3.82)$$

Fig. 3.9 displays this solution compared to the linearized and the exact nonlinear solution for two levels of initial amplitude. For $\tilde{\theta}_0 = 80°$ the multiple scales solution closely matches the exact solution, whereas for $\tilde{\theta}_0 = 150°$ there are noticeable discrepancies, due to the neglected higher-order terms of the expansion.

3.5.5   The Method of Averaging

Several perturbation methods are based on *averaging* (e.g., Mitropolsky 1965; Mitropolskii and Nguyen 1997; Nayfeh 1973; Sanders and Verhulst 1985; Levi 1999). We describe here a commonly applied variant, the so-called *Krylov-Bogoliubov* method.

For illustrating this approach we first consider the unforced and undamped pendulum (the forced case is treated in Sect. 3.7.4), whose equation of motions is given by (3.54):

$$\ddot{\theta} + \theta = \frac{1}{6}\varepsilon\theta^3, \quad \theta = \theta(\tau). \qquad (3.83)$$

When $\varepsilon = 0$ the problem is linear, and the solution is

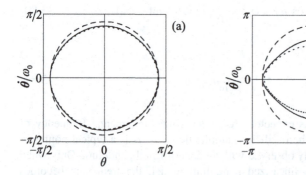

Fig. 3.9. Phase plane orbits for the undamped, unforced pendulum. Pendulum started from rest with **(a)** $\tilde{\theta}_0 = 80°$ or **(b)** $\tilde{\theta}_0 = 150°$. (—) exact nonlinear solution, (– – –) exact linear solution, and (......) multiple scales approximation

$$\theta = a_0 \cos(\tau + \varphi_0), \tag{3.84}$$

where a_0 and φ_0 are constants.

When $\varepsilon \neq 0$ the problem is nonlinear, and (3.84) does not solve it. However, we assume the nonlinear solution to be similar to the linear one, with the difference that the amplitude and the phase are allowed to vary in time, that is:

$$\theta = a \cos(\tau + \varphi); \quad a = a(\tau), \quad \varphi = \varphi(\tau). \tag{3.85}$$

The unknown functions $a(\tau)$ and $\varphi(\tau)$ are now considered dependent variables of the problem, by employing a shift of variables from the original dependent variable $\theta(\tau)$ to the new variables a and φ. In performing this transformation there are three unknowns (θ, a, φ), but only two equations ((3.83) and (3.85)). Thus an additional and independent equation is required for ensuring uniqueness of the transformation. For this additional equation it is typically useful to pose the condition that the velocity has a functional form which is similar to that of the linear case. A transformation of variables so defined is called a *Van der Pol transformation*. By (3.84) the linear pendulum velocity is $\dot{\theta} = -a_0 \sin(\tau + \varphi_0)$. Thus, as a third equation for the transformation, we state the requirement that

$$\dot{\theta} = -a \sin(\tau + \varphi). \tag{3.86}$$

To express the equation of motion in terms of the new variables a and φ we need to eliminate θ from two of the three equations (3.83), (3.85) and (3.86). By differentiating (3.85) with respect to τ, one can eliminate θ from (3.85) and (3.86) to obtain:

$$\dot{a} \cos \psi - a\dot{\varphi} \sin \psi = 0, \tag{3.87}$$

where

$$\psi \equiv \tau + \varphi. \tag{3.88}$$

Similarly, by differentiating (3.86) with respect to τ, one can eliminate θ from (3.86) and (3.83) to obtain:

$$\dot{a} \sin \psi + a\dot{\varphi} \cos \psi + \frac{1}{6}\varepsilon a^3 \cos^3 \psi = 0. \tag{3.89}$$

Solving then (3.87) and (3.89) for $\dot{a}$ and $\dot{\varphi}$ it is found that:

$$\dot{a} = -\frac{1}{6}\varepsilon a^3 \sin \psi \cos^3 \psi,$$
$$\dot{\varphi} = -\frac{1}{6}\varepsilon a^2 \cos^4 \psi . \tag{3.90}$$

Note that no approximations have yet been introduced. Equations (3.90) with (3.85) simply restate the original equation of motion (3.83) in the form of two first-order equations, with a and φ as the dependent variables instead of θ.

We now come to the approximations. From (3.90) it appears that $\dot{a}$ and $\dot{\varphi}$ are small quantities, since $\varepsilon \ll 1$. This implies that the variables a and φ change much more slowly with τ than does $\psi = \tau + \varphi$. Thus a and φ will hardly change during one period of oscillation, from $\psi = 2\pi j$ to $\psi = 2\pi (j + 1)$, $j = 0, 1, \ldots$. The approximation involved in Krylov–Bogoliubov averaging, then, is to replace the equations for $\dot{a}$ and $\dot{\varphi}$ with their average values during one period of oscillation. In other words: for determining the slow variations of a and φ, the rapid variations in ψ-terms can be neglected for a first approximation.

We average Eq. (3.90) by integrating both sides of each equation from $\psi = 0$ to $\psi = 2\pi$ (one period of oscillation), while considering a, φ, $\dot{a}$, and $\dot{\varphi}$ to be constants during that period. The result becomes:

$$\dot{a} \approx -\frac{1}{6}\varepsilon a^3 \frac{1}{2\pi} \int_0^{2\pi} \sin\psi \cos^3\psi \, d\psi = 0,$$

$$\dot{\varphi} \approx -\frac{1}{6}\varepsilon a^2 \frac{1}{2\pi} \int_0^{2\pi} \cos^4\psi \, d\psi = -\frac{1}{16}\varepsilon a^2. \tag{3.91}$$

These *averaged equations* are readily solved for a and φ to yield:

$$a(\tau) = a_0,$$

$$\varphi(\tau) = -\frac{1}{16}\varepsilon a_0^2 \tau + \varphi_0, \tag{3.92}$$

where a_0 and φ_0 are constants to be determined by the initial conditions. Returning to the original variable $\theta(\tau)$ one obtains, by (3.85):

$$\theta(\tau) = a\cos(\tau + \varphi)$$

$$= a_0 \cos(\tau - \frac{1}{16}\varepsilon a_0^2 \tau + \varphi_0), \tag{3.93}$$

or, letting $\varepsilon = 1$ and re-introducing the original time-variable $t = \tau/\omega_0$:

$$\theta(t) = a_0 \cos(\hat{\omega}t + \varphi_0), \quad \hat{\omega} \equiv (1 - \frac{1}{16}a_0^2)\omega_0. \tag{3.94}$$

This approximate solution is identical to the zero order part of the multiple scales expansion (*cf.* (3.80)).

Higher-order averaging approximations require extensions of the Krylov-Bogoliubov method; for example, the *generalized method of averaging* is useful for this purpose (Nayfeh 1973). Thus a well-defined procedure exists for systematically increasing the accuracy of solutions. This is a significant strength of the method, as is its solid basis of mathematical proofs concerning convergence and accuracy (e.g. Sanders and Verhulst 1985).

An obstacle for the general application of averaging can be to find a transformation of variables that will render a system suitable for averaging, e.g. a system of the form $\dot{\mathbf{x}} = \varepsilon\mathbf{f}(\mathbf{x}, t)$, $\mathbf{x} \in R^n$, of which (3.90) for the pendulum is a specific example. One trick that typically works well is to first find the solution of the unperturbed system ($\varepsilon = 0$) in terms of unknown constants, and then take these constants as new dependent variables of the perturbed system; this is exactly what is accomplished by the *Van der Pol transformation* in (3.85).

We shall return to the method of averaging in Sect. 3.7.4, showing how to use averaging with forced vibrations.

3.5.6 The Method of Harmonic Balance

With the *method of harmonic balance* (also called the *describing function method*) one assumes a periodic solution in the form of a harmonic series:

$$\theta(t) = \sum_{j=0}^{N} A_j \cos(j\omega t + j\varphi_0), \qquad (3.95)$$

where ω, φ_0 and A_j, $j = 0, N$, are unknown constants to be determined.

The series is substituted into the equation of motion, and the coefficient of each of the lowest $N + 1$ harmonics required to vanish. This yields a system of $N + 1$ algebraic equations in the $N + 3$ unknowns ω, φ_0 and A_j, $j = 0, N$. Usually one solves the algebraic system for the $N + 1$ unknowns ω and A_0, A_2, A_3, ..., expressing all unknowns in terms of A_1 and φ_0. This will leave A_1 and φ_0 for determination by the initial conditions.

Often something is known in advance about a periodic solution to be determined, e.g., from experimental observations or computer simulations. If this is the case, the method of harmonic balance may quickly yield an accurate result. On the other hand, if little is known about the character of the solution, then a large number of terms may be required to ensure an accurate approximation, and this accuracy is not easily quantified.

As for the example problem of the free pendulum, we assume the third-order nonlinearity to produce only odd-ordered time harmonics, that is, $A_{2i} = 0$, $i = 0, 1,$ .... Choosing $N = 3$ will thus provide a two-term series:

$$\theta(\tau) = A_1 \cos\psi + A_3 \cos 3\psi, \quad \psi \equiv \omega\tau + \varphi_0. \qquad (3.96)$$

Substituting this into the pendulum equation (3.54) we obtain (omitting ε):

$$
\begin{aligned}
(-\omega^2 A_1 \cos \psi &- 9\omega^2 A_3 \cos 3\psi) + (A_1 \cos \psi + A_3 \cos 3\psi) \\
&= \frac{1}{6}(A_1 \cos \psi + A_3 \cos 3\psi)^3 \\
&= \frac{1}{8}A_1 \left(A_1^2 + A_1 A_3 + 2A_3^2\right) \cos \psi \\
&\quad + \frac{1}{24}(A_1^3 + 6A_1^2 A_3 + 3A_3^3) \cos 3\psi + hh(5),
\end{aligned}
\tag{3.97}
$$

where $hh(5)$ denotes fifth and higher harmonics. Equating to zero the coefficients of $\cos\psi$ and $\cos 3\psi$ result in, respectively:

$$
\begin{aligned}
(1 - \omega^2)A_1 &= \frac{1}{8}A_1 \left(A_1^2 + A_1 A_3 + 2A_3^2\right), \\
(1 - 9\omega^2)A_3 &= \frac{1}{24}\left(A_1^3 + 6A_1^2 A_3 + 3A_3^3\right),
\end{aligned}
\tag{3.98}
$$

which are not readily solved for A_1 and A_3. However, assuming A_1 to be small and A_3 to be even smaller, it appears from the second equation that $A_3 = O(A_1{}^3)$. In keeping terms to order $A_1{}^3$ the two equations may then be approximated by

$$
\begin{aligned}
(1 - \omega^2)A_1 &= \frac{1}{8}A_1 \left(A_1^2 + O(A_1^4)\right), \\
(1 - 9\omega^2)A_3 &= \frac{1}{24}A_1^3 + O(A_1^5).
\end{aligned}
\tag{3.99}
$$

Solving for A_3 one finds that:

$$
A_3 = \frac{\frac{1}{24}A_1^3 + O(A_1^5)}{-8 + \frac{9}{8}A_1^2 + O(A_1^4)} = -\frac{1}{192}A_1^3 + O(A_1^5) \quad \text{for } A_1 \ll 1,
\tag{3.100}
$$

and, solving for ω that:

$$
\omega = \left(1 - \frac{1}{8}A_1^2 + O(A_1^4)\right)^{1/2} = 1 - \frac{1}{16}A_1^2 + O(A_1^4) \quad \text{for } A_1 \ll 1.
\tag{3.101}
$$

Re-substituting into (3.96) the following approximate solution is then obtained:

$$
\theta(t) \approx A_1 \cos(\hat{\omega}t + \varphi_0) - \frac{1}{192}A_1^3 \cos(3(\hat{\omega}t + \varphi_0)),
\tag{3.102}
$$

where the original time-variable $t = \tau/\omega_0$ has been substituted for τ, the constants A_1 and φ_0 are determined by the initial conditions, and

$$
\hat{\omega} \equiv \left(1 - \frac{1}{16}A_1^2\right)\omega_0.
\tag{3.103}
$$

Comparing this harmonic balance result to the multiple scales solution (3.80) and to the method of averaging solution (3.94) (where a_0 plays the role of A_1), we note that the three methods agree to the lowest order of approximation.

From the above example it may seem that the method of harmonic balance involves considerably less algebra than the methods of multiple scales and averaging. However, this is the case only because the assumed harmonic series captures the essential features of the true solution. If nothing was known about the true solution, then a large number of terms had to be included and the resulting number of algebraic manipulations would quickly outperform those required for the multiple scales method.

The method of harmonic balance, by contrast to perturbation methods, does not assume weak nonlinearity, and may thus work even with strongly nonlinear terms. Mickens (1984) discussed the conditions under which harmonic balance could be used with good results, and recommended it mainly for strongly nonlinear problems; for weakly nonlinear problems perturbation methods (like the method of multiple scales and the method of averaging) should be preferred. More recently variants of the method of harmonic balance have been suggested for improving its efficiency and accuracy (Wu et al. 2018; Zhou et al. 2020; Sorokin et al. 2015b).

3.6 The Forced Response – Multiple Scales Analysis

3.6.1 Posing the Problem

We now consider obtaining approximate nonlinear solutions for the pendulum equation with damping and forcing included (*cf.* (3.11)):

$$\ddot{\theta} + 2\beta\omega_0\dot{\theta} + \left(\omega_0^2 - q\Omega^2 \cos(\Omega t)\right) \sin\theta = 0. \qquad (3.104)$$

As for the corresponding unforced problem we assume rotations to be finite but not very large, so that $\sin\theta \approx \theta - \frac{1}{6}\theta^3$ remains a reasonable approximation. Introducing nondimensional time τ and nondimensional frequency of excitation ω::

$$\tau = \omega_0 t, \ \omega = \Omega /\omega_0, \qquad (3.105)$$

the equation of motion becomes:

$$\ddot{\theta} + 2\beta\dot{\theta} + \left(1 - q\omega^2 \cos(\omega\tau)\right)\left(\theta - \frac{1}{6}\theta^3\right) = 0, \qquad (3.106)$$

where now $\dot{\theta} \equiv d\theta/d\tau$.

When using perturbation analysis one needs to decide upon the terms to be considered small or weak. For the present example we consider the damping term, the parametric excitation term and the nonlinearity to be of magnitude order $\varepsilon \ll 1$. Substituting $2\beta \rightarrow \varepsilon\, 2\beta$, $q \rightarrow \varepsilon q$ and $\frac{1}{6}\theta^3 \rightarrow \varepsilon\, \frac{1}{6}\theta^3$ into (3.106) we obtain, upon neglecting terms of order ε^2 and higher:

$$\ddot{\theta} + 2\varepsilon\beta\dot{\theta} + \left(1 - \varepsilon q\omega^2 \cos(\omega\tau)\right)\theta + \varepsilon\gamma\theta^3 = 0, \tag{3.107}$$

where a parameter γ has been introduced to quantify the coefficient of the nonlinear term (for the pendulum $\gamma = -\frac{1}{6}$). Note again that ε has no physical interpretation, but merely serves to indicate the assumed smallness of terms.

Ordering terms according to magnitude is an important and decisive step of any perturbation analysis. The above ordering reflects that the terms describing linear inertia and linear stiffness are supposed to be 'strong'. They will determine the response at the lowest level of approximation. The remaining terms are assumed to modify this response at a higher level of approximation, in this case the ε-level.

Equations similar to (3.107) arise in numerous problems of nonlinear mechanics. It may be recognized as a *Duffing equation* with parametric excitation, or as a nonlinear *Mathieu equation*. For example, this equation also models transverse single-mode vibrations of a column subject to base excitation, with the linear restoring term representing bending forces, and the cubic nonlinearity representing nonlinear stretching of the neutral axis. Or single-mode transverse vibrations of the measurement pipe of a Coriolis type flowmeter with a pulsating fluid (Enz and Thomsen 2011a). The feature making (3.107) particular to the pendulum problem is the specific value of the nonlinear coefficient, $\gamma = -\frac{1}{6}$. By allowing γ to take on arbitrary values, including positive and zero, the below analysis will be much broader in scope.

3.6.2 Perturbation Equations

We seek a first-order approximate solution to Eq. (3.107). Using the method of multiple scales, a solution is assumed in the form of a uniformly valid expansion:

$$\theta(\tau, \varepsilon) = \theta_0(T_0, T_1) + \varepsilon\theta_1(T_0, T_1) + O(\varepsilon^2), \tag{3.108}$$

where the two independent time-scales T_0 and T_1 are given by

$$\begin{aligned} T_0 &= \tau, \\ T_1 &= \varepsilon\tau. \end{aligned} \tag{3.109}$$

The expansion is substituted into (3.107) by noting that:

$$\dot{\theta} = (D_0 + \varepsilon D_1)\theta,$$
$$\ddot{\theta} = (D_0^2 + 2\varepsilon D_0 D_1 + O(\varepsilon^2))\theta, \qquad (3.110)$$

where $D_i^j \equiv \partial^j/\partial T_i^j$. Performing the substitution, and equating to zero the coefficients to like powers of ε one finds that, to order ε^0:

$$D_0^2\theta_0 + \theta_0 = 0, \qquad (3.111)$$

and that, to order ε^1:

$$D_0^2\theta_1 + \theta_1 = -2D_0D_1\theta_0 - 2\beta D_0\theta_0 - \gamma\theta_0^3 + q\omega^2 \cos(\omega T_0)\theta_0. \qquad (3.112)$$

The general solution of the zero order problem (3.111) can be written:

$$\theta_0 = A(T_1)e^{iT_0} + \overline{A}(T_1)e^{-iT_0}, \qquad (3.113)$$

where A is an arbitrary function of the slow scale T_1, which is determined so as to make the solution of the first-order problem (3.112) free of secular terms.

Substituting the zero order solution (3.113) into the first-order problem (3.112) we obtain:

$$D_0^2\theta_1 + \theta_1 = \left(-i2A' - i2\beta A - 3\gamma A^2\overline{A}\right)e^{iT_0} - \gamma A^3 e^{i3T_0}$$
$$+ \frac{1}{2}q\omega^2 \left(Ae^{i(\omega+1)T_0} + \overline{A}e^{i(\omega-1)T_0}\right) + cc, \qquad (3.114)$$

where $A' \equiv dA/dT_1$, the term $\cos(\omega T_0)$ has been expressed in complex exponential form, and cc denotes complex conjugates of the preceding terms.

The function $A(T_1)$ is determined by the requirement that the solution for θ_1 in (3.114) should be free of secular terms, that is, free of terms containing $T_0 e^{iT_0}$ as a factor. Since secular solution terms are caused by resonant excitation terms, we examine the right side of (3.114) for the presence of resonant terms. Terms oscillating at a frequency equal to the natural frequency of the homogeneous system are resonant, and their sum should be equated to zero to get rid of secular solution terms. The natural frequency of the homogeneous system here equals unity (square root of the coefficient to θ_1), and so we should equate to zero the sum of all terms containing e^{iT_0} as a factor, since this term oscillates at unit frequency (a term $e^{irt} \equiv \cos(rt) + i\sin(rt)$ oscillates at frequency r). The first parenthesis of (3.114) is readily seen to make up a resonant term. However, the term $\frac{1}{2}q\omega^2 e^{i(\omega-1)T_0}$ will also be resonant in case $\omega = 2$. We shall refer to the case $\omega = 2$ as the case of *primary parametric resonance*[4].

The case $\omega \approx 2$ (as opposed to $\omega = 2$) is referred to as *near-resonance*. Near-resonant excitation causes *small divisor* terms to appear in the solution for θ_1, and small divisor terms are just as unacceptable as secular terms. To understand this we consider the particular solution of the equation $D_0^2\theta_1 + \theta_1 = e^{i(\omega-1)T_0}$, which is $\theta_1 = (1 - (\omega - 1)^2)^{-1}e^{i(\omega-1)T_0}$. Hence, when $\omega \approx 2$ the divisor $(1 - (\omega - 1)^2)$ will be small, and θ_1 will be correspondingly large. This violates the assumption of θ_1 being small, as compared to θ_0. Consequently, small divisor terms should be eliminated from the response, by requiring the sum of *near-resonant* excitation terms to vanish identically.

As a strength of the method of multiple scales we note that the perturbation equations, when set up, automatically reveal which special cases to consider. For the pendulum example it *might* (dependent on experience) be obvious that there is something special about the value $\omega = 2$ for the excitation frequency: When $\omega = 2$ the pendulum support oscillates up and down at twice the linear natural frequency of free pendulum oscillations. By simple physical reasoning one realizes that this is just the right frequency ratio for generating large responses using little effort (as would be $\omega = 1$, if an oscillating torque was applied at the pendulum hinge instead). For other problems the special case(s) to consider may not be similarly obvious, but setting up the multiple scales perturbation equations would reveal them anyway.

Thus, for the pendulum example there are two cases to consider below: the *non-resonant case* (ω away from 2), and the *near-resonant case* ($\omega \approx 2$).

3.6.3 The Non-resonant Case

When ω is away from 2 the only resonant term of (3.114) is the one containing e^{iT_0} as a factor. Requiring the coefficient to e^{iT_0} to vanish the following *solvability condition* is obtained:

$$i2A' + i2\beta A + 3\gamma A^2\overline{A} = 0. \tag{3.115}$$

With this condition fulfilled, a particular solution to equation (3.114) is obtained by adding in turn the responses to each of the remaining harmonic excitation-terms, that is:

---

[4]More generally, for linear parametrically excited systems with n degrees of freedom and distinct natural frequencies ω_j, $j = 1,n$, one can show (Nayfeh & Mook 1979) that parametric resonance occurs when the excitation frequency $\Omega = |\omega_i \mp \omega_j|/n$ for $i,j,n = 1,2,\ldots$, where n is the *order* of the parametric resonance; this is an example of *combination resonances*. Typically the first-order ($n = 1$) primary ($i = j = 1$) resonance is the one of main interest.

$$\theta_1 = \frac{1}{8}\gamma A^3 e^{i3T_0} - \frac{1}{2}q\omega\left(\frac{1}{\omega+2}Ae^{i(\omega+1)T_0} + \frac{1}{\omega-2}\overline{A}e^{i(\omega-1)T_0}\right) + cc. \qquad (3.116)$$

Usually, when the excitation is stationary, one is primarily interested in stationary (post-transient) behavior. The homogeneous part of the solution describes the transient response is therefore not needed.

The function $A(T_1)$ is determined from the solvability condition (3.115) by first expressing it in polar form:

$$A = \frac{1}{2}ae^{i\varphi}, \quad a(T_1), \ \varphi(T_1) \in R, \qquad (3.117)$$

Substituting into (3.115) we obtain, when separating real and imaginary parts:

$$a' = -\beta a,$$
$$\varphi' = \frac{3}{8}\gamma a^2. \qquad (3.118)$$

Solving for a and φ one finds that

$$a = a_0 e^{-\beta T_1},$$
$$\varphi = \varphi_0 - \frac{3}{8}a_0^2\frac{\gamma}{2\beta}e^{-2\beta T_1}, \qquad (3.119)$$

where a_0 and φ_0 are arbitrary constants. The first equation shows that $a \to 0$ as $T_1 \to \infty$, since $\beta > 0$. Then by (3.117) also $A \to 0$, and by (3.113, 3.116) and (3.108) the stationary response becomes $\theta(\tau) = 0$. Consequently the state $\theta(\tau) = 0$ is a stable equilibrium when ω is away from 2, so that oscillations of the pendulum damp out in time no matter how wildly the pendulum support is shaken up and down.

Equations (3.118) are the *modulation equations* of this perturbation analysis. They govern modulations in slow time T_1 of the amplitudes and phases of terms oscillating harmonically in fast time T_0. For determining stationary responses one may attempt solving the modulation equations to see what happens with amplitudes and phases as time approaches infinity. This approach was used above. Typically, however, the modulation equations are not readily solved – and in fact do not need to be solved for obtaining stationary responses. Instead one seeks the singular points, i.e. the equilibriums, of the modulation equations – and examines their stability with respect to slight disturbances. For the singular points it holds by definition that $a' = \varphi' = 0$, which describes a stationary response in which amplitudes and phases do not change in time. The following section will illustrate this technique.

3.6.4 The Near-Resonant Case

When $\omega \approx 2$ we are faced with the case of primary parametric resonance. Some measure of the nearness of ω to the resonant value 2 will be in need, so we let

$$\omega = 2 + \varepsilon\sigma \tag{3.120}$$

where σ is the so-called *detuning parameter*. This definition expresses the assumption that ω differs from the value 2 only by a small quantity $\varepsilon\sigma$. Substituting (3.120) into the ω-dependent oscillatory terms of (3.114), near-resonant terms will convert into resonant terms. We do not substitute into ω-dependent terms that are non-oscillatory (such as $\frac{1}{2}q\omega^2$), since this is unnecessary for the purpose. By noting that

$$
\begin{aligned}
(\omega - 1)T_0 &= ((2+\varepsilon\sigma) - 1)T_0 = T_0 + \varepsilon\sigma T_0 = T_0 + \sigma T_1, \\
(\omega + 1)T_0 &= ((2+\varepsilon\sigma) + 1)T_0 = 3T_0 + \varepsilon\sigma T_0 = 3T_0 + \sigma T_1,
\end{aligned}
\tag{3.121}
$$

the substitution of (3.120) into (3.114) yields:

$$
\begin{aligned}
D_0^2\theta_1 + \theta_1 ={}& \left(-i2A' - i2\beta A - 3\gamma A^2\overline{A}\right)e^{iT_0} - \gamma A^3 e^{i3T_0} \\
& + \frac{1}{2}q\omega^2\left(Ae^{i3T_0} + \overline{A}e^{iT_0}\right)e^{i\sigma T_1} + cc.
\end{aligned}
\tag{3.122}
$$

Resonant terms (proportional to e^{iT_0}) appear here, that will feed off secular terms in the solution for θ_1. To eliminate secular terms we require the sum of all resonant terms to vanish, that is, the solvability condition becomes:

$$-i2A' - i2\beta A - 3\gamma A^2\overline{A} + \frac{1}{2}q\omega^2\overline{A}e^{i\sigma T_1} = 0. \tag{3.123}$$

This condition differs from that of the non-resonant case (3.115) only by an added excitation term (last term of (3.123)). With the condition fulfilled, a particular solution to (3.122) becomes

$$\theta_1 = \frac{1}{8}\gamma A^3 e^{i3T_0} - \frac{1}{16}q\omega^2 A e^{i(3T_0 + \sigma T_1)} + cc. \tag{3.124}$$

For determining $A(T_1)$ we let $A = \frac{1}{2}ae^{i\varphi}$, insert into (3.123), separate real and imaginary parts, and obtain the modulation equations for the *resonant* case:

$$
\begin{aligned}
a' &= -\beta a + \frac{1}{4}q\omega^2 a \sin(\sigma T_1 - 2\varphi), \\
a\varphi' &= \frac{3}{8}\gamma a^3 - \frac{1}{4}q\omega^2 a \cos(\sigma T_1 - 2\varphi).
\end{aligned}
\tag{3.125}
$$

Considering only stationary oscillations of the pendulum, it will suffice to determine the conditions on which the amplitudes and phases will not change in time, that is, to locate singular points. However, the notion of singular points is tied to autonomous systems of equations, which (3.125) is not (due to the explicit dependence on T_1). To eliminate T_1 we introduce a new dependent variable $\psi = \psi(T_1)$ by

$$\psi \equiv \sigma T_1 - 2\varphi. \tag{3.126}$$

On substituting into (3.125) an autonomous pair of equations is then obtained:

$$a' = -\beta a + \frac{1}{4}q\omega^2 a \sin \psi,$$
$$a\psi' = -\frac{3}{4}\gamma a^3 + \frac{1}{2}q\omega^2 a \cos \psi + \sigma a. \tag{3.127}$$

We now seek the singular points of (3.127). These are given by the condition that $a' = \psi' = 0$, since then a and ψ do not change in (slow) time and the pendulum will either be at rest or oscillate at constant amplitude and phase. Letting $a' = \psi' = 0$ in (3.127 and solving for a and ψ, one finds that there are two possibilities. The first is that $a = 0$, while $\cos\psi \to -2\sigma/(q\omega^2)$ and $\sin\psi \to 4\beta/(q\omega^2)$ as $a \to 0$ (and undefined when $a = 0$). This implies that $\theta(\tau) = 0$ so that the pendulum is at rest. The other possibility is that $a \neq 0$, the more interesting case. With $a' = \psi' = 0$ in (3.127 we obtain, dividing both equations by $a \neq 0$, that

$$\frac{1}{2}q\omega^2 \sin \psi = 2\beta,$$
$$\frac{1}{2}q\omega^2 \cos \psi = \frac{3}{4}\gamma a^2 - \sigma. \tag{3.128}$$

Solving for a and ψ gives (square and add the two equations to find a; then divide the two equations to find ψ):

$$a^2 = \frac{4}{3\gamma}\left(\sigma \pm \sqrt{C_1}\right),$$
$$\tan \psi = \frac{2\beta}{\frac{3}{4}\gamma a^2 - \sigma} = \frac{\pm 2\beta}{\sqrt{C_1}}, \tag{3.129}$$

where

$$C_1 \equiv \left(\frac{1}{2}q\omega^2\right)^2 - (2\beta)^2. \tag{3.130}$$

These nonlinear solutions are seen to exist only when $C_1 \geq 0$, and $\sigma + \sqrt{C_1} > 0$ or $\sigma - \sqrt{C_1} > 0$. We are now able to write down the near-resonant solution:

$$\theta(\tau) = \theta_0(T_0, T_1) + \varepsilon\theta_1(T_0, T_1) + O(\varepsilon^2), \tag{3.131}$$

where:

$$
\begin{aligned}
\theta_0 &= A(T_1)e^{iT_0} + \bar{A}(T_1)e^{-iT_0} \quad \text{(by (3.113))} \\
&= \frac{1}{2}a\left(e^{i(T_0+\varphi)} + e^{-i(T_0+\varphi)}\right) \quad \text{(by (3.117))} \\
&= a\cos(T_0 + \varphi) \\
&= a\cos\left(T_0 + \frac{1}{2}(\sigma T_1 - \psi)\right) \quad \text{(by (3.126))} \\
&= a\cos\left(\tau + \frac{1}{2}\left(\frac{\omega - 2}{\varepsilon}\varepsilon\tau - \psi\right)\right) \quad \text{(by (3.109, 3.120))} \\
&= a\cos\left(\frac{1}{2}(\omega\tau - \psi)\right),
\end{aligned}
\tag{3.132}
$$

and, by (3.124) and the same substitutions as above:

$$
\begin{aligned}
\theta_1 &= \frac{1}{8}\gamma\left(A^3 e^{i3T_0} + \bar{A}^3 e^{-i3T_0}\right) \\
&\quad - \frac{1}{16}q\omega^2\left(Ae^{i(3T_0+\sigma T_1)} + \bar{A}e^{-i(3T_0+\sigma T_1)}\right) \\
&= \frac{1}{32}\gamma a^3 \cos\left(\frac{3}{2}(\omega\tau - \psi)\right) \\
&\quad - \frac{1}{16}q\omega^2 a \cos\left(\frac{3}{2}(\omega\tau - \frac{1}{3}\psi)\right).
\end{aligned}
\tag{3.133}
$$

The amplitude a and phase ψ is governed by the modulation equations (3.127), for which the stationary values are given by (3.129).

In terms of the original frequency and time variables (Ω, t) of the pendulum problem, the response becomes:

$$
\begin{aligned}
\theta(t) &= a\cos\left(\frac{1}{2}(\Omega t - \psi)\right) \\
&\quad + \frac{1}{32}\varepsilon\gamma a^3 \cos\left(\frac{3}{2}(\Omega t - \psi)\right) - \frac{1}{16}\varepsilon q\omega^2 a \cos\left(\frac{3}{2}(\Omega t - \frac{1}{3}\psi)\right) \\
&\quad + O(\varepsilon^2),
\end{aligned}
\tag{3.134}
$$

where now ε merely serves to indicate the approximation order of terms. The response is seen to be *frequency-locked* at half the excitation frequency and higher harmonics (integer multiples of $\frac{1}{2}\Omega$) hereof.

3.6.5   Stability of Stationary Solutions

Stationary solutions may be stable or unstable. The stable solutions may serve as predictions of states in which the system may be observed in, e.g., laboratory experiments or computer simulations. The *un*stable solutions predict states that will *not* be observed – since the system will either be in a stable state or on the way to one. Thus, knowing a number of stationary solutions for a problem is of rather limited value without knowing the stability of these solutions. In this section we consider the stability of the solutions obtained in Sections 3.6.3–3.6.4 for the pendulum problem.

Stability of Non-resonant Solutions If ω is away from 2 (the non-resonant case), the only stationary solution is $\theta(\tau) = 0$, corresponding to the constant amplitude of oscillation $a = 0$ (*cf.* Section 3.6.3). This solution is stable for $\beta > 0$, as follows readily from the first equation in (3.118), since when $\beta > 0$ then $a' = -\beta a$ implies that $a \to 0$ for $T_1 \to \infty$ for any values of q, ω and initial conditions. (The Jacobian eigenvalues could also be used, as in the next paragraph.)

Stability of Near-resonant Solutions If ω is close to 2 (the near-resonant case) there are three stationary solutions, according to Section 3.6.4. The first is $a = 0$ as for the non-resonant case, and the two others are given by (3.129). The three solutions were obtained as singular points for the modulation equations (3.127). The determination of solution stability, therefore, reduces to determining the stability of singular points for a system having the form $\dot{\mathbf{x}} = \mathbf{f}(\mathbf{x})$. In Sect. 3.4.4 it was shown how to employ Jacobian eigenvalues for this.

The Jacobian of the modulation equations (3.127) is given by:

$$\mathbf{J}(a, \psi) = \begin{bmatrix} \frac{\partial a'}{\partial a} & \frac{\partial a'}{\partial \psi} \\ \frac{\partial \psi'}{\partial a} & \frac{\partial \psi'}{\partial \psi} \end{bmatrix} = \begin{bmatrix} -\beta + \frac{1}{4}q\omega^2 \sin \psi & \frac{1}{4}q\omega^2 a \cos \psi \\ -\frac{3}{2}\gamma a & -\frac{1}{4}q\omega^2 \sin \psi \end{bmatrix}. \tag{3.135}$$

We first examine the stability of the singular point $a = 0$, where the limit values of ψ are given by $\cos \psi = -2\sigma/(q\omega^2)$, $\sin \psi = \pm\sqrt{1 - \cos^2 \psi} = \pm\sqrt{1 - (2\sigma/(q\omega^2))^2}$. At this point the Jacobian becomes:

$$\mathbf{J}(a, \psi) = \begin{bmatrix} -\beta \pm \frac{1}{2}\sqrt{C} & 0 \\ 0 & \pm\sqrt{C} \end{bmatrix}, \quad C \equiv \left(\frac{1}{2}q\omega^2\right)^2 - \sigma^2. \tag{3.136}$$

where $\pm$ indicates an undetermined sign. The Jacobian governs the linearized motion near the singular point, and so the diagonal form of (3.136) implies the linearized time evolution of a to be uncoupled with that of ψ. Since when $a = 0$ we do not care about the stability of ψ, we only consider the eigenvalue associated with a-motions:

$$\lambda_1 = -\beta \pm \frac{1}{2}\sqrt{\left(\tfrac{1}{2}q\omega^2\right)^2 - \sigma^2}. \tag{3.137}$$

The solution $a = 0$ is unstable if $\mathrm{Re}(\lambda_1) > 0$, that is, if:

$$q > \frac{2}{\omega^2}\sqrt{\sigma^2 + (2\beta)^2}. \tag{3.138}$$

Since γ does not appear in this condition, the stability of the zero solution does not depend on the nonlinearity.

We then examine the stability of the two nonlinear solutions given by (3.129), denoted here by a_1 and a_2:

$$a_1^2 = \frac{4}{3\gamma}(\sigma + \sqrt{C_1}), \quad a_2^2 = \frac{4}{3\gamma}(\sigma - \sqrt{C_1}), \tag{3.139}$$

where

$$C_1 \equiv \left(\tfrac{1}{2}q\omega^2\right)^2 - (2\beta)^2. \tag{3.140}$$

The Jacobian (3.135) evaluated at these two singular points becomes:

$$\mathbf{J}(\tilde{a}, \tilde{\psi}) = \begin{bmatrix} 0 & \frac{1}{2}\tilde{a}(\tfrac{3}{4}\gamma\tilde{a}^2 - \sigma) \\ -\frac{3}{2}\gamma\tilde{a} & -2\beta \end{bmatrix}, \tag{3.141}$$

where $\tilde{a}$ denotes either a_1 or a_2, and where (3.128) has been used for eliminating $\sin\psi$ and $\cos\psi$. The eigenvalues of $\mathbf{J}(\tilde{a}, \tilde{\psi})$ are determined by the requirement that $|\mathbf{J}(\tilde{a}, \tilde{\psi}) - \lambda\mathbf{I}| = 0$, which leads to a polynomial equation in λ:

$$\lambda^2 + 2\beta\lambda + C_2 = 0, \tag{3.142}$$

with solutions

$$\lambda = -\beta \pm \sqrt{\beta^2 - C_2}, \tag{3.143}$$

where

$$C_2 \equiv \frac{3}{4}\gamma\tilde{a}^2\left(\frac{3}{4}\gamma\tilde{a}^2 - \sigma\right). \tag{3.144}$$

Unstable solutions are characterized by having at least one eigenvalue λ with a positive real part. By (3.143) a positive real part requires:

$$\beta^2 - C_2 > 0 \quad \text{and} \quad -\beta + \sqrt{\beta^2 - C_2} > 0, \tag{3.145}$$

which is fulfilled (for $\beta > 0$) if and only if:

$$C_2 < 0. \tag{3.146}$$

For any *softening* nonlinearity one has $\gamma < 0$ (recall that for the pendulum $\gamma = -\frac{1}{6}$), and the condition for instability becomes, upon inserting (3.144):

$$\frac{3}{4}\gamma\tilde{a}^2 - \sigma > 0. \tag{3.147}$$

For the two solutions $\tilde{a} = a_1$ and $\tilde{a} = a_2$ we obtain, by inserting (3.140):

$$\frac{3}{4}\gamma\tilde{a}^2 - \sigma = \begin{cases} (\sigma + \sqrt{C_1}) - \sigma = \sqrt{C_1} > 0 \text{for } \tilde{a} = a_1, \\ (\sigma - \sqrt{C_1}) - \sigma = -\sqrt{C_1} < 0 \text{for } \tilde{a} = a_2. \end{cases} \tag{3.148}$$

Thus, by condition (3.147) we conclude that when $\gamma < 0$ the a_1-solution is *unstable*, whereas the a_2-solution is *stable*. (Conversely, for a *hardening* nonlinearity $\gamma > 0$, the a_1-solution is stable whereas the a_2-solution is unstable).

3.6.6   Discussing Stationary Responses

A Typical Response Fig. 3.10 shows a typical near-resonant *frequency response* for the pendulum system. The curves describe stationary values of the pendulum amplitudes a as a function of excitation frequency ω, as given by (3.129) and the trivial response $a = 0$. Stable solution branches are indicated by solid lines and unstable branches by dashed lines, according to the results of Section 3.6.5. Some typical nonlinear features appear:

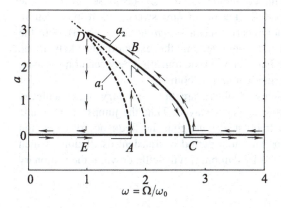

Fig. 3.10. Frequency response $a(\omega)$ for a pendulum subjected to resonant parametric excitation. (—) Stable, (– – –) unstable, (– • –) backbone. ($\gamma = -1/6$, $q = 0.2$, $\beta = 0.05$)

First, *the response peak bends over to the left*. This feature characterizes systems with nonlinear restoring forces of the *softening* type ($\gamma < 0$). A hardening restoring force ($\gamma > 0$) would bend the curve to the right. This contrasts the 'straight-up' peak characteristic of linear systems.

Second, for certain values of the excitation frequencies ω there are *multiple stationary solutions*. This is a consequence of the bent peak: Left of point E in the figure, the only possible solution is the trivial one, a = 0, which is stable. Between E and A there are *three* possible solutions: $a = 0$ (stable), $a = a_1$ (unstable), and $a = a_2$ (stable). The unstable a_1-solution will not be observed in experiments, though it will influence the transient response by repelling nearby states. Now, which of the two stable solutions will be observed in experiments? It depends on the initial conditions. This peculiarity is special to nonlinear systems – since linear systems have at most a single stable state to approach, and will go there however started. Between A and C there are *two* possible solutions: $a = 0$ (which is now unstable) and $a = a_2$ (still stable). Right of C we leave the region $\omega \approx 2$ of primary parametric resonance, and so the response turns linear with the stable solution $a = 0$ as the only one possible.

Third, there are discontinuous *jumps* in the response, also a consequence of the bent peak. Think of an experiment in which the frequency of excitation ω is increased from zero, very slowly to allow for stationary states to settle. The system will remain at rest, until at point A the zero solution turns unstable in favor of the stable a_2 solution. A sudden upward jump in stationary amplitude will then occur to B, beyond which the amplitude decreases smoothly to zero at C. Then consider the experiment reversed, by decreasing ω slowly from a value higher than C. Amplitudes will then increase smoothly until D is reached, at which the amplitude abruptly jumps to E and the system returns to a state of rest.

Thus, if an experimental frequency-sweep was performed for a physical model of the system, one would obtain *two* frequency–response curves: one for the upward sweep (resembling $EABC$ in Fig. 3.10), and one for the downward sweep (resembling $CBDE$). Having no other information than these measured frequency-responses, three fundamental features of the system could be inferred: 1) The presence of amplitude-jumps implies the system to be *nonlinearly affected*, with the jumps reflecting a bent, continuous frequency–response. 2) With the maximum amplitude being largest for the downward sweep, the response-curve must bend to the left, and the nonlinearity has a *softening* character. 3) With the system being at rest outside the resonant region, the excitation is likely to be *parametric*, since for purely external excitation the transition between non-resonant and resonant behavior would be gradual and smooth.

Notice that the responses discussed above are the *stationary* ones, achieved when the effects of initial disturbances has decayed. Thus the jumps in amplitude does not occur 'in time', but in frequency: Changing the frequency of excitation across a jump value, there will be an initial period of transient oscillations, which after some time (mainly determined by damping) will settle down at the stationary amplitudes for the new frequency.

Response Backbone Fig. 3.10 also shows, in dash-dotted line, the *backbone* (or *skeleton*) curve of the frequency response; it is obtained by zeroing the forcing and damping parameters in the frequency response equation. For the present example this means letting $q = \beta = 0$ in (3.129)–(3.130), and inserting $\sigma = \omega - 2$ from (3.120), which gives $C_1 = 0$ and, considering only the positive branch of the backbone:

$$a = \sqrt{\frac{4(\omega - 2)}{3\gamma}} \quad \text{(on backbone).} \tag{3.149}$$

The radical must be non-negative, which means that for $\gamma > 0$ the backbone curve is defined only for $\omega \geq 0$ (i.e. it bends to the *right*, towards higher frequencies), while for $\gamma < 0$ the backbone curve is defined only for $\omega \leq 0$ (i.e. it bends *left*, towards lower frequencies). In the linear case $\gamma \to 0$ which gives $a \to \pm \infty$, i.e. the backbone points straight up, reflecting that for linear systems the free oscillation frequency does not depend on amplitude. The backbone is an important curve, and more easy to calculate than the forced frequency response. It organizes all frequency responses of the system under study, independent of input forcing and damping, and tells which way a nonlinear frequency response bends and how much. It also describes the relation between oscillation amplitude and frequency, a very characteristic feature for nonlinear systems. In the present example, with the parametrically excited pendulum, we have $\gamma = -1/6$ and (3.149) gives $\omega = 2(1 - \frac{1}{16}a^2)$, which is just twice (since the excitation is parametric) the nonlinear natural frequency of the freely oscillating undamped pendulum, *cf.* (3.81).

Response backbones can also be determined from numerical models, using, e.g., free-decay ("ring down") simulation or pseudo-arclength continuation, and from experimental measurements by, e.g., free-decay measurement or control-based continuation techniques; see, e.g., Denis et al. (2017), Givois et al. (2020), Londoño et al. (2015), Peter et al. (2016), Renson et al. (2016).

Peak Response The maximum response is often of interest in applications. It can often be rather easily calculating by noting that the *peak point* of the response curve is located where the backbone intersects the response curve. This point can be found by first solving (3.130) with $C_1 = 0$ to give the *peak response frequency*, ω^*:

$$\omega^* = 2\sqrt{\frac{\beta}{q}}, \tag{3.150}$$

which is independent on the nonlinearity parameter γ – and then substitute this frequency into in (3.129), along with $\sigma = \omega - 2$ from (3.120), to give the corresponding *peak response amplitude*, a^*:

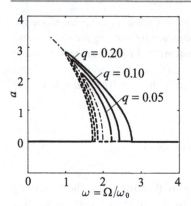

Fig. 3.11. Frequency responses $a(\omega)$ for three different levels of system excitation q. ($\gamma = -1/6$, $\beta = 0.05$)

$$a^* = \sqrt{\frac{4(\omega^* - 2)}{3\gamma}} = \sqrt{\frac{8(\sqrt{\beta/q} - 1)}{3\gamma}}, \gamma \neq 0. \qquad (3.151)$$

Here the radical must be non-negative, so a peak point so defined only exists when either $\gamma > 0$ and $\beta \geq q$, or when $\gamma < 0$ and $\beta \leq q$. In the linear case the response under parametric resonance is unlimited, i.e. $a^* \to \infty$, even with damping. With the example parameters given in the legend of Fig. 3.10 inserted into (3.150) and (3.151), one finds the peak response frequency $\omega^* = 1$, and the corresponding peak response amplitude $a^* \approx 2.8$, agreeing with the values that can be read from the axis values.

Influence of System Parameters Fig. 3.11 shows a set of frequency-responses obtained for varying levels of excitation q. It appears that stronger excitation implies larger response amplitudes, and a wider resonant region. Note too that even small levels of excitation, if resonant, may feed off large amplitudes, even though the level of damping is significant (here 5%). This is also a characteristic associated with parametric excitation: small inputs may create large outputs, even in the presence of damping[5].

Fig. 3.12 depicts the influence of the level of (small) damping. Increased damping tends to lower the response-peak, whereas the width of the resonant region is virtually unaffected.

Fig. 3.13 shows the frequency responses for three values of the coefficient of nonlinearity. The response-peaks bend to the left when $\gamma < 0$ (softening), to the right when $\gamma > 0$ (hardening), and straight up when $\gamma = 0$ (linear case). Note that the linear, near-resonant response is *unbounded*. Generally, for a system in

---

[5]This also means parametric excitation can be used for *amplifying* or 'pumping' motions or signals (Rhoads et al. 2008; Rhoads and Shaw 2010; Thomas et al. 2013; Neumeyer et al. 2017, 2019), and even for *attenuating* or damping vibrations (Dohnal 2008).

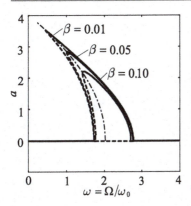

Fig. 3.12. Frequency responses $a(\omega)$ for three different levels of viscous damping β. ($\gamma = -1/6$, $q = 0.2$)

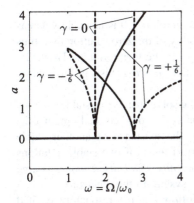

Fig. 3.13. Frequency responses $a(\omega)$ for three different coefficients of nonlinearity: $\gamma = -1/6$ (softening), $\gamma = +1/6$ (hardening) and $\gamma = 0$ (linear). ($q = 0.2$, $\beta = 0.05$)

parametric resonance, some nonlinear mechanism is required for limiting the response, even in the presence of damping. For real, physical systems such limiting factors are always present. Once again we are reminded that linear theory merely predicts the *onset* of critical behavior. Post-critical analysis inherently requires consideration to nonlinear effects.

Stability Diagram Fig. 3.14 shows a *stability diagram*. A stability diagram indicates the existence and stability of stationary solutions in a plane spanned by the excitation parameters, here ω and q. As appears from (3.139) and (3.148) the number and the stability of solutions are governed by the signs of C_1 and $\sigma \pm \sqrt{C_1}$. The boundary curves $C_1 = 0$ and $\sigma \pm \sqrt{C_1} = 0$ of the diagram do not depend on the nonlinear coefficient γ, though the stability and magnitude of solutions in the different domains certainly do.

Unhatched regions of the stability diagram correspond to ranges of excitation parameters ω and q for which the only solution is the linear one, $a = 0$, which is stable.

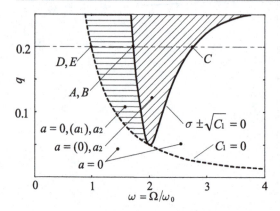

Fig. 3.14. The existence and stability of stationary amplitudes $a = 0$, $a = a_1$ and $a = a_2$. Solutions keyed in parentheses are unstable. Points $(A$-$E)$ refer to Fig. 3.10. $(\gamma < 0, \beta = 0.05)$

In the diagonally hatched domain, the linear solution $a = 0$ is unstable whereas the nonlinear solution $a = a_2$ is stable. The instability of the zero solution (and thus the boundary curve) could be predicted by analyzing the linear system. However, linear analysis would not reveal that another solution $(a = a_2)$ exists in this region, and is stable.

In the horizontally hatched region there are three solutions: $a = 0$ (stable), $a = a_1$ (unstable), and $a = a_2$ (stable). Here, linearized analysis would correctly predict the zero solution to exist and be stable. However, the existence of another stable solution $(a = a_2)$ would not be revealed. The coexistence of two stable solutions implies that each of them is stable *only to sufficiently small disturbances*. Large disturbances of either stable state may throw the system to the other one.

Nonlinearity as a Creator of Periodic Motion For the resonantly excited parametric pendulum a linearized analysis predicts the zero solution to be unstable, and the pendulum rotation amplitudes to grow unbounded. With growing angles of rotation, however, nonlinearities come into play so that θ ceases to be an adequate approximator of $\sin\theta$. One effect of the nonlinearity is to limit the growing response to a finite value, as was illustrated above.

But how does the nonlinearity limit the response? Re-examining the modulation equations (3.127), it appears that the nonlinear term $-\frac{3}{4}\gamma a^2$ affects the amplitude a only indirectly, through changing the phase ψ. For small values of a this influence is negligible, since then $\left| -\frac{3}{4}\gamma a^2 \right| \ll 1$. If the phase is such that $q\omega^2\sin\psi > 4\beta$ then $a' > 0$. This corresponds to a situation in which more energy is pumped into the system than can be dissipated through damping. Consequently, vibration amplitudes will build up, a feature captured by linear analysis as well $(\gamma = 0)$. However, while a is growing, the nonlinear term $-\frac{3}{4}\gamma a^2$ comes into play. It will cause the phase ψ to change, until at some point the energy supplied to the system exactly balances the energy being dissipated by damping. When this occurs there is no energy left over for further increasing the amplitude a. Then, in the absence of external disturbances (additional providers of energy), the response will stabilize into stationary periodic motion – and a so-called *limit cycle* is born.

There are other ways in which nonlinearities may create periodic oscillations out of linearly unbounded motions. One is associated with *dry friction*, which is responsible for the unpleasant squeals from, e.g., activated disk-breaks, train-wheels in curved tracks, the teacher's chalk across the blackboard or rusty door hinges – or the more pleasant sounds from a bow and violin. And there is *flutter* and *galloping*, that is, structural oscillations caused by fluids flowing across plates or strings, inside tubes, etc. Here, linear theory predicts oscillations of the structure to grow unbounded when the flow-speed exceeds a critical value. Typically however, nonlinearity will limit the response – and you hear the ringing of airborne power cables, the sound of an oboe or a human voice, see vibrations of the aircraft wing, or have difficulties in controlling self-sustained oscillations of the garden water hose. Nature disposes on a rich source of nonlinearities for turning nonperiodic input into periodic oscillations.

Bifurcations As parameters are varied in a dynamical system, the stability of equilibrium solutions may change, as may the number and character of solutions. Such changes, representing marked qualitative shifts in system behavior, are called *bifurcations*. The specific values of system parameters at which bifurcations take place are then *bifurcation values*, or *critical values*. Approaching a bifurcation value, the system approaches a *critical state*, at which arbitrarily small changes in the parameters of the system may cause the magnitude and/or character of the response to change drastically.

For example, in Fig. 3.10 the frequencies $\omega_A = \omega_B$, ω_C, and $\omega_D = \omega_E$ are bifurcation values. Marked changes of possible responses take place upon crossing any of these frequencies, whereas in between only trivial changes in response-magnitude occur. The corresponding *points* $(\omega_A, 0)$, $(\omega_C, 0)$, and (ω_D, a_D) at the response curve are called *bifurcation points*.

In decreasing ω from above ω_C in Fig. 3.10, a stable limit cycle is born at ω_C and the zero solution becomes unstable. This type of bifurcation is rather common in problems of nonlinear dynamics. It is the *Hopf bifurcation*, named after the famous mathematician who established conditions for its existence.

The Hopf bifurcation at $(\omega_C, 0)$ involves the transition *stable equilibrium* $\rightarrow$ *unstable equilibrium + stable limit cycle*. This is called a *supercritical* or *soft* Hopf bifurcation. The term 'soft' here serves to indicates that no drastic change of response occurs across the bifurcation value. For the present case the amplitude simply changes from zero to a non-zero but small value. Supercritical/soft bifurcations represent smooth and peaceable changes in system behavior.

The bifurcation at $(\omega_A, 0)$, by contrast, involves the transition *unstable equilibrium* $\rightarrow$ *stable equilibrium + unstable limit cycle*. This too is a Hopf bifurcation, since a limit cycle is created from an equilibrium, though, the Hopf bifurcation is in this case *subcritical* or *hard*. Subcritical/hard Hopf bifurcations are more spectacular than their supercritical/soft counterparts, since a finite disturbance to the stable equilibrium may throw the system beyond the unstable limit cycle. A real structure might not survive the 'hard' effect of a subcritical Hopf bifurcation. We shall return to Hopf- and several other bifurcations in Chap. 5.

A Note on the 'Stability of Systems' Some older texts on linear vibrations and stability briefly touch upon the effects of nonlinearity. One may here find statements like: *'When the linearized system is stable, then so is the corresponding nonlinear system'*. By this one could be misled into thinking that linear analysis would be fully sufficient for checking that a physical system will always return to rest, however nonlinear, and however disturbed. This is not so, as illustrated above for the pendulum example: Just below resonance, when the state of rest is stable, a strong disturbance may cause the pendulum to escape to a coexistent stable state of large amplitude oscillations.

Nevertheless, the cited statement is true, when one conceives 'system' as an equilibrium solution of a system of differential equations (not the underlying physical system), and 'stable' as stable to sufficiently small disturbances.

For systems that are known to possess at most a single equilibrium – e.g. all linear systems – this dissection of terms is rather irrelevant. If the 'system is stable', then the system (mathematical or physical) will be either at *the* state of equilibrium, or on the way to it, irrespective of the magnitude and character of disturbances applied. Since the question on stability can be decided on by linearization, then one can just as well neglect nonlinearities from the very beginning.

The lesson to be learnt is that it makes little sense to characterize a system (physical or mathematical) as being 'stable' or 'unstable'. *States* can be stable or unstable. Systems may possess zero, one, or any number of static and dynamic states of equilibrium. Some states may be stable and others unstable, according to the specific values of system parameters. A system will be either in a stable state, or on its way to one. Upon disturbing the system in some particular state of equilibrium, it may return to this state or escape to another one, depending on the magnitude and character of the disturbance. And finally: linear analysis is sufficient for determining the stability of a given state of equilibrium, but nonlinear analysis may be required for determining the states to linearize around.

3.7 Externally Excited Duffing Systems

In this section we consider the harmonically excited *Duffing's equation*:

$$\ddot{u} + 2\beta\omega_0\dot{u} + \omega_0^2 u + \gamma u^3 = q\cos\Omega t, \tag{3.152}$$

where for mechanical systems $\omega_0{}^2$ is the linear stiffness parameter (if positive, then ω_0 is the linear natural frequency), β the damping ratio, γ the coefficient of a cubic nonlinearity, and q and Ω the amplitude and frequency, respectively, of an externally applied harmonic excitation. Duffing's equation appears so often in applied mechanics that a basic knowledge of its possible solutions is indispensable. The equation looks simple – basically just a linear damped single-DOF oscillator with harmonic excitation, and an additional cubic nonlinearity. Nevertheless, the solutions of this equation exhibit an extremely rich variation (e.g. see Kovacic and

Brennan 2011). Dependent on the signs of ω_0^2 and γ there can be one or three static equilibria, each of which can be stable or unstable. And dependent on the balance between input forcing amplitude q, dissipation coefficient β, and the frequency ratio Ω/ω_0, the response can be anything from simple mono-frequency harmonic oscillations, over nonlinearly distorted oscillations with multiple amplitudes possible, to chaotic/random-like oscillations (see Chap. 6), and bursting oscillations (Rakaric and Kovacic 2016). In this section we focus on oscillations that occur near the static equilibria of (3.152), while chaotic solutions are considered in Chap. 6.

Comparing (3.152) to the equation of motion (3.107) for the pendulum, one sees that they are similar except for the excitation term. For the pendulum the excitation is parametric (multiplying a dependent variable), whereas for the Duffing system the excitation is external (to the free system).

The analysis to follow will provide us with the opportunity to further exercise the method of multiple scales, to enlighten the significance of ordering terms, and to introduce the phenomena of superharmonic and subharmonic resonance. Nayfeh and Mook (1979) and Kovacic and Brennan (2011) may be consulted for further reference on Duffing systems.

To provide an opportunity for viewing results in the light of real systems, we first present two physical models for which the Duffing system (3.152) apply.

3.7.1 Two Physical Examples

Both examples involve a hinged-hinged, transversely loaded beam.

In the first example, one support is allowed to move freely in the horizontal direction. Considering large rotations, a problem of *nonlinear elasticity* arises, with a nonlinear restoring force of the hardening type.

In the second example, the supports of the beam are separated a fixed horizontal distance apart. Rotations are assumed to be small and the nonlinearity involved is due to *midplane stretching*, causing a nonlinear restoring force of the hardening type. In this example the linear stiffness parameter can be positive, negative or zero, depending on whether the support-separation is larger than, less than, or equal to the undeformed length of the beam.

Nonlinear Elasticity The hinged-hinged elastic beam in Fig. 3.15(a) has bending stiffness EI and mass per unit length ρA, and is subjected to harmonic forcing halfway along its length l. The equation of motion is set up by using Newton's second law for a beam-element (Fig. 3.15(b)), as follows:

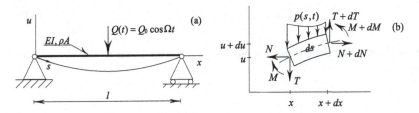

Fig. 3.15. (a) Beam prone to nonlinear elasticity; (b) Differential beam-element

$$(N + dN) - N = 0 \quad \Rightarrow \quad N' = 0,$$
$$(T + dT) - T - pds = \rho A ds \ddot{u} \quad \Rightarrow \quad T' = p + \rho A \ddot{u}, \tag{3.153}$$
$$(M + dM) - M + Tds - Ndu = 0 \quad \Rightarrow \quad M' + T - Nu' = 0,$$

where for this case we need to use a curvilinear axial coordinate s, and primes denote partial differentiation wrt s. Longitudinal and rotational inertia has been neglected, and the beam is assumed to be slender. Differentiating the third equation with respect to s we obtain, upon inserting the two first equations, that

$$M'' + p + \rho A \ddot{u} = 0, \tag{3.154}$$

where N has been eliminated, since $N' = 0$ and $N(l) = 0$ implies that $N = 0$.

No nonlinearities have entered so far, at least not explicitly so. For the linearized version of the problem the rotations are assumed to be small, $(u')^2 \ll 1$, so that $M = EI\kappa \approx EIu'' \approx EId^2 u/dx^2$, where κ is the curvature, is a satisfactory approximation for the internal bending moment. For the present example, however, the rotations are supposed to be finite and we need to consider a better approximation to the true nonlinear curvature κ, which is given by (Pearson 1974; Hodges 1984):

$$
\begin{aligned}
\kappa &= \frac{u''(s)}{\sqrt{1 - (u'(s))^2}} \\
&\approx u''(s)\left(1 + \frac{1}{2}(u'(s))^2\right) \quad \text{for} \quad (u'(s))^2 \ll 1,
\end{aligned}
\tag{3.155}
$$

where the last approximation is a truncated Taylor expansion, valid for rotations that are finite but not very large.

Substituting $M = EI\kappa$ and the approximation for κ into (3.154), one obtains the following equation of motion:

$$EI\left[\left(1 + \frac{1}{2}\left(u'\right)^2\right)u''\right]'' + Q_0 \cos(\Omega t)\delta(s - \tfrac{1}{2}l) + \rho A \ddot{u} = 0, \tag{3.156}$$

where $p(s,t) = \delta(s - \tfrac{1}{2}l)Q_0\cos(\Omega t)$ has also been substituted, with δ denoting Dirac's delta function. Aiming towards a single-mode approximation to the response u, we assume that

$$u(s,t) = a(t)\varphi(s), \tag{3.157}$$

where $a(t)$ is the time-dependent amplitude of the lowest linear mode $\varphi(s)$ of free vibrations:

$$\varphi(s) = \varphi_1(s) = \sin(\pi s / l). \tag{3.158}$$

Then substitute the assumed solution (3.157) into the equation of motion (3.156), multiply by $\varphi(s)$, and integrate along the length of the beam:

$$EI \int_0^l \left[\left(1 + \tfrac{1}{2}(a\varphi')^2 \right) a\varphi'' \right]'' \varphi \, ds + Q_0 \cos(\Omega t) \int_0^l \delta(s - \tfrac{1}{2}l)\varphi \, ds$$
$$+ \rho A \ddot{a} \int_0^l \varphi^2 \, ds = 0. \tag{3.159}$$

Substituting (3.158) for φ and performing the integrations then yields:

$$\frac{\pi^4 EI}{2l^3} \left(1 + \frac{3\pi^2}{8l^2} a^2 \right) a + Q_0 \cos(\Omega t) + \tfrac{1}{2}\rho A l \ddot{a} = 0, \tag{3.160}$$

which is similar to what can be found in, e.g., Bolotin (1964). The modal amplitude a is seen to be governed by a Duffing equation of the form (3.152), with parameters:

$$u(t) = a(t), \quad \omega_0^2 = \frac{EI}{\rho A} \left(\frac{\pi}{l} \right)^4, \quad \gamma = \frac{1}{8} \left(\frac{\pi \omega_0}{l} \right)^2, \quad q = \frac{-2Q_0}{\rho A l}. \tag{3.161}$$

We note that the linear stiffness term is always *positive* (since $\omega_0^2 > 0$), and that the nonlinearity is of the *hardening* type ($\gamma > 0$). For an in-depth treatment of many problems involving nonlinear elasticity see Antman (1995).

Damping was excluded from this example. One could add mass-proportional damping ($-\rho A c \, \dot{u} ds$) to the left side of the second equation in (3.153), or equivalently add modal damping ($2\beta\omega_0\dot{a}$)) directly to (3.160).

Neglecting longitudinal inertia in (3.153) simplifies the calculations, and is often justified in beam vibration problems. But in this particular case, where nonlinear elasticity is of concern, there are good reasons to include it; the reason we did not was mostly to keep the example simple enough to be illustrative. The situation is that *if* beam deformations (or rather beam slopes, $|u'|$) are large enough for nonlinear elasticity to be important, then nonlinear *inertia* will likely *also* be important. The effect of nonlinear inertia is *softening* (Atluri 1973), while as we saw the isolated effect of nonlinear elasticity is hardening. Thus the combined effect of nonlinear elasticity can be softening or hardening, depending mainly on the slenderness of the beam (Atluri 1973; Anderson et al. 1996; Lacarbonara and Yabuno 2006; Sayag and Dowell 2016).

Midplane Stretching The beam in Fig. 3.16 differs from that of the previous example (Fig. 3.15), in that the right support cannot move horizontally as the beam deforms transversely. The supports are fixed a prescribed distance $(1 + \eta)l$ apart, where $|\eta| \ll 1$ describes a small initial stretch or compression of the beam. When $\eta > 0$ the initial separation between supports is larger than the undeformed length

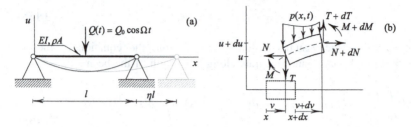

Fig. 3.16. (a) Beam prone to midplane stretching; (b) Differential beam-element

l of the beam, and the beam is pre-*tensioned*. If $\eta < 0$ the separation is less than the undeformed beam-length, and the beam pre-*compressed*.

To deform this beam transversely one must supply considerably more energy than if one support was free to slide, at least when $\eta \geq 0$. This is because the beam cannot deform solely in bending but will need to stretch axially as well, that is, to become longer, since the supports cannot slide. Thus, for this problem one can expect stretching of the beam midplane to contribute significantly to the transverse stiffness of the beam. Midplane stretching implies that longitudinal deformations v should be accounted for in addition to transverse deformations u.

On the other hand, since so much energy is required for deforming the beam transversely, one can expect transverse deformations and cross-sectional rotations to be small. This allows for the linear measure of curvature to be adopted, by contrast to the previous example where large curvatures were considered the key source of nonlinearity.

For the beam-element in Fig. 3.16(b) we obtain, using Newton's second law, the following equations, employing usual slender-beam assumptions and neglecting rotational and longitudinal inertia:

$$(N + dN) - N = 0 \Rightarrow N' = 0,$$
$$(T + dT) - T - p dx = \rho A dx \ddot{u} \Rightarrow T' = \rho A \ddot{u} + p, \tag{3.162}$$
$$(M + dM) - M + T(dx + dv) - N du = 0 \Rightarrow M' + T(1 + v') - Nu' = 0.$$

As is common practice for slender beams we assume that $v = O(u^2)$, that is, longitudinal deformations are second in order as compared to transverse deflections. This implies that v' can be neglected when compared to unity, and to u, but not when compared to u^2.

Differentiating the third equation in (3.162) with respect to x we obtain, inserting the two first equations and neglecting v' in the term $(1 + v')$, that:

$$M'' + \rho A \ddot{u} + p - Nu'' = 0. \tag{3.163}$$

The first of the equations (3.162) implies N to be a constant. To find its value we apply Hooke's law,

$$N = EA\Lambda, \tag{3.164}$$

with Λ denoting the longitudinal strain:

$$\Lambda = \frac{\sqrt{(dx+dv)^2 + du^2} - dx}{dx} = \sqrt{(1+v')^2 + (u')^2} - 1 \approx v' + \frac{1}{2}(u')^2, \tag{3.165}$$

where, as explained above, v' should *not* be neglected in comparison with $(u')^2$. Inserting into (3.164) and re-arranging one finds that:

$$v' = \frac{N}{EA} - \frac{1}{2}(u')^2, \tag{3.166}$$

so that the longitudinal extension v is:

$$v(x,t) = \int_0^x \left(\frac{N}{EA} - \frac{1}{2}(u')^2\right) dx = \frac{Nx}{EA} - \frac{1}{2}\int_0^x (u')^2 dx. \tag{3.167}$$

Imposing the boundary condition $v(l,t) = \eta l$ on (3.167) yields:

$$\eta l = \frac{Nl}{EA} - \frac{1}{2}\int_0^l (u')^2 dx, \tag{3.168}$$

from which N can be determined:

$$N = EA\left(\eta + \frac{1}{2l}\int_0^l (u')^2 dx\right). \tag{3.169}$$

Substituting (3.169), $M = EIu''$ and $p = Q_0\delta(x - \frac{1}{2}l)\cos(\Omega t)$ into (3.163) then yields the partial differential equation of motion for the beam:

$$EIu'''' + \rho A\ddot{u} + Q_0\cos(\Omega t)\delta(x - \frac{1}{2}l) - EA\left(\eta + \frac{1}{2l}\int_0^l (u')^2 dx\right)u'' = 0. \tag{3.170}$$

As for the previous example we seek an approximate solution in terms of the lowest mode $\varphi(x)$ of linear free vibrations, that is:

$$u(x,t) = a(t)\varphi(x), \quad \varphi(x) = \sin(\pi x/l). \tag{3.171}$$

This approximation is adequate when the frequency of excitation Ω is well below the second linear natural frequency of the beam. Other functions could be assumed, as could a complete expansion $\sum_{i=1} a_j(t)\varphi_j(x))$. Inserting (3.171) into

(3.170) we obtain, on multiplying by φ, integrating over the beam length and rearranging, that:

$$\frac{1}{2}\rho A l \ddot{a} + \left(\frac{\pi^4 EI}{2l^3} + \frac{\pi^2 EA}{2l}\eta\right)a + \frac{\pi^4 EA}{8l^3}a^3 = -Q_0 \cos(\Omega t). \tag{3.172}$$

This is a Duffing equation of the form (3.152), with parameters:

$$u(t) = a(t), \quad \omega_0^2 = \frac{EI}{\rho A}\left(\frac{\pi}{l}\right)^4\left(1+\eta\left(\frac{l}{\pi}\right)^2\frac{A}{I}\right), \quad \gamma = \frac{1}{4}\frac{E}{\rho}\left(\frac{\pi}{l}\right)^4, \quad q = \frac{-2Q_0}{\rho A l}. \tag{3.173}$$

The nonlinearity appears to be *hardening* ($\gamma > 0$). The linear stiffness coefficient ω_0^2 equals the squared lowest natural frequency $EI\pi^4/\rho A l^4$ of an ordinary hinged-hinged beam, though, with an additional term that accounts for a stiffening or weakening effect when $\eta \neq 0$.

Note that the linear stiffness can be *negative*, since $\omega_0^2 < 0$ whenever:

$$\eta < -\eta_{crit}; \quad \eta_{crit} = \left(\frac{\pi}{l}\right)^2\frac{I}{A} = \left(\frac{\pi}{s}\right)^2, \tag{3.174}$$

where $s = l/r$ is the beams slenderness ratio, and $r = \sqrt{I/A}$ the radius of gyration of the beam cross section. This condition is met when an initial contraction of supports implies a compressive load of the beam that exceeds the static buckling load $EI\pi^2/l^2$. In that case, it appears from (3.172), there are three possible configurations of static equilibrium: The unstable, straight configuration $a = 0$, and two stable, symmetrical states of buckling:

$$a = \pm\sqrt{\frac{-\omega_0^2}{\gamma}} = \pm 2\sqrt{-\left(\frac{I}{A}\right) - \left(\frac{l}{\pi}\right)^2\eta}. \tag{3.175}$$

It appears that $a \to \infty$ when $\gamma \to 0$, so that the buckling amplitude is bounded only in the presence of nonlinearity ($\gamma \neq 0$). This is yet an illustration that even though critical states may be determined by linearization, the prediction of *post-critical* states requires consideration to nonlinear effects.

Nonlinear Elasticity or Midplane Stretching – Which is the Stronger? If axial motion is unrestricted, then midplane stretching is usually not relevant; nonlinear elasticity is obviously the dominating nonlinearity, of the two. But what if boundary conditions are such that both could be at play? We can estimate the relative effect by comparing the coefficients of the cubic nonlinearity for the two cases. By (3.161) and (3.173) we have, with subscripts NE and MS denoting the coefficients of nonlinear elasticity and midplane stretching, respectively:

$$\gamma_{NE} = \frac{1}{8}\left(\frac{\pi\omega_0}{l}\right)^2 = \frac{1}{8}\left(\frac{\pi}{l}\right)^6\frac{EI}{\rho A},$$

$$\gamma_{MS} = \frac{1}{4}\frac{E}{\rho}\left(\frac{\pi}{l}\right)^4,$$

(3.176)

which gives the following simple ratio of the two nonlinear coefficients:

$$\frac{\gamma_{NE}}{\gamma_{MS}} = \frac{1}{2}\left(\frac{\pi}{s}\right)^2 = \frac{1}{2}\eta_{crit},$$

(3.177)

where s and η_{crit} are defined with (3.174). For slender beams, where Bernoulli–Euler theory applies, one has $s \geq 20 - 200$, giving ratios of the nonlinear coefficients in the range 10^{-2}–10^{-4} or smaller. Thus at similar levels of modal response amplitude u, the midplane stretching nonlinearity is much stronger.

For the nonlinear *forces* (γu^3) to be of similar magnitude for the two types of nonlinearity, the deformations u should thus be of order magnitude $(10^{-2}$–$10^{-4})^{-1/3}$ $\approx$ 4–20 times larger for nonlinear elasticity than for midplane stretching.

So, if axial boundary motion is restricted so as to create midplane stretching, at least for slender beams one can usually neglect nonlinear elasticity in comparison with midplane stretching nonlinearity.

3.7.2 Primary Resonance, Weak Excitations

We now turn to obtaining approximate solutions for the Duffing system (3.152). When $\Omega \approx \omega_0$, then the excitation frequency is close to the linear natural frequency of the system, and we are dealing with a case of *primary external resonance*. At primary resonance one can expect weak excitations to cause comparatively large responses. Consequently we assume for (3.152) that if $q = O(\varepsilon)$ where $\varepsilon \ll 1$, then $u = O(1)$. Further, the influence of the damping term and the nonlinear term is supposed to be weak as compared to the terms describing linear inertia and stiffness. Assuming damping and nonlinearity to be both $O(\varepsilon)$, the corresponding terms will enter the analysis at the same level of approximation as the external forcing. This magnitude ordering is expressed by writing (3.152) in the form:

$$\ddot{u} + \omega_0^2 u = \varepsilon\left(q\cos\Omega t - 2\beta\omega_0\dot{u} - \gamma u^3\right).$$

(3.178)

Heading on for a multiple scales approximation, we assume a solution in form of a uniformly valid expansion:

$$u(t;\varepsilon) = u_0(T_0, T_1) + \varepsilon u_1(T_0, T_1) + O(\varepsilon^2),$$

(3.179)

where $T_0 = t$ and $T_1 = \varepsilon t$ are the fast and the slow time-scale, respectively, and u_0 and u_1 are the unknown functions to be determined. On substituting the expansion into (3.178) one obtains:

$$
\begin{aligned}
\big(D_0^2 + 2\varepsilon D_0 D_1 &+ O(\varepsilon^2)\big)\big(u_0 + \varepsilon u_1 + O(\varepsilon^2)\big) + \omega_0^2\big(u_0 + \varepsilon u_1 + O(\varepsilon^2)\big) \\
&= \varepsilon\big[q\cos\Omega T_0 - 2\beta\omega_0(D_0 + \varepsilon D_1)\big(u_0 + \varepsilon u_1 + O(\varepsilon^2)\big) \\
&\quad - \gamma\big(u_0 + \varepsilon u_1 + O(\varepsilon^2)\big)^3\big],
\end{aligned}
\tag{3.180}
$$

where $D_i{}^j \equiv \partial^j/\partial T_i{}^j$. Equating to zero the coefficients to like powers of ε yields, to order ε^0:

$$
D_0^2 u_0 + \omega_0^2 u_0 = 0,
\tag{3.181}
$$

and to order ε^1:

$$
D_0^2 u_1 + \omega_0^2 u_1 = q\cos\Omega T_0 - 2D_0 D_1 u_0 - 2\beta\omega_0 D_0 u_0 - \gamma u_0^3.
\tag{3.182}
$$

The general solution to the zero order problem (3.181) is:

$$
u_0 = A(T_1)e^{i\omega_0 T_0} + \overline{A}(T_1)e^{-i\omega_0 T_0},
\tag{3.183}
$$

where $A(T_1) \in C$ is the unknown function to be determined. Substituting the solution for u_0 into the first-order problem (3.182) we obtain that:

$$
\begin{aligned}
D_0^2 u_1 + \omega_0^2 u_1 = &\tfrac{1}{2}qe^{i\Omega T_0} - i2\omega_0 A'e^{i\omega_0 T_0} \\
&- i2\beta\omega_0^2 A e^{i\omega_0 T_0} - \gamma\big(A^3 e^{i3\omega_0 T_0} + 3A^2\overline{A}e^{i\omega_0 T_0}\big) + cc,
\end{aligned}
\tag{3.184}
$$

where the term $\cos(\Omega T_0)$ has been expressed in exponential form.

Terms proportional to $e^{i\omega_0 T_0}$ are resonant to the left-hand side of the equation. These will cause secular terms (proportional to $T_0 e^{i\omega_0 T_0}$) to appear in the particular solution for u_1. Further, since by assumption $\Omega \approx \omega_0$, the term $\tfrac{1}{2}qe^{i\Omega T_0}$ will be near-resonant, causing small divisor terms to appear in the particular solution for u_1. To convert near-resonant terms to resonant terms a detuning parameter σ is introduced, measuring the nearness to resonance by:

$$
\Omega = \omega_0 + \varepsilon\sigma.
\tag{3.185}
$$

Substituting this into (3.182) we obtain, upon noting that $\Omega T_0 = \omega_0 T_0 + \varepsilon\sigma T_0 = \omega_0 T_0 + \sigma T_1$ and rearranging terms:

$$D_0^2 u_1 + \omega_0^2 u_1 = \left[\frac{1}{2}qe^{i\sigma T_1} - i2\omega_0(A' + \beta\omega_0 A) - 3\gamma A^2\overline{A}\right]e^{i\omega_0 T_0}$$
$$- \gamma A^3 e^{i3\omega_0 T_0} + cc. \tag{3.186}$$

Any particular solution u_1 to this equation will contain secular terms, unless the function $A(T_1)$ meets the solvability condition:

$$\frac{1}{2}qe^{i\sigma T_1} - i2\omega_0(A' + \beta\omega_0 A) - 3\gamma A^2\overline{A} = 0. \tag{3.187}$$

With this condition fulfilled a particular solution of (3.186) becomes:

$$u_1 = \frac{\gamma}{8\omega_0^2}A^3 e^{i3\omega_0 T_0} + cc, \tag{3.188}$$

so that, by (3.179) and (3.183):

$$u(t; \varepsilon) = Ae^{i\omega_0 T_0} + \varepsilon\frac{\gamma}{8\omega_0^2}A^3 e^{i3\omega_0 T_0} + O(\varepsilon^2) + cc. \tag{3.189}$$

To solve (3.187) for $A(T_1)$ we let

$$A = \frac{1}{2}ae^{i\varphi}, \quad a(T_1), \ \varphi(T_1) \in R. \tag{3.190}$$

Substituting into (3.187) one obtains, upon separating real and imaginary parts, the following pair of modulation equations:

$$a' = -\beta\omega_0 a + \frac{q}{2\omega_0}\sin(\sigma T_1 - \varphi),$$
$$a\varphi' = \frac{3\gamma}{8\omega_0}a^3 - \frac{q}{2\omega_0}\cos(\sigma T_1 - \varphi). \tag{3.191}$$

To transform these into autonomous equations (i.e. with no explicit dependence on T_1) a new dependent variable ψ is introduced, by

$$\psi = \sigma T_1 - \varphi \quad (\Rightarrow \psi' = \sigma - \varphi'). \tag{3.192}$$

Slow modulations of amplitudes $a(T_1)$ and phases $\psi(T_1)$ are then governed by:

$$a' = -\beta\omega_0 a + \frac{q}{2\omega_0}\sin\psi,$$
$$a\psi' = \sigma a - \frac{3\gamma}{8\omega_0}a^3 + \frac{q}{2\omega_0}\cos\psi. \tag{3.193}$$

Stationary solutions are defined by having constant-valued amplitudes and phases. Thus, for seeking stationary solutions we locate the singular points of

(3.193). Letting $a' = \psi' = 0$, we first note that $a = 0$ is *not* a solution, by contrast to the case of the parametrically excited pendulum (*cf.* Section 3.6.4). Squaring and adding the two equations (with $a' = \psi' = 0$) we find that a is given by the implicit solution to the following *frequency response equation*:

$$\left[\left(2\beta\omega_0^2\right)^2 + \left(2\sigma\omega_0 - \tfrac{3}{4}\gamma a^2\right)^2 \right] a^2 = q^2. \tag{3.194}$$

The corresponding phase ψ is then given by

$$\tan\psi = \frac{-\beta\omega_0}{\sigma - \frac{3\gamma}{8\omega_0} a^2}. \tag{3.195}$$

Having determined a from (3.194) and ψ from (3.195), the function $A(T_1)$ is given by (3.190) and (3.192) as:

$$A(T_1) = \tfrac{1}{2} a e^{i\varphi} = \tfrac{1}{2} a e^{i(\sigma T_1 - \psi)}, \tag{3.196}$$

and so the approximate solution (3.189) becomes

$$\begin{aligned}
u(t; \varepsilon) &= \tfrac{1}{2} a e^{i(\sigma T_1 - \psi)} e^{i\omega_0 T_0} + \varepsilon \frac{\gamma}{64\omega_0^2} a^3 e^{i3(\sigma T_1 - \psi)} e^{i3\omega_0 T_0} + O(\varepsilon^2) + cc \\
&= a\cos(\sigma T_1 + \omega_0 T_0 - \psi) \\
&\quad + \varepsilon \frac{\gamma}{32\omega_0^2} a^3 \cos(3(\sigma T_1 + \omega_0 T_0 - \psi)) + O(\varepsilon^2).
\end{aligned} \tag{3.197}$$

Inserting (3.185), $T_0 = t$ and $T_1 = \varepsilon t$, we find that the stationary periodic solutions $u(t)$ for the primary resonant Duffing equation are given by:

$$\begin{aligned}
u(t) &= a\cos(\Omega t - \psi) \\
&\quad + \varepsilon \frac{\gamma}{32\omega_0^2} a^3 \cos(3(\Omega t - \psi)) + O(\varepsilon^2),
\end{aligned} \tag{3.198}$$

where the constants a and ψ are given by (3.194) and (3.195) with $\sigma = \Omega - \omega_0$, and where ε now merely serves the purpose of indicating the level of approximation. Note that the oscillations occur at the frequency of excitation Ω, and a higher harmonic 3Ω of this.

Nonstationary (transient) solutions are also given by (3.198), though, with time varying amplitudes and phases $a(t)$ and $\psi(t)$ as determined by the solutions to the modulation equations (3.193) with relevant initial conditions.

Plotting Frequency Responses To see how the stationary amplitude a varies with the excitation frequency Ω we plot the frequency response. Equation (3.194) is not readily solved for a, since one has to locate the roots of a third order polynomial in a^2. Instead we compute data for the response curves by considering Ω, rather than a, as the dependent variable. Inserting $\sigma = \Omega - \omega_0$ into (3.194) and solving for Ω, we obtain the response curves in the form $\Omega(a)$, as follows:

$$\frac{\Omega}{\omega_0} = 1 + \frac{3\gamma}{8\omega_0^2}a^2 \pm \sqrt{\left(\frac{q}{2\omega_0^2 a}\right)^2 - \beta^2}. \tag{3.199}$$

Fig. 3.17 shows a typical frequency response, as given by (3.199) with $\gamma = \frac{1}{2}$. The response-peak is bent towards the right, as expected for a hardening nonlinearity. This causes a jump-down in stationary amplitudes to occur from A to B, and an jump-up from C to D.

Other aids in plotting frequency responses may be software to determine numerical solutions of the algebraic frequency response equations. However, if this is performed simply for a uniformly increasing or decreasing sequence of frequency (or amplitude) values, some non-trivial post-processing will be in need to connect the solution points (Ω, a) by line segments to form the response curve. Here numerical *arclength continuation* or *path following algorithms* may be better alternatives, in using a lengthwise curve parameter rather than frequency (or amplitude) as the independent variable (see Chap. 5 and, e.g., Nayfeh and Balachandran 1995; Kaas-Petersen 1989). Such methods take as their starting point a single solution point (Ω, a), and find the next curve point based on local gradient information, tracing the curve even through possible turning points and branches.

Stability of Solutions The stability of stationary solutions (indicated by line type in Fig. 3.17) is determined by evaluating Jacobian eigenvalues. At a singular point $(\tilde{a}, \tilde{\psi})$, the Jacobian of the modulation equations (3.193) becomes:

$$\mathbf{J}(\tilde{a}, \tilde{\psi}) = \begin{bmatrix} -\beta\omega_0 & \frac{q}{2\omega_0}\cos\tilde{\psi} \\ -\frac{3\gamma}{4\omega_0}\tilde{a} - \frac{q}{2\omega_0}\frac{1}{\tilde{a}^2}\cos\tilde{\psi} & -\frac{q}{2\omega_0}\frac{1}{\tilde{a}}\sin\tilde{\psi} \end{bmatrix}, \tag{3.200}$$

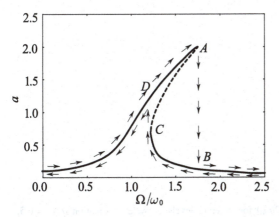

Fig. 3.17. Frequency response of the resonant Duffing equation. (——) stable, (– – –) unstable. ($\gamma = 0.5$, $q = 0.2$, $\beta = 0.05$, $\omega_0 = 1$)

where the singular points $\tilde{a}$ and $\tilde{\psi}$ are given by the solutions to (3.194) and (3.195). Using (3.193) with $a' = \psi' = 0$ to eliminate $\cos\tilde{\psi}$ and $\sin\tilde{\psi}$, we obtain:

$$\mathbf{J}(\tilde{a},\tilde{\psi}) = \begin{bmatrix} -\beta\omega_0 & -\left(\sigma\tilde{a} - \frac{3\gamma}{8\omega_0}\tilde{a}^3\right) \\ \frac{1}{\tilde{a}^2}\left(\sigma\tilde{a} - \frac{9\gamma}{8\omega_0}\tilde{a}^3\right) & -\beta\omega_0 \end{bmatrix}. \tag{3.201}$$

Computing eigenvalues as the solutions of $|\mathbf{J}(\tilde{a},\tilde{\psi}) - \lambda\mathbf{I}| = 0$ it is found that:

$$\lambda = -\beta\omega_0 \pm \sqrt{-\left(\sigma - \frac{3\gamma}{8\omega_0}\tilde{a}^2\right)\left(\sigma - \frac{9\gamma}{8\omega_0}\tilde{a}^2\right)}. \tag{3.202}$$

A solution $\tilde{a}$ is unstable if at least one eigenvalue has a positive real part, otherwise it is stable. Hence, for $\tilde{a}$ to be unstable it is required that:

$$\left(\sigma - \frac{3\gamma}{8\omega_0}\tilde{a}^2\right)\left(\sigma - \frac{9\gamma}{8\omega_0}\tilde{a}^2\right) + (\beta\omega_0)^2 < 0. \tag{3.203}$$

This condition is met on the part A-C of the frequency response in Fig. 3.17, which is thus unstable and drawn in dashed line. All other parts of the response are stable.

Influence of System Parameters Fig. 3.18 depicts the variation in resonant response due to variations in external load-level q. Naturally, as q is reduced then so is the height of the resonance peak, and consequently the nonlinear bend of the peak. Also, the width of the resonant region shrinks when q is reduced.

Fig. 3.19 shows the influence of the level of viscous damping β. Increased damping appears to be similar in effect to that of decreased loading, in that the resonance peak is lowered and the nonlinear bend is reduced. However, the width of the resonant region does not shrink, as for the case of varying q.

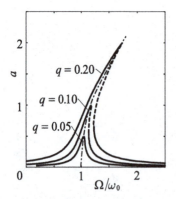

Fig. 3.18. Frequency responses $a(\Omega/\omega_0)$ for different levels of system excitation q. ($\gamma = 0.5$, $\beta = 0.05$, $\omega_0 = 1$)

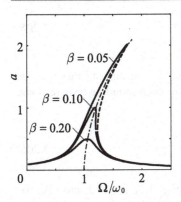

Fig. 3.19. Frequency responses $a(\Omega/\omega_0)$ for different levels of viscous damping β. ($\gamma = 0.5$, $q = 0.2$, $\omega_0 = 1$)

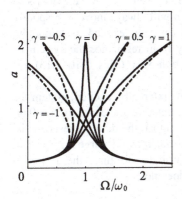

Fig. 3.20. Frequency responses $a(\Omega/\omega_0)$ for varying levels of nonlinearity γ. ($q = 0.2$, $\beta = 0.05$, $\omega_0 = 1$)

Fig. 3.20 shows the influence of the coefficient of nonlinearity γ. The resonance peaks bend right when $\gamma > 0$ and left when $\gamma < 0$. The effect increases with the magnitude of γ. When $\gamma = 0$ the response is linear, and the peak is straight up.

Backbones The backbone curves in Fig. 3.18 and Fig. 3.19 are obtained by letting $q = \beta = 0$ in (3.199), using (3.185) to substitute σ, and solving for a, which gives:

$$a = \omega_0 \sqrt{\frac{8}{3\gamma}\left(\frac{\Omega}{\omega_0} - 1\right)} \quad \text{(on backbone).} \tag{3.204}$$

Peak Response The peak response amplitude a^* is found by zeroing the radical in in (3.199) and solving for $a = a^*$, giving:

$$a^* = \frac{q}{2\beta\omega_0^2},$$ (3.205)

which is independent on the nonlinearity parameter γ. Substituting this for a in (3.199), along with $\sigma = \Omega - \omega_0$ from (3.185), gives the corresponding peak response frequency ratio:

$$\frac{\Omega^*}{\omega_0} = 1 + \frac{3\gamma}{32}\left(\frac{q}{\beta\omega_0^3}\right)^2.$$ (3.206)

With the example parameters in the legend of Fig. 3.17 inserted into (3.205)–(3.206) one finds $\Omega^*/\omega_0 = 1.75$, $a^* = 2$, agreeing with what can be read from the axis values.

Responses for External and Parametrical Excitation Compared The frequency responses of Figs. 3.17–3.20 have no line of zero amplitude. This is a general feature of externally excited systems, which will always move in response to excitation, however small. By contrast, parametrically excited system only respond to near-resonant excitation, whereas away from resonance the system is unaffected by the excitation, as exemplified in Section 3.6.6 (Figs. 3.10–3.13) for the pendulum with an oscillating support.

Further, the damped resonant response of an externally excited system is bounded even in the absence of nonlinearity, as illustrated by the curve for $\gamma = 0$ in Fig. 3.20. For parametrically excited systems, by contrast, the damped response is bounded only in the presence of nonlinearity, as illustrated by the curve for $\gamma = 0$ in Fig. 3.13. With external excitation damping contributes to limit the resonant response, whereas with parametrical excitation it does not.

3.7.3   Non-resonant Hard Excitations

In Sect. 3.7.2 the excitation amplitude q was assumed to be small, $q = O(\varepsilon)$. If the frequency of excitation Ω is away from the linear natural frequency ω_0, then a small excitation will cause only a weak and essentially linear response. So, when Ω is away from ω_0 it takes a *hard* excitation to drive the system into the nonlinear regime. To study the nonlinear effect of hard excitation, we assume the excitation amplitude to be similar in magnitude to the terms describing linear restoring force and inertia, that is, $q = O(1)$. The equation of motion (3.152) then reads:

$$\ddot{u} + \omega_0^2 u = q\cos\Omega t - \varepsilon\left(2\beta\omega_0\dot{u} + \gamma u^3\right),$$ (3.207)

where the damping and the nonlinearity is still assumed to be small. Substituting for u the multiple scales expansion:

$$u(t; \varepsilon) = u_0(T_0, T_1) + \varepsilon u_1(T_0, T_1) + O(\varepsilon^2), \qquad (3.208)$$

one obtains, to order ε^0:

$$D_0^2 u_0 + \omega_0^2 u_0 = q \cos \Omega T_0, \qquad (3.209)$$

and to order ε^1:

$$D_0^2 u_1 + \omega_0^2 u_1 = -2D_0 D_1 u_0 - 2\beta \omega_0 D_0 u_0 - \gamma u_0^3. \qquad (3.210)$$

As a consequence of the magnitude reordering of terms, the external forcing now appears at the ε^0 level of approximation. The general solution of (3.209) is:

$$u_0 = A(T_1) e^{i\omega_0 T_0} + Q e^{i\Omega T_0} + cc, \qquad (3.211)$$

where

$$Q \equiv \frac{q/2}{\omega_0^2 - \Omega^2} \qquad (3.212)$$

Substituting the solution for u_0 into (3.210), the first-order problem becomes:

$$
\begin{aligned}
D_0^2 u_1 + \omega_0^2 u_1 = & -\left[i2\omega_0 (\beta \omega_0 A + A') + 3\gamma A(A\overline{A} + 2Q^2) \right] e^{i\omega_0 T_0} \\
& - Q\left[i2\beta \omega_0 \Omega + \gamma(6A\overline{A} + 3Q^2) \right] e^{i\Omega T_0} - \gamma\left(A^3 e^{i3\omega_0 T_0} + Q^3 e^{i3\Omega T_0} \right) \\
& - 3\gamma Q\Big[AQ e^{i(2\Omega + \omega_0)T_0} + A^2 e^{i(\Omega + 2\omega_0)T_0} \\
& \quad + \overline{A}Q e^{i(2\Omega - \omega_0)T_0} + \overline{A}^2 e^{i(\Omega - 2\omega_0)T_0} \Big] + cc.
\end{aligned}
\qquad (3.213)
$$

Here the first bracketed term is seen to be resonant to the left-hand side of the equation. Further, it appears, near-resonant terms that will produce small divisor terms in u_1 arise whenever $\Omega \approx \frac{1}{3}\omega_0$, $\Omega \approx 3\omega_0$, $\Omega \approx \omega_0$, or $\Omega \approx 0$. The first two cases are examples of *secondary resonances*, with $\Omega \approx \frac{1}{3}\omega_0$ called a *superharmonic resonance*, and $\Omega \approx 3\omega_0$ a *subharmonic resonance*. The case $\Omega \approx \omega_0$ corresponds to primary resonance, which is irrelevant here by the assumption that Ω is away from ω_0. Thus, there are four cases to consider:

1. Non-resonant excitation: Ω is away from $\frac{1}{3}\omega_0$, $3\omega_0$, ω_0 and 0
2. Superharmonic resonance: $\Omega \approx \frac{1}{3}\omega_0$
3. Subharmonic resonance: $\Omega \approx 3\omega_0$
4. Quasi-static excitation: $\Omega \approx 0$

The first three cases are treated below, whereas the fourth is left as an exercise.

Non-resonant Excitation If Ω is away from $\frac{1}{3}\omega_0$, $3\omega_0$, ω_0 and 0, then only the first bracketed term of (3.213) will produce secular terms. We require this term to vanish identically:

$$i2\omega_0(\beta\omega_0 A + A') + 3\gamma A(A\bar{A} + 2Q^2) = 0. \tag{3.214}$$

Letting $A = \frac{1}{2}ae^{i\varphi}$, where a and φ are real functions of T_1, one arrives at the modulation equations:

$$a' = -\beta\omega_0 a,$$
$$a\varphi' = \frac{3\gamma}{\omega_0}\left(Q^2 + \frac{1}{8}a^2\right)a. \tag{3.215}$$

Integrating the first equation yields $a(T_1) = a_0 e^{-\beta\omega_0 T_0}$, so that $a \to 0$ for $T_1 \to \infty$ and the stationary value of a is zero. Thus, to first order, the stationary response is governed by:

$$u(t) = u_0 + O(\varepsilon) = A(T_1)e^{i\omega_0 T_0} + Qe^{i\Omega T_0}$$
$$+ cc + O(\varepsilon) = \frac{1}{2}ae^{i\varphi}e^{i\omega_0 T_0} + \frac{q/2}{\omega_0^2 - \Omega^2}e^{i\Omega T_0} \tag{3.216}$$
$$+ cc + O(\varepsilon) = \frac{q}{\omega_0^2 - \Omega^2}\cos(\Omega t) + O(\varepsilon).$$

This is essentially the linear, off-resonant response, corresponding to small amplitude vibrations (virtually independent of damping) at the frequency of excitation Ω.

Superharmonic Resonance If $\Omega \approx \frac{1}{3}\omega_0$, then the term proportional to $e^{i3\Omega T_0}$ in (3.213) will be near-resonant. To measure the nearness of Ω to $\frac{1}{3}\omega_0$, we define a detuning parameter σ through

$$3\Omega = \omega_0 + \varepsilon\sigma, \tag{3.217}$$

whereby $3\Omega T_0 = \omega_0 T_0 + \sigma T_1$. On substituting into (3.213) it is found that secular terms are eliminated by the following solvability condition:

$$i2\omega_0(\beta\omega_0 A + A') + 3\gamma A(A\bar{A} + 2Q^2) + \gamma Q^3 e^{i\sigma T_1} = 0. \tag{3.218}$$

Letting $A = \frac{1}{2}ae^{i\varphi}$ where $a(T_1)$, $\varphi(T_1) \in R$ we obtain, upon separating real and imaginary parts, the modulation equations for superharmonic resonance:

$$a' = -\beta\omega_0 a - \frac{\gamma Q^3}{\omega_0}\sin\psi,$$
$$a\psi' = \left(\sigma - \frac{3\gamma Q^2}{\omega_0}\right)a - \frac{3\gamma}{8\omega_0}a^3 - \frac{\gamma Q^3}{\omega_0}\cos\psi, \tag{3.219}$$

where, to obtain autonomous equations, a new phase variable ψ has been substituted for φ:

$$\psi = \sigma T_1 - \varphi. \tag{3.220}$$

Stationary oscillations correspond to singular points of the modulation equations, that is, to those (a, φ) for which $a' = \psi' = 0$. Letting $a' = \psi' = 0$ in (3.219) we obtain, by squaring and adding the two equations, the frequency response equation for the superharmonic case:

$$\left[\left(\beta \omega_0^2 \right)^2 + \left(\sigma \omega_0 - 3\gamma (Q^2 + \tfrac{1}{8}a^2) \right)^2 \right] a^2 = \left(\gamma Q^3 \right)^2, \tag{3.221}$$

with corresponding phase ψ given by (divide the two equations):

$$\tan \psi = \frac{-\beta \omega_0^2}{\sigma \omega_0 - 3\gamma (Q^2 + \tfrac{1}{8}a^2)}. \tag{3.222}$$

The frequency response equation (3.221) is not readily solved for a or for Ω (which is hidden in Q, cf. (3.212)). However, numerical solutions are rather easily obtained.

We may now write down the response for the superharmonic case. To a first approximation it becomes, by (3.208), (3.211), (3.220) and (3.217):

$$u(t) = u_0 + O(\varepsilon) = A(T_1)e^{i\omega_0 T_0} + Qe^{i\Omega T_0} + cc + O(\varepsilon) = \cdots$$
$$= a\cos(3\Omega t - \psi) + \frac{q}{\omega_0^2 - \Omega^2} \cos \Omega t + O(\varepsilon). \tag{3.223}$$

For stationary oscillations the constant amplitude a and phase ψ is given by (3.221)–(3.222), whereas for nonstationary (transient) motions the time-varying amplitudes $a(t)$ and phases $\psi(t)$ are given by the modulation equation (3.219).

As appears from (3.223) the superharmonic response consists of oscillations at three times the excitation frequency, overlaying the linear response at frequency Ω. Note that the frequency 3Ω approximately equals ω_0 for this case.

Typical superharmonic frequency responses are shown in Fig. 3.21 for different values of excitation level q, and in Fig. 3.22 for different levels of damping β. The response curves were obtained by solving (3.221) numerically, with (3.217) and (3.212) substituted for σ and Q. The backbone curves are obtained by letting $Q = \beta = 0$ in (3.221, using (3.217) to substitute σ, and solving for a, which gives:

$$a = \omega_0 \sqrt{\frac{8}{3\gamma} \left(3\frac{\Omega}{\omega_0} - 1 \right)} \quad \text{(on backbone)}. \tag{3.224}$$

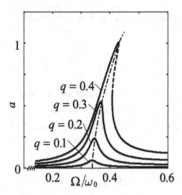

Fig. 3.21. Superharmonic frequency response $a(\Omega/\omega_0)$ for varying excitation levels q. ($\gamma = 0.5$, $\beta = 0.05$, $\omega_0 = 1$)

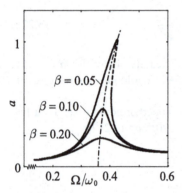

Fig. 3.22. Superharmonic frequency response $a(\Omega/\omega_0)$ for varying damping levels β. ($\gamma = 0.5$, $q = 0.4$, $\omega_0 = 1$)

As for the case of primary resonance, the response curves bend to the right when $\gamma > 0$. The response magnitude seems highly sensitive to the magnitude of excitation: with hard excitation the resonance peak is pronounced, whereas with non-hard excitations ($q = 0.1$, e.g.) there is virtually no superharmonic response.

Imagine an experimental frequency-sweep to be carried out for a system prone to superharmonic resonance. Increasing the frequency of excitation Ω, one will observe resonantly growing oscillations when $\Omega \approx \frac{1}{3}\omega_0$, that is, far below the region of linear resonance ($\Omega \approx \omega_0$). Further, the system will oscillate at $3\Omega \approx \omega_0$, that is, at a frequency far above that of the excitation.

Subharmonic Resonance If $\Omega \approx 3\omega_0$, then the term proportional to $e^{i(\Omega - 2\omega_0)T_0}$ in (3.213) will be near-resonant. To express this nearness we choose a detuning parameter σ through

$$\Omega = 3\omega_0 + \varepsilon\sigma, \tag{3.225}$$

whereby $\Omega T_0 = 3\omega_0 T_0 + \sigma T_1$. On substituting into (3.213), one finds that secular terms are eliminated if:

$$i2\omega_0(\beta\omega_0 A + A') + 3\gamma A(A\bar{A} + 2Q^2) + 3\gamma Q\bar{A}^2 e^{i\sigma T_1} = 0. \tag{3.226}$$

Letting $A = \frac{1}{2} ae^{i\varphi}$ where $a(T_1), \varphi(T_1) \in R$, and separating real and imaginary parts, the modulation equations for the case of subharmonic resonance is obtained:

$$a' = -\beta\omega_0 a - \frac{3\gamma Q}{4\omega_0} a^2 \sin\psi,$$

$$a\psi' = \left(\sigma - \frac{9\gamma Q^2}{\omega_0}\right)a - \frac{9\gamma}{8\omega_0}a^3 - \frac{9\gamma Q}{4\omega_0}a^2 \cos\psi, \tag{3.227}$$

where

$$\psi = \sigma T_1 - 3\varphi. \tag{3.228}$$

For locating stationary solutions we let $a' = \psi' = 0$ and obtain the subharmonic frequency response equation:

$$\left[(3\beta\omega_0^2)^2 + \left(\sigma\omega_0 - 9\gamma(Q^2 + \tfrac{1}{8}a^2)\right)^2\right]a^2 = \left(\tfrac{9}{4}\gamma Qa^2\right)^2. \tag{3.229}$$

As appears $a = 0$ is a solution. Solving for $a \neq 0$ it is found that

$$a^2 = C_1 \pm \sqrt{C_1^2 - C_2}, \tag{3.230}$$

where

$$C_1 \equiv \frac{8\omega_0\sigma}{9\gamma} - 6Q^2, \quad C_2 \equiv \left(\frac{8\omega_0}{9\gamma}\right)^2 \left[(3\beta\omega_0)^2 + \left(\sigma - \frac{9\gamma Q^2}{\omega_0}\right)^2\right]. \tag{3.231}$$

Since C_2 is always positive, nontrivial solutions $a \neq 0$ can occur only when $C_1 > 0$ and $C_1^2 - C_2 > 0$.

To a first approximation the solution $u(t)$ governing subharmonic response becomes, by (3.208), (3.211), (3.228) and (3.225):

$$u(t) = u_0 + O(\varepsilon^2) = A(T_1)e^{i\omega_0 T_0} + Qe^{i\Omega T_0} + cc + O(\varepsilon) = \cdots$$

$$= a\cos(\tfrac{1}{3}\Omega t - \tfrac{1}{3}\psi) + \frac{q}{\omega_0^2 - \Omega^2}\cos(\Omega t) + O(\varepsilon). \tag{3.232}$$

The subharmonic response is seen to consist of oscillations at a frequency three times lower than the frequency of excitation, overlaying the linear response at frequency Ω. For this case the frequency $\frac{1}{3}\Omega$ approximately equals the linear natural frequency ω_0. Thus, in experiments one will observe the system performing strong oscillations at its natural frequency, as in linear resonance, but at a frequency far above that of linear resonance.

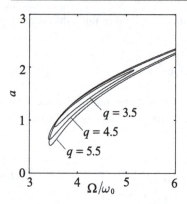

Fig. 3.23. Subharmonic frequency response for different levels of excitation q. ($\gamma = 0.5$, $\beta = 0.05$, $\omega_0 = 1$)

Fig. 3.23 shows typical frequency responses, as given by (3.230) with (3.225) and (3.212) substituted for σ and Q. As appears, the stationary amplitude can become quite large – in particular when compared to the very weak response predicted by linear theory (the q term of (3.232)).

Fig. 3.24 depicts regions of the (Ω, q) plane where subharmonic resonances can be excited. The boundary curves enclosing such regions are defined by the condition $C_1^2 - C_2 = 0$ and $C_1 > 0$, which by (3.231) expands to:

$$q^2 = \frac{16\omega_0^6}{63\gamma}\left(1 - \left(\frac{\Omega}{\omega_0}\right)^2\right)^2\left(\frac{\Omega}{\omega_0} - 3 \pm \sqrt{\left(\frac{\Omega}{\omega_0} - 3\right)^2 - 63\beta^2}\right) . \qquad (3.233)$$

As the nonlinearity γ is decreased, it appears from Fig. 3.24, a stronger loading is required for exciting superharmonic resonances.

For the case of subharmonic resonance, we conclude that large responses can be excited at driving frequencies much higher than the natural frequency of the system. Nayfeh and Mook (1979) refer to a case in which the propellers of a commercial airplane excited a subharmonic of order $\frac{1}{2}$ in the wings, which in turn excited a subharmonic of order $\frac{1}{4}$ in the rudder. The airplane broke up.

3.7.4 Obtaining Forced Responses by Averaging

For the above analysis of Duffing's equation we employed multiple scales expansions. To illustrate a possible alternative we here employ *averaging analysis* for obtaining the primary resonant response to this equation. This will also demonstrate how to apply averaging for *forced* systems, which was excluded from the brief presentation of the averaging technique given in Section 3.5.5.

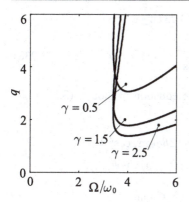

Fig. 3.24. Regions of the loading-plane where subharmonic resonances may appear. ($\beta = 0.05$, $\omega_0 = 1$)

Thus, we reconsider the system (3.152), subjected to near-resonant weak excitation, weak damping, and weak nonlinearity:

$$\ddot{u} + 2\varepsilon\beta\omega_0\dot{u} + \omega_0^2 u + \varepsilon\gamma u^3 = \varepsilon q \cos(\Omega t),$$
$$\Omega^2 - \omega_0^2 = O(\varepsilon), \quad \varepsilon \ll 1. \tag{3.234}$$

When $\gamma = 0$ the problem is linear, and the stationary solution has the form $u = a\sin(\Omega t + \varphi)$, where a and φ are constants. With the averaging method one assumes the solution of the weakly nonlinear problem to be similar in character to the linear one, though, with the amplitude a and phase φ allowed to vary in time. Thus we let $u(t) = a(t)\sin(\Omega t + \varphi(t))$, and write this in the form:

$$u = a(t) \sin \psi, \quad \psi(t) \equiv \Omega t + \varphi(t). \tag{3.235}$$

The unknown amplitudes $a(t)$ and $\varphi(t)$ are thereby considered dependent variables of the problem, and a variable shift is performed from the original dependent variable $u(t)$ to the new variables a and φ. In performing this transformation there are three unknowns (u, a, φ), but only two Eqs. ((3.234) and (3.235)). For a third equation we impose the arbitrary but convenient restriction that the velocity $\dot{u}$ should take a form similar to that of the linear case, that is:

$$\dot{u} = a\Omega \cos \psi, \tag{3.236}$$

which by (3.235) requires that $\dot{a}\sin\psi + a\,\dot{\varphi}\cos\psi = 0$, that is:

$$a\dot{\varphi} = -\dot{a} \tan \psi. \tag{3.237}$$

The acceleration $\ddot{u}$ becomes, by (3.236) and (3.237):

$$\ddot{u} = \dot{a}\Omega\cos\psi - a\Omega(\Omega + \dot{\varphi})\sin\psi = \frac{\dot{a}\Omega}{\cos\psi} - a\Omega^2\sin\psi, \tag{3.238}$$

Inserting (3.235), (3.236) and (3.238) into the equation of motion (3.234) one obtains an equation governing $a(t)$:

$$\dot{a}\Omega = \varepsilon\big[a(\Omega^2 - \omega_0^2)\sin\psi\cos\psi - 2\beta\omega_0 a\Omega\cos^2\psi \\ -\gamma a^3\sin^3\psi\cos\psi + q\cos(\psi - \varphi)\cos\psi\big]. \tag{3.239}$$

Inserting this into (3.237) an equation for $\varphi(t)$ is obtained:

$$\Omega a\dot{\varphi} = \varepsilon\big[-a(\Omega^2 - \omega_0^2)\sin^2\psi + 2\beta\omega_0 a\Omega\sin\psi\cos\psi \\ + \gamma a^3\sin^4\psi - q\cos(\psi - \varphi)\sin\psi\big]. \tag{3.240}$$

The pair of equations (3.239)–(3.240) just restate the equation of motion (3.234), which is a second-order ODE in u, in form of two first-order equations in a and ψ.

We now turn to the averaging approximations. As appears from (3.239)–(3.240) $\dot{a}$ and $\dot{\varphi}$ will be small, since $\varepsilon 1$, so that a and φ change much more slowly with t than does $\psi = \Omega t + \varphi$. Hence a and φ will hardly change during one period of the oscillating terms. As an approximation we then replace the equations for $\dot{a}$ and $\dot{\varphi}$ with their average values during one period of oscillation. This corresponds to assuming that, for the determination of the slow variations of a and φ, the rapid variations in ψ-terms can be neglected. The average of a particular term $f(a, \psi)$ is given by $\int_0^{2\pi} f(a, \psi)\,d\psi$, where a and ψ are treated as constants during the period of integration. Averaging the right-hand terms of (3.239)-(3.240), the following pair of *averaged equations of motion* is obtained (now omitting ε):

$$\dot{a}\Omega = -\beta\omega_0 a\Omega + \frac{1}{2}q\cos\varphi, \\ \Omega a\dot{\varphi} = -\frac{1}{2}a(\Omega^2 - \omega_0^2) + \frac{3}{8}\gamma a^3 - \frac{1}{2}q\sin\varphi. \tag{3.241}$$

The averaged equations of motion take the form of *modulation equations*, governing slow modulations in time of the amplitude and phase of a periodic oscillation.

For obtaining the approximate response one has to solve (3.241) for $a(t)$ and $\varphi(t)$, and then substitute into (3.235) for obtaining $u(t)$. This is not straightforward, since Eq. (3.241) are still nonlinear.

For obtaining the *stationary response*, one simply lets $\dot{a} = 0$ and $\dot{\varphi} = 0$ in (3.241), and solves the resulting pair of algebraic equations for the constant values of a and φ. This yields:

$$\left[(2\beta\omega_0\Omega)^2 + \left((\Omega^2 - \omega_0^2) - \frac{3}{4}\gamma a^2\right)^2\right]a^2 = q^2,$$

$$\tan\varphi = \frac{-(\Omega^2 - \omega_0^2) + \frac{3}{8}\gamma a^2}{2\beta\omega\Omega}, \tag{3.242}$$

where the first equation determines the frequency response, that is, the stationary values of oscillation amplitude a as a function of the frequency of excitation Ω.

The stability of stationary solutions is examined by evaluating Jacobian eigenvalues of the modulation equations (3.241), as illustrated in Section 3.7.2. for the method of multiple scales. Plotting then the frequency response $a(\Omega/\omega_0)$, figures similar to those obtained by using the method of multiple scales appear (*cf.* Sections 3.7.2, Figs. 3.17–3.20).

We can compare the frequency response equation (3.242), obtained by averaging analysis, with the similar equation (3.194) obtained by multiple scales perturbation analysis. Expanding the equations and comparing terms, one finds the two equations to differ only in terms describing the linear ($\gamma = 0$) part of the response. The differences vanish as $\Omega \to \omega_0$, but increase as $|\Omega - \omega|$ or β grow large. With both methods, however, we have assumed that Ω is close to ω_0 and that damping is weak. Hence, on the assumptions employed, the two methods yield similar results.

Alternatively to the transform in (3.235) one could use instead $u = a(t)\sin\psi$ where $\psi(t) = \omega_0 t + \varphi(t)$, i.e. with ω_0 instead of Ω as the constant part of the frequency parameter. This would give similar same approximate results, differing only a little when $\Omega \approx \omega_0$ and β is small, as assumed. It might seemed strange to use the free oscillation parameter ω_0 for a harmonically forced problem with known excitation frequency Ω. However, this is actually consistent with the generally workable approach for using the method of averaging, which is to use, as a basis for the variable transform, the solution for the unperturbed problem ($\varepsilon = 0$), and then letting the *constants* of the $\varepsilon = 0$ be time-dependent *variables* in the $\varepsilon \neq 0$ solution. In the present example the unperturbed problem corresponding to (3.234) is $\ddot{u} + \omega_0^2 u = 0$, with solution $u = a\sin(\omega_0 t + \varphi)$, with a and φ constants; thus $u = a(t)\sin\psi$ with $\psi(t) = \omega_0 t + \varphi(t)$ would be a suitable transform, that would produce modulation equations similar to (3.241), suitable to averaging.

For the above example the method of averaging seems more straightforward than the method of multiple scales. It mostly is, in particular when one has some prior knowledge of the kind of solution to expect, and when higher-order approximations are not in need. With averaging the special cases to consider do not readily and automatically reveal themselves during the analysis, as with the method of multiple scales. And to obtain a higher-order correction (such as the ε-term in (3.198)) one needs to employ a more involved version of the averaging technique (e.g., Mitropolsky 1965; Sanders and Verhulst 1985).

3.7.5 Multiple Scales Analysis with Strong Nonlinearity

Sometimes perturbation methods can be used even when nonlinearities are *not* weak, as has been otherwise assumed in this chapter so far. The quality of a perturbation-based prediction depend on the degree to which the underlying assumptions are fulfilled. In the previous sections, when calculating resonant frequency responses, an essential assumption was on the nearness to resonance. This was expressed as an assumed smallness of a detuning parameter $\sigma = \Omega - \omega_0$, with Ω being the excitation frequency and ω_0 the resonance frequency in question, e.g. a linear natural frequency. Even with the assumptions of weak damping and excitation fulfilled, an assumption based on small σ will gradually worsen with the increase of $|\sigma|$, i.e. as the deviation of Ω from the constant frequency ω_0 increases. With a stronger (stiffness-)nonlinearity the resonance peak bends more over at larger amplitudes, and even more so with smaller damping. This means that further up the resonance peak σ becomes too large for a prediction based on small σ to have acceptable accuracy. Thus, when comparing with numerical simulation, one will typically note a marked drop in accuracy at the upper parts of a strongly nonlinearly bent resonance peak.

But this can sometimes be cured by considering the detuning σ instead as a deviation from the *nonlinear* free oscillation frequency, i.e. the frequency on the nonlinear response backbone, which depends on oscillation amplitude a (but not on damping and excitation). Thus instead of $\sigma = \Omega - \omega_0$, with a constant ω_0, one can use $\sigma = \Omega - \omega(a)$, where ω is the nonlinear free and undamped oscillation frequency at oscillation amplitude a. So, if one can set up a reasonably simple expression for $\omega(a)$, a multiple scales analysis can be conducted that has good accuracy as long as the excitation frequency is close to the *backbone* of the forced frequency response. Considering a typical nonlinear frequency response, such as Fig. 3.17, you will note that actually σ will be *smaller* the further up the resonance peak you come, to become eventually zero at the very top of the peak (which intersects the backbone).

Burton and Rahman 1986 suggested and tested such an approach to derive approximate analytical expressions for the stationary frequency response of a Duffing oscillator with arbitrarily weak or strong nonlinearity, i.e. (3.152) with an arbitrary ratio of γ/ω_0^2 (including zero and infinity). To find $\omega(a)$ for this case amounts to find the nonlinear oscillation frequency for (3.152) when $\beta = q = 0$. This frequency can be expressed as a function of oscillation amplitude by a complete elliptic integral (Polyanin and Zaitsev 2003; Kovacic and Brennan 2011; Kovacic 2020). But one can also for $\omega(a)$ use a series solution in terms of elementary functions, as given by Burton and Hamdan (1983), and used in, e.g. Thomsen (2008b). In any case, multiple scales approximate solutions for the corresponding forced, damped frequency response can be obtained that agree excellently with numerical simulation, even for very high ratios γ/ω_0^2.

3.8 Two More Classical Nonlinear Oscillators

Next we consider the *van der Pol oscillator* and the *Rayleigh oscillator*, which are classical in the sense that they have a long history, and still serve to illustrate and understand essential phenomena occurring with nonlinear oscillators. Both are basically unforced and linearly damped single-DOF oscillators, with an additional cubically nonlinear damping term. For both oscillators the composite (linear + nonlinear) damping term is such that energy is fed into the system at lower oscillation amplitudes (i.e. the linearized damping is *negative*), but drained at higher amplitudes, so that oscillations will grow or shrink in amplitude until some balance of stationary oscillations is established. With no external forcing involved such oscillations are *self-excited* or self-excit*ing* (Nayfeh and Mook 1979; Tondl 1991). They play a significant role in numerous technical and natural contexts, such as with friction-generated noise and vibrations, flow-induced vibrations, biological clocks (e.g. animal heart beats and brainwaves), electrical circuits, and economical/sociological processes.

A self-excited oscillator is *not* driven primarily by an explicitly time-dependent forcing term, but rather by some energy source internal to the system. An example is friction-induced vibrations, where the typically negative slope of the velocity versus friction force characteristic near zero velocity corresponds to *negative damping*; thus energy is fed *into* the system, rather than being drained out (*cf.* Sects. 3.9.5–3.9.6). This causes oscillations to grow, as is commonly experienced with, e.g., squeaking car and bike brakes, door hinges and train tracks, and with bowed musical instruments. However, with growing oscillation amplitudes other phenomena will always start to outbalance the energy input, e.g. with sliding friction surfaces the negative slope in the friction versus velocity curve will turn positive for higher velocities, corresponding to 'normal' positive damping.

Another example of self-excitation is *flow-induced vibrations*, such as with fluttering aircraft wings, singing air cables, wind musical instruments, and human or animal voicing and whistling: Explicit time-dependent input forces are absent, while instead some physical mechanism allows turning energy of the fluid/air/gas flow into mechanical vibrations, effectively corresponding to negative damping which will make oscillations grow. Again, with any real physical system this growth is sooner or later be limited by other forces, whose significance increase with oscillation amplitude, and the oscillations may stabilize at a certain level (or turn chaotic, *cf.* Chap. 6, or the device may eventually break or disintegrate).

3.8.1 The Van Der Pol Oscillator

A nondimensional form of the *van der Pol oscillator* for the variable $u = u(t)$ is:

$$\ddot{u} - \varepsilon(1 - u^2)\dot{u} + u = 0, \tag{3.243}$$

which is the *van der Pol equation*, named after the Dutch electrical engineer and physicist Balthasar van der Pol (1889–1959). It is an autonomous ordinary differential equation, characterized by its nonlinear damping term, with a coefficient ε which is not necessarily small. The equation arose within vacuum tube electrical engineering, but serves to illustrate, or directly model, the essence of many phenomena also in mechanical engineering, those characterized by self-excitation.

The case of practical interest is $\varepsilon > 0$, where the zero solution is unstable, while there is a stable limit cycle attracting solutions from every initial condition. To see this consider starting the system with zero velocity near $u = 0$. The term u^2 in (3.243) is then small and insignificant compared to unity, so the system is effectively $\ddot{u} - \varepsilon\dot{u} + u = 0$. This is a linear system with *negative* damping, meaning that initially the oscillations will be sinusoidal with unit frequency, and exponentially growing amplitude. But then u^2 will grow so large, that it cannot be ignored compared to unity, and eventually the term $-(1 - u^2)$ in (3.243) will change sign, meaning that the damping turns *positive*. This will reduce the oscillation amplitude, until the term again changes sign. After a while at state of stationary oscillations is attained, with $u(t)$ oscillating with constant amplitude and frequency components.

For *small* ε we can calculate approximate solutions to (3.243) using perturbation analysis. Using the averaging method (Sect. 3.5.5) we first employ a transformation of variables based on the solution to the unperturbed ($\varepsilon = 0$) problem, i.e. $\ddot{u} + u = 0$, with solution $u(t) = a_0\sin(t + \varphi_0)$, where a_0 and φ_0 are constants. Considering these constants for the $\varepsilon = 0$ problem as new time-dependent *variables* for the $\varepsilon \neq 0$ problem, the transform to use is:

$$u(t) = a\sin\psi; \quad \dot{u}(t) = a\cos\psi, \quad \psi(t) = t + \varphi(t). \qquad (3.244)$$

Calculating $\ddot{u}$ from this, and inserting into (3.243), one arrives, following the same procedure as in Sect. 3.5.5, at a pair of first-order differential equations for a and φ:

$$\dot{a} = \varepsilon\left(a\cos^2\psi - a^3\sin^2\psi\cos^2\psi\right),$$
$$a\dot{\varphi} = \varepsilon\left(-a\sin\psi\cos\psi + a^3\sin^3\psi\cos\psi\right). \qquad (3.245)$$

For small ε the evolution in a and φ is slow in time t, so that the right-hand sides can be averaged over a period 2π of the rapidly oscillating terms $\sin\psi$ and $\cos\psi$, giving (use e.g. the list of averaging integrals in App. C.4):

$$\dot{a} = \varepsilon\frac{1}{2}a\left(1 - \frac{1}{4}a^2\right),$$
$$a\dot{\varphi} = 0. \qquad (3.246)$$

The second equation admits the trivial solution $a = 0$, which also solves the first equation; thus $a = 0$ is a solution. The linearization for $\varepsilon 1$ of the first equation is $\dot{a} = \frac{1}{2}\varepsilon a$, from which readily appears that $a = 0$ is unstable for any positive value of ε (one could also check stability using Jacobian eigenvalues as in Sect. 3.4.4).

When $a \neq 0$ the second Eq. in (3.246) gives $\dot{\varphi} = 0$, with a constant solution $\varphi(t) = \varphi_0$, while letting $\dot{a} = 0$ in the first equation gives a nontrivial stationary solution $a = a_0 = 2$. Substituting this into (3.244), the approximate solution to (3.243) for small ε then becomes:

$$u(t) = 2\sin(t + \varphi_0); \quad \dot{u}(t) = 2\cos(t + \varphi_0), \quad \varepsilon \ll 1, \tag{3.247}$$

which is seen to be independent on ε (but still assumes small ε), and is stable for all positive values of ε. It forms a limit cycle in the $(u, \dot{u})$ phase plane, to which orbits starting from any nonzero initial condition will be attracted, the quicker the higher the value of ε. It appears from (3.247) that $u^2 + \dot{u}^2 = 2^2$, meaning that in the $(u, \dot{u})$ phase plane the limit cycle forms a circle of radius 2.

Fig. 3.25 shows solutions to the van der Pol equation, as obtained by direct numerical simulation of (3.243). The upper left figure depicts a post-transient time series for a small value of $\varepsilon = 0.1$. It has a time-harmonic/sinusoidal shape with amplitude 2, while the corresponding stationary phase plane orbits in the upper right figure approaches the close-to-circular limit cycle of radius 2, i.e. as predicted by the approximate perturbation solution (3.247) for small ε.

The two lower figures in Fig. 3.25 shows numerical solutions when $\varepsilon = 1$, i.e. *not* small. With this much stronger level of damping nonlinearity the time series appears with clear *harmonic distortion*, i.e. not as a pure sinusoidal; a Fourier transform of the time series would show non-zero components not only at the fundamental linear oscillation frequency $\omega = 1$, but as well at the higher harmonics, $\omega = 3, 5, \ldots$ (odd harmonics only, due to the u-symmetry of (3.243), which is invariant to the transformation $u \rightarrow -u$). The effect of nonlinearity is even more visible in the phase plane plot (lower right figure), where the stable limit cycle (thick line) is clearly non-circular. The nonlinear distortion, increasing with increasing ε, is completely absent in the approximate perturbation solution (3.247). Nevertheless, the approximate predictions, with an oscillation amplitude of 2, are at least roughly agreeing with numerical simulation even for the evidently non-small value of $\varepsilon = 1$.

For even larger values of the damping parameter, $\varepsilon \gg 1$, the oscillations change shape towards resembling a train of rectangular pulses (not shown), that is: For most of the oscillation period u changes very slowly, while the change from positive to negative values occurs almost instantaneously; this is called *relaxation oscillations*.

3.8.2 The Rayleigh Oscillator

A nondimensional form of the *Rayleigh oscillator* equation $u = u(t)$ is:

$$\ddot{u} - \varepsilon(1 - \frac{1}{3}\dot{u}^2)\dot{u} + u = 0, \tag{3.248}$$

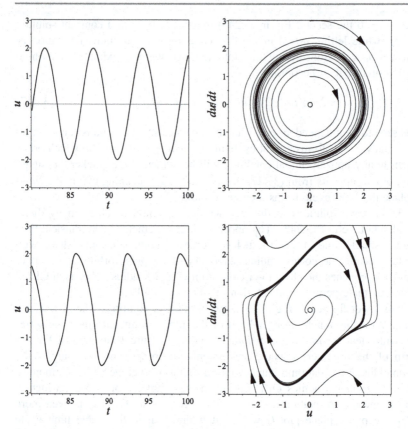

Fig. 3.25. Post-transient time traces (left) and phase plane orbits (right) obtained by numerical simulation of the van der Pol oscillator Eq. (3.243) for $\varepsilon = 0.1$ (top) and $\varepsilon = 1$ (bottom). In the phase planes thick lines are stable limit cycles, and thin lines orbits approaching the stable limit cycle from different initial conditions

which differs from the van der Pol Eq. (3.243) only by the nonlinear coefficient to $\dot{u}$ being proportional to $\dot{u}^2$ rather than to u^2. That is, rather than changing sign at a certain level of $|u|$, the damping changes sign at a certain level of $|\dot{u}|$. This makes no difference to the qualitative behavior of the oscillators, who both exhibit the characteristic of an unstable trivial solution and a single stable limit cycle. Also, the time series and phase plane plot for the Rayleigh oscillator qualitatively resembles those for the van der Pol oscillator in Fig. 3.25. The two oscillators can be used to model the same kind of phenomena, though one of them may fit better than the other for a given task (e.g., Veskos and Demiris 2006).

The equivalence between the Rayleigh and the van der Pol oscillator equation is so close that the one can be obtained from the other by a simple transformation of variables. To see this differentiate the Rayleigh oscillator Eq. (3.248) once with respect to time, and substitute $\dot{u} = x$, with x as a new independent variable; then the resulting equation for x is identical to the van der Pol Eq. (3.243), just with

x instead of u. Hence, for a given ε, solutions to the Rayleigh oscillator equation can be obtained simply by integrating in time the solutions for the van der Pol equation. This includes the approximate solution for small ε (3.247), which for the Rayleigh oscillator equation becomes:

$$u(t) = -2\cos(t + \varphi_0); \quad \dot{u}(t) = 2\sin(t + \varphi_0), \quad \varepsilon \ll 1, \tag{3.249}$$

which just corresponds to a phase shift of $\pi/2$ compared to the van der Pol equation.

3.9 Vibro-Impact Analysis Using Discontinuous Transformations

Here we show how near-elastic vibro-impact problems can be conveniently analyzed by a discontinuity-reducing transformation of variables, combined with an extended averaging procedure; the presentation is based mainly on Fidlin (2006) and Thomsen and Fidlin (2008).

Vibro-impact systems are characterized by repeated impacts. Applications include devices to crush, grind, forge, drill, punch, tamp, pile, and cut a variety of objects, and vibrating machinery or structures with stops or clearances (Babitsky 1998). Also, vibro-impact is involved in noise- and wear-producing processes, as with rattling gear boxes and heat exchanger tubes. The analysis of such systems is often difficult, mainly due to the inherent presence of strong nonlinearity: Even if a vibro-impact system can be considered linear or weakly nonlinear in-between impacts, the impacts correspond to a force–displacement relation which cannot be linearized at the point of impact contact, or has a dominating nonlinear component. Using kinematic impact formulations, this nonlinear effect is taken into account though the prescription of a discontinuous change in relative velocity at impact times. With near-elastic impacts, these velocity discontinuities are "large", i.e. of the order of magnitude of the impact velocities themselves. Such problems are not natural candidates for approximate perturbation analysis. But sometimes they can be transformed to be so.

In this section we present a general procedure and some application examples, showing how near-elastic vibro-impact problems, linear or nonlinear in-between impacts, can be conveniently analyzed by a discontinuity-reducing transformation of variables, combined with an extended averaging procedure that allows presence of the resulting small discontinuities of the transformed system. Only first-order analytical predictions are derived, while the more elaborate extension to second order is described in Thomsen and Fidlin 2008. We consider only *near-elastic* vibro-impact, i.e. with a coefficient of restitution close to unity, though averaging may be applied also for the inelastic case (Fidlin 2006). A recent application example of the technique is provided in Rebouças et al. (2019), using it to calculate the vibro-impact response for a cantilever beam with a one-sided stop, and comparing with laboratory experiments.

The classical approach for analyzing vibro-impact problems is *stitching* (Kobrinskii 1969), where the equations of motions are integrated in-between impacts, kinematic impact conditions are used to switch between time intervals of solution, and solutions over times involving several impacts are obtained by 'gluing' together a suitable number of such partial solutions. Variants of this involve setting up and analyzing *discrete maps* between successive impacts (Guckenheimer and Holmes 1983). For numerical simulation, stitching and its variants can be simple and effective, with appropriate numerical algorithms (e.g., Wiercigroch 2000). However, for obtaining purely analytical solutions stitching is elaborate, and any nonlinearity in-between impacts makes it difficult to apply. Furthermore, for typical applications it is not necessary to obtain solutions at the level of detail provided by exact methods. Of more interest may be condensed measures such as oscillation frequencies, stationary amplitudes, and the stability of stationary motions.

Thus, for vibro-impact problems approximate analytical methods are necessary and useful. Among these are the methods of *harmonic linearization* (Babitsky 1998), *averaging* (Fidlin 2006), and *direct separation of motions* (Blekhman 2000), each with their particular strengths. In particular the method of harmonic linearization, as introduced by Babitsky (1998) in the 1960s, has been used successfully to solve many vibro-impact problems (e.g. Rebouças et al. 2017). This approach is convenient in use, but mathematically not well supported; results need careful validation by e.g. numerical simulation, in particular for systems operating away from resonance.

The approach presented in this section – discontinuous transformation combined with extended averaging – is founded on original ideas by in particular Zhuravlev (1976), Pilipchuk (1988), and Ivanov (1997). In Thomsen and Fidlin (2008) the approach is mathematically supported by a theorem similar to the standard averaging theorem (Sanders and Verhulst 1985), thus providing estimates of the accuracy of approximation, and a systematic procedure for increasing the accuracy to any desired level. By contrast to harmonic linearization, it assumes a *kinematic* rather than a kinetic impact formulation; that is: the impact process is described simply by a coefficient of restitution relating relative velocities just before and after impact, while the details of impact process itself is not involved. Compared to classical or semi-analytical stitching, it provides analytical solutions valid at all times, i.e. free of switching conditions, and also works for systems that are nonlinear in-between impacts; the latter nonlinearities can be weak or strong or even essential (as e.g. u^3 or $\text{sgn}(u)$), if just the solution of the unperturbed system is known. Compared to the averaging approach described by Ivanov (1997), the discontinuous transformations used with the present approach need not to eliminate the impact discontinuities completely. This is a considerable advantage, since setting up (and physically interpreting) transformations that eliminate discontinuities completely is far from trivial and requires quite some ingenuity. Still, there is no general rule for suggesting workable transformations, but the examples provided may work directly or serve as inspiration for other cases.

The purpose of the presentation is to demonstrate the practical applicability of a method for analyzing vibro-impact problems. The mathematical models used are chosen for their simplicity; we are not concerned with the accuracy of these models for describing physical reality, but only with the accuracy of the method we suggest for analyzing them. No attempt is given to cover abundant literature on the modeling of particular vibro-impact systems and the study of their dynamical behavior.

Sect. 3.9.1 illustrates in a simple setting the basic idea of employing an unfolding transformation to eliminate discontinuities for purely elastic vibro-impact systems. With inelastic impacts such transformations will not eliminate the discontinuities, but for near-elastic impacts they will be reduced to a value which is small as compared to the impact velocities. This motivates the following Sect. 3.9.2, which shows how to apply asymptotic first-order averaging for general systems of ordinary differential equations containing small discontinuities. Finally Sects. 3.9.3–3.9.6 presents four application examples, where discontinuous unfolding transformation and averaging are combined, resulting in approximate analytical expressions for key properties such as oscillation amplitudes and frequencies. Problem 3.23 exercises the key techniques in a simple setting.

3.9.1 The Unfolding Discontinuous Transformation: Basic Idea

We begin by considering a simple vibro-impact system, just to show how an unfolding transformation can be used to eliminate impact discontinuities. Assuming purely elastic impacts (coefficient of restitution $R = 1$), the motions of the harmonic oscillator in Fig. 3.26(a), bouncing against a rigid stop at $s = 0$, are governed by:

$$\ddot{s} + s = 0 \quad \text{for} \quad s > 0,$$
$$s_+ = s_-, \dot{s}_+ = -\dot{s}_- \quad \text{for} \quad s = 0, \tag{3.250}$$

where overdots denote differentiation with respect to time t, and subscripts plus and minus indicate states immediately before and after impact, respectively. The general solution is:

$$s(t) = A|\sin(t + \theta)|, \tag{3.251}$$

which is shown in Fig. 3.26(b) (solid line) for initial conditions corresponding to $(A, \theta) = (1, 0)$; it is nothing but a folded or full-wave rectified sine function. But this in turn means that the discontinuity in (3.250) can be removed by a discontinuous transformation defined by:

$$s = |z|; \quad z_- z_+ < 0, \tag{3.252}$$

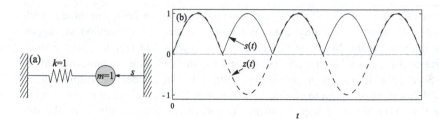

Fig. 3.26. (a) Harmonic oscillator with a stop at $s = 0$. (b) A solution $s(t)$ to (3.250) (in solid line), and its unfolding $z(t)$ (dashed) as given by the transformation (3.252)

where the inequality condition implies that the new dependent variable $z(t)$ changes sign at every impact; this removes the non-uniqueness of the transformation due to $|z|$ in the first equation. Inserting (3.252) into (3.250), the system transforms into:

$$\ddot{z} + z = 0, \quad \text{for} \quad z \neq 0,$$
$$z_+ = z_-, \dot{z}_+ = \dot{z}_- \quad \text{for} \quad z = 0, \tag{3.253}$$

where the second line is given only to emphasize that the transformed system is now continuous at $z = 0$. This also appears from the graph of $z(t)$ in Fig. 3.26(b) (dashed line), where the 'unfolding' or mirroring of every other oscillation of $s(t)$ due to (3.252) is seen to eliminate the discontinuity at zero. Hence the transformation (3.252) turns (3.250) into a simple linear oscillator equation, with solution $z = A\sin(t + \theta)$, corresponding to $s = |z| = A|\sin(t + \theta)|$.

Supposing we did not already know the exact solution (3.251), it could be easily derived by first using the discontinuous transformation (3.252), then solve the transformed system for z, and finally back-transform to obtain s. Even if impacts were not purely elastic, and others sources of energy dissipation or energy input or nonlinearities were present, a transformation similar to (3.252) might possibly eliminate or reduce the discontinuity of the original system, as will be illustrated with the examples in Sects. 3.9.3–3.9.6.

With many applications a system so transformed will be weakly nonlinear, and with velocity discontinuities that are small (as compared to impact velocities). Weak nonlinearities can be handled by perturbation methods such as averaging, but even small discontinuities implies obstacles for this. Next we describe how to employ averaging for systems with small discontinuities.

3.9.2 Averaging for Vibro-Impact Systems: General Procedure

We summarize a technique, presented in Zhuravlev and Klimov (1988) and developed in Fidlin (2006), for averaging differential equations with small discontinuities, recalling that such differential equations may appear after a discontinuity-reducing transformation for a near-elastic vibro-impact system.

Common perturbation methods such as averaging, multiple scales, harmonic linearization, and direct separation of motions all have difficulties in handling discontinuities such as occur with vibro-impact systems. Typically the coordinates of the impacting objects remain continuous, while the discontinuous change of velocities during impact may be of the order of magnitude of the velocities themselves. An effective approach, which enables the application of averaging to systems with impacts, works by eliminating or reducing the discontinuity from the equations by employing a variable transformation that contains or unfolds the essential discontinuity, as was illustrated for a simple case in Sect. 3.9.1. This idea was suggested by Zhuravlev (1976), and developed for different applications in e.g. Ivanov (1997), Pilipchuk (1988, 2002), Fidlin (1991, 2004b), Dimentberg and Iourtchenko (2004). The original objective was to eliminate the discontinuity completely and to apply standard averaging. This is quite successful for perfectly elastic impacts, but in the case of even small energy dissipation during impacts, it becomes difficult to find suitable transformations that are reasonably simple and with a clear physical interpretation. This seems to impede general use of discontinuity-removing transformations.

However, it is actually not necessary to eliminate the discontinuities completely, but only to reduce them to a sufficiently small level, comparable to other sources of energy dissipation, and to generalize the averaging technique for that case. This combination – of a discontinuity-reducing transformation and extended averaging – leads to an efficient approach for asymptotic analysis of impacting oscillators, as will be described below and exemplified in the following sections.

Standard Averaging – for Smooth Systems Standard averaging applies to systems of the form:

$$\frac{d\mathbf{x}}{d\varphi} = \varepsilon \mathbf{f}(\varphi, \mathbf{x}), \tag{3.254}$$

where $\mathbf{x}(\varphi) \in D \subset R^n$, φ is the independent variable (not necessarily time or even related to time), and $\varepsilon \ll 1$ is a parameter indicating the smallness of the right-hand side. According to *the averaging theorem* (Sanders and Verhulst 1985; Fidlin 2006), if $\mathbf{f}$ is 2π-periodic (but not necessarily continuous) in φ and bounded and Lipschitz-continuous (explained below) in $\mathbf{x}$ on D, then on the scale $\varphi \leq O(1/\varepsilon)$, $\mathbf{x}$ is asymptotically close to the solution $\mathbf{x}_1$ of the averaged system:

$$\frac{d\mathbf{x}_1}{d\varphi} = \varepsilon \langle \mathbf{f}(\varphi, \mathbf{x}_1) \rangle, \tag{3.255}$$

where $\langle\rangle$ denotes averaging over φ during a single period of $\mathbf{f}$:

$$\langle g(\varphi)\rangle_\varphi = \frac{1}{2\pi} \int_0^{2\pi} g(\varphi)d\varphi. \tag{3.256}$$

For application of the method it is important to recall that a function $\mathbf{f}(\mathbf{x}, \varphi)$, $\mathbf{x}(\varphi)$ $\in D \subset R^n$, is bounded and *Lipschitz-continuous* in $\mathbf{x}$ on D if, and only if, non-negative constants $M_{\mathbf{f}}$ and $\lambda_{\mathbf{f}}$ exist such that, for all $(\mathbf{x}_1, \mathbf{x}_2) \in D$:

$$\|\mathbf{f}(\mathbf{x}, \varphi)\| \leq M_{\mathbf{f}} \text{ and } \|\mathbf{f}(\mathbf{x}_2, \varphi) - \mathbf{f}(\mathbf{x}_1, \varphi)\| \leq \lambda_{\mathbf{x}}\|\mathbf{x}_2 - \mathbf{x}_1\|. \tag{3.257}$$

In other words, a Lipschitz-continuous function $\mathbf{f}(\mathbf{x}, \varphi)$ is limited in how sudden it can change with $\mathbf{x}$: The slope of a line (in R^n) joining any two points on the graph (in R^n) of the function will never exceed its Lipschitz constant λ.

Extended First-order Averaging – for Discontinuous Systems For vibro-impact problems formulated with kinematic impact conditions, the inherent discontinuities in velocity preclude using standard averaging. However, a special form of the averaging theorem can be proved (Zhuravlev and Klimov 1988), that holds for systems with *small* (i.e. $O(\varepsilon)$) discontinuities in the state variables, that is, instead of (3.254):

$$\frac{d\mathbf{x}}{d\varphi} = \varepsilon\mathbf{f}(\mathbf{x}, \varphi) \quad \text{for} \quad \varphi \neq j\pi, j = 0, 1, \ldots, \mathbf{x}(\varphi) \in D \subset R^n, \varepsilon \ll 1,$$

$$\mathbf{x}_+ - \mathbf{x}_- = \varepsilon\mathbf{g}(\mathbf{x}_-) \quad \text{for} \quad \varphi = j\pi, \tag{3.258}$$

where $\mathbf{f}$ is 2π-periodic (not necessarily continuous) in φ, $\mathbf{f}$ and $\mathbf{g}$ are bounded and Lipschitz-continuous in $\mathbf{x}$ on $\mathbf{x} \in D \subset R^n$, and $\mathbf{x}_-$ and $\mathbf{x}_+$ are the states $\mathbf{x}$ immediately before and after the j'th impact, corresponding to the passage of φ through the value $j\pi$. For that case the averaged system becomes, instead of (3.255) (Zhuravlev 1988; Fidlin 2005, 2006):

$$\frac{d\mathbf{x}_1}{d\varphi} = \varepsilon\left(\langle\mathbf{f}(\mathbf{x}_1, \varphi)\rangle_\varphi + \pi^{-1}\mathbf{g}(\mathbf{x}_1)\right), \tag{3.259}$$

where again the subscript 1 indicates the first level of approximation to $\mathbf{x}$, i.e. $\mathbf{x} = \mathbf{x}_1 + \varepsilon\mathbf{x}_2 + O(\varepsilon^2)$, so that the error $\|\mathbf{x}_1(t) - \mathbf{x}(t)\|$ is $O(\varepsilon)$ on the time-scale $1/\varepsilon$. The averaged motions $\mathbf{x}_1$ may themselves be determined at different levels of accuracy, i.e. $\mathbf{x}_1 = \mathbf{x}_{11} + \varepsilon\mathbf{x}_{12} + O(\varepsilon^2)$, of which (3.259) gives the first order approximation $\mathbf{x}_{11}$, while the second-order approximation(s) $\mathbf{x}_{12}$ (and $\mathbf{x}_2$) are derived in Thomsen and Fidlin 2008.

To transform vibro-impact problems into the form (3.258) can be a challenge, in particular since for near-elastic vibro-impact the discontinuity in velocity is not small. At least two transformations are then required: One for transforming large discontinuities into small ones, and another one for transforming the impact-free part of the equations of motion into the form of the first equation in (3.258). Below we illustrate how this can be accomplished with four specific examples of vibro-impact systems.

3.9.3   Example 1: Damped Harmonic Oscillator Impacting a Stop

Consider the system in Fig. 3.27, which is an extension of the harmonic oscillator with a stop in Section 3.9.1. Here we assume inelastic impacts with a coefficient of restitution R slightly less than unity, linear viscous damping in-between impacts with small coefficient β, and a stop situated a small distance Δ away from the equilibrium (without stop) of the spring at $s = 0$. The motions are governed by:

$$\ddot{s} + 2\beta\dot{s} + s = 0 \quad \text{for } s > -\Delta,$$
$$s_+ = s_-, \dot{s}_+ = -R\dot{s}_- \quad \text{for } s = -\Delta, \tag{3.260}$$

with $s = s(t)$, given initial conditions $s(0)$ and $\dot{s}(0)$, and parameter ranges:

$$0 < \beta \ll 1, \quad 0 < (1 - R) \ll 1, \quad |\Delta| \ll 1. \tag{3.261}$$

The unfolding transformation (3.252) can be used also here, slightly modified:

$$s = |z| - \Delta, \quad z_- z_+ < 0. \tag{3.262}$$

Inserting this into (3.260) gives a transformed system in the new variable z:

$$\ddot{z} + 2\beta\dot{z} + z = \Delta \text{sgn} z \quad \text{for} \quad z \neq 0,$$
$$z_+ - z_- = 0; \quad \dot{z}_+ - \dot{z}_- = -(1 - R)\dot{z}_- \quad \text{for} \quad z = 0. \tag{3.263}$$

The main difference between (3.260) and (3.263) is, that in (3.263) the discontinuous jump in velocity at impact ($s = -\Delta$ or $z = 0$) has been reduced to a small value (proportional to $1 - R$, i.e. small when $R \approx 1$), while a small nonlinearity

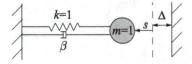

Fig. 3.27. Harmonic oscillator with near-elastic impacts, viscous damping, and a stop offset a distance Δ from the equilibrium $s = 0$ of the linear spring

(proportional to Δ) has emerged in the equation of motion valid in-between impacts ($z \neq 0$). Though (3.263 is still nonlinear, it is only weakly so, which means that perturbation methods can be applied to calculate approximate solutions as follows:

Considering terms in (3.263) having small coefficients as perturbations, the unperturbed system corresponding to (3.263) is a linear undamped and unforced harmonic oscillator. This means that the method of averaging can be used, after having applied a standard van der Pol transformation:

$$z = A \sin \psi, \quad \dot{z} = A \cos \psi, \quad \psi = t + \theta, \tag{3.264}$$

from which directly follows:

$$A \dot{\theta} = -\dot{A} \tan \psi, \quad \dot{\psi} = 1 + \dot{\theta}, \quad \ddot{z} = \dot{A} / \cos \psi - A \sin \psi, \tag{3.265}$$

where $A = A(t) > 0$ and $\theta = \theta(t)$ denote the slowly changing amplitude and phase, respectively. Inserting (3.264) and (3.265) into (3.263) gives the following system in the new dependent variables A and θ:

$$
\left.
\begin{aligned}
\dot{A} &= -2\beta A \cos^2 \psi + \Delta \cos \psi \operatorname{sgn}(\sin \psi) \\
\dot{\theta} &= 2\beta \sin \psi \cos \psi - \frac{\Delta}{A} |\sin \psi|
\end{aligned}
\right\} \quad \text{for } \psi \neq j\pi, \quad j = 0, 1, \ldots,
$$

$$
\left.
\begin{aligned}
A_+ - A_- &= -(1 - R)A_- \\
\theta_+ - \theta_- &= 0
\end{aligned}
\right\} \quad \text{for } \psi = j\pi,
\tag{3.266}
$$

where impacts occur at $\psi = j\pi$. With small parameters as assumed in (3.261), and assuming also $A \gg |\Delta|$, we note from (3.266) that $\dot{A}$ and $\dot{\theta}$ are $O(\Delta, \beta)$, so that:

$$
\begin{aligned}
\frac{dA}{d\psi} &= \frac{\dot{A}}{\dot{\psi}} = \frac{\dot{A}}{1 + \dot{\theta}} = \dot{A}\left(1 + O(\dot{\theta})\right) = \dot{A} + O(\dot{A})O(\dot{\theta}) = \dot{A} + O\left((\Delta, \beta)^2\right), \\
\frac{d\theta}{d\psi} &= \frac{\dot{\theta}}{\dot{\psi}} = \frac{\dot{\theta}}{1 + \dot{\theta}} = \dot{\theta}\left(1 + O(\dot{\theta})\right) = \dot{\theta} + O(\dot{\theta})O(\dot{\theta}) = \dot{\theta} + O\left((\Delta, \beta)^2\right).
\end{aligned}
\tag{3.267}
$$

Thus, to first order of accuracy in the small parameters, we can replace $\dot{A}$ and $\dot{\theta}$ with $dA/d\psi$ and $d\theta/d\psi$, respectively. Then (3.266) has the general form (3.258), with φ, $\mathbf{x}$, $\mathbf{g}$, and $\mathbf{f}$ defined by, respectively:

$$
\varphi = \psi; \quad \mathbf{x}(\varphi) = \left\{ \begin{aligned} A(\psi) \\ \theta(\psi) \end{aligned} \right\}; \quad \varepsilon\mathbf{g}(\mathbf{x}) = \left\{ \begin{aligned} -(1 - R)A \\ 0 \end{aligned} \right\};
$$

$$
\varepsilon\mathbf{f}(\mathbf{x}, \varphi) = \left\{ \begin{aligned} -2\beta A \cos^2 \psi + \Delta \cos \psi \operatorname{sgn}(\sin \psi) \\ 2\beta \sin \psi \cos \psi - (\Delta/A)|\sin \psi| \end{aligned} \right\}.
\tag{3.268}
$$

The condition under (3.258), that $\mathbf{f}$ and $\mathbf{g}$ should be bounded and Lipschitz-continuous in $(A, \theta) \in D \subset R^2$, is seen to be fulfilled when Δ/A is bounded and Lipschitz-continuous in A; this holds true under the already stated assumption A $|\Delta|$.

With the conditions fulfilled, application of (3.259) with (3.256) to (3.266) gives the averaged system

$$
\begin{aligned}
\dot{A}_1 &= -\tilde{\beta} A_1, \\
\dot{\theta}_1 &= -\frac{2\Delta}{\pi A_1}.
\end{aligned}
\tag{3.269}
$$

where

$$
\tilde{\beta} = \beta + \frac{1-R}{\pi},
\tag{3.270}
$$

and where also here the subscript 1 indicates a first-order approximate solution, asymptotically valid under assumptions (3.261). The effective damping constant $\tilde{\beta}$ is seen to integrate the two dissipative effects present: inelastic restitution during impacts (parameter R), and viscous damping in-between impacts (parameter β). The linear system (3.269) can readily be integrated to give:

$$
\begin{aligned}
A_1 &= C_1 e^{-\tilde{\beta} t}, C_1 > 0, \\
\theta_1 &= -\frac{2\Delta}{\pi C_1 \tilde{\beta}} e^{\tilde{\beta} t} + C_2,
\end{aligned}
\tag{3.271}
$$

where the constants C_1 and C_2 are determined by the initial conditions. Back-substituting this into (3.264) and (3.262) then gives the approximate solution s_1 for the original variable s:

$$
\begin{aligned}
s_1 &= C_1 e^{-\tilde{\beta} t} \left| \sin\left(t - \frac{2\Delta}{\pi C_1 \tilde{\beta}} e^{\tilde{\beta} t} + C_2 \right) \right| - \Delta, 0 < t < t_*, \\
t_* &= O\left(\min\left\{ \tilde{\beta}^{-1}, |\Delta|^{-1}, \tilde{\beta}^{-1} \ln\left(C_1 |\Delta|^{-1} \right) \right\} \right),
\end{aligned}
\tag{3.272}
$$

where the time-horizon t_* ensures the error in s_1 to be of the same magnitude order as the small parameters (*cf.* the error estimate under (3.259)), and that $A_1 > |\Delta|$ (to ensure boundedness and Lipschitz continuity, *cf.* the remark under (3.268)).

Fig. 3.28 compares this approximate solution with results of numerical simulation of the original equation of motion (3.260) (using the MATLAB-function *ode23*, with impacts handled by the *event function* feature), for parameters as indicated in the figure legend. The equilibrium mass-to-stop distance Δ is positive

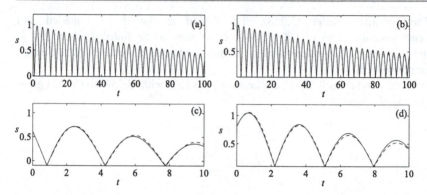

Fig. 3.28. Impact oscillator position $s(t)$, $t \in [0; t_*]$, as obtained by the approximate analytical prediction (3.272) (in solid line), and by numerical simulation of the original equation of motion (3.260) (dashed line), for initial conditions corresponding to $(C_1, C_2) = (1, 0)$ Parameter values $(\Delta, 1 - R, \beta)$: In (a): $(\varepsilon, \varepsilon, \varepsilon)$, in (b) $(-\varepsilon, \varepsilon, \varepsilon)$, in (c): $(10\varepsilon, 10\varepsilon, 10\varepsilon)$, in (d): $(-10\varepsilon, 10\varepsilon, 10\varepsilon)$, where $\varepsilon = 0.01$

for the left figures (a,c) (corresponding to a unilateral clearance), and negative for the right figures (b,d) (i.e. the mass is pre-compressed against the stop). For the upper figures (a,b) the small parameters are of the order $\varepsilon = 0.01$, and there is no discernible difference between approximate and simulated response. For the lower figures (c,d) the small parameters are 10 times larger, and at the end $t = t_*$ (*cf.* (3.272)) of the time series there are visible discrepancies of order magnitude 0.1 between the approximate and the simulated response. These findings are consistent with the accuracy estimate given under (3.259), according to which the error at $t = t_* = O(\tilde{\beta}^{-1}, \Delta^{-1})$ is of the same magnitude order as the small parameters, i.e. 0.01 for figures (a,b), and 0.1 for figures (c,d). This highlights the asymptotic nature of the analytical prediction (3.272), which implies that the quality of predictions decline as the assumption of small parameters (and Lipschitz continuity) fails to hold. Note, however, that the quality of predictions at a given level of parameter-smallness can be systematically improved by using higher order approximations (*cf.* Thomsen and Fidlin 2008).

It should be recalled that the small displacement-related parameters should be "small" as compared to the oscillation amplitude A, *cf.* the discussion below (3.268). When $|s|$ drops below $|\Delta|$, which occurs for $t \gg t_*$, then the parameters cannot longer be considered "small", and (3.272) will not provide a good approximation. This holds even for $\Delta > 0$, where $|s| < |\Delta|$ implies the mass does not hit the stop, and the system becomes linear with a simple exponentially decaying harmonic solution; the nonlinear asymptotic solution (3.272) is incapable of reproducing this.

We have demonstrated how a discontinuous transformation followed by averaging can provide simple, analytical predictions of the response of a strongly nonlinear vibro-impact system, provided the dissipation is weak, and the stop is situated near the equilibrium. Here the system was considered linear in-between impacts, but this is not necessary, weak nonlinearities could be added to (3.260)

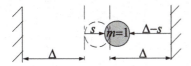

Fig. 3.29. Simple impact oscillator: Point mass in a clearance 2Δ

without causing additional complications in the procedure. This is illustrated in Sects. 3.9.5 and 3.9.6.

3.9.4   Example 2: Mass in a Clearance

The motions of a free point mass in a clearance of measure 2Δ (Fig. 3.29) is governed by:

$$\ddot{s} = 0 \quad \text{for} \quad |s| < \Delta,$$
$$s_+ = s_-; \quad \dot{s}_+ = -R\dot{s}_- \quad \text{for} \quad |s| = \Delta, \tag{3.273}$$

where the impacts at $|s| = \Delta$ are assumed to be near-elastic, $0 < (1 - R)\,1$, while Δ can be arbitrarily large. Discontinuous changes in velocity occur at the times of impact. To set up a suitable transformation for eliminating or reducing these, we first note that for purely elastic impacts ($R = 1$) the speed $|\dot{s}|$ remains unchanged between impacts, i.e. the coordinate s will trace out a regular zigzag line in time. Such a folded line can be described by standard trigonometric functions, for example:

$$\Pi(z) = \arcsin(\sin z), \tag{3.274}$$

which is depicted in Fig. 3.30(a). Thus, if a curve $z(t)$ is a straight line, then $\Pi(z(t))$ is a zigzag line, 2π-periodic in z (but not necessarily in t). Consequently, if $z(t)$ is *almost* a straight line, i.e. $z(t) = c_0 + c_1 t + \varepsilon(t)$, $\varepsilon \ll 1$, then $\Pi(z(t))$ can be considered a slightly perturbed zigzag line. Conversely, if $\Pi(z(t))$ is (close to) a periodic zigzag line, then $z(t)$ is (close to) a straight line, that is, all the "zags" of $\Pi(z(t))$ are mirrored about a line parallel to the time axis, to create an unfolded line that is (close to) straight.

Now, when $R = 1$ the variable s in (3.273) describes a zigzag line in time with period $4\Delta/|\dot{s}|$, where $|\dot{s}|$ is the constant speed. With near-elastic impacts, $0 < 1 - R$ 1, $|\dot{s}|$ will decrease at bit at every impact, and so will the slopes of the zigs and zags of $s(t)$, which becomes a close-to-periodic zigzag line. As an unfolding transformation for s we may thus use

$$s = \frac{2\Delta}{\pi} \Pi(z), \tag{3.275}$$

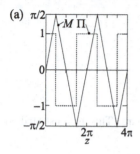

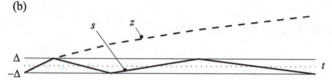

Fig. 3.30. (a) Function $\Pi(z)$ in (3.274) and its derivative $M(z)$; (b) motions $s(t)$ of the mass in a clearance for $0 < 1 - R \ll 1$, and its unfolding $z(t)$ as given by the transformation (3.275)

which was suggested by Zhuravlev and Klimov (1988) for $R = 1$, but here we use it even for $R \approx 1$. From (3.274) and (3.275) it follows that:

$$\dot{s} = \frac{2\Delta}{\pi} M(z)\dot{z}, \quad \ddot{s} = \frac{2\Delta}{\pi}(M'(z)\dot{z} + M(z)\ddot{z}) \quad \text{for } z \neq \frac{\pi}{2} + j\pi, \quad j = 0, 1, \dots, \quad (3.276)$$

where $M(z) = d\Pi/dz = \mathrm{sgn}(\cos z)$. With (3.274)–(3.276), the system (3.273) then transforms into:

$$\ddot{z} = 0, \quad z \neq \frac{\pi}{2} + j\pi \quad \text{for } j = 0, 1, \cdots,$$
$$z_+ = z_-, \quad \dot{z}_+ - \dot{z}_- = -(1 - R)\dot{z}_- \quad \text{for } z = \frac{\pi}{2} + j\pi. \qquad (3.277)$$

where it appears that the discontinuity in the velocity $\dot{z}$ of the new dependent variable z has been reduced to a small value proportional to $(1 - R)$. This also appears from Fig. 3.30(b), where $z(t)$ describes a polygon with small angles between the straight line segments, i.e. the changes in slope of $z(t)$ are much smaller than those in the slope of $s(t)$.

To transform (3.277) further into the general form (3.258), we first reduce (3.277) to the required first-order form by introducing a new dependent variable $v = \dot{z}$, giving:

$$\left.\begin{array}{l} \dot{z} = v \\ \dot{v} = 0 \end{array}\right\} \quad \text{for } z \neq \frac{\pi}{2} + j\pi, \quad j = 0, 1, \cdots,$$

$$\left.\begin{array}{l} z_+ - z_- = 0 \\ v_+ - v_- = -(1 - R)v_- \end{array}\right\} \quad \text{for } z = \frac{\pi}{2} + j\pi, \qquad (3.278)$$

or, eliminating time as the independent variable by dividing the second equation with the first:

$$\frac{dv}{dz} = 0 \quad \text{for} \quad z \neq \frac{\pi}{2} + j\pi,$$
$$v_+ - v_- = -(1 - R)v_- \quad \text{for} \quad z = \frac{\pi}{2} + j\pi,$$

(3.279)

where instead z now takes the role of the *in*dependent variable. This system has the general form (3.258), and can thus be averaged using (3.259) into:

$$\frac{dv_1}{dz_1} = -\frac{1-R}{\pi}v_1, \quad v_1 = \frac{dz_1}{dt},$$

(3.280)

which is a linear system with solution:

$$z_1 = \frac{\pi}{1-R}\ln(C_1 t + C_2), \quad v_1 = \frac{\pi}{1-R}\frac{C_1}{C_1 t + C_2},$$

(3.281)

where the constants $C_1 > 0$ and C_2 are determined by initial conditions. The corresponding motion in terms of the original variable $s(t)$ is then, by (3.275):

$$s = \frac{2\Delta}{\pi}\Pi\left(\frac{\pi}{1-R}\ln(C_1 t + C_2)\right).$$

(3.282)

Fig. 3.31 compares results of using the approximate first-order prediction (3.282) to results of numerical simulation of the original system (3.273). As appears there is no discernible difference in results for the position variable $s(t)$. For the velocity $\dot{s}(t)$ the difference is small but visible, and in fact very illustrative of the effect of discontinuous averaging of systems with near-elastic impacts: The vertical lines for $\dot{s}$ in Fig. 3.31 correspond to impacts, where the velocity changes sign and its magnitude decreases by the small value $(1 - R)v_-$. The horizontal lines in the numerical solution (dashed) correspond to time intervals of free flight with constant velocity. In the approximate solution, by contrast, there is no change in velocity magnitude after impacts; instead the change is distributed over the time interval between impacts, keeping the average energy loss the same as for the numerical solution. Thus the inaccuracy of the approximate solution for the velocity $\dot{s}$ is of the order of magnitude of the small parameters, while the inaccuracy in position s is even smaller.

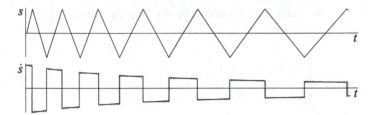

Fig. 3.31. Position $s(t)$ and velocity $\dot{s}(t)$ of a mass in a clearance 2Δ, as predicted by the approximate expression (3.282) (solid line), and by numerical simulation of (3.273) (dashed line), for the case $R = 0.9$

3.9.5   Example 3: Self-excited Friction Oscillator with a One-Sided Stop

Fig. 3.32(a) shows the classical 'mass on moving belt' model (Panovko and Gubanova 1965), though extended with a stop at the right, which restricts motions to $s < \Delta$ (the left stop is ignored in this section but included in the following Sect. 3.9.6). Without stop(s) this system is often used for illustrating friction-induced oscillation, e.g., Thomsen and Fidlin (2003) used averaging to derive stationary amplitudes for pure slip and stick–slip oscillations. With stop(s), the system models rubbing objects with slipping parts, e.g., loosely mounted brake pads.

A typical nondimensional model formulation of the equations of motion is:

$$\ddot{s} + s = -h(\dot{s}) \quad \text{for} \quad s < \Delta, \quad \dot{s} < v_b,$$
$$s_+ = s_-, \quad \dot{s}_+ = -R\dot{s}_- \quad \text{for} \quad s = \Delta, \tag{3.283}$$

where $s(t)$ is the displacement of the mass from the static equilibrium $s_0 = \mu(-v_b)$ at belt speed v_b, R is the coefficient of impact restitution, and the friction law is a cubic Stribeck model with friction coefficient $\mu(v_r) = \mu_s \text{sgn}(v_r) - k_1 v_r + k_3 v_r^3$ depending on relative interface velocity $v_r = \dot{s} - v_b$ (Ibrahim 1992b; Thomsen and Fidlin 2003, and Fig. 3.32(b)). Here μ_s is the static coefficient of friction, k_1 the slope of the friction-velocity curve at zero relative velocity, k_3 the coefficient governing increased friction at higher velocities. The dissipation function h in (3.283) is then:

$$h(\dot{s}) = h_1 \dot{s} + h_2 \dot{s}^2 + h_3 \dot{s}^3, \tag{3.284}$$

where $h_1 = 2\beta - k_1 + 3k_3 v_b^2 < 0$, $h_2 = -3k_{3b}$, $h_3 = k$, and β is the linear viscous damping ratio. Of interest here is the effect of impacts, so $\dot{s} < v_b$ (i.e. $v_r < 0$) is assumed to avoid unnecessary complications connected with sticking motions (stick-slip is considered in Thomsen and Fidlin 2003).

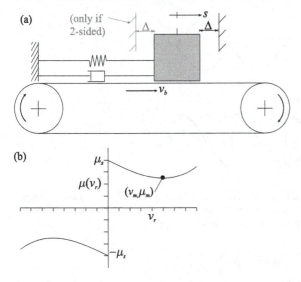

Fig. 3.32. (a) Self-excited friction-oscillator with one or two stops; (b) Friction coefficient μ as a function of relative interface velocity $v_r = \dot{s} - v_b$

Assuming weak dissipation the parameters h_1, h_2, and h_3 are small. In addition we assume near-elastic impacts, and a small distance from the equilibrium of the unstrained spring, i.e.:

$$0 < (1 - R) \ll 1, \quad |h_{1,2,3}| \ll 1; \quad |\Delta| \ll 1. \tag{3.285}$$

To turn (3.283)–(3.284) into the form (3.258), we repeat the procedure from Sect. 3.9.3, and even start with the same discontinuous transformation (3.262), though with a shift in sign (since the stop is now situated $s = +\Delta$):

$$s = \Delta - |z|, \quad z_+ z_- < 0. \tag{3.286}$$

In the z-variable, then, every other oscillation of s and $\dot{s}$ will be mirrored, so that if $R = 1$ the velocity-discontinuity at impact is eliminated, while for near-elastic impacts $0 < (1 - R) 1$ the discontinuity will be small, as illustrated in Fig. 3.33. Inserting (3.286) into (3.283), the transformed system becomes:

$$\ddot{z} + z = \Delta \operatorname{sgn} z - h_1 \dot{z} + h_2 \dot{z}^2 \operatorname{sgn} z - h_3 \dot{z}^3 \quad \text{for} \quad z \neq 0, |\dot{z}| < v_b,$$
$$\dot{z}_+ - \dot{z}_- = -(1 - R)\dot{z}_- \quad \text{for} \quad z = 0. \tag{3.287}$$

Next, to turn (3.287) further into the general form (3.258), we note that the first equation in (3.287) is quasi-linear, so that the standard van der Pol transformation (3.264)–(3.265) can be used, giving:

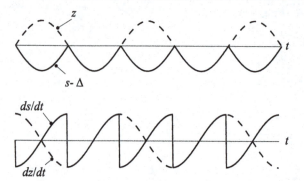

Fig. 3.33 Displacements (top) and velocities (bottom) for a friction-oscillator impacting a single near-elastic stop. In the transformed variable z the discontinuity in velocity dz/dt is small, while for the original variable s the discontinuity in velocity ds/dt is large

$$
\left.
\begin{aligned}
\dot{A} &= \left(\Delta + h_2 A^2 \cos^2 \psi\right) \cos \psi \, \mathrm{sgn}(\sin \psi) \\
&\quad - \left(h_1 + h_3 A^2 \cos^2 \psi\right) A \cos^2 \psi \\
\dot{\theta} &= -\left(\frac{\Delta}{A} + h_2 A \cos^2 \psi\right) |\sin \psi| \\
&\quad + \left(h_1 + h_3 A^2 \cos^2 \psi\right) \cos \psi \sin \psi
\end{aligned}
\right\}
\quad \text{for} \quad \psi \neq j\pi, j = 0, 1, \ldots,
$$

$$
\left.
\begin{aligned}
A_+ - A_- &= -(1 - R)A_- \\
\theta_+ - \theta_- &= 0, \psi = j\pi
\end{aligned}
\right\}
\quad \text{for} \quad \psi = j\pi.
\tag{3.288}
$$

Under assumptions (3.285) and $A \gg |\Delta|$ we find, using order analysis as in (3.267), that $dA/d\psi = \dot{A} + O\left((\Delta, h_{1,2,3})^2\right)$ and $d\theta/d\psi = \dot{\theta} + O\left((\Delta, h_{1,2,3})^2\right)$, so that to first order of accuracy of the small parameters, we can replace $\dot{A}$ and $\dot{\theta}$ with $dA/d\psi$ and $d\theta/d\psi$, respectively. Then (3.288) has the general form (3.258), and (3.259) can be used to calculate the averaged system:

$$
\begin{aligned}
\dot{A}_1 &= -\hat{\beta} A_1 - \frac{3 h_3 A_1^3}{8}, \\
\dot{\theta}_1 &= -\frac{2}{\pi}\left(\frac{\Delta}{A_1} + \frac{h_2 A_1}{3}\right),
\end{aligned}
\tag{3.289}
$$

where

$$
\hat{\beta} = \frac{h_1}{2} + \frac{1 - R}{\pi}.
\tag{3.290}
$$

This system is rather similar to (3.269)–(3.270) for the viscously damped simple impact oscillator, though with the important difference that in (3.289) – (3.290) the linear damping parameter $\hat{\beta}$ can be also negative (if h_1 is sufficiently negative, cf. the definition of h_1 below (3.283)). In that case energy from the running belt is transferred to the mass and the spring, leading to self-excited oscillations, that may be stabilized by nonlinearities such as the h_3-term.

Though the system (3.289) is (weakly) nonlinear, we can easily determine the stationary solutions, which are those of primary interest. By (3.286) and (3.264) we have:

$$s_1 = \Delta - A_1|\sin(t+\theta)|, \tag{3.291}$$

and thus constant-amplitude solutions can be identified by letting $\dot{A}_1 = 0$ in (3.289). This gives a trivial solution $A_1 = 0$ (where the mass does not move), and a non-trivial solution $A_1 = A_{1\infty}$ corresponding to stationary oscillations,

$$A_{1\infty} = \sqrt{-\frac{8\hat{\beta}}{3h_3}}. \tag{3.292}$$

This solution is stable (i.e. $\partial\dot{A}_1/\partial A_1 < 0$ from (3.289) with $A_1 = A_{1\infty}$) only when it exists, i.e. when $\hat{\beta} < 0$ for $h_3 > 0$, and in that case the *trivial* solution $A_1 = 0$ is unstable. In (3.292) the inelasticity of impacts is present only through the parameter $\hat{\beta}$, which combines the dissipative effects of friction and impact, cf. (3.290). In the absence of impacts $\hat{\beta} = h_1/2$, and expression (3.292) for the oscillation amplitude reduces to the known expression for the no-stick oscillation amplitude of a cubic friction oscillator (Thomsen and Fidlin 2003). Hence the effect of near elastic impacts on oscillation amplitude is equivalent to that of linear viscous damping with coefficient $\frac{2}{\pi}(1 - R)$. cf. (3.289)–(3.290). Examining this equivalence further, it turns out, that most of the (non-sticking) analytical results derived in Thomsen and Fidlin (2003) for the system without stop also hold for the system with a one-sided stop – provided the linear dissipation parameter β is replaced everywhere by the effective viscous damping coefficient $\hat{\beta}$ as defined in (3.289), and the low-speed slope h_1 of the friction curve is replaced by the effective value $\hat{h}_1 = 2\hat{\beta} - k_1 + 3k_3v_b^2$.

Substituting (3.292) into the second equation in (3.289), and solving the linear equation for θ_1, one finds:

$$\theta_{1\infty} = \frac{-2}{\pi A_{1\infty}}\left(\Delta + \frac{1}{3}h_2 A_{1\infty}^2\right)t + \theta_0, \tag{3.293}$$

so that by (3.291) the oscillating stationary solution $s_{1\infty}$ can be written:

$$s_{1\infty} = \Delta - A_{1\infty}|\sin(\omega_\infty t + \theta_0)|, \tag{3.294}$$

where θ_0 is an arbitrary constant phase, and the oscillation period is π/ω_∞, where

$$\omega_\infty = 1 - \frac{2}{\pi A_{1\infty}}\left(\Delta + \tfrac{1}{3}h_2 A_{1\infty}^2\right). \tag{3.295}$$

These simple expressions for the oscillations of the strongly nonlinear vibro-impact problem with friction provides good agreement with numerical simulation, even for larger values of $|1 - R|$, as long as Δ is small. Fig. 3.34 shows an example, with parameters (given in the figure legend) resulting in $\hat{\beta} = -0.0068$ and $A_{1\infty} = 0.60$, and good agreement between the approximation (3.294) (in solid line) and numerical simulation of the original equation of motion (3.283) (dashed). With other parameters, the errors (i.e. deviations from numerical simulation), will decrease or increase as the parameters assumed small in (3.285) becomes smaller or larger, respectively. Also, for fixed parameters, the errors can be reduced by using a more accurate second order analysis, the result of which is shown dotted in Fig. 3.34, and derived and discussed in detail in Thomsen and Fidlin 2008.

As appears from (3.292) and Fig. 3.35 (solid line), the first-order approximation to the stationary oscillation amplitude $A_{1\infty}$ does not depend on the equilibrium-to-stop distance Δ while numerical simulation shows a clear increase of $A_{1\infty}$ with Δ (Fig. 3.35, circles). For small Δ the error resulting from this is $O(\Delta)$, which is consistent with the error estimate under (3.259). For larger Δ the accuracy can be improved by using second-order averaging (*cf.* Fig. 3.35 dotted line, and Thomsen and Fidlin 2008).

3.9.6 Example 4: Self-excited Friction Oscillator in a Clearance

With a *two*-stop friction-oscillator (Fig. 3.32(a)), we let the static equilibrium of the mass on the running belt be in the middle of the clearance of measure 2Δ, but do not assume the clearance to be small. Motions $s(t)$ are then governed by the nondimensional system:

$$\begin{aligned}\ddot{s} + s &= -h(\dot{s}) \quad \text{for} \quad |s| < \Delta, \dot{s} < v_b, \\ s_+ &= s_-, \dot{s}_+ = -R\dot{s}_- \quad \text{for} \quad |s| = \Delta,\end{aligned} \tag{3.296}$$

where the function h is given by (3.284). The system differs from the one-stop system (3.283) only in the impact condition, which is $|s| = \Delta$ instead of $s = \Delta$. However, this changes the character of solutions, so that the mirror-transformation (3.286), used for the one-stop system, will *not* reduce the velocity-discontinuity to a small value. Instead we re-employ the transformation (3.275), used for the problem of a free mass in a clearance:

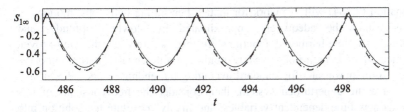

Fig. 3.34. Stationary friction oscillator displacement $s_{1\infty}$ as a function of time t, comparing the analytic first-order prediction (3.294) (solid line), the second-order approximation (Thomsen and Fidlin 2008) (dotted), and numerical simulation of the original equation of motion (3.283) (dashed). Parameters: $h_1 = -0.02$, $h_2 = -0.1$, $h_3 = 0.05$, $R = 0.99$, $\Delta = 0.05$

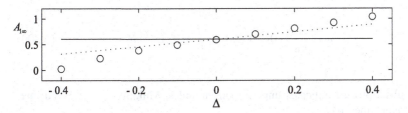

Fig. 3.35. Stationary amplitude $A_{1\infty}$ as a function of the distance Δ from the (unstable) equilibrium of the mass on a moving belt to the hard stop: The analytic first-order prediction (3.292) (solid line), the analytical second-order prediction (Thomsen and Fidlin 2008) (dotted), and numerical simulation of the original equation of motion (3.283) (circles). Parameters as for Fig. 3.34

$$s = \frac{2\Delta}{\pi} \Pi(z), \qquad (3.297)$$

with Π defined by (3.274). Inserting this into (3.296) gives:

$$\ddot{z} + M(z)\Pi(z) = -\bar{h}_1 \dot{z} - \bar{h}_2 \dot{z}^2 M(z) - \bar{h}_3 \dot{z}^3 \quad \text{for} \quad z \neq \frac{\pi}{2} + j\pi, \quad j = 0, 1, \ldots,$$

$$\dot{z}_+ - \dot{z}_- = -(1-R)\dot{z}_- \quad \text{for} \quad z = \frac{\pi}{2} + j\pi,$$

$$(3.298)$$

where $\bar{h}_1 = h_1, \bar{h}_2 = 2\Delta h_2 / \pi$, $\bar{h}_3 = 4\Delta^2 h_3 / \pi^2$, and $M(z) = d\Pi/dz$. We assume a small energy input (corresponding to a small negative slope of the friction characteristics), and small energy dissipation due to friction and impact, i.e.:

$$0 < (1-R) \ll 1; \quad \left| \bar{h}_{1,2,3} \right| \ll 1; \quad \Delta = O(1). \qquad (3.299)$$

Next we should transform (3.298) into the general form (3.258) applicable for averaging. But contrary to the one-sided oscillator of Sect. 3.9.5, the van der Pol transformation (3.264) will not work for that: When hitting two stops, motions of the mass cannot be adequately approximated by slowly amplitude- and phase-modulated time-harmonic functions. Instead we rely on the well-proven general method of transforming weakly nonlinear differential equations into the first-order form appropriate for standard averaging (*cf.* ending of Sect. 3.5.5), that is: First solve the unperturbed system, then consider the free constants of this solutions as new time-dependent variables, and finally substitute the solution into the nonlinear equations of motions to obtain the equations governing the slow evolution of these variables. For classical weakly nonlinear oscillators this approach results in the van der Pol transformation (3.264). Here, to find a workable transformation, we consider the unperturbed system corresponding to (3.298), which can be written in terms of a potential V:

$$\ddot{z}_0 + \frac{dV}{dz_0} = 0, \quad V(z_0) = \int_0^{z_0} M(\zeta)\Pi(\zeta)d\zeta, \tag{3.300}$$

where subscript zero indicates unperturbed variables. Multiplying by $\dot{z}_0$ and integrating over time gives:

$$E = \frac{1}{2}\dot{z}_0^2 + V(z_0), \tag{3.301}$$

where the potential energy $V(z_0)$ is π-periodic with each period defined by a positive parabola, and the constant of integration E is the mechanical energy of the unperturbed system. This energy is limited by two conditions: First, the condition $\dot{z}_0^2 > 0$ with (3.301) gives $E > V(z_0)$, which implies $E > \max(V(z_0)) = V(\pi/2) = \pi^2/8$. Second, we consider only slipping motions, i.e. $\dot{s} < v_b$, which with (3.297), (3.301), and $M^2 = 1$ implies that $E < \frac{\pi^2}{8}(1 + (v_b/\Delta)^2)$. Hence:

$$1 < \frac{8E}{\pi^2} < 1 + (v_b/\Delta)^2. \tag{3.302}$$

With E thus restricted, (3.301) gives the velocity of z_0:

$$\dot{z}_0 = \sqrt{2(E - V(z_0))}. \tag{3.303}$$

Then we use this solution for the unperturbed system as a basis for a *transformation* of variables, where E is considered the new dependent variable, i.e. we let:

$$\dot{z} = \sqrt{2(E - V(z))}, \tag{3.304}$$

which gives:

$$E = \frac{1}{2}\dot{z}^2 + V(z) \quad \Rightarrow \quad \dot{E} = \left(\ddot{z} + \frac{dV}{dz}\right)\dot{z}, \tag{3.305}$$

or, substituting V from (3.300) and inserting into (3.296):

$$\dot{E} = -\bar{h}_1\dot{z}^2 - \bar{h}_2\dot{z}^3 M(z) - \bar{h}_3\dot{z}^4 \quad \text{for} \quad z \neq \frac{\pi}{2}+j\pi, \quad j = 0, 1, \ldots,$$

$$E_+ - E_- = -\left(1-R^2\right)\left(E_- - \frac{\pi^2}{8}\right) \quad \text{for} \quad z = \frac{\pi}{2}+j\pi. \tag{3.306}$$

Dividing the first equation with $\dot{z}$ and using (3.304), we obtain an equation where time is eliminated, while instead z takes the role of the independent variable:

$$\frac{dE}{dz} = -\bar{h}_1\sqrt{2(E-V(z))} - 2\bar{h}_2(E-V(z))M(z)$$

$$\qquad - \bar{h}_3(2(E-V(z)))^{3/2} \quad \text{for} \quad z \neq \frac{\pi}{2}+j\pi, \tag{3.307}$$

$$E_+ - E_- = -\left(1-R^2\right)\left(E_- - \frac{\pi^2}{8}\right) \quad \text{for} \quad z = \frac{\pi}{2}+j\pi.$$

Under assumptions (3.299) the system has the general form (3.258), and can thus be averaged using (3.259), giving:

$$\frac{dE_1}{dz} = q(E_1), \tag{3.308}$$

where

$$q(E_1) = -\bar{h}_1\left\{\frac{1}{2}\sqrt{2E_1 - \frac{\pi^2}{4}} + \frac{2}{\pi}E_1\arcsin\left(\frac{\pi}{2\sqrt{2E_1}}\right)\right\}$$

$$\quad - \frac{1-R^2}{\pi}\left(E_1 - \frac{\pi^2}{8}\right)$$

$$\quad - \bar{h}_3\left\{\left(2E_1 - \frac{\pi^2}{4}\right)^{3/2} + \frac{3}{4}E_1\sqrt{2E_1 - \frac{\pi^2}{4}} + \frac{3}{\pi}E_1^2\arcsin\left(\frac{\pi}{2\sqrt{2E_1}}\right)\right\}. \tag{3.309}$$

Stationary solutions to (3.308) determines periodic oscillations of the mass on the belt with two impacts per oscillation period. The physical meaning can be illustrated by considering the case $h_3 = \bar{h}_3 = 0$ (which implies $h_2 = \bar{h}_2 = 0$), i.e. with friction monotonically decreasing with increased interface velocity. In that case increasing amplitudes of self-excited oscillations cannot be limited by increased friction, but only by the hard stops. Then the mass gains energy during slipping and dissipates it during impacts. If the energy obtained during slipping exceeds what is lost during impact, the total energy will increase. But the oscillation amplitudes cannot exceed the fixed clearance width 2Δ, so increased system energy can only go into increased velocity, and correspondingly increased oscillation frequency. Thus, when friction-induced oscillations are first initiated, their amplitude will build up until the mass starts hitting the stops, whereafter the frequency of oscillations increase until a balance between gained and dissipated energy is attained.

To calculate the frequency of stationary oscillations we first determine the corresponding stationary energy $\tilde{E}_1$ as a solution of $q(\tilde{E}_1) = 0$. Then the corresponding oscillation period $\tilde{T}$ (equal to twice the free-flight time between two successive impacts) and frequency $\tilde{\omega} = 2\pi/\tilde{T}$ is calculated from (3.304), which is valid in-between two impacts, $z \in \;] - \pi/2; \pi/2\; [$:

$$\frac{dz}{dt} = \sqrt{2\left(\tilde{E} - V(z)\right)}$$

$$\Rightarrow \tilde{T} = 2 \int_{z=-\frac{\pi}{2}}^{z=\frac{\pi}{2}} dt = 2 \int_{-\frac{\pi}{2}}^{\frac{\pi}{2}} \frac{dz}{\sqrt{2\left(\tilde{E} - V(z)\right)}}, \tag{3.310}$$

which, by inserting (3.300) and (3.274) and integrating, gives:

$$\tilde{\omega} = \frac{2\pi}{\tilde{T}} = \frac{\pi/2}{\arcsin\left(\frac{\pi/2}{\sqrt{2\tilde{E}}}\right)}. \tag{3.311}$$

Fig. 3.36 illustrates how the stationary oscillation frequency computed by this expression increases with the input energy parameter $(-h_1)$ for parameters as given in the legend. As appears the approximate results (in solid line) agree asymptotically with numerical simulation (circles) of the original system (3.296) for small values of $|h_1|$, i.e. as the assumptions (3.299) are better fulfilled.

3.9.7   Second-Order Analysis

It is a common experience with nonlinear systems, demonstrated also by the examples in Sects. 3.9.3–3.9.6, that essential system behavior is revealed by approximate analysis to lowest (i.e. first) order. However, higher order analysis may be necessary when better numerical accuracy is required, or if parameters that are assumed small are not really so, or if certain phenomena are only revealing themselves at higher order. The general averaging method for vibro-impact analysis

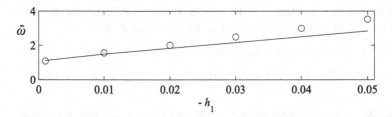

Fig. 3.36. Stationary oscillation frequency for a friction oscillator hitting two stops, as given by
the analytical approximation (3.311) (solid line), and by numerical simulation of the original
system (3.296) (circles). Parameters: $\Delta = 1$, $R = 0.95$, $h_2 = h_3 = 0$

described in Sect. 3.9.2 can be systematically extended to any order of accuracy
required, i.e. $\mathbf{x} = \mathbf{x}_1 + \varepsilon\mathbf{x}_2 + \ldots$, though the computational burden increases rapidly
with accuracy order. Of primary interest is typically to improve the accuracy of the
slowly changing variable $\mathbf{x}_1$, while increased accuracy in the small and rapidly
oscillating motions $\varepsilon\mathbf{x}_2$ may be interesting only by its effect on $\mathbf{x}_1$. A procedure for
extended second-order averaging for discontinuous systems is provided in Fidlin
(2006), and applied in Thomsen and Fidlin (2008) to the friction oscillator with a
one-sided stop of Sect. 3.9.5. As appears from Fig. 3.35 (dotted line) the
second-order analysis increases accuracy. However, it is also considerably more
elaborate.

3.10 Summing Up

Hopefully some basic understanding of nonlinear phenomena and perturbation and
other approximate methods has by now emerged. Only near-equilibrium behavior
of mostly weakly nonlinear one-dimensional systems has been dealt with, and in
most of the elaborated examples the nonlinearity involved took the form of a cubic
restoring term. All of these limitations will be explored in the following chapters.

3.11 Problems

Problem 3.1 Consider a fixed-free elastic rod, subjected to viscous damping,
distributed axial loading and a nonlinear material law. The axial vibrations are
governed by:

$$\rho A\ddot{u} = \sigma' A - 2\rho A\beta\dot{u} + \rho AF(x,t), \quad \sigma = Eu'(1+\gamma u'),$$
$$x \in [0;1], u = u(x,t), \quad u(0,t) = \sigma(1,t) = 0, \tag{3.312}$$

where ρA is the mass per unit length, A the cross-sectional area, $F(x,t)$ the loading, β the coefficient of damping, σ the stress and u' the longitudinal strain.

a) Eliminate σ to obtain the following equation of motion:

$$\ddot{u} - c^2 u'' = 2\gamma c^2 u' u'' - 2\beta\dot{u} + F(x,t), \, u(0,t) = u'(1,t) = 0. \qquad (3.313)$$

b) Compute the linear undamped natural frequencies ω_j and corresponding mode shapes $\varphi_j(x)$ of the rod.

c) Assume a solution in form of the eigenfunction expansion:

$$u(x,t) = \sum_{j=1}^{N} a_j(t)\varphi_j(x). \qquad (3.314)$$

Insert into the equation of motion, utilize the orthogonality of mode shapes, and show that the modal amplitudes a_i are governed by an equation of the form

$$\ddot{a}_i + \omega_i^2 a_i = -2\beta\dot{a}_i + \gamma \sum_{j,k=1}^{N} B_{ijk} a_j a_k + f_i(t), \quad i = 1,N, \qquad (3.315)$$

where B_{ijk} is defined through an integral of mode shapes.

Problem 3.2 For a fixed–fixed wire subjected to transverse excitation, the non-linear planar transverse vibrations are governed by:

$$\ddot{w} - c_0^2 w'' = \frac{c_1^2}{2l} w'' \int_0^l (w')^2 \, dx - 2\mu\dot{w} + F(x,t), \, \, w(0,t) = w(l,t) = 0. \quad (3.316)$$

a) Show that the undamped linear natural frequencies and mode shapes are:

$$\omega_j = j\pi c_0/l, \quad \varphi_j(x) = \sqrt{2}\sin(j\pi x/l). \qquad (3.317)$$

b) Assume a solution in form of the eigenfunction expansion:

$$w(x,t) = \sqrt{2}\sum_{j=1}^{N} u_j(t)\varphi_j(x). \qquad (3.318)$$

Insert into the equation of motion, utilize the orthogonality of mode shapes, and show that the modal amplitudes u_j are governed by an equation of the form

$$\ddot{u}_j + \omega_j^2 u_j = -2\beta_j\dot{u}_j - \left(jc_1(\pi/l)^2\right)^2 u_j \sum_{k=1}^{N} k^2 u_k^2 + f_j(t), \qquad (3.319)$$

where β_j denote the modal damping factors.

Problem 3.3 For each of the systems below, determine the singular points, determine their type and stability, and sketch the phase plane orbits near the singular points when $0 < \beta < 1$.

a) $\ddot{u} + 2\beta \dot{u} + u + u^3 = 0$
b) $\ddot{u} + 2\beta \dot{u} + u - u^3 = 0$
c) $\ddot{u} + 2\beta \dot{u} - u + u^3 = 0$
d) $\ddot{u} + 2\beta \dot{u} - u - u^3 = 0$

Problem 3.4 Consider the equation of motion for a pendulum with a constant torque M at the supporting hinge:

$$\ddot{\theta} + 2\beta\dot{\theta} + \omega^2 \sin \theta = M. \tag{3.320}$$

Letting $x_1 = \theta$ and $x_2 = \dot{\theta}$, determine the singular points in the (x_1, x_2) phase plane. Discuss the character of nonlinear motions near the singular points.

Problem 3.5 Consider the equation of motion governing arbitrarily large oscillations of a pendulum with quadratic damping:

$$\ddot{\theta} + 2\beta\dot{\theta}|\dot{\theta}| + \omega^2 \sin \theta = 0. \tag{3.321}$$

Letting $x_1 = \theta$ and $x_2 = \dot{\theta}$, determine the singular points in the (x_1, x_2) phase plane. Discuss the character of nonlinear motions near the singular points and sketch the corresponding phase plane orbits.

Problem 3.6 Consider the system shown in Fig. P3.6, consisting of a pair of linked and guided rigid masses m_1 and m_2 in a gravity field g. The horizontal motions $x(t)$ of m_1 are restricted by linear springs having total stiffness k.

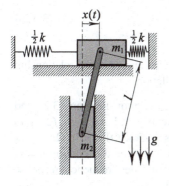

Fig. P3.6

a) Show that the motions of m_1 are governed by:

$$\left(m_1 + \frac{m_2 x^2}{l^2 - x^2}\right)\ddot{x} + \frac{m_2 l^2 x \dot{x}^2}{(l^2 - x^2)^2} + kx + m_2 g \frac{x}{\sqrt{l^2 - x^2}} = 0. \qquad (3.322)$$

b) Let $R = m_2/m_1$ and $u = x/l$. Expand the equation of motion to order three, assuming $|u| \ll 1$, and show that small but finite oscillations are governed by:

$$\left(1 + Ru^2\right)\ddot{u} + Ru\dot{u}^2 + \omega_0^2 u + \frac{Rg}{2l} u^3 = 0, \qquad (3.323)$$

where $\omega_0^2 = k/m_1 + Rg/l$.

c) Using the method of multiple scales, obtain a two-term approximate relationship between the amplitude and the frequency of the motion.

Problem 3.7 Consider a system having a fifth-order nonlinear restoring term:

$$\ddot{u} + \alpha u^5 = 0. \qquad (3.324)$$

Using the method of harmonic balance, show that to a first approximation the free oscillations of this system occur at the frequency $\omega = a^2\sqrt{5\alpha/8}$, where a is the amplitude of oscillation.

Problem 3.8 Consider the transversely vibrating wire described in Problem 3.2, for the case of no damping and no external excitation.

a) Assume that a single-mode approximation is appropriate ($N = 1$), and show that the free response is given by:

$$w(x, t) = 2u(t) \sin\frac{\pi x}{l}, \qquad (3.325)$$

where the modal amplitude $u(t)$ is governed by:

$$\ddot{u} + \omega^2 u + \gamma u^3 = 0, \qquad (3.326)$$

where $\omega = \pi c_0/l$ and $\gamma = (c_1 \pi^2 / l^2)^2$.

b) Use the method of averaging to obtain a two-term approximate relationship between the amplitude and the frequency of oscillation.

Problem 3.9 Consider the transversely vibrating wire of Problem 3.2, and assume a single-mode approximation to be appropriate, $N = 1$. Let $u \equiv u_1, f \equiv f_1, \omega \equiv \omega_1, \beta \equiv \beta_1, \gamma \equiv c_1(\pi/l)^2$, and assume harmonic excitation $f(t) = p\sin(\Omega t)$, weak damping $\beta = O(\varepsilon)$, $\varepsilon \ll 1$, weak nonlinearity $\gamma = O(\varepsilon)$, and hard excitation $p = O(1)$. Using the method of multiple scales, obtain a first-order frequency–response equation for the case of superharmonic resonance, $\Omega \approx \frac{1}{3}\omega$.

Problem 3.10 Consider the undamped Duffing system with negative linear stiffness:

$$\ddot{u} - \frac{1}{2}(u - u^3) = K \cos \Omega t. \tag{3.327}$$

Using the method of multiple scales, determine the first term of a uniformly valid expansion for u, describing the response near the center at $u = 1$ when, respectively:

a) $\Omega \approx 1$
b) $\Omega \approx 2$
c) $\Omega \approx \frac{1}{2}$
d) $\Omega \approx 3$
e) $\Omega \approx \frac{1}{3}$

Problem 3.11 Consider a system governed by the following equation of motion:

$$\ddot{u} + u + \alpha u^5 = K \cos(\Omega t) - 2\beta \dot{u}. \tag{3.328}$$

a) Determine the frequency response equation when $\Omega \approx 1$, and sketch the amplitude of the response as a function of Ω for different values of K.
b) Show that secondary resonances exist when Ω is near 5, 3, 2, $\frac{1}{2}$, $\frac{1}{3}$ or $\frac{1}{5}$.
c) For each case of secondary resonance, set up the equations governing modulations of phases and amplitudes.

Problem 3.12 Consider the following system with a linear-quadratic nonlinearity:

$$\ddot{u} + 2\beta\omega\dot{u} + \omega^2 u - \gamma u^2 = K \cos \Omega t, \tag{3.329}$$

where the damping and the nonlinearity is assumed to be small $\beta, \gamma = O(\varepsilon)$, $\varepsilon \ll 1$.

a) Using the method of multiple scales, determine a first-order uniformly valid expansion for $u(t)$ for the case of primary resonance ($\Omega \approx \omega$) and weak excitation, $K = O(\varepsilon)$. Determine the stationary response. At which level of approximation does the nonlinearity affect the response? (Compare to the Duffing equation).
b) For the case of strong excitation, $K = O(1)$, identify all possible secondary resonances (Ω away from ω).
c) For conditions as in b), obtain a first approximation for the stationary response in the case of superharmonic resonance, $\Omega \approx \frac{1}{2}\omega$. How does the nonlinearity affect the response?
d) For the case of very small damping and excitation ($\beta, K = O(\varepsilon^2)$), determine the stationary response for the case of primary resonance ($\Omega \approx \omega$). Compare the change in the character of the response as compared to case a). [Tip: this requires an extra time scale in the analysis, $T_2 = \varepsilon^2 t$.]

Problem 3.13 Consider a system with a quadratic-cubic restoring force:

$$\ddot{u} + \varepsilon^2 2\beta\omega\dot{u} + \omega^2 u + \varepsilon\eta u^2 + \varepsilon^2\gamma u^3 = \varepsilon^2 q\cos\Omega t, \tag{3.330}$$

where $\varepsilon \ll 1$, and η and γ are positive parameters.

a) Using the method of multiple scales, determine a first-order uniformly valid expansion for $u(t)$ for the case of primary resonance, $\Omega \approx \omega$.
b) Determine the stationary frequency response, and discuss whether this has a softening or hardening character, depending on the ratio γ/η.

Problem 3.14 Consider a system with nonlinear restoring force and nonlinear damping, subjected to harmonic forcing:

$$\ddot{\theta} + \sin\theta + 2\beta\theta^2\dot{\theta} = q\cos\Omega t. \tag{3.331}$$

a) Consider the case of primary resonance, $\Omega \approx 1$. Using the method of multiple scales, show that small but finite oscillations are approximately governed by:

$$\theta(t;\varepsilon) = a(T_1)\cos(T_0 + \varphi(T_1)) + O(\varepsilon), \tag{3.332}$$

where $a(T_1)$ and $\varphi(T_1)$ are solutions of:

$$\begin{aligned} 2a' &= -\frac{1}{2}\beta a^3 + q\sin\psi, \\ 2a\psi' &= 2\sigma a + \frac{1}{8}a^2 + q\cos\psi, \end{aligned} \tag{3.333}$$

where $\psi = \sigma T_1 - \varphi$, and $\varepsilon\sigma = \Omega - 1$.
b) Obtain the frequency response equation. Show that the maximum stationary amplitude is $(27/\beta)^{1/3}$, and sketch the frequency response.

Problem 3.15 The forced response of a self-excited system to a slowly varying external excitation $f(\varepsilon t) = O(1)$ is governed by

$$\ddot{u} + \omega_0^2 u = \varepsilon(1 - u^2)\dot{u} + \omega_0^2 f(\varepsilon t). \tag{3.334}$$

a) Using the method of multiple scales, show that

$$u = A(T_1)e^{i\omega_0 T_0} + cc + f(T_1) + O(\varepsilon), \tag{3.335}$$

where $T_0 = t$, $T_1 = \varepsilon t$, and the slow modulations of $A(T_1)$ are governed by:

$$2A' = (1 - A\bar{A} - f^2)A. \tag{3.336}$$

b) Express $A(T_1)$ in polar form and determine the equations governing slow modulations of amplitudes and phases.
c) Show that when $f^2 < 1$ a nontrivial stationary solution exists:

$$u(t) = \sqrt{1 - f(t)^2} \cos(\omega_0 t + \varphi_0) + f(t) + O(\varepsilon). \tag{3.337}$$

Problem 3.16 The first-mode transverse response of an axially spring-loaded beam subjected to harmonic base excitation (Fig. P3.16) is given by

$$\ddot{u} + 2\beta\dot{u} + \omega^2 u + \gamma u^3 = p\Omega^2 \cos \Omega t, \tag{3.338}$$

where $u(t)$ denotes the modal amplitude of the beam with respect to the straight, undeformed configuration, and where the damping β, the nonlinearity γ and the load parameter p are all assumed to be small and positive. The linear stiffness coefficient ω^2 may become negative. This is the case when the axial spring is pre-compressed by a force that is larger than the buckling load of the beam. The static equilibrium $u = 0$ then turns unstable in favor of two buckled equilibriums $u = \pm \tilde{u}$, where $\tilde{u} = \sqrt{-\omega^2/\gamma}$.

a) Rewrite the equation of motion to describe small but finite oscillations near the buckled equilibrium $u = \tilde{u}$, by introducing $\eta(t) = u(t) - \tilde{u}$ and $\tilde{\omega}^2 = -2\omega^2$. What is the frequency of linearized free vibrations of the buckled beam?
b) Determine a first-order multiple scales expansion for $u(t) = \eta(t) + \tilde{u}$ for the case of primary resonant excitation of the buckled beam. Determine the stationary response.

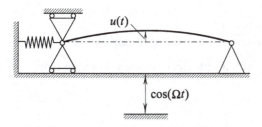

Fig. P3.16

Problem 3.17 Consider a modified Mathieu equation with quadratic damping:

$$\ddot{u} + (\omega_0^2 + 2\varepsilon \cos 2t)u + \varepsilon\beta\dot{u}|\dot{u}| = 0. \tag{3.339}$$

a) Using the method of multiple scales, show that a first-order uniformly valid expansion for the response is:

$$u = a\cos(\omega_0 T_0 + \varphi) + O(\varepsilon). \tag{3.340}$$

b) For the case of primary resonance ($\omega_0 \approx 1$), show that the slow modulations of amplitudes and phases are governed by:

$$a' = -\frac{a}{2\omega_0}\sin\psi - \frac{4}{3\pi}\beta\omega_0 a^2,$$
$$a\psi' = 2\sigma a - \frac{a}{\omega_0}\cos\psi, \tag{3.341}$$

where $\varepsilon\sigma = 1 - \omega_0$ and $\psi = 2\sigma T_1 - 2\varphi$.

c) Obtain the frequency response equation and sketch the frequency response (stationary amplitude versus σ).

d) Examine the stability of solutions.

Problem 3.18 (Back to basics ...) You have been beached on a distant sandbank, with no textbooks surviving. Written in the sand you find the below puzzle. With nothing else to do, you find yourself solving it, using a very basic approach.

Given the equation of motion for a linear single-degree-of-freedom oscillator:

$$\ddot{u} + 2c\dot{u} + ku = p, \tag{3.342}$$

where

$$0 < c < \sqrt{|k|}, \quad k \in R, \quad p \in R. \tag{3.343}$$

a) Show that $u = p/k$ is a static equilibrium

b) Show that this equilibrium is stable when $k > 0$, and unstable when $k < 0$

Problem 3.19 Consider a system with a fifth-order nonlinear restoring term:

$$\ddot{u} + u + u^5 = 0. \tag{3.344}$$

Using the method of multiple scales, obtain a two-term approximate solution for the motions $u(t)$, and set up a relationship between the amplitude and the frequency of the motion.

Problem 3.20 Make a MATLAB (or similar) program for numerical simulation of a pendulum on a vertically vibrating support, cf. Fig. 3.5 and eqs. (3.11)–(3.12). The program should be able to display time series and animate solutions. Extra options could be, e.g., phase plane plots and frequency spectra. It should be possible to easily change pendulum system parameters and simulation/solution parameters and watch the effect – very much as with a laboratory pendulum system with excitation control and measurement instruments.

Problem 3.21 The vertical column AB in Fig. P3.21 has length L and is infinitely rigid and massless. It is hinged at A, while at B there is a transverse spring of stiffness k. At B there is also a concentrated mass m and a vertical time-harmonic pulsating load P with amplitude $\tilde{P}$, frequency Ω, and average (i.e. "static") value $\bar{P}$. Gravity can be ignored.

a) Show that small, undamped, transverse vibrations are governed by:

$$\ddot{\theta} + (\omega_0^2 - \bar{p} + \tilde{p}\cos\Omega t)\theta = 0. \qquad (3.345)$$

where $\theta(t)$ is the angle of AB wrt vertical, $\omega_0^2 = k/m$, $\bar{p} = \bar{P}/mL$, and $\tilde{p} = \tilde{P}/mL$.

b) Classify the forces and the system (cf. Sect. 1.9) when $\tilde{P} = 0$ and $\tilde{P} \neq 0$, respectively.

c) When $\tilde{P} = 0$, determine the value of $\bar{P}$ at which the vertical equilibrium $\theta = 0$ becomes unstable.

d) When $\tilde{P} \neq 0$, $\bar{P} < kL$, and $\Omega > 2\sqrt{\omega_0^2 - \bar{p}}$, use a Strutt diagram (see e.g. App. C.2.3) for Mathieu's equation, and known analytical (first order) approximations for the stability borders, to determine the value of $\tilde{P}$ at which $\theta = 0$ becomes unstable. What is the physical interpretation of the quantity $\sqrt{\omega_0^2 - \bar{p}}$?

e) How would small linear viscous damping affect the above results, in principle?

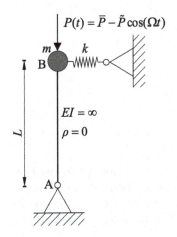

Fig. P3.21

Problem 3.22 Fig. P3.22a shows a simple model of a rotor of free length L, suspended vertically in a gravity field g, and with small internal damping. The shaft is elastic with Young's modulus E, has principal moments of inertia I_1 and $I_2 \le I_1$, and is massless except for a concentrated mass m at the free end B. Bearings prevent transverse movements and slopes of the shaft at point A.

a) Determine the range of rotor speeds Ω for which the straight configuration of the rotor is unstable in the absence of gravity (i.e. for a horizontal rotor).
b) Determine the instability range for Ω in the presence of gravity, and show that its width decreases with increasing mass m.

[Tip: You may need the following expression for the transverse deformation δ at the free end of a clamped-free beam (Fig. P3.22(b)):

$$\delta = \frac{QL^3}{3EI} F(\gamma), \quad \text{where} \quad \gamma = \frac{NL^2}{EI}, \tag{3.346}$$

where Q and N are, respectively, the transverse and axial loads at the free end, and γ is a nondimensional parameter describing axial load. For $\gamma > 0$, $F(\gamma)$ is a monotonically decreasing function satisfying $F(0) = 1, F(\gamma) \to 0$ for $\gamma \to \infty$, and with monotonically increasing slope. (One can show that $F(\gamma) = 3\gamma^{-1}(1 - \tanh(\sqrt{\gamma})/\sqrt{\gamma})$, but to answer the question this expression is not needed.)].

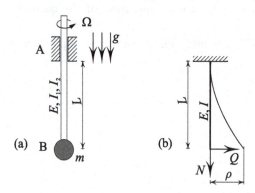

Fig. P3.22

Problem 3.23 Fig. P3.23 shows a model for a linear oscillator performing free vibroimpact against a rigid stop. The oscillator has mass m, linear stiffness k, linear viscous damping coefficient c, and the spring is undeformed at $x = 0$. The stop is positioned at $x = \Delta$ (which can be negative or positive or zero), and impacts are considered purely elastic. The equation of motion is:

$$\ddot{x} + 2\beta\dot{x} + \omega^2 x = 0 \quad \text{for} \quad x < \Delta,$$
$$\dot{x}_+ = -\dot{x}_- \qquad\qquad \text{for} \quad x = \Delta,$$

$$(3.347)$$

where $\omega = \sqrt{k/m}$ is the natural frequency of the unimpacting oscillator, $\beta = c/(2\sqrt{km})$ is the viscous damping ratio, and the initial conditions are $x(0) = x_0$, $\dot{x}(0) = v_0$.

a) Use a discontinuous transformation of variables, $x = \Delta - |z|$, $z_+ z_- < 0$, to transform the equation of motion into a system with a discontinuity which is small when $\omega^2 \Delta$ is small.

b) For the case $\Delta = 0$, solve the transformed system for $z(t)$, and use this to calculate the exact solution for $x(t)$ when $x_0 = 0$ and $v_0 < 0$.

c) Compute and graph the exact solution when $\beta = 0.1$, $\omega = 1$, $\Delta = 0$, and $(x_0, v_0) = (0, -0.1)$. Compare to results of numerical simulation, using e.g. MATLAB and *event-handling* to solve the ODE numerically (here "an event" is an impact, i.e. the condition $x = \Delta$ and $\dot{x} > 0$.

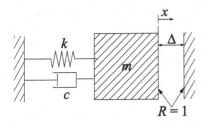

Fig. P3.23

4 Nonlinear Multiple-DOF Systems: Local Analysis

4.1 Introduction

Certain nonlinear phenomena can occur only with systems having multiple degrees of freedom. Thus, *nonlinear interaction* requires the presence of at least two components that can interact. In this chapter we shall consider only 2-DOF systems, since these will suffice for illustrating important properties of nonlinear multiple-DOF systems in general. A 2-DOF model typically arises as an approximation to a real multiple-DOF or continuous system, for which a single-DOF model fails to describe the behavior to be studied.

To accustom ourselves to an additional degree of freedom the first example is worked out in some detail, whereas for the remaining examples we focus on main results. Nonlinearities are assumed to be weak, and perturbation methods (multiple scales or averaging) are used to obtain first-order approximate solutions.

By contrast to a single degree of freedom system – which has only a single linear natural frequency and a single mode of motion – a system having n degrees of freedom system has n linear natural frequencies and n corresponding modes: the linear normal modes. For linear systems these modes are *uncoupled*, that is, energy imparted externally to any one mode remains in that mode; there is normally[1] no mechanism by which energy can be shared with other modes. Nonlinearities, however, may provide such a mechanism, so that energy imparted into one mode may under certain conditions be exchanged with other modes. This is called *nonlinear mode-coupling*, or *modal interaction*.

---

[1]With parametric/internal excitation this may be different: With time-varying coefficients even linear systems may exhibit mode coupling, e.g., under conditions of parametric resonance or anti-resonance (Dohnal 2008). The presence of asymmetrical phenomena like unidirectional flow may also couple modes; this applies, e.g., to Coriolis flowmeters, where the measurement pipe is typically excited to vibrate resonantly at the lowest mode of vibration, while fluid flowing through the pipe excites motion in also the *second* mode of vibration (Thomsen and Fuglede 2020; Thomsen and Dahl 2010; Enz et al. 2011b).

© The Author(s), under exclusive license to Springer Nature Switzerland AG 2021
J. J. Thomsen, *Vibrations and Stability*,
https://doi.org/10.1007/978-3-030-68045-9_4

Modal interactions can be especially pronounced when two or more of the linear natural frequencies of a system are *commensurate* or *near*-commensurate, that is, when the natural frequencies are related by integers or near-integers – i.e. $\omega_2 \approx 2\omega_1$, $\omega_2 \approx 3\omega_1$, $\omega_3 \approx \omega_2 \pm \omega_1$, $\omega_3 \approx 2\omega_2 \pm \omega_1$, etc. Depending on the order of the nonlinearity, such frequency relationships can cause the corresponding modes to be strongly coupled, and *internal resonance* is said to exist. For example, if the system has quadratic nonlinearities, then internal resonances may arise when $\omega_i \approx 2\omega_j$ or $\omega_i \approx \omega_j \pm \omega_k$, whereas for cubic nonlinearities the possible conditions are $\omega_i \approx 3\omega_j$, $\omega_i \approx 2\omega_j \pm \omega_k$, or $\omega_i \approx \omega_j \pm \omega_k \pm \omega_l$. Internal resonance is responsible for several interesting phenomena, especially when combined with *external* resonance. Nayfeh and co-workers at the Virginia Polytechnic Institute and State University have produced an impressive number of significant contributions within this area (e.g., Nayfeh and Balachandran 1989; Nayfeh 1989; Balachandran and Nayfeh 1991; and in particular the monograph Nayfeh 1997 and references cited there). See also Cartmell's textbook (1990), and the surveys on autoparametric resonance by Verhulst (1996b) and Tondl et al. (2000).

Sometimes some of the system frequencies ore stiffnesses are not fixed but may vary in time, e.g. with the state of the system. Then some special and sometimes useful variants of nonlinear action can take place, in particular with weakly coupled nonlinear systems, such as, e.g., *energy pumping/targeted energy transfer* (Gendelman et al. 2001; Vakakis et al. 2001; Jian et al.), and *resonance/resonant capture* (Quinn et al. 1995; Vakakis et al. 2001; Jian et al. 2003).

Section 4.2 discusses the *autoparametric vibration absorber* – a device purposefully designed to operate at conditions of combined internal and external resonance. Section 4.3 deals with the nonlinear dynamics of the *non-shallow arch*, an example of a structure for which internal resonance is more or less unavoidable. Brief mentions of other internal resonant structures are given in Sect. 4.4. In Sect. 4.5 we turn to *the follower-loaded double pendulum*, which besides displaying rich dynamics allows us to demonstrate how to employ multiple scales perturbation for systems written in first-order matrix form. Finally, in Sects. 4.6–4.8, we consider nonlinear interactions associated with *vibration induced sliding of mass*.

4.2 The Autoparametric Vibration Absorber

4.2.1 The System

Consider the system in Fig. 4.1, having two degrees of freedom $x(t)$ and $\theta(t)$. Essentially this is the vibration absorber suggested by Haxton and Barr (1972) (see also Cartmell and Lawson (1994)), which for some applications can be a suitable alternative to the traditional tuned mass damper (Hunt 1979; den Hartog 1985; Korenev and Reznikov 1993; Krenk and Høgsberg 2014). The spring-supported mass m_1, which is externally excited by a force $F(t)$, represents the system to be

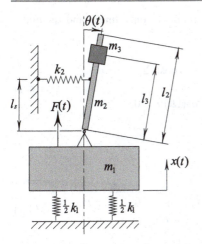

Fig. 4.1 Schematic of an autoparametric vibration absorber (after Cartmell 1990)

damped. The pendulum part constitutes the vibration absorber. This part could be replaced by an elastic beam, as in the experimental absorber described in Haxton and Barr (1972), or by any other elastic component. The essential feature is that nonlinear effects are used to suppress vibrations.

The kinetic and potential energies of the system are, respectively:

$$
\begin{aligned}
T &= \tfrac{1}{2}m_1\dot{x}^2 + \tfrac{1}{2}m_3\left(\dot{x}^2 + l_3^2\dot{\theta}^2 - 2l_3\dot{x}\dot{\theta}\sin\theta\right) \\
&\quad + \tfrac{1}{2}m_2\left(\dot{x}^2 + \tfrac{1}{3}l_2^2\dot{\theta}^2 - l_2\dot{x}\dot{\theta}\sin\theta\right),
\end{aligned}
\tag{4.1}
$$

$$
V = \tfrac{1}{2}k_1x^2 + \tfrac{1}{2}k_2l_s^2\tan^2\theta.
$$

Using Lagrange's equations,

$$
\frac{d}{dt}\left(\frac{\partial L}{\partial \dot{x}}\right) - \frac{\partial L}{\partial x} = F(t); \quad \frac{d}{dt}\left(\frac{\partial L}{\partial \dot{\theta}}\right) - \frac{\partial L}{\partial \theta} = 0; \quad L \equiv T - V,
\tag{4.2}
$$

the equations of motion become:

$$
(m_1 + m_2 + m_3)\ddot{x} + k_1x - \tfrac{1}{2}(m_2l_2 + 2m_3l_3)\left(\dot{\theta}^2\cos\theta + \ddot{\theta}\sin\theta\right) = Q\cos(\Omega t),
$$

$$
(\tfrac{1}{3}m_2l_2^2 + m_3l_3^2)\ddot{\theta} - \tfrac{1}{2}(m_2l_2 + 2m_3l_3)\ddot{x}\sin\theta + k_2l_s^2\tan\theta\left(1 + \tan^2\theta\right) = 0,
$$

$$
\tag{4.3}
$$

where harmonic forcing has been assumed, $F(t) = Q\cos(\Omega t)$. The equations can be rearranged by eliminating θ from the first equation and x from the second.

Expanding the nonlinearities we then obtain, retaining only linear and quadratic terms:

$$\ddot{x} + \omega_1^2 x + \gamma_1 \theta^2 - \frac{\gamma_1}{\omega_2^2} \dot{\theta}^2 = q \cos(\Omega t),$$

$$\ddot{\theta} + \omega_2^2 \theta + \gamma_2 x \theta - \frac{\gamma_2}{\omega_1^2} q \cos(\Omega t)\theta = 0,$$

(4.4)

where the constants are given by:

$$\omega_1^2 = \frac{k_1}{m_1 + m_2 + m_3}, \quad \omega_2^2 = \frac{k_2 l_s^2}{\frac{1}{3} m_2 l_2^2 + m_3 l_3^2}, \quad q = \frac{Q}{m_1 + m_2 + m_3},$$

$$\gamma_1 = \frac{\frac{1}{2}(m_2 l_2 + 2 m_3 l_3)}{m_1 + m_2 + m_3} \omega_2^2, \quad \gamma_2 = \frac{\frac{1}{2}(m_2 l_2 + 2 m_3 l_3)}{\frac{1}{3} m_2 l_2^2 + m_3 l_3^2} \omega_1^2.$$

(4.5)

To illustrate the essential properties of the system, without getting lost in algebra, we assume that the absorber masses m_2 and m_3 are much smaller than the primary mass m_1. This implies that, to a first approximation, the terms containing θ^2 and $q \cos(\Omega t)\theta$ can be ignored in comparison to the remaining linear and nonlinear terms. Finally, on assuming linear viscous damping, we are left with the following equations of motion for the pendulum absorber:

$$\ddot{x} + 2\beta_1 \dot{x} + \omega_1^2 x + \gamma_1 \theta^2 = q \cos \Omega t,$$

(4.6)

$$\ddot{\theta} + 2\beta_2 \dot{\theta} + \omega_2^2 \theta + \gamma_2 x \theta = 0.$$

(4.7)

If the equations are linearized (by letting $\gamma_{1,2} = 0$), one finds that the x and the θ-motions are uncoupled; these motions only influence one another through the nonlinear coupling terms $\gamma_{1,2}$.

The system (4.6)–(4.7) is said to be *autoparametric*, because the solution $x(t)$ of (4.6) acts as a parametric excitation in the $x\theta$ term of (4.7). To see this more clearly we ignore the nonlinearity of (4.6) (let $\gamma_1 = 0$), which then has a linear solution of the form $x(t) = B \cos(\Omega t + \varphi)$. In this case (4.7) becomes

$$\ddot{\theta} + 2\beta_2 \dot{\theta} + \left(\omega_2^2 + \gamma_2 B \cos(\Omega t + \varphi)\right)\theta = 0,$$

(4.8)

which is a linear equation subjected to parametric excitation. It is a *Mathieu* equation, for which it is known that $\theta \to \infty$ when $\Omega \approx 2\omega_2$, even in the presence of damping (cf. App. C, or Nayfeh and Mook 1979). Thus, restricting ourselves to linear analysis, we would doom the vibration absorber useless for practical purposes, due to potential dynamic instabilities of the pendulum part.

However, retaining the nonlinearity ($\gamma_1 > 0$), it appears from (4.6) that any increase in θ will tend to lower the amplitude of x, so that the parametric excitation in (4.7) will decrease in magnitude. Thus we may expect the nonlinearity to limit the linearly unbounded θ solution.

The idea underlying the autoparametric absorber, then, is that energy imparted to the primary mass m_1 may be transferred into motions of the pendulum part through a nonlinear coupling between the primary mass and the absorber. To see how this works we now perform a perturbation analysis for the condition $\Omega \approx \omega_1$, which is the most important case corresponding to primary resonance of the main system m_1. The system (4.6)–(4.7) is a special case of a slightly more general quadratic system analyzed in Nayfeh and Mook (1979).

4.2.2 First-Order Approximate Response

When $\Omega \approx \omega_1$ we expect small levels of excitation q to cause comparatively large responses $x(t)$. Assuming weak nonlinearities and small damping, we order the terms in (4.6)–(4.7) so that excitation, nonlinearities and damping will all appear at the same level of approximation:

$$
\begin{aligned}
\ddot{x} + \omega_1^2 x &= \varepsilon\big(q\cos(\Omega t) - \gamma_1\theta^2 - 2\beta_1\dot{x}\big), \\
\ddot{\theta} + \omega_2^2\theta &= \varepsilon\big(-\gamma_2 x\theta - 2\beta_2\dot{\theta}\big),
\end{aligned}
\tag{4.9}
$$

where $\varepsilon \ll 1$ serves to indicate the assumed smallness of terms. Using the method of multiple scales, we seek a first-order uniformly valid solution:

$$
\begin{aligned}
x &= x_0(T_0, T_1) + \varepsilon x_1(T_0, T_1) + O(\varepsilon^2), \\
\theta &= \theta_0(T_0, T_1) + \varepsilon\theta_1(T_0, T_1) + O(\varepsilon^2),
\end{aligned}
\tag{4.10}
$$

where x_0, x_1, θ_0, θ_1 are functions to be determined, and $T_0 = t$, $T_1 = \varepsilon t$. Substituting into (4.9) and equating to zero coefficients of like powers of ε it is found that, to order ε^0:

$$
\begin{aligned}
D_0^2 x_0 + \omega_1^2 x_0 &= 0, \\
D_0^2 \theta_0 + \omega_2^2 \theta_0 &= 0,
\end{aligned}
\tag{4.11}
$$

and that, to order ε^1:

$$
\begin{aligned}
D_0^2 x_1 + \omega_1^2 x_1 &= -2D_0 D_1 x_0 + q\cos(\Omega T_0) - \gamma_1\theta_0^2 - 2\beta_1 D_0 x_0, \\
D_0^2 \theta_1 + \omega_2^2 \theta_1 &= -2D_0 D_1 \theta_0 - \gamma_2 x_0\theta_0 - 2\beta_2 D_0\theta_0,
\end{aligned}
\tag{4.12}
$$

where $D_i^j \equiv \partial^j / \partial T_i^j$. The general solutions of the zero order equations (4.11) are:

$$
\begin{aligned}
x_0 &= A(T_1)e^{i\omega_1 T_0} + \bar{A}(T_1)e^{-i\omega_1 T_0}, \\
\theta_0 &= B(T_1)e^{i\omega_2 T_0} + \bar{B}(T_1)e^{-i\omega_2 T_0},
\end{aligned}
\tag{4.13}
$$

where A and B are unknown complex-valued functions of the slow time-scale T_1, and overbars denote complex conjugation. Substituting x_0 and θ_0 into (4.12), the first-order problem becomes:

$$
\begin{aligned}
D_0^2 x_1 + \omega_1^2 x_1 &= -i2\omega_1(A' + \beta_1 A)e^{i\omega_1 T_0} + \frac{1}{2}q e^{i\omega_1 T_0}e^{i\sigma_1 T_1} \\
&\quad - \gamma_1\left(B^2 e^{i2\omega_2 T_0} + B\bar{B}\right) + cc, \\
D_0^2 \theta_1 + \omega_2^2 \theta_1 &= -i2\omega_2(B' + \beta_2 B)e^{i\omega_2 T_0} \\
&\quad - \gamma_2\left(ABe^{i(\omega_1 + \omega_2)T_0} + A\bar{B}e^{i(\omega_1 - \omega_2)T_0}\right) + cc,
\end{aligned}
\tag{4.14}
$$

where the term $\cos(\Omega t)$ has been expressed in exponential form, cc denote complex conjugates of preceding terms, and where a *detuning parameter* σ_1 has been introduced to indicate the nearness to *primary external resonance*:

$$
\Omega = \omega_1 + \varepsilon\sigma_1 \quad (\Rightarrow \Omega T_0 = \omega_1 T_0 + \sigma_1 T_1),
\tag{4.15}
$$

Next we determine the requirements for $A(T_1)$ and $B(T_1)$ that ensure the solutions x_1 and θ_1 to be free of secular terms. It appears from Eq. (4.14) that one needs to distinguish between two cases: $\omega_1 \neq 2\omega_2$ and $\omega_1 = 2\omega_2$.

If ω_1 is away from $2\omega_2$, then none of the nonlinear terms (those with $\gamma_{1,2}$) in (4.14) will produce secular terms. The response is then essentially that of the corresponding linear problem, that is: excitation of the primary mass causes this to oscillate ($x(t) \neq 0$) whereas the pendulum part of the absorber is at rest ($\theta(t) = 0$).

If ω_1 is close to $2\omega_2$ we are faced with the more interesting case of *internal resonance*. This is the condition for which the vibration absorber is specifically designed. We analyze it by introducing an additional detuning parameter σ_2, by:

$$
\begin{aligned}
\omega_1 &= 2\omega_2 + \varepsilon\sigma_2 \\
&(\Rightarrow 2\omega_2 T_0 = \omega_1 T_0 - \sigma_2 T_1 \text{ and } (\omega_1 - \omega_2)T_0 = \omega_2 T_0 + \sigma_2 T_1).
\end{aligned}
\tag{4.16}
$$

Note that when $\Omega \approx \omega_1$ and $\omega_1 \approx 2\omega_2$, as assumed, then also $\Omega \approx 2\omega_2$. This is the condition under which the response of the corresponding linear system may be unbounded, as discussed above.

Substituting (4.16) into (4.14), and equating to zero the sum of terms proportional to $e^{i\omega_1 T_0}$ in the first equation and to $e^{i\omega_2 T_0}$ in the second, we arrive at the following conditions for the elimination of secular terms:

$$- i2\omega_1(A' + \beta_1 A) - \gamma_1 B^2 e^{-i\sigma_2 T_1} + \frac{1}{2}qe^{i\sigma_1 T_1} = 0,$$

$$- i2\omega_2(B' + \beta_2 B) - \gamma_2 A\bar{B}e^{i\sigma_2 T_1} = 0. \tag{4.17}$$

When these conditions are fulfilled the particular solutions of (4.14) become:

$$x_1 = -\frac{\gamma_1}{\omega_1^2}B\bar{B} + cc,$$

$$\theta_1 = \frac{\gamma_2}{\omega_1(\omega_1 + 2\omega_2)}ABe^{i(\omega_1 + \omega_2)T_0} + cc, \tag{4.18}$$

from which little can be inferred without knowing the functions $A(T_1)$ and $B(T_1)$. To determine A and B we let

$$A = \frac{1}{2}ae^{i\alpha}, \quad a(T_1), \alpha(T_1) \in R,$$

$$B = \frac{1}{2}be^{i\beta}, \quad b(T_1), \beta(T_1) \in R. \tag{4.19}$$

Substituting this into the solvability conditions (4.17) one obtains, upon separating real and imaginary parts, the following set of modulation equations:

$$a' = -\beta_1 a + \frac{q}{2\omega_1}\sin\psi_1 + \frac{\gamma_1}{4\omega_1}b^2\sin\psi_2,$$

$$b' = -\beta_2 b - \frac{\gamma_2}{4\omega_2}ab\sin\psi_2, \tag{4.20}$$

$$a\alpha' = -\frac{q}{2\omega_1}\cos\psi_1 + \frac{\gamma_1}{4\omega_1}b^2\cos\psi_2,$$

$$b\beta' = \frac{\gamma_2}{4\omega_2}ab\cos\psi_2, \tag{4.21}$$

where

$$\psi_1 \equiv \sigma_1 T_1 - \alpha,$$

$$\psi_2 \equiv \sigma_2 T_1 + \alpha - 2\beta. \tag{4.22}$$

Substituting into (4.10) the equations (4.13), (4.18), (4.19), (4.22), (4.15) and (4.16), $T_0 = t$ and $T_1 = \varepsilon t$, one finds the following first-order approximate solution:

$$x(t) = a\cos(\Omega t - \psi_1)$$
$$+ \varepsilon\frac{-\gamma_1}{2\omega_1^2}b^2 + O(\varepsilon^2),$$
$$\theta(t) = b\cos\left(\frac{1}{2}\Omega t - \frac{1}{2}(\psi_1 + \psi_2)\right)$$
$$+ \varepsilon\frac{\gamma_2}{2\omega_1(\omega_1 + 2\omega_2)}ab\cos\left(\frac{3}{2}\Omega t - \frac{1}{2}(3\psi_1 + \psi_2)\right) + O(\varepsilon^2),$$

(4.23)

where then functions a, b, ψ_1 and ψ_2 are solutions of (4.20)–(4.21) with (4.22).

In seeking stationary solutions we let $a' = b' = y_1' = y_2' = 0$ in (4.20)–(4.21). Stationary values of a, b, ψ_1 and ψ_2 then become solutions of an algebraic system of equations:

$$0 = -\beta_1 a + \frac{q}{2\omega_1}\sin\psi_1 + \frac{\gamma_1}{4\omega_1}b^2\sin\psi_2,$$
$$0 = -\beta_2 b - \frac{\gamma_2}{4\omega_2}ab\sin\psi_2,$$
$$\sigma_1 a = -\frac{q}{2\omega_1}\cos\psi_1 + \frac{\gamma_1}{4\omega_1}b^2\cos\psi_2,$$
$$\frac{1}{2}(\sigma_1 + \sigma_2)b = \frac{\gamma_2}{4\omega_2}ab\cos\psi_2.$$

(4.24)

In solving this system for a and b two possibilities appear. The first is that:

$$a = a_l \equiv \frac{q/(2\omega_1)}{\sqrt{\beta_1^2 + \sigma_1^2}},$$
$$b = b_l \equiv 0,$$

(4.25)

so that, according to (4.23):

$$x(t) = a_l\cos(\Omega t - \psi_1) + O(\varepsilon^2),$$
$$\theta(t) = O(\varepsilon^2).$$

(4.26)

This is essentially the linear solution (as indicated by the subscript 'l'), predicting no motions $\theta(t)$ of the absorber pendulum. The other possibility is that:

$$a = a_n \equiv 2\left(\frac{\omega_2}{\gamma_2}\right)\sqrt{(2\beta_2)^2 + (\sigma_1 + \sigma_2)^2},$$
$$b = b_n \equiv 2\sqrt{-\mu_1 \pm \sqrt{\left(\frac{q}{2\gamma_1}\right)^2 - \mu_2^2}},$$

(4.27)

where

$$
\begin{aligned}
\mu_1 &= \frac{2\omega_1\omega_2}{\gamma_1\gamma_2}\left(2\beta_1\beta_2 - \sigma_1(\sigma_1+\sigma_2)\right), \\
\mu_2 &= \frac{2\omega_1\omega_2}{\gamma_1\gamma_2}\left(2\beta_2\sigma_1 + \beta_1(\sigma_1+\sigma_2)\right),
\end{aligned}
\tag{4.28}
$$

which upon insertion into (4.23) yields that, to order ε:

$$
\begin{aligned}
x(t) &= a_n \cos(\Omega t - \psi_1) + O(\varepsilon), \\
\theta(t) &= b_n \cos\left(\tfrac{1}{2}\Omega t - \tfrac{1}{2}(\psi_1+\psi_2)\right) + O(\varepsilon).
\end{aligned}
\tag{4.29}
$$

In what follows Eq. (4.29), with a and b as given by (4.27), will be termed the 'nonlinear solution'. The nonlinear solution exists only for combinations of system parameters rendering all radicals in (4.27) positive. The stability of the nonlinear solutions is determined by examining eigenvalues of the Jacobian of the modulation equations (4.20)–(4.21), with (4.27) inserted for a and b.

From (4.29) one notice a somewhat peculiar feature of the nonlinear solution: The motion $x(t)$ of the directly excited primary mass is independent of the magnitude q of the excitation (a does not depend on q). We also note from the θ–equation in (4.29) that the absorber mass becomes locked at a frequency exactly half the frequency of excitation Ω.

4.2.3 Frequency and Force Responses

Fig. 4.2 shows *force response* curves, as given by (4.25) and (4.27), for the design-case of perfectly tuned external and internal resonance, that is: $\Omega = \omega_1$ and $\omega_1 = 2\omega_2$ ($\Rightarrow \sigma_1 = \sigma_2 = 0$). Dashed parts of the responses are unstable.

As appears, for low levels of excitation, $q \leq q_1$, the amplitude a of the primary mass increases linearly, following the linear solution (4.25). In this range of loading there is no motion of the absorber mass ($b = 0$). When $q > q_1$ the solution $b = 0$ turns unstable, and the oscillation amplitude b of the absorber pendulum starts to grow (through a supercritical Hopf bifurcation; *cf.* Sect. 3.6.6 and Chap. 5). Also, the linearly increasing branch of the a-amplitude becomes unstable in favor of the nonlinear branch, which is constant-valued. Hence, beyond $q = q_1$ the oscillation amplitude of the primary mass does not change – it becomes *saturated* – and all additional energy supplied to the primary mass spills over into the pendulum mode (b-amplitude) of motion. The nonlinear vibration absorber limits the response amplitude of the primary system to a finite value, soaking up energy that would otherwise tend to increase this amplitude.

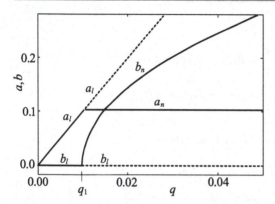

Fig. 4.2 Stationary system amplitudes a and b as functions of excitation level q. Perfectly tuned external and internal resonance; $x(t) \propto a$, $\theta(t) \propto b$. ($\omega_1 = 2\omega_2 = 1$, $\Omega = \omega_1 = 1$, $\gamma_1 = \gamma_2 = 1$, $\beta_1 = \beta_2 = 0.05$)

Fig. 4.3 shows the consequence of slight external detuning, $\sigma_1 = 0.2$; That is, the primary mass is not driven exactly at resonance $\Omega = \omega_1$. As appears, the absorber still limits the a-response, though at a somewhat higher level. Comparing to the case of perfect tuning the most noticeable difference is that *jumps* in the stationary response now appear: Upon increasing the load-level q from below q_1, the stationary absorber amplitude b raises abruptly from zero at $q = q_2$. Similarly, upon decreasing q from above q_2 there is a down-jump to zero at $q = q_1$.

Fig. 4.5 is a frequency response, showing the vibration amplitude a of the primary mass as a function of excitation frequency Ω, for three levels of loading magnitude q. Note how the high amplitude linear resonance peaks near $\Omega = \omega_1$ are destabilized in the presence of nonlinear interaction. Effectively the peaks are 'cut off', and the linear resonance at ω_1 turns into a nonlinear anti-resonance.

Fig. 4.4 depicts the influence of slight *internal* detuning, that is, $\sigma_2 = \omega_1 - 2\omega_2$ differs slightly from zero so that the system is slightly offset from perfect internal resonance. The frequency responses still limit the amplitude a of the primary mass, albeit less efficiently than for perfect internal tuning, $\sigma_2 = 0$.

4.2.4 Concluding Remarks on the Vibration Absorber

Some of the nonlinear phenomena encountered in this section appear with single-DOF systems as well. This concerns *frequency locking, amplitude jumps* and the presence of *multiple solution branches*. Others are particular to multiple-DOF systems: *modal interaction, internal resonance* and *mode saturation*. None of them appear with linear systems. Hence, for the system considered – and others whose mathematical model are similar to it – experimental observations cannot be meaningfully explained by considering only the linearized equations of motion.

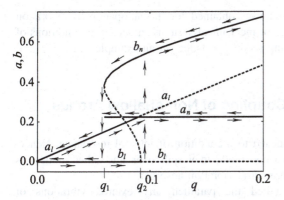

Fig. 4.3 Stationary amplitudes as functions of excitation level q. Perfectly tuned internal resonance, but slightly detuned external resonance. ($\Omega = 1.2$, and other parameters as for Fig. 4.2.)

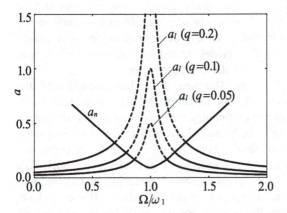

Fig. 4.4 Stationary amplitude a of the primary mass as a function of excitation frequency Ω for three levels of absorber detuning $\sigma_2 = \omega_1 - 2\omega_2$. ($q = 0.1$, $\omega_1 = 1$, $\gamma_1 = \gamma_2 = 1$, $\beta_1 = \beta_2 = 0.05$)

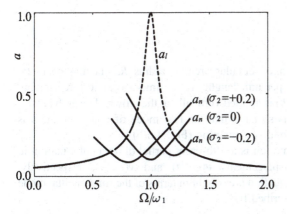

Fig. 4.5 Stationary amplitude a of the primary mass as a function of excitation frequency Ω for three levels of excitation level q. Perfectly tuned internal resonance. ($\omega_1 = 2\omega_2 = 1$, $\gamma_1 = \gamma_2 = 1$, $\beta_1 = \beta_2 = 0.05$)

The force and frequency-responses obtained for the autoparametric vibration absorber are common to a range of mechanical systems subjected to conditions of internal resonance. The following section provides yet an example.

4.3 Nonlinear Mode-Coupling of Non-shallow Arches

Non-shallow arches are predisposed to a condition of internal resonance. Hence, excitation energy imparted into one mode may spill over into other modes, as illustrated above for the autoparametric vibration absorber.

Bolotin, in the 1950s, studied the parametrically excited vibrations of non-shallow arches (Bolotin 1964). Through laboratory experiments he succeeded in validating many of his own theoretical predictions concerning regions of dynamic instability and magnitudes of vibration amplitude. However, he also reported the interesting experimental observation, that '... *with the approach to resonance, the picture of vibration is more complex and difficult to understand*' (Bolotin 1964, p. 331). Bolotin hypothesized that these irregularities were consequences of non-idealities in the physical model. At certain conditions, he believed, the mathematical model ceased to be a valid descriptor of reality.

This took place in the pre-chaos era. Thomsen (1992) later reconsidered this same system, under the hypothesis that the model of Bolotin was perfectly adequate, but that the strange experimental findings were signs of chaotic dynamics. Indeed, it was found that the simple mathematical arch-model of Bolotin displayed chaotic vibrations under conditions similar to those of the laboratory experiments.

We summarize below a multiple scales analysis of Bolotin's arch-model. This analysis will shed further light on the phenomenon of modal interaction, and will provide clues for explaining the strange experimental findings. However, local perturbation analysis only predicts those of the possible responses that are smooth and non-chaotic. To study chaotic responses we need to rely on numerical simulation; for this we return to the non-shallow arch in Chap. 6.

4.3.1 The Model

Fig. 4.6(a) shows a hinged-hinged circular arch of radius R, opening span 2α, bending stiffness EI and mass per unit length ρA. A time-harmonic force having magnitude $P(t) = P_0 + P_t \cos(\theta t)$ is applied vertically at the crown. The arch is *non-shallow*, that is, the angle 2α is so large that the first mode of linear vibration is antisymmetric whereas the second is symmetric (Fig. 4.6(b)).

The linearized equation of motion is a fifth-order partial differential equation in the radial and tangential displacements $u(\varphi, t)$ and $v(\varphi, t)$, respectively. Approximate solutions to this equation are sought in terms of the fundamental mode of antisymmetric vibration, described by:

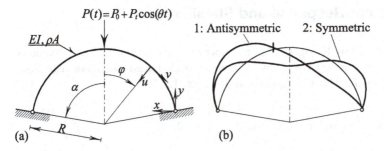

Fig. 4.6 The double-hinged crown-loaded arch. (a) Geometry and loading; (b) Fundamental modes of vibration

$$u(\varphi,t) = \tilde{f}(t)\sin\left(\frac{\pi\varphi}{\alpha}\right); \quad v(\varphi,t) = -\frac{\alpha}{\pi}\tilde{f}(t)\left(1 + \cos\left(\frac{\pi\varphi}{\alpha}\right)\right), \qquad (4.30)$$

where $\tilde{f}(t)$ is the unknown modal amplitude. Nonlinearities arise when incorporating the second-order effects of vertical displacements of the crown, as well as the influence of symmetric components of displacement. The resulting two-mode approximation for the vibrating arch becomes (Thomsen 1992):

$$\ddot{f} + 2\beta\dot{f} + (1 - m\omega^2 u)f = 0, \qquad (4.31)$$

$$\ddot{u} + 2\beta\omega\dot{u} + \omega^2 u + \kappa(f\ddot{f} + \dot{f}^2) = \frac{q}{m}\cos(\Omega\tau), \qquad (4.32)$$

where the nondimensional quantities are defined in terms of the linear natural frequencies $\tilde{\omega}_a$ and ω_s of antisymmetric and symmetric vibrations, as follows:

f	Horizontal crown-displacement
u	Vertical crown-displacement (due to symmetric vibrations)
$\omega = \omega_s/\tilde{\omega}_a$	Frequency ratio (of symmetric to antisymmetric vibrations)
m, κ	Nonlinear coefficients, $m = m(\alpha)$, $\kappa = \kappa(\alpha)$
$q = P_t/(P^* - P_0)$	Magnitude of external load (P^* is the static buckling load)
$\Omega = \theta/\tilde{\omega}_a$	Frequency of external load
β	Viscous damping ratio
τ	Nondimensional time, $\tau = \tilde{\omega}_a t$

The arch equations (4.31)–(4.32) resemble those of the autoparametric vibration absorber, Eq. (4.4). Thus, interesting dynamics may arise when $\Omega \approx \omega \approx 2$, corresponding to combined external and internal resonance. For non-shallow arches of a rather wide range of opening spans 2α, curiously, the natural frequency of symmetric vibrations is just about twice the frequency of antisymmetric vibration. For example, an opening span $2\alpha = 160°$ yields $\omega = \omega_s/\tilde{\omega}_a \approx 2.44$. Non-shallow arches are thus inherently near to internal resonance.

4.3.2 Linear Response and Stability

Linearizing (4.32) by letting $\kappa = 0$, one notices that the symmetric arch vibrations components $u(t)$ is governed by a conventional forced oscillator equation. A particular solution for this equation is (cf. Sect. 1.2.3):

$$u(\tau) = A_0 \frac{q}{m\omega^2} \cos(\Omega\tau - \psi); \quad A_0 \equiv \left(\left(1 - (\Omega/\omega)^2\right)^2 + (2\beta\Omega/\omega)^2 \right)^{-1/2}.$$

$$(4.33)$$

Substituting this into (4.31), we see that antisymmetric vibrations are then governed by:

$$\ddot{f} + 2\beta\dot{f} + (1 - qA_0 \cos(\Omega\tau - \psi))f = 0. \tag{4.34}$$

Obviously $f = 0$ is a solution. This corresponds to the case where the force acting vertically at the crown creates only symmetric vibrations of the arch. However, under certain conditions the zero solution $f = 0$ may become unstable. Since (4.34) can be transformed into a standard Mathieu equation, one can calculate approximations to the boundaries of dynamic instability (see e.g., App. C, or Nayfeh and Mook 1979). Sample results are depicted in Fig. 4.7, in planes spanned by the loading parameters Ω and q. In hatched regions the zero solution $f = 0$ is unstable. If ω is far from 2 (Fig. 4.7(a)), there are two distinct regions of dynamic instability: One near $\Omega = 2$ (primary parametric resonance), and one near $\Omega = \omega$ (primary external resonance). As ω approaches the value 2 (Fig. 4.7(b)) the two regions merge into one, and approach the $q = 0$ axis. This is the so-called *autoparametric* case, which is relevant for non-shallow arches where inherently $\omega \approx 2$ as discussed above. In real service such structures can be dangerous, since unstable vibrations may be excited by extremely small loads within a rather broad frequency band, as appears from Fig. 4.7(b).

Thus, from the above *linearized* analysis one finds that, when $\Omega \approx \omega \approx 2$, even small levels of symmetric loading q may create unstable antisymmetric vibrations. In this linear setting unstable solutions will grow unbounded with time, since no mechanism exists for limiting the growth. To see what actually happens beyond the initial exponential growth, one needs to carry out a *nonlinear* analysis, as described in the next section.

4.3.3 Nonlinear Response and Stability

For the system (4.31)–(4.32) we may examine either the case of primary external resonance with weak excitation ($\Omega \approx \omega$, $q = O(\varepsilon)$), or the case of primary parametric resonance with hard excitation ($\Omega \approx 2$, $q = O(1)$). Since $\omega \approx 2$ the results

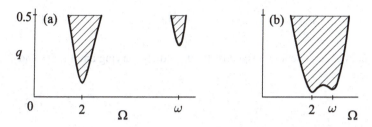

Fig. 4.7 Primary regions of linear instability (hatched) when **(a)** $\omega \gg 2$ and **(b)** $\omega \approx 2$

become about the same, though, it turns out, with the latter case producing a slightly closer fit to numerical results.

Considering the case $\Omega \approx 2$, $q = O(1)$, the terms of (4.31)–(4.32) are ordered accordingly, so that the loading terms and linear terms appear at the lowest level of approximation, whereas terms describing damping and nonlinearities appear at the next, higher level:

$$\ddot{f} + \varepsilon 2\beta\dot{f} + (1 - \varepsilon m\omega^2 u)f = 0, \tag{4.35}$$

$$\ddot{u} + \varepsilon 2\beta\omega\dot{u} + \omega^2 u + \varepsilon\kappa(f\ddot{f} + \dot{f}^2) = \frac{q}{m}\cos(\Omega\tau), \tag{4.36}$$

where ε denotes the usual bookkeeper of small terms. Using the method of multiple scales we assume for the solution a uniformly valid expansion:

$$f = f_0(T_0, T_1) + \varepsilon f_1(T_0, T_1) + O(\varepsilon^2), \tag{4.37}$$

$$u = u_0(T_0, T_1) + \varepsilon u_1(T_0, T_1) + O(\varepsilon^2). \tag{4.38}$$

The analysis proceeds just as for the autoparametric vibration absorber (cf. Sect. 4.2). A pair of detuning parameters, σ_1 and σ_2, are introduced to indicate the nearness to internal and parametric resonance, respectively:

$$\begin{aligned} \omega &= 2 + \varepsilon\sigma_1, \\ \Omega &= 2 + \varepsilon\sigma_2. \end{aligned} \tag{4.39}$$

Performing then the usual operations one ends up with the following two-term approximation for the nonlinear arch response:

$$\begin{aligned} f(\tau) = a_1 \cos(\tfrac{1}{2}\Omega\tau - \tfrac{1}{2}\psi_1) \\ + \frac{2\Lambda_1 a_1 a_2}{(\tfrac{3}{2}\Omega)^2 - 1}\cos(\tfrac{3}{2}\Omega\tau - \tfrac{1}{2}\psi_1 + \psi_2) + O(\varepsilon^2), \end{aligned} \tag{4.40}$$

$$u(\tau) = a_2 \cos(\Omega\tau + \gamma_2) + O(\varepsilon^2), \tag{4.41}$$

where (a_1, a_2) and (ψ_1, ψ_1) denote, respectively, the slowly varying amplitudes and phases, and where

$$\Lambda_1 \equiv -\tfrac{1}{4}m\omega^2. \tag{4.42}$$

Comparing to the similar Eq. (4.23) for the autoparametric vibration absorber, we see that symmetric vibrations u of the arch correspond to primary-mass motions x of the damper, and that antisymmetric vibrations f of the arch correspond to pendulum motions θ of the damper.

For Eqs. (4.40)–(4.41) it turns out that the amplitudes $a_1(T_1)$ and $a_2(T_1)$ are governed by the modulation equations:

$$
\begin{aligned}
a_1' &= -\beta a_1 - \Lambda_1 a_1 a_2 \sin(\psi_1 + \psi_2),\\
a_2' &= -\beta\omega a_2 + \Lambda_2 a_1^2 \sin(\psi_1 + \psi_2) + q^*(\beta\cos\psi_2 - \sin\psi_2),\\
a_1\psi_1' &= \sigma_2 a_1 - 2\Lambda_1 a_1 a_2 \cos(\psi_1 + \psi_2),\\
a_2\psi_2' &= (\sigma_1 - \sigma_2)a_2 + \Lambda_2 a_1^2 \cos(\psi_1 + \psi_2) - q^*(\cos\psi_2 + \beta\sin\psi_2),
\end{aligned}
\tag{4.43}
$$

where primes denote derivatives with respect to slow time T_1, and

$$\Lambda_2 \equiv -\frac{1}{2}\frac{\kappa}{\omega}, \qquad q^* \equiv \frac{q/m}{\Omega + \omega}. \tag{4.44}$$

In seeking stationary solutions we let $a_1' = a_2' = y_1' = y_2' = 0$, and find that there are two possibilities. The first essentially corresponds to the linear solution, where the symmetric load excites only symmetric vibrations:

$$
\begin{aligned}
a_1 &= 0,\\
a_2^2 &= \frac{(q^*)^2(1 + \beta^2)}{(\sigma_1 - \sigma_2)^2 + (\beta\omega)^2}.
\end{aligned}
\tag{4.45}
$$

The other possibility is that the symmetric load excites antisymmetric vibrations through a nonlinear interaction between modes:

$$
\begin{aligned}
a_1^2 &= \frac{1}{\Lambda_1\Lambda_2}\left(\chi_1 \pm \sqrt{(\Lambda_1 q^*)^2(1 + \beta^2) - \chi_2^2}\right),\\
a_2^2 &= \frac{\tfrac{1}{4}\sigma_2^2 + \beta^2}{\Lambda_1^2},
\end{aligned}
\tag{4.46}
$$

where

$$\chi_1 \equiv -\frac{1}{2}\sigma_2(\sigma_1 - \sigma_2) - \beta^2\omega, \quad \chi_2 \equiv (\sigma_1 - \sigma_2)\beta - \frac{1}{2}\beta\omega\sigma_2. \quad (4.47)$$

The stability of solutions is governed by the real part and sign of the Jacobian eigenvalues of (4.43), evaluated at the solution points.

Figs. 4.8, 4.9 and 4.10 depict aspects of the stationary first-order multiple scales solutions, when $\omega = 2.44$, $\kappa = 2.63$, $m = 3.32$ and $\beta = 0.03$. The parameter values correspond to a weakly damped arch of opening span 160°.

Fig. 4.8 shows the amplitudes of antisymmetric (a_1) and symmetric (a_2) vibrations versus load q for the near-resonant case $\Omega = 2.6$. The linear solution (4.45) is indexed by 'l' and the nonlinear solution (4.46) by 'n'. Possible jumps in amplitude are indicated by arrows, and dashed parts of the solutions are unstable. The curves appear very similar to the corresponding ones for the vibration absorber (Fig. 4.3). Increasing the load q from below, it is seen, the amplitude a_2 of symmetric vibrations initially increases on the linear branch, whereas there is no antisymmetric motion ($a_{1l} = 0$). For $q_1 < q < q_2$ there are two stable solutions: the linear and the nonlinear. The one actually reached depends on the initial conditions. For $q > q_2$ the nonlinear solution is the only stable one. In this range the directly excited symmetric mode a_{2n} becomes saturated, that is, its amplitude is independent of the loading q. Any additional supply of energy associated with increasing q is then transferred to the antisymmetric mode a_{1n}.

Fig. 4.9 shows the amplitudes a_1 and a_2 versus excitation frequency Ω, for a constant magnitude of the load $q = 0.3$. This value of q corresponds to 30% of the static buckling load, a quite hard excitation. It appears that when Ω is increased beyond Ω_1, or decreased below Ω_2, the linear solution becomes unstable and the solution jumps to the nonlinear branches. As an interesting feature one may notice that for $\Omega_3 < \Omega < \Omega_4$ both the linear and the nonlinear solution are unstable. Hence, within this range of excitation frequencies the system will neither be at rest nor perform small amplitude periodic motions – even though the excitation is periodic. Where will it go then? It will turn further out in state space, exploring regions that are inaccessible by the local methods on which we have relied. It might settle down into stable periodic motion with very large amplitude. Or it may wander restlessly around on a chaotic attractor. The laboratory experiments of Bolotin, as well as numerical simulations, indicate that the system behaves chaotically for this range of excitation frequencies. Figure also shows the results ($\bigcirc$, $\square$) of numerically integrating the original model equations (4.35)–(4.36), showing reasonable agreement with the theoretical results. The largest discrepancies are associated with the nonlinear branch a_{1n}, for which the (ignored) second term of Eq. (4.40) contributes significantly to the response.

Fig. 4.10 depicts regions of stability and instability of the stationary solutions in the plane of the loading parameters Ω and q. The curve separating regions B and D constitutes the classical boundary of linear parametric instability (compare to Fig. 4.7). Linear theory predicts that outside this border (i.e. in regions D and A) the

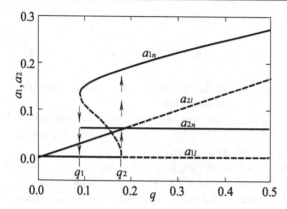

Fig. 4.8 Stationary modal arch amplitudes a_1 and a_2 as functions of load magnitude q. l: 'linear' solution, n: nonlinear solution. Dashed parts unstable. ($\Omega = 2.6$, $\omega = 2.44$, $\kappa = 2.63$, $m = 3.32$, $\beta = 0.03$)

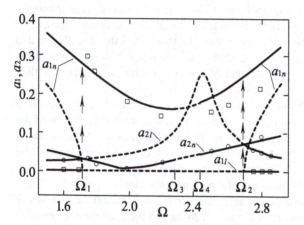

Fig. 4.9 Stationary modal arch amplitudes a_1 and a_2 as functions of excitation frequency Ω. l: 'linear' solution, n: nonlinear solution. Dashed parts unstable. $\bigcirc$,$\square$: numerical solution. ($q = 0.3$, $\omega = 2.44$, $\kappa = 2.63$, $m = 3.32$, $\beta = 0.03$)

zero solution is stable, whereas within the border (regions B and C) it is unstable. The nonlinear analysis enables us to extend and sharpen the conclusions drawn from linear analysis, as follows. As for region D, it is correct that the zero solution is stable. However, the nonlinear (non-zero) solution is stable as well. This means that the zero solution is stable only to sufficiently small perturbations, since a large perturbation may carry the system to the stable nonlinear branch. As for region B, it is correct that the zero solution is unstable. However, this does not mean that the amplitudes of the system grow without bounds. Rather, they become limited at a finite value, perhaps very small. As for the knife-shaped region C, the nonlinear local analysis revealed that no stable periodic motions of small amplitude exist here.

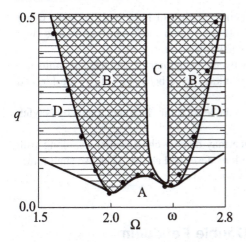

Fig. 4.10 Stability of stationary solutions as a function of loading parameters Ω and q. Region A: only the *linear* solution is stable; B: only the *nonlinear* solution is stable; C: *neither* solutions are stable; D: *both* solutions are stable. •: Numerical simulation. ($\omega = 2.44$, $\kappa = 2.63$, $m = 3.32$, $\beta = 0.03$)

Numerical simulations show that quasiperiodic and chaotic motions prevail in this region (cf. Chap. 6).

4.4   Other Systems Possessing Internal Resonance

The below examples are meant to be just indicative of other areas where internal resonance may dominate the response of a system.

A system consisting of two orthogonally clamped beams with a two-to-one internal resonance have been studied analytically and experimentally in a number of papers (e.g., Nayfeh and Zavodney 1988; Nayfeh and Balachandran 1989; Nayfeh et al. 1989; Balachandran and Nayfeh 1991).

Cylindrical shells may exhibit nonlinear coupling between breathing and flexural modes due to internal two-to-one resonances (Nayfeh and Raouf 1987; Nayfeh and Balachandran 1989).

The dynamics of ships rolling in sea waves (Nayfeh and Khdeir 1986) are also governed by coupled equations with quadratic nonlinearities, thus being a candidate for complicated motions.

The nonlinear dynamics of the Golden Gate Bridge in San Francisco have been considered with a special emphasis to the occurrence of internal resonance (Rossikhin and Shitikova 1995).

Consideration to modal interactions caused by a three-to-one internal resonance was included in a study of vibration-induced fluid flow in pipes (Jensen 1997).

Langthjem (1995b) studied the dynamic stability of immersed fluid-conveying tubes. Seemingly chaotic coupling between flutter and whirling modes was observed for an experimental tube immersed in water.[2]

Chaotic vibrations due to coupling between the symmetric and the antisymmetric mode of an elastically buckled two-bar linkage have been studied by Sorokin and Terentiev (1998).

These are just examples; Many more can be found in the monographs by Nayfeh (1997) and Tondl et al. (2000).

Most systems with internal resonance may display chaotic motions, along with the regular and predictable kinds of motions described above. Thus, we shall return to internal resonant systems in Chap. 6 on chaos.

4.5   The Follower-Loaded Double Pendulum

In this section we examine possible periodic motions of a well-known archetype of a *flutter*-prone structure (cf. Sect. 2.4.2), the double pendulum of Fig. 4.11. The system differs from those already described in this chapter in that it is *autonomous*, that is, the external load is not an explicit function of time. This case will provide us with an opportunity to exercise the method of multiple scales on models written in first-order matrix form. But first a brief account of the background.

Some structures are prone to flutter-instability due to the presence of non-conservative follower-type forces. Examples are compressors, turbines, fluid conveying pipes, aircraft wings, rockets, suspension bridges, and shafts with controlled speed of revolution. The load-limits of stability and the initial unstable behavior of such systems may be established through linearization, whereas the prediction of long-time behavior requires a nonlinear analysis. Traditionally, long-time behavior has been characterized as either soft flutter (finite amplitude periodic motion, possibly causing fatigue failure), explosive flutter (perhaps causing ultimate failure), or escape to a different state of equilibrium. However, as pointed out in Thomsen (1995), in some cases the final state may be none of the above but rather stationary chaotic.

Theoretical interest in non-conservative problems of stability was stimulated by a number of flutter-related structural failures, the most renowned undoubtedly being the collapse of the Tacoma Narrows Bridge in 1940 (e.g., Billah and Scanlan 1991; Koughan J 1996; Peterson 1990; Ross 1984). It has persevered ever since. Most flutter research concentrates on certain archetypal flutter-prone models that are *generic*, that is, they illustrate the phenomena in question in the simplest possible setting. One such model is the *Beck's column* – a fixed-free column subject to a tangential load at the free end (e.g., Ziegler 1968; Langthjem and Sugiyama 2000;

---

[2]Langthjem kindly demonstrated this experiment to the author, who got pretty wet in the attempt to count the periods of the modes involved; they were nearly two-to-one.

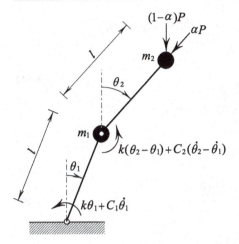

Fig. 4.11 The partially follower-loaded elastic double pendulum

Andersen and Thomsen 2002). Another one is the *partially follower-loaded elastic double pendulum*, which is to be considered here. In fact, this double pendulum can be viewed as a two-mode approximation to Beck's column, at least for small deflections.

4.5.1 The Model

Consider the system of Fig. 4.11, composed of two rigid and weightless rods of equal length l that carry concentrated masses m_1 and m_2. A tangential follower load αP and a non-following (constant-directional) load $(1-\alpha P)$ is applied at the free end. Here, P is an indicator of load-*level* whereas the parameter $\alpha \in R$ characterizes the loading-*arrangement*. The special case $\alpha = 0$ corresponds to purely conservative loading, e.g., as provided by gravity. Another special case is $\alpha = 1$ corresponding to a perfectly following load, e.g., as provided by fluid flowing through the bars in a horizontal or zero-gravity arrangement (e.g. Langthjem and Sugiyama 1999). At the frictionless hinges there are linear torsional springs with stiffness k, and linear dampers with viscous damping coefficients C_1 and C_2. The bar-angles $\theta_1(t)$ and $\theta_2(t)$ define the instantaneous state of the system, and are thus generalized coordinates.

The equations of motion are conveniently set up by using Lagrange's equations. The kinetic energy T, potential energy V, Rayleigh dissipation function R, and generalized forces Q_1 and Q_2 are, respectively:

$$T = \frac{1}{2}m_1 l^2 \dot{\theta}_1^2 + \frac{1}{2}m_2 l^2 \left(\dot{\theta}_1^2 + \dot{\theta}_2^2 + 2\dot{\theta}_1 \dot{\theta}_2 \cos(\theta_2 - \theta_1) \right),$$

$$V = \frac{1}{2}k\theta_1^2 + \frac{1}{2}k(\theta_2 - \theta_1)^2,$$

$$R = \frac{1}{2}C_1 \dot{\theta}_1^2 + \frac{1}{2}C_2 \left(\dot{\theta}_2 - \dot{\theta}_1 \right)^2, \tag{4.48}$$

$$Q_1 = Pl((1 - \alpha) \sin \theta_1 + \alpha \sin(\theta_1 - \theta_2)),$$

$$Q_2 = Pl(1 - \alpha) \sin \theta_2.$$

Using Lagrange's equations:

$$\frac{d}{dt}\left(\frac{\partial T}{\partial \dot{\theta}_i} \right) - \frac{\partial(T - V)}{\partial \theta_i} + \frac{\partial R}{\partial \dot{\theta}_i} = Q_i, \quad i = 1, 2, \tag{4.49}$$

the following equations of motion are obtained:

$$(1 + m)\ddot{\theta}_1 + \cos(\theta_2 - \theta_1)\ddot{\theta}_2 + (c_1 + c_2)\dot{\theta}_1$$
$$- c_2\dot{\theta}_2 + 2\theta_1 - \theta_2 + \dot{\theta}_2^2 \sin(\theta_1 - \theta_2)$$
$$= p((1 - \alpha) \sin \theta_1 + \alpha \sin(\theta_1 - \theta_2)),$$
$$\cos(\theta_2 - \theta_1)\ddot{\theta}_1 + \ddot{\theta}_2 + c_2(\dot{\theta}_2 - \dot{\theta}_1)$$
$$- \theta_1 + \theta_2 - \dot{\theta}_1^2 \sin(\theta_1 - \theta_2)$$
$$= p(1 - \alpha) \sin \theta_2, \tag{4.50}$$

where nondimensional quantities have been introduced as follows:

$$\theta_i = \theta_i(\tau), \quad \tau = \tilde{\omega}t, \quad \tilde{\omega}^2 = \frac{k}{m_2 l^2},$$

$$m = \frac{m_1}{m_2}, \quad p = \frac{Pl}{k}, \quad c_i = \frac{C_i}{\tilde{\omega}m_2 l^2}, \quad i = 1, 2, \tag{4.51}$$

To keep the number of parameters at a manageable level we fix the mass ratio at $m = 2$ (this corresponds to an equivalent continuous beam having uniform mass distribution), and let $c_1 = c_2 = c$.

The governing equations (4.50) are nonlinear in the state variables θ_1 and θ_2. For studying small-amplitude motions of the pendulum near $(\theta_1, \theta_2) = (0, 0)$ we may expand the equations into a Taylor-series. Keeping terms to order three one arrives at the following approximate system, written in matrix form as a system of four first-order differential equations:

$$\dot{\mathbf{x}} = \mathbf{A}(\alpha, p)\mathbf{x} + \mathbf{f}(\alpha, p, \mathbf{x}), \qquad (4.52)$$

where $\mathbf{x} = \{\theta_1, \theta_2, \dot{\theta}_1, \dot{\theta}_2\}^T$ is a vector of state variables, $\mathbf{f}$ is a vector containing cubic nonlinearities,

$$\mathbf{f}(\alpha, p, \mathbf{x}) = \sum_{j,k,l=1}^{4} \mathbf{b}_{jkl}(\alpha, p) x_j x_k x_l, \qquad (4.53)$$

where the coefficients $\mathbf{b}_{jkl}(\alpha, p)$ can be found in Thomsen (1995), and the matrix $\mathbf{A}$ describes the linear part of the system:

$$\mathbf{A}(\alpha, p) = \begin{bmatrix} \mathbf{0} & \mathbf{I} \\ -\mathbf{M}^{-1}\mathbf{K} & -\mathbf{M}^{-1}\mathbf{C} \end{bmatrix}, \qquad (4.54)$$

wherein the sub-matrices are given by

$$\mathbf{M} = \begin{bmatrix} 3 & 1 \\ 1 & 1 \end{bmatrix}, \quad \mathbf{C} = \begin{bmatrix} 2c & -c \\ -c & c \end{bmatrix}, \quad \mathbf{K} = \begin{bmatrix} 2-p & p\alpha - 1 \\ -1 & 1 - p(1-\alpha) \end{bmatrix}. \qquad (4.55)$$

For finite pendulum motions near the upright position $(\theta_1, \theta_2) = (0, 0)$, the approximate Eq. (4.52) captures the dynamics of the full Eq. (4.50).

4.5.2 The Zero Solution and Its Stability

It appears from (4.50) and (4.52) that the $(\theta_1, \theta_2) = (0, 0)$ or $\mathbf{x}(\tau) = \mathbf{0}$ is a possible solution. Is it stable? And, if unstable, then *how* do unstable solutions evolve in time?

For the analysis of stability one calculates the eigenvalues of the Jacobian $\mathbf{J}(\mathbf{x})$ of (4.52), evaluated at the singular point $\tilde{\mathbf{x}} = \mathbf{0}$. Since by (4.52) $\mathbf{J}(\tilde{\mathbf{x}}) = \mathbf{A}$ where $\mathbf{A}$ is given in (4.54), the Jacobian eigenvalues λ become solutions of the determinant equation $|\mathbf{A}(\alpha, p) - \lambda\mathbf{I}| = 0$. This expands to a polynomial equation:

$$H_0\lambda^4 + H_1\lambda^3 + H_2\lambda^2 + H_3\lambda + H_4 = 0, \qquad (4.56)$$

where

$$\begin{aligned} H_0 &= 2, \quad H_1 = 7c, \quad H_2 = 2p(\alpha - 2) + c^2 7, \\ H_3 &= c(3p(\alpha - 1) + 2), \quad H_4 = p(3 - p)(\alpha - 1) + 1. \end{aligned} \qquad (4.57)$$

The zero solution $\mathbf{x} = \mathbf{0}$ is asymptotically stable (locally) if all roots of the characteristic equation (4.56) have negative real parts. According to the *Routh-Hurwitz criterion* (see App. C) this is the case when the following four conditions are all met:

$$H_1 > 0, \quad D_2 = H_1 H_2 - H_0 H_3 > 0,$$
$$D_3 = D_2 H_3 - H_1^2 H_4 > 0, \quad H_4 > 0, \tag{4.58}$$

where H_j, $j = 0$, 4 are the simple functions of loading parameters (α, p) and damping coefficient c given in (4.57).

Fig. 4.12 depicts domains of stability and instability of the zero solution in the plane of loading parameters (α, p), for a small value of the damping coefficient $(c = 0.1)$. The unhatched white domain corresponds to loadings for which the zero solution is stable. Since a nonlinear system is considered, we should here be more strict and say that the zero solution is *locally* stable, that is: Disturbing the upright position of the double pendulum, it will return to that position if the disturbance is 'small', though, a strong disturbance might cause it to escape to another state of static or dynamic equilibrium.

The unstable (hatched) domains in Fig. 4.12 can be classified according to the nature and sign of the four roots (eigenvalues) of the characteristic polynomial (4.56). For small positive damping the local behavior near $\theta_1 = \theta_2 = 0$ can be summarized as follows:

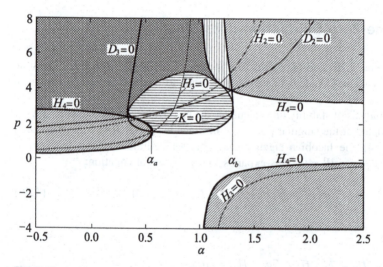

Fig. 4.12 Stability of the zero solution. Legends (e.g., $K = 0$) attached to the 'positive' side of curves (e.g., where $K > 0$). Domains: ☐ *stable*, ▦ *pure divergence*, ▤ *pure flutter*, ▨ *decayed-oscillation divergence*, ▥ *fluttering divergence*. (Thomsen 1995)

- *Pure divergence* (cross-hatched in Fig. 4.12): four real roots, two positive and two negative.
- *Pure flutter* (horizontally-hatched): two pairs of complex conjugate roots, one pair with positive and one pair with negative real part.
- *Decayed-oscillation divergence* (diagonally-hatched): two real roots of opposite sign, and a pair of complex conjugate roots with negative real part.
- *Fluttering divergence.* (vertically-hatched): two real roots of opposite sign, and a pair of complex conjugate roots with positive real part.

The notions of *flutter* and *divergence* are tied to the nature of the eigenvalues this way: Consider a linearized system $\dot{\mathbf{x}} = \mathbf{Ax}$ with a Jacobian eigenvalue $\lambda = a + ib$, $a, b \in R$. The solution $\mathbf{x}(t)$ consists of a sum of terms having the form $\mathbf{x}(t) = \varphi e^{\lambda t} = \varphi e^{(a+ib)t} = \varphi e^{at}(\cos(bt) + i \sin(bt))$, one term for each eigenvalue. Thus, $e^{\lambda t}$ describes the time-evolution of a particular component of the solution. With a real and positive eigenvalue ($a > 0$, $b = 0$) the term $e^{\lambda t}$ will escape towards infinity at exponential rate (*divergence*). With a complex-valued eigenvalue having positive real part ($a > 0$, $b \neq 0$) the solution will oscillate around zero with amplitudes increasing expontially in time (*flutter*). When several eigenvalues are involved, flutter and divergence may combine to create damped oscillations around a solution that diverges exponentially from zero (*decayed-oscillation divergence*), or exponentially growing oscillations around a solution diverging exponentially from zero (*fluttering divergence*).

We have now established conditions under which the upright position of the double pendulum may become unstable. And we have described the kinds of motion accompanying unstable behavior (flutter, divergence, etc.). For a truly linear system ($\mathbf{f} = \mathbf{0}$ in (4.52)) this would end the discussion. There would be no equilibrium states other than the zero solution, and so any unstable solution would approach infinity as predicted by the local analysis of stability. However, in the presence of nonlinearities ($\mathbf{f} \neq \mathbf{0}$) the system may possess any number of stable and unstable states of static or dynamic equilibrium. So, saying that the zero solution for the nonlinear double pendulum is unstable does not reveal where the system will then be instead, as in the linear setting. We have to calculate the stable state(s) to which the system will go in place of the zero. In the following section we consider the existence and stability of periodic solutions near the zero solution.

4.5.3 Periodic Solutions

When for the linear system ($\mathbf{f} = \mathbf{0}$) the zero solution is unstable, the state (θ_1, θ_2) of the double pendulum will approach infinity. Usually the presence of nonlinearities ($\mathbf{f} \neq \mathbf{0}$) will limit this growth of state variables, perhaps into the form of finite amplitude periodic oscillations. One can expect such periodic solutions to occur near regions of the loading parameter space where the so-called *Hopf conditions* are met. That is, where the characteristic equation (4.56) has a pair of purely imaginary roots $\lambda = \pm i\omega$, while the remaining roots have negative real parts. The Hopf conditions establish conditions for a *Hopf bifurcation* to occur, and this is just the kind of bifurcation we are looking for – turning a stable equilibrium unstable in favor of stable periodic oscillations (we shall return to Hopf in Chap. 5).

One can rather easily show that the Hopf conditions are met on the part of the curve $D_3 = 0$ that is restricted by $H_3 > 0$ and $D_2 > 0$. In Fig. 4.12 the Hopf bifurcation set is located on $D_3 = 0$, $\alpha_a < \alpha < \alpha_b$.

To examine periodic motions we perform a perturbation analysis for values of the loading parameters (α, p) near the Hopf bifurcation set. First introduce into (4.52) a nondimensional parameter ε to indicate the assumed smallness of nonlinear terms:

$$\dot{\mathbf{x}} = \mathbf{A}(\alpha, p)\mathbf{x} + \varepsilon \mathbf{f}(\alpha, p, \mathbf{x}). \tag{4.59}$$

Then assume that, for a fixed value of loading 'conservativeness' α, the loading magnitude p is given a slight perturbation εp_1 from the critical value p_0:

$$p = p_0 + \varepsilon p_1, \tag{4.60}$$

where (α, p_0) belongs to the Hopf bifurcation set, and p_1 is a free perturbation parameter (playing a role similar to that of the *detuning parameters* in previous examples).

Using the method of multiple scales we seek an approximate solution to (4.59) in form of a uniformly valid expansion:

$$\mathbf{x}(\tau; \varepsilon) = \mathbf{x}_0(T_0, T_1) + \varepsilon \mathbf{x}_1(T_0, T_1) + O(\varepsilon^2), \tag{4.61}$$

where $\mathbf{x}_0$ and $\mathbf{x}_1$ are the unknown functions to be determined, and where $T_0 = \tau$ and $T_1 = \varepsilon \tau$ are the fast and slow time-scales, respectively. By virtue of (4.60) and (4.61) we may Taylor-expand $\mathbf{f}$ and $\mathbf{A}$ in terms of ε in the vicinity of the bifurcation point (α, p_0), as follows:

$$\begin{aligned} \mathbf{f}(\alpha, p, \mathbf{x}) &= \mathbf{f}(\alpha, p_0, \mathbf{x}_0) + O(\varepsilon), \\ \mathbf{A}(\alpha, p) &= \mathbf{A}_0 + \varepsilon p_1 \mathbf{A}_1 + O(\varepsilon^2), \end{aligned} \tag{4.62}$$

where

$$\mathbf{A}_0 \equiv \mathbf{A}(\alpha, p_0), \quad \mathbf{A}_1 \equiv \left. \frac{\partial \mathbf{A}}{\partial p} \right|_{p=p_0}. \tag{4.63}$$

Substituting (4.61), (4.62) and (4.53) into (4.59), by noting that $d/d\tau = \partial/\partial T_0 + \varepsilon \partial/\partial T_1$, and equating then coefficients to like powers ε one obtains, to order ε^0:

$$\frac{\partial \mathbf{x}_0}{\partial T_0} - \mathbf{A}_0 \mathbf{x}_0 = \mathbf{0}, \tag{4.64}$$

and to order ε^1:

$$\frac{\partial \mathbf{x}_1}{\partial T_0} - \mathbf{A}_0 \mathbf{x}_1 = -\frac{\partial \mathbf{x}_0}{\partial T_1} + p_1 \mathbf{A}_1 \mathbf{x}_0 + \sum_{j,k,l=1}^{4} \mathbf{b}_{jkl}(\alpha, p_0) x_{0j} x_{0k} x_{0l}, \tag{4.65}$$

where x_{0j} denotes the j'th component of $\mathbf{x}_0$.

The definition of $\mathbf{A}_0$ ensures that two of its eigenvalues are purely imaginary, while the two others have negative real parts. Disregarding damped terms (corresponding to those eigenvalues having negative real parts) the solution of the zero order problem (4.64) can therefore be written:

$$\mathbf{x}_0 = a(T_1)\mathbf{u}e^{i\omega_0 T_0} + \bar{a}(T_1)\bar{\mathbf{u}}e^{-i\omega_0 T_0}, \tag{4.66}$$

where $a(T_1)$ is a yet unknown function, and where $(i\omega_0)$ and $\mathbf{u}$ are, respectively, the purely imaginary eigenvalue and corresponding eigenvector of the eigenvalue problem:

$$(\mathbf{A}_0 - i\omega_0 \mathbf{I})\mathbf{u} = 0. \tag{4.67}$$

One can easily show that $\omega_0^2 = H_3(\alpha, p_0)/H_1 = \frac{3}{7} p_0(\alpha - 1) + \frac{2}{7}$, which is independent of the coefficient of damping c.

Substituting (4.66) into the first-order problem (4.65) gives:

$$\frac{\partial \mathbf{x}_1}{\partial T_0} - \mathbf{A}_0 \mathbf{x}_1 = \mathbf{q}_1 e^{i\omega_0 T_0} + \mathbf{q}_3 e^{i3\omega_0 T_0} + cc, \tag{4.68}$$

where

$$\mathbf{q}_1 = -\frac{da}{dT_1}\mathbf{u} + p_1 a \mathbf{A}_1 \mathbf{u} + a^2 \bar{a} \sum_{j,k,l=1}^{4} \mathbf{b}_{jkl}(\alpha, p_0)\left(\bar{u}_j u_k u_l + u_j \bar{u}_k u_l + u_j u_k \bar{u}_l\right),$$

$$\mathbf{q}_3 = a^3 \sum_{j,k,l=1}^{4} \mathbf{b}_{jkl}(\alpha, p_0) u_j u_k u_l, \tag{4.69}$$

where u_j denotes the j'th component of the eigenvector $\mathbf{u}$, and $\bar{u}_j$ is the complex conjugate of u_j. The first term on the right-hand side of (4.68) is resonant to the homogeneous equation, since $(i\omega_0)$ is an eigenvalue of $\mathbf{A}_0$. The solution of (4.68) will then contain secular terms proportional to $T_0 e^{i\omega_0 T_0}$, unless the function $\mathbf{q}_1$ is restricted so as to prevent this. One can show (Kuo et al. 1972, Nayfeh and Balachandran 1995) that secular terms are eliminated when $\mathbf{q}_1$ is orthogonal to the left eigenvector $\mathbf{v}$ of $\mathbf{A}_0$, that is, we require $\mathbf{v}^T\mathbf{q}_1 = 0$ where $\mathbf{v}$ is a solution of $\mathbf{v}^T(\mathbf{A}_0 - i\omega_0\mathbf{I}) = \mathbf{0}^T$, or equivalently, of $(\mathbf{A}_0^T - i\omega_0\mathbf{I})\mathbf{v} = \mathbf{0}$. Substituting $\mathbf{q}_1$ from (4.69) into the condition $\mathbf{v}^T\mathbf{q}_1 = 0$ one obtains the following solvability condition:

$$\frac{da}{dT_1} - \beta a - \gamma a^2\bar{a} = 0, \tag{4.70}$$

where

$$\beta = \frac{p_1\mathbf{v}^T\mathbf{A}_1\mathbf{u}}{\mathbf{v}^T\mathbf{u}} \equiv \beta_R + i\beta_I,$$

$$\gamma = \frac{\mathbf{v}^T\sum_{j,k,l=1}^4\mathbf{b}_{jkl}(\alpha,p_0)\left(\bar{u}_j u_k u_l + u_j\bar{u}_k u_l + u_j u_k\bar{u}_l\right)}{\mathbf{v}^T\mathbf{u}} \tag{4.71}$$

$$\equiv \gamma_R + i\gamma_I,$$

where (β_I, β_R) and (γ_I, γ_R) denote real and imaginary parts of β and γ. For the determination of the function $a(T_1)$ we let

$$a = Ae^{i\varphi}, \quad A(T_1), \varphi(T_1) \in R, \tag{4.72}$$

whereby (4.70) converts into a pair of real-valued modulation equations:

$$\frac{dA}{dT_1} = \beta_R A + \gamma_R A^3 \equiv G(A), \tag{4.73}$$

$$A\frac{d\varphi}{dT_1} = \beta_I A + \gamma_I A^3. \tag{4.74}$$

In seeking stationary solutions it appears that there are two possibilities. The first is that $A = 0$, which corresponds to the zero solution $\mathbf{x} = \mathbf{0}$. The other is that $A = \sqrt{-\beta_R/\gamma_R}$, corresponding to nonlinear periodic motion as given by:

$$\begin{aligned}
\mathbf{x} &= \mathbf{x}_0 + \varepsilon\mathbf{x}_1 + O(\varepsilon^2)\\
&= a(T_1)\mathbf{u}e^{i\omega_0 T_0} + cc + O(\varepsilon)\\
&= Ae^{i\varphi}\mathbf{u}e^{i\omega_0 T_0} + cc + O(\varepsilon)\\
&= 2\text{Re}\left(Ae^{i\varphi}\mathbf{u}e^{i\omega_0 T_0}\right) + O(\varepsilon)\\
&= 2|\mathbf{u}|(-\beta_R/\gamma_R)^{1/2}\cos\omega\tau + O(\varepsilon),
\end{aligned} \tag{4.75}$$

where the fundamental frequency of oscillation is $\omega \equiv \omega_0 + \beta_I - \beta_R\gamma_I/\gamma_R$, and $|\mathbf{u}|$ holds the moduli of the complex-valued elements in $\mathbf{u}$.

Table 4.1 Stability of zero and limit cycle solutions

β_R	<0	<0	>0	>0
γ_R	<0	>0	<0	>0
Limit cycle	(Nonexistent)	Unstable	Stable	(Nonexistent)
Zero solution	Stable	Stable	Unstable	Unstable

Evaluating $G'(A) \equiv dG/dA$ of (4.73) (i.e. the Jacobian of G) at the stationary points one finds $G'(0) = \beta_R$ and $G'(\sqrt{-\beta_R/\gamma_R}) = -2\beta_R$, respectively. Consequently, the zero solution is unstable for $\beta_R > 0$ (and otherwise stable), whereas the limit cycle (4.75) exists for $\beta_R/\gamma_R < 0$ and is unstable for $\beta_R < 0$ (and otherwise stable); These results are summarized in Table 4.1. Note that if $\beta_R < 0$ and $\gamma_R > 0$ the zero solution is stable only to *small* initial disturbances, since a sufficiently strong disturbance may throw the state of the system beyond the unstable limit cycle.

For the actual double pendulum a numerical evaluation of β and γ from (4.71) reveals that in the range $\alpha_a < \alpha < \alpha_b$ (cf. Fig. 4.12) one has $\beta_R > 0$ for $p > p_0$, and $\beta_R < 0$ for $p < p_0$, while $\gamma_R < 0$ for all p. This implies that for α between α_a and α_b, and p above $D_3 = 0$, a stable limit cycle coexists with the unstable zero solution (soft flutter), whereas for p below $D_3 = 0$ the stable zero is the only solution. Thus, the Hopf bifurcation occurring at the curve $D_3 = 0$, $\alpha \in [\alpha_a]$, is *supercritical* (cf. Sect. 3.6.6 and Chap. 5).

Fig. 4.13 depicts amplitudes of the stable limit cycles given by (4.75), as a function of loading magnitude p. It appears that for p near $p_0 \approx 1.397$ the predicted amplitudes agree with those obtained by numerical simulation of the full nonlinear equations (4.50). The increasing discrepancies for larger values of p and θ reflect that (4.75) is only a zero order approximation to the solution of (4.52), which in turn is only an approximation to the original equations (4.50).

4.5.4 Non-periodic and Non-zero Static Solutions

We have been concerned with locating Hopf bifurcations and periodic solutions near $(\theta_1, \theta_2) = (0, 0)$. Perturbation analysis can be used too for examining *divergence bifurcations* of this system, leading to non-zero static solutions (Thomsen 1995). Quasiperiodic and non-periodic chaotic motions of the system will be discussed in Chap. 6.

4.5.5 Summing Up

For the non-conservative double pendulum a linearized analysis predicts that certain values of loading parameters will destabilize the zero solution, causing motions of

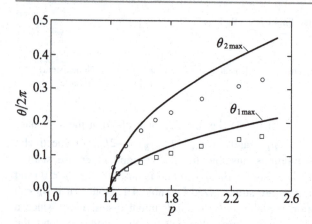

Fig. 4.13 Oscillation amplitudes of the double pendulum for $\alpha = 0.8$, $c = 0.1$, $m = 2$ and varying p. (———) Theory (Eq. (4.75)); $\bigcirc, \square$: numerical simulation (Eq. (4.50)). (Thomsen 1995)

the system to grow unbounded with time. To examine the actual post-critical behavior (which cannot be unbounded), we employed a multiple scales perturbation analysis to the nonlinear system, written in first-order matrix form. This analysis proved the zero solution to become unstable in favor of stable finite amplitude periodic motion. Also, we showed that for some parameters the linearly stable zero solution can be destabilized by a strong disturbance. Some observations have appeared that hold generally:

- A linearized analysis of stability only reveals the consequences of applying a *small* disturbance to a *specific state of equilibrium*.
- Linear analysis is generally incapable of predicting *post-critical behavior*.
- A linearized analysis predicts unstable solutions to grow exponentially in time. This prediction is valid only *initially*, that is, to the time where motions have become too large for the linearization to remain adequate.

4.6   Pendulum with a Sliding Disk

4.6.1   Introduction

Now we turn to studying nonlinear interactions between vibrating structures and sliding solid bodies. As with internal resonance, the kind of interaction considered is *nonlinear*. With linearization the interaction disappears, and the motions of structure and solid body become completely independent.

To put focus on the nonlinear effects of sliding in a simple setting, we consider in this section a vibrated pendulum with a disk sliding on its rod. Sections 4.7–4.8 then consider similar effects associated with flexible structures.

4.6.2   The System

The Pendulum (Fig. 4.14) consists of a rigid rod that is free to rotate around a rapidly vibrating support, and a solid disk that is free to slide along the rod. Gravity acts parallel to the direction of support motion. The arrangement differs from *Kapitza's pendulum* (e.g., Kapitza 1951, 1965; Panovko and Gubanova 1965; Stephenson 1908b) only by the presence of the movable disk. Kapitza's pendulum can be stabilized in the upright position when the support is oscillating at small amplitude and high frequency.

The model in Fig. 4.14 was suggested and studied by Chelomei (1983), with the aim to explain his seemingly gravity defying experimental observations: Under certain circumstances the pendulum rod would stabilize in the inverted (up-pointing) position with the disk floating near a fixed position along the rod. Actually, as was shown by Blekhman and Malakhova (1986), this phenomenon cannot be explained by the simple model in Fig. 4.14, but requires consideration to flexural vibrations of the rod as well as imperfections in the excitation (cf. Chap. 7, and Thomsen and Tcherniak 2001). But the model serves very well to illustrate the basic mechanism by which mass can be caused to move by effects of nonlinear interaction. We here set up the equations of motion, and employ Krylov-Bogoliubov averaging for locating possible equilibriums.

4.6.3   Equations of Motion

In Fig. 4.14 $U(t)$ denotes the position of the solid disk having mass M and rotary inertia I, sliding without slip along the rod. The rod has length l and mass m, and is assumed to be perfectly rigid and uniform. Its rotation with respect to a vertical

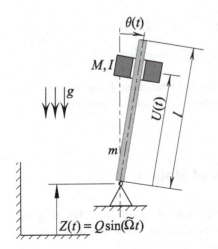

Fig. 4.14 Pendulum on a vibrating support, with a disk sliding on the rigid rod

gravity field g is $\theta(t)$. The rod-hinge performs prescribed vertical harmonic oscillations, $Z(t) = Q\sin(\tilde{\Omega}t)$. The kinetic and potential energies of the system are, respectively:

$$T = \frac{1}{2}M\left(\dot{U}^2 + U^2\dot{\theta}^2 + \dot{Z}^2 + 2\dot{Z}(\dot{U}\cos\theta - U\dot{\theta}\sin\theta)\right)$$
$$+ \frac{1}{2}I\dot{\theta}^2 + \frac{1}{2}m\left(\frac{1}{3}l^2\dot{\theta}^2 + \dot{Z}^2 - l\dot{\theta}\dot{Z}\sin\theta\right), \tag{4.76}$$
$$V = Mg(U\cos\theta + Z) + mg\left(\frac{1}{2}l\cos\theta + Z\right).$$

Using Lagrange's equations, and adding linear viscous damping, one obtains the following equations of motion:

$$(\tfrac{1}{3}ml^2 + MU^2 + I)\ddot{\theta} + 2c_1\dot{\theta} + 2MU\dot{U}\dot{\theta}$$
$$- (MU + \tfrac{1}{2}ml)\left(g - Q\tilde{\Omega}^2\sin(\tilde{\Omega}t)\right)\sin\theta = 0, \tag{4.77}$$

$$\ddot{U} + 2c_2\dot{U} - \dot{\theta}^2 U + \left(g - Q\tilde{\Omega}^2\sin(\tilde{\Omega}t)\right)\cos\theta = 0, \quad U \in [0; L], \tag{4.78}$$

or, in nondimensional form:

$$(1 + \gamma + \alpha u^2)\ddot{\theta} + 2\beta_1\dot{\theta} + 2\alpha u\dot{u}\dot{\theta}$$
$$- (1 + \tfrac{2}{3}\alpha u)\left(1 - q\Omega^2\sin(\Omega\tau)\right)\sin\theta = 0, \tag{4.79}$$

$$\ddot{u} + 2\beta_2\dot{u} - \dot{\theta}^2 u + \tfrac{2}{3}\left(1 - q\Omega^2\sin(\Omega\tau)\right)\cos\theta = 0, \quad u \in [0; 1], \tag{4.80}$$

where

$$\tau = \omega t, \quad \omega^2 = \tfrac{3}{2}g/l, \quad u = U/l, \quad q = \tfrac{3}{2}Q/l, \quad \Omega = \tilde{\Omega}/\omega,$$
$$\alpha = 3M/m, \quad \gamma = I/(\tfrac{1}{3}ml^2), \quad \beta_1 = c_1/(\tfrac{1}{3}ml^2\omega), \quad \beta_2 = c_2/\omega. \tag{4.81}$$

Here ω is the linear natural frequency of the rod when there is no disk, Ω is the ratio of excitation frequency to natural frequency, and α and γ represents the mass and rotary inertia, respectively, of the disk.

4.6.4   Inspecting the Equations of Motion

Equation (4.80) governs motions $u(\tau)$ of the disk along the rod. Ignoring the excitation and the damping, one sees that the acceleration of the disk along the rod is given by $\ddot{u} = \dot{\theta}^2 u - \frac{2}{3}\cos\theta$. The first term is of the centrifugal type, forcing the disk towards the tip of the rod if $\dot{\theta} \neq 0$. The second term represents the effect of gravity, forcing the disk towards the rod-tip when the rod points downwards ($\cos\theta < 0$) or towards the hinge when the rod points upwards ($\cos\theta > 0$).

With the rod at rest ($\dot{\theta} \neq 0$) the disk will move at constant acceleration $\frac{2}{3}\cos\theta$ towards $u = 0$ or $u = 1$. With the rod oscillating around $\theta = \pi$ (pointing downwards) the disk falls off the rod, since the centrifugal term and the gravity co-operate in driving the disk against the rod-tip. However, with the rod oscillating around $\theta = 0$ (pointing upwards) the two terms counteract, and may possibly balance each other out, on the average. If this equilibrium state is stable, the disk will be at rest or perform small oscillations somewhere along the rod.

But how should the rod be brought to oscillate in the upright position, against gravity? To see this we consider (4.79), governing rotations $\theta(\tau)$ of the rod. For a fixed value of $u(\tau)$ (i.e., the disk is fixed to the rod) the Coriolis term $2a u \dot{u}\dot{\theta}$ drops out, and when further $\sin\theta \approx \theta$ for small θ the equation becomes that of an ordinary linear pendulum, subjected to parametric excitation. It is a *Mathieu* equation, for which it is known that the equilibrium $\theta = 0$ can be stabilized for certain conditions of (small) amplitude q and (high) frequency Ω (cf. App. C; Nayfeh and Mook 1979; Panovko and Gubanova 1965). On the other hand, when $\theta \equiv 0$ the centrifugal forces cannot hold the disk in position. So, we are seeking conditions under which the rod can be stabilized in performing small but rapid oscillations near $\theta = 0$.

4.6.5 Seeking Quasi-statical Equilibriums by Averaging

We consider here the existence and stability of the following configuration of the system: the rod performs small amplitude vibrations near $\theta = 0$, and the disk performs small amplitude vibrations near some fixed value of $u \in]0; 1[$. We shall fail in this respect, as did Blekhman and Malakhova (1986) using quite other methods, simply because there *are* no stable equilibriums. Still, the analysis will be illustrative of the phenomenon of vibration induced sliding.

Assuming $\theta \approx 0$ implies $\sin\theta \approx \theta$ and $\cos\theta \approx 1$. Assuming further that movements of the disk are much slower than those of the rod, the Coriolis term $2a u \dot{u}\dot{\theta}$ of (4.79) can be neglected for a first approximation. Since the rod equation (4.79) is parametrically excited we assume vibrations of the rod to occur at half the excitation frequency. Thus, we employ to (4.79) a Van der Pol transformation $(\theta, \dot{\theta})$ $\rightarrow (a\sin\psi, \frac{1}{2}a\Omega\cos\psi)$ with $a = a(\tau)$, $\psi = \psi(\tau) = \frac{1}{2}\Omega\tau + \varphi(\tau)$ and $\dot{a}\sin\psi + a\,\dot{\varphi}$ $\cos\psi \equiv 0$. Averaging out the rapid variations in ψ one arrives at the following equations, which govern slow modulations of the rod amplitude a and phase φ:

$$(1+\gamma+\alpha u^2)2\Omega\dot{a} = -2\beta_1\Omega a - (1+\tfrac{2}{3}\alpha u)q\Omega^2 a\cos 2\varphi,$$

$$(1+\gamma+\alpha u^2)2\Omega a\dot{\varphi} = -\tfrac{1}{2}(1+\gamma+\alpha u^2)\Omega^2 a - (1+\tfrac{2}{3}\alpha u)(2-q\Omega^2\sin 2\varphi)a.$$

$$(4.82)$$

Similarly, substituting $\dot{\theta} = \tfrac{1}{2}a\Omega\cos\psi$ into (4.80) and averaging out variations in ψ (assumed to be much faster than those of u), an averaged equation governing slow motions of the disk is obtained:

$$\ddot{u} + 2\beta_2\dot{u} - \tfrac{1}{8}\Omega^2 a^2 u + \tfrac{2}{3} = 0, \tag{4.83}$$

which clearly shows how the averaged centrifugal term $\tfrac{1}{8}\Omega^2 a^2 u$ forces the disk to slide against gravity ($\ddot{u}$ increases) when the rod is oscillating, $a \neq 0$. As appears, the disk will attain quasi-statical equilibrium on the rod at the position

$$u = \frac{16}{3\Omega^2 a^2} \quad \text{for} \quad a = \tilde{a} > \frac{4}{\sqrt{3}\Omega}, \tag{4.84}$$

where $\tilde{a}$ is the stationary amplitude of rod vibrations (obtained by letting $\dot{a} = \dot{\varphi} = 0$ in (4.82) and inserting (4.84)), and the requirement on $\tilde{a}$ ensures that $u < 1$. However, this equilibrium is *unstable*, as appears too from (4.83): When u is moved beyond the equilibrium the positive acceleration $\ddot{u}$ will increase, which in turn further increases u.

Thus, on the assumptions made there are no stable quasi-statical equilibriums for $u \in {]}0; 1[$. An equilibrium does exist, where the forces driving the disk to slide along the rod balances gravity, but any slight disturbance of this will cause the disk to escape towards the tip of the rod. This, of course, does not contradict the experimental observations of Chelomei; these just cannot be explained by the simple model analyzed here. We shall return to Chelomei's pendulum in Chap. 7, using a proper model that explains the experimental observations.

In returning to the purpose of this section, we note that (4.83) remains valid as a simple descriptor of how vibrations may induce sliding of mass. In the following sections we shall consider more useful examples of this.

4.7 String with a Sliding Point Mass

We here consider vibration-induced sliding of mass as a means for damping out structural vibrations. The principle was suggested by Babitsky and Veprik (1993), and later extended and further investigated (Thomsen 1996a, b; Thomsen and Miranda 1998); see also the work on using sliding mass and internal resonance for vibration damping, by Golnaraghi and co-workers (e.g. Khalily et al. 1994;

Siddiqui et al. 2003), for energy harvesting (Bukhari et al. 2020, Shin et al. 2020), or for general self-tuning into resonance (Krack et al. 2017, Müller and Krack 2020a, b).

The present section contains an extract of the study by Thomsen (1996a), emphasizing the method of analysis and the nonlinear phenomena involved. Equations of motion for a string with a sliding point mass are set up using Hamilton's Principle, discretized using mode shape expansion, and analyzed using averaging. Responses to near-resonant external and parametric excitations are obtained and discussed, as are responses to slow frequency-sweeps across resonance.

4.7.1   Model System and Equations of Motion

Fig. 4.15 shows the model. The string has undeformed length ℓ, axial stiffness EA, and mass per unit length ρA. It performs transverse vibrations with configuration W (X, t) at time t in a fixed reference frame (X, W). External excitations are provided by prescribed displacements $W_0(t)$ of the string base along W, and displacements Q $(t) = Q_0 + Q_1(t)$ of the right string-end along X. The point mass m at $X = Y(t)$ is confined to move on the string, acted upon by restoring forces $F_r(Y)$ parallel to the X-axis, and by frictional forces between mass and string.

The string is assumed to behave linearly elastic, to have negligible rotary and axial inertia, and to perform transverse vibrations at small but finite rotations, $W'(X, t))^2 \ll 1$. The base excitation is assumed to be small, $|W_0(t)| \ll \ell$, as is the axial excitation, $|Q_1(t)| \ll Q_0 \ll \ell$. Energy dissipation is assumed to be sufficiently small for a linear viscous damping model to apply. The restoring force $F_r(Y)$ is assumed to be weak, and may represent, e.g., gravity or spring-loading.

The equations of motion are set up using Hamilton's principle (cf. Sect. 1.5.2). On the above assumptions the kinetic and potential energies of the system are, respectively:

$$T = \int_0^\ell \frac{1}{2}\rho A \dot{W}^2 dX + \frac{1}{2}m\left(\dot{Y}^2 + \left(\dot{W} + W'\dot{Y}\right)^2\right)\Big|_{X=Y(t)},$$

$$V = \int_0^\ell \frac{1}{2}EA\left(\eta' + \frac{1}{2}(W')^2\right)^2 dX + \int_{Y_0}^Y F_r(X)dX,$$

(4.85)

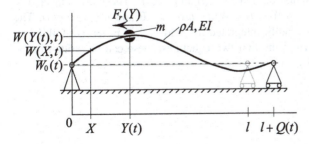

Fig. 4.15 Model system: a string with sliding point mass and transverse base excitation

where dots and primes denote partial differentiation with respect to t and X, respectively. The quantity $(\dot{W} + W'\dot{Y})$ equals $dW(Y(t), t)/dt$, that is, the instantaneous velocity of the point mass parallel to the W-axis. The quantity $(\eta' + \frac{1}{2}(W')^2)$ represents the axial strain Λ of the deformed string, with $\eta = \eta(X, t)$ denoting displacements of material string-points in the X-direction, and the strain energy density at stress σ defined by $\frac{1}{2}\sigma\Lambda = \frac{1}{2}EA\Lambda^2$.

The Lagrangian L of the system is, with $\hat{\delta}(X)$ denoting Dirac's delta function:

$$L = T - V = \int_0^\ell h\, dX,$$

$$h = \frac{1}{2}\rho A \dot{W}^2 + \frac{1}{2}m\hat{\delta}(X - Y)\left(\dot{Y}^2 + \left(\dot{W} + W'\dot{Y}\right)^2\right) \tag{4.86}$$

$$- \frac{1}{2}EA\left(\eta' + \frac{1}{2}(W')^2\right)^2 - \hat{\delta}(X - Y)\int_{Y_0}^X F_r(\xi)\, d\xi.$$

Stationarity of the action integral $I = \int L\, dt$ requires (with δ denoting variation):

$$\delta I = \int_{t_1}^{t_2}\int_0^\ell \delta h\, dX\, dt$$

$$= \int_{t_1}^{t_2}\int_0^\ell \left(\frac{\partial h}{\partial \dot{W}}\delta\dot{W} + \frac{\partial h}{\partial W'}\delta W' + \frac{\partial h}{\partial Y}\delta Y + \frac{\partial h}{\partial \dot{Y}}\delta\dot{Y} + \frac{\partial h}{\partial \eta'}\delta\eta'\right) dX\, dt = 0.$$

$$\tag{4.87}$$

Employing integration by parts and requiring δI to vanish for arbitrary admissible variations δW, δY and $\delta\eta$, the requirement of stationarity (4.87) imply:

$$\frac{\partial}{\partial t}\frac{\partial h}{\partial \dot{W}} + \frac{\partial}{\partial X}\frac{\partial h}{\partial W'} = 0; \quad \int_0^\ell \left(\frac{\partial}{\partial t}\frac{\partial h}{\partial \dot{Y}} - \frac{\partial h}{\partial Y}\right)dX = 0; \quad \frac{\partial}{\partial X}\frac{\partial h}{\partial \eta'} = 0. \tag{4.88}$$

With the functional h given by (4.86), the expressions in (4.88) provide three equations for the determination of $W(X, t)$, $Y(t)$ and $\eta(X, t)$. These are subjected to boundary conditions $W(0, t) = W(\ell, t) = W_0(t)$, $\eta(0, t) = 0$ and $\eta(\ell, t) = Q(t)$. The third equation in (4.88) is readily integrated to yield $\eta(X, t)$ in terms of $W'(x, t)$. Using this to eliminate η from the first two equations, these can be written in the following nondimensional form:

$$(\ddot{w} + \ddot{w}_0) + c_1(\dot{w} + \dot{w}_0)$$
$$+ \alpha\hat{\delta}(x - y)(\ddot{w} + \ddot{w}_0 + 2\dot{w}'\dot{y} + w''\dot{y}^2 + w'\ddot{y})$$
$$- \pi^{-2}\left(1 + q(t) + 2\pi^{-2}\mu\int_0^1 (w')^2 d\xi\right)w'' = 0, \tag{4.89}$$
$$\ddot{y} + c_2\dot{y} + f_r(y) = -\left((\ddot{w} + \ddot{w}_0 + 2\dot{w}'\dot{y} + w''\dot{y}^2 + w'\ddot{y})w'\right)\big|_{x=y(\tau)},$$

with boundary conditions $w(0, \tau) = w(1, \tau) = 0$. Linear viscous damping has been added, and nondimensional variables and parameters are defined by:

$$x = \frac{X}{\ell}, \quad \tau = \tilde{\omega}t, \quad \tilde{\omega}^2 = \frac{EAQ_0}{\rho A\ell}\left(\frac{\pi}{\ell}\right)^2,$$

$$w(x,\tau) = \frac{W(X,t) - W_0(t)}{\ell}, y(\tau) = \frac{Y(t)}{\ell},$$

$$\alpha = \frac{m}{\rho A\ell}, \quad c_{1,2} = \frac{C_{1,2}}{\tilde{\omega}}, \quad w_0(\tau) = \frac{W_0(t)}{\ell}, \tag{4.90}$$

$$q(t) = \frac{Q_1(t)}{Q_0}, \quad f_r(y) = \frac{F_r(Y)}{m\tilde{\omega}^2\ell}, \quad \mu = \frac{\pi^2\ell}{4Q_0}.$$

Note that time is normalized by the fundamental natural frequency of a string with no point mass and constant tension, that α denotes the fraction of point mass to string mass, and that string deformations are measured with respect to the moving base (so that boundary conditions become homogeneous).

The first equation in (4.89) governs transverse motions $w(x, \tau)$ of the string on $x \in [0; 1]$. The terms describe in turn linear inertia, linear damping, added inertial effect due to the point mass, and axial force in the string. The added inertia of the point mass equals $d^2w(y(\tau), \tau)/dt^2$, and is seen to contain nonlinear terms due to the non-straight motion of the mass. The term describing axial force adds the linear effect of the externally applied displacement of the string-end to the nonlinear effect of axial stretching. With the point mass fixed at $x = y$ (implying $\dot{y} = \ddot{y} = 0$) the equation reduces to that of a tensioned string with a lumped mass.

The second equation in (4.89) governs motions $y(\tau)$ of point mass along the string. The left-hand side represents the dynamics of an unexcited oscillator with general restoring force $f_r(y)$. Movements of the point mass are seen to be excited by string motions w through the nonlinear interaction terms on the right-hand side. These represent the total differential $d^2w/d\tau^2$ times the rotation w' of the string at the position of the point mass $x = y(\tau)$. For the classical problems involving externally prescribed movements of a mass along a structure (e.g. cars passing bridges or loads moving along crane booms), one can safely neglect these terms. This implies that positions of the point mass are unaffected by vibrations of the structure. For the present problem, by contrast, the point mass is driven to slide solely by nonlinear interaction terms originating from transverse vibrations of the string.

We aim towards a simple model representing essential features. For this the nonlinear terms are ordered by the following physical assumptions, with $\varepsilon \ll 1$ serving as the scaling parameter: String motions are small, $w = O(\varepsilon)$; The base excitation is small compared to string motion, $w_0 = O(\varepsilon^2)$; Sliding of the point mass is slow compared to string motion, $y = y(\varepsilon\tau) \Rightarrow \dot{y} = O(\varepsilon)$, $\ddot{y} = O(\varepsilon^2)$; Damping is weak, $c_{1,2} = O(\varepsilon)$; The restoring force f_r is weak, comparable in magnitude to that of nonlinear interaction, $f_r(y) = O(\varepsilon^2)$; The constant part of the axial excitation is small compared to string length, $(Q_0/\ell) = O(\varepsilon)$, and the time-varying part is small compared to the constant part, $q = (Q_1/Q_0) = O(\varepsilon)$. On these assumptions the definition of μ implies that $\mu = O(\ell/Q_0) = O(\varepsilon^{-1})$. Substituting then $w \to \varepsilon w$, $w_0 \to \varepsilon^2 w_0$, $\dot{y} \to \varepsilon \dot{y}$, etc., into Eqs. (4.89) one obtains, neglecting terms of order ε^3 and higher and omitting ε notation, the following approximate equations with leading nonlinearities preserved:

$$
\ddot{w} + c_1 \dot{w} + \alpha \hat{\delta}(x - y)(\ddot{w} + 2\dot{w}'\dot{y})
$$

$$
- \pi^{-2}\left(1 + q(t) + 2\pi^{-2}\mu \int_0^1 (w')^2 \, d\xi\right)w''
$$

$$
= -\left(1 + \alpha\hat{\delta}(x - y)\right)\ddot{w}_0, \tag{4.91}
$$

$$
\ddot{y} + c_2\dot{y} + f_r(y) = -(\ddot{w}w')|_{x=y(\tau)}.
$$

An N-mode expansion is now assumed for the string deflections w, that is: $w(x, \tau) = \sum_{j=1}^{N} u_j(\tau)\varphi_j(x)$, where u_j are unknown time-functions and $\varphi_j = \sin(j\pi x)$ are the linear mode shapes of a tensioned string with no point mass. On substitution into (4.91), multiplication by $\varphi_j(x)$ and subsequent integration over $x \in [0; 1]$, a set of ordinary differential equations in $u_j(\tau)$ and $y(\tau)$ is obtained:

$$
\sum_{j=1,N} \left(\delta_{ij} + 2\alpha\varphi_i(y)\varphi_j(y)\right)\ddot{u}_j + \sum_{j=1,N} \left(c_1\delta_{ij} + 4\alpha\dot{y}\varphi_i(y)\varphi_j'(y)\right)\dot{u}_j
$$

$$
+ i^2(1 + q(\tau))u_i = -2\left(\frac{1}{i\pi}(1 - (-1)^i) + \alpha\varphi_i(y)\right)\ddot{w}_0, \quad i = 1, N, \tag{4.92}
$$

$$
\ddot{y} + c_2\dot{y} + f_r(y) = -\sum_{j=1,N} \varphi_j(y)\ddot{u}_j \sum_{j=1,N} \varphi_j'(y)u_j \quad \varphi_j(x) = \sin j\pi x,
$$

where δ_{ij} is the Kronecker delta. For a dominant j'th mode the equations reduce to the single-mode approximation:

$$
\left(1 + 2\alpha \sin^2(j\pi y)\right)\ddot{u}_j + (c_1 + 2j\pi\alpha\dot{y}\sin(2j\pi y))\dot{u}_j
$$

$$
+ j^2\left(1 + q(\tau) + \mu j^2 u_j^2\right)u_j = -2\left(\frac{1}{j\pi}(1 - (-1)^j) + \alpha\sin(j\pi y)\right)\ddot{w}_0, \tag{4.93}
$$

$$
\ddot{y} + c_2\dot{y} + f_r(y) = -\frac{1}{2}j\pi \sin(2j\pi y)u_j\ddot{u}_j, \quad j = 1, N,
$$

that is, for the fundamental mode ($j = 1$, $u \equiv u_1$):

$$\left(1 + 2\alpha \sin^2(\pi y)\right)\ddot{u} + \left(c_1 + 2\pi \alpha \dot{y} \sin(2\pi y)\right)\dot{u}$$

$$+ \left(1 + q(\tau) + \mu u^2\right)u = -\left(\frac{4}{\pi} + 2\alpha \sin(\pi y)\right)\ddot{w}_0, \qquad (4.94)$$

$$\ddot{y} + c_2 \dot{y} + f_r(y) = -\frac{1}{2}\pi \sin(2\pi y)\ddot{u}u.$$

For the single-mode approximations (4.93) and (4.94) to be adequate near a j'th dominant mode, the mass α must be small enough that $\varphi_j(x) = \sin(j\pi x)$ remains a valid approximation to the actual shape of string vibrations. Equations (4.94) constitute the system model for the analysis to follow.

4.7.2 Illustration of System Behavior

The system displays a rich variety of dynamic behavior. Just a few examples illustrating fundamental features are given here. The results were obtained by numerical integration of Eq. (4.94).

Fig. 4.16(a) shows the response to resonant transverse base excitation of the fundamental mode. The top figure depicts the string response $u(\tau)$ when the mass is fixed to the string at $y = y_0 = 0.1$. For a fixed mass the response is linear, with harmonic vibrations building until limited by viscous damping. The middle figure shows the response obtained when the mass is free to slide against a linear spring having static equilibrium at $y = y_0$. Initially, string vibrations build up as for the case with fixed mass. However, as appears from the bottom figure, this triggers sliding of the mass towards the middle of the string, and thus changes the resonance frequency of the combined system. A stationary state arises, with string vibrations being reduced, and the mass in equilibrium nearer the middle of the string. By contrast to the case with no spring, the mass will return to y_0 if the excitation is removed or shifted out of resonance. The effects of resonant base excitation are further analyzed in Sects. 4.7.3–4.7.4.

Fig. 4.16(b) shows the response to resonant axial excitation of the fundamental parametric resonance near $\Omega = 2$. A small value of the initial condition $u(0)$ was applied to perturb the unstable zero solution. The top figure depicts the string response $u(\tau)$ when the mass is fixed to the string at $y = y_0 = 0.1$. As shown, the response grows exponentially (figure clipped at $|a| = 0.15$), since axial stretching and other response-limiting factors have been neglected. The fixed mass implies the system to be linear, and thus to respond unboundedly to resonant parametric excitation. When the mass is free to slide against a linear spring, the string response initially grows exponentially (middle figure). However, string vibrations initiate movements of the mass (bottom figure). At some position of the mass along the string, the nonlinear coupling forces pulling the mass towards $y = \frac{1}{2}$ balances the tension of the spring pulling it towards $y = y_0$. The displacement of the mass in turn

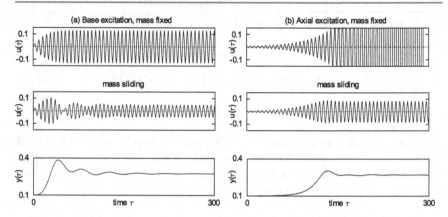

Fig. 4.16 System response to (**a**) resonant transverse base excitation, and (**b**) resonant axial excitation. Numerical integration of Eqs. (4.94). Top: string motion $u(\tau)$ with point mass fixed at $y = y_0$; Middle: string motion with point mass sliding (spring-loaded); Bottom: point mass-position $y(\tau)$ associated with middle figures. Non-zero parameters for part (a): $w_0(\tau) = w_A\sin(\Omega\tau)$, $w_A = 0.01$, $\Omega = 1$, $c_1 = 0.1$, $c_2 = 0.35$, $\alpha = 0.3$, $f_r(y) = \kappa(y - y_0)$, $\kappa = 0.01$, $y_0 = 0.1$; part (b): q $(\tau) = q_A\sin(\Omega\tau)$, $q_A = 0.3$, $\Omega = 2$, $c_1 = 0.1$, $c_2 = 0.05$, $\alpha = 0.1$, $f_r(y) = \kappa(y - y_0)$, $\kappa = 0.03$, $y_0 = 0.1$, $u(0) = 0.005$

affects vibrations of the string, which level off into stationary finite-amplitude motion. Thus, sliding of the mass effectively limits this otherwise unlimited response. The effect of axial resonant excitation is analyzed further in Sect. 4.7.5.

Other choices of parameters may cause stationary responses that qualitatively differ from those in Fig. 4.16. For example, the point mass may end up by oscillating back and forth around an unstable equilibrium, causing a *beating* (i.e. quasiperiodic) response of the string[3]. Further parameter perturbations may turn the beating into chaos, as described in Chap. 6.

4.7.3   Response to Near-Resonant Base Excitation

We here consider obtaining analytical frequency responses for the case of near-resonant base excitation. Away from resonance the string amplitudes are too small to excite sliding of the point mass, and the dynamics of the system are essentially linear. So, a near-resonant (or very hard) excitation is required for

---

[3]Beating vibrations change their amplitudes periodically in time at a slow rate. Beats may occur when two modes are summed that are close in frequency. Consider as an example $\sin(\omega t + \omega t) + \sin(\omega t - \varepsilon\omega t) = 2 \sin(\varepsilon\omega t) \cos(\omega t)$, which is beating when $\varepsilon \ll 1$. Guitarists may rely on audible beats for tuning their instruments: As two tones come closer in frequency one hear the beats slowing down, and finally disappearing at perfect tune.

significant nonlinear effects to show up. We assume the point mass to be small, $\alpha \ll$ 1, and employ averaging for obtaining first-order approximations to the response.

With a base excitation near-resonant to the fundamental mode, Eq. (4.94) apply with $q(\tau) = 0$ and $w_0(\tau) = w_A \sin(\Omega\tau)$. Here w_A is the constant amplitude of excitation, and $\Omega \approx 1$ is the constant frequency of excitation which is near to the natural frequency of the string with no point mass. The effect of nonlinear stretching is easily included, though to keep the analysis simple it is neglected in this section ($\mu = 0$). Equations (4.94) thus become:

$$\left(1 + 2\alpha \sin^2(\pi y)\right)\ddot{u} + \left(c_1 + 2\pi\alpha\dot{y}\sin(2\pi y)\right)\dot{u} + u$$
$$= \Omega^2 w_A \left(\frac{4}{\pi} + 2\alpha \sin(\pi y)\right) \sin(\Omega\tau), \tag{4.95}$$

$$\ddot{y} + c_2\dot{y} + f_r(y) = -\frac{1}{2}\pi \sin(2\pi y)\ddot{u}u. \tag{4.96}$$

Averaged System For the string equation (4.95), a Van der Pol transformation $(u, \dot{u}) \to (a \sin\psi, a\Omega \cos\psi)$, with $a = a(\tau)$, $\psi = \psi(\tau) = \Omega\tau + \theta(\tau)$ and the constraint $\dot{a} \sin\psi + a\dot{\theta} \cos\psi \equiv 0$, yields a pair of first-order equations in the time-varying amplitude $a(\tau)$ and phase $\theta(\tau)$:

$$\left(1 + 2\alpha \sin^2(\pi y)\right)\Omega\dot{a} = \left(\Omega^2 - 1 + 2\alpha\Omega^2 \sin^2(\pi y)\right)a \sin\psi \cos\psi$$
$$- \left(c_1 + 2\pi\alpha\dot{y}\sin(2\pi y)\right)\Omega a \cos^2\psi$$
$$+ \Omega^2 w_A \left(\frac{4}{\pi} + 2\alpha \sin(\pi y)\right)\sin(\psi - \theta)\cos\psi,$$
$$\left(1 + 2\alpha \sin^2(\pi y)\right)\Omega a\dot{\theta} = - \left(\Omega^2 - 1 + 2\alpha\Omega^2 \sin^2(\pi y)\right)a \sin^2\psi$$
$$+ \left(c_1 + 2\pi\alpha\dot{y}\sin(2\pi y)\right)\Omega a \sin\psi \cos\psi$$
$$- \Omega^2 w_A \left(\frac{4}{\pi} + 2\alpha \sin(\pi y)\right)\sin(\psi - \theta)\sin\psi. \tag{4.97}$$

For the point mass equation (4.96) the Van der Pol transformation yields

$$\ddot{y} + c_2\dot{y} + f_r(y) = -\frac{1}{2}\pi \sin(2\pi y)\left(\Omega a\dot{a}\tan\psi - \Omega^2 a^2 \sin^2\psi\right). \tag{4.98}$$

With $(\Omega^2 - 1, \alpha, c_1, w_A, a) \ll 1$, all right-hand terms of (4.97) will be small. Consequently $\dot{a}$ and $\dot{\theta}$ are small. This implies that $a(\tau)$ and $\theta(\tau)$ are slowly varying as compared to $\psi(\tau) = \Omega\tau + \theta(\tau)$, so that right-hand terms can be approximated by their averages. Replacing right-hand terms of form $G(\psi)$ with $\frac{1}{2\pi}\int_0^{2\pi} G(\psi)\,d\psi$ one obtains, treating a and θ as constants during the period of integration:

$$
\begin{aligned}
\left(1+2\alpha \sin^2(\pi y)\right)\Omega \dot{a} &= -\frac{1}{2}\left(c_1 + 2\pi\alpha\dot{y}\sin(2\pi y)\right)\Omega a \\
&\quad - \Omega^2 w_A\left(\frac{2}{\pi}+\alpha\sin(\pi y)\right)\sin\theta, \\
\left(1+2\alpha \sin^2(\pi y)\right)\Omega a\dot{\theta} &= -\frac{1}{2}\left(\Omega^2 - 1 + 2\alpha\Omega^2 \sin^2(\pi y)\right)a \\
&\quad - \Omega^2 w_A\left(\frac{2}{\pi}+\alpha\sin(\pi y)\right)\cos\theta.
\end{aligned}
\tag{4.99}
$$

Similarly, with $(c_2, f_r, a, \dot{a}) \ll 1$, Eq. (4.98) can be averaged to yield:

$$
\ddot{y} + c_2\dot{y} + f_r(y) = \frac{1}{4}\pi\Omega^2 a^2 \sin(2\pi y).
\tag{4.100}
$$

The right-hand term of this equation clearly shows how a string vibrating at instantaneous amplitude $a(\tau)$ will cause sliding $y(\tau)$ of the point mass (we recognize a similar term in (4.83) for the pendulum with a sliding disk). When there is no external restoring forces ($f_r = 0$), vibrations of the string will drive the point mass towards $y = \frac{1}{2}$, where $\sin(2\pi y)$ shifts sign. This in turn affects string vibrations, according to (4.99). However, even if the string vibrations decay to zero, the point mass will remain at the stable equilibrium $y = \frac{1}{2}$. The presence of external restoring forces ($f_r \neq 0$) generally destroys the equilibrium at $y = \frac{1}{2}$, in favor of equilibriums depending on string amplitude and frequency.

Stationary Solutions Stationary solutions for the averaged system (4.99)–(4.100) are determined by the condition that $\dot{a} = \dot{\theta} = \dot{y} = \ddot{y} = 0$, that is, by:

$$
\begin{aligned}
\Omega^2 w_A\left(\frac{2}{\pi}+\alpha\sin(\pi y)\right)\sin\theta &= -\frac{1}{2}c_1\Omega a, \\
\Omega^2 w_A\left(\frac{2}{\pi}+\alpha\sin(\pi y)\right)\cos\theta &= -\frac{1}{2}(\Omega^2 - 1 + 2\Omega^2\alpha \sin^2(\pi y))a, \\
f_r(y) &= \frac{1}{4}\pi\Omega^2 a^2 \sin(2\pi y).
\end{aligned}
\tag{4.101}
$$

Eliminating θ this implies:

$$
\begin{aligned}
\left[\Omega^2 w_A\left(\frac{2}{\pi}+\alpha\sin(\pi y)\right)\right]^2 &= \left[(c_1\Omega)^2 + (\Omega^2 - 1 + 2\Omega^2\alpha \sin^2(\pi y))^2\right]\frac{1}{4}a^2, \\
a^2 = \frac{4}{\pi\Omega^2}\frac{f_r(y)}{\sin(2\pi y)}, \quad &\tan\theta = \frac{c_1\Omega}{\Omega^2 - 1 + 2\Omega^2\alpha \sin^2(\pi y)}.
\end{aligned}
\tag{4.102}
$$

When $f_r(y) \equiv 0$ the second equation yields $y \in \{0, \frac{1}{2}, 1\}$. Substituting this into the first equation, one obtains the string amplitudes a for these limit-cases, corresponding to linear vibrations of the string with no point mass ($y = 0$, $y = 1$), or with a point mass at the middle of the string ($y = \frac{1}{2}$).

When $f_r(y) \neq 0$, substitution of the second equation into the first gives an equation in y that can be solved numerically. The y-solution is then substituted into the second and third equation to yield the string amplitude a and corresponding phase θ. Frequency-responses are given by the curves $y(\Omega)$ and $a(\Omega)$ so obtained. The stability of solutions is determined by evaluating eigenvalues of the Jacobian of the system (4.99)–(4.100), written as a system of four first-order equations. A solution of (4.102) having one or more Jacobian eigenvalues with positive real part is unstable.

Backbone curves for the frequency responses are given by the solutions to (4.102) for which $w_A = c_1 = 0$, that is, by:

$$y_{bbone} = \frac{1}{\pi} \arcsin \left(\frac{\Omega^{-2} - 1}{2\alpha} \right)^{1/2}, \quad \frac{1}{1+2\alpha} < \Omega^2 < 1,$$

$$a_{bbone}^2 = \frac{4f_r(y_{bbone})}{\pi \Omega^2 \sin(2\pi y_{bbone})} = \frac{4\alpha f_r(y_{bbone})}{\pi \left[(1 - \Omega^2)(\Omega^2(1+2\alpha) - 1) \right]^{1/2}}. \tag{4.103}$$

The slope of $y_{bbone}(\Omega)$ is negative for all Ω in the range defined. This implies a frequency response of the softening type (peak bent towards lower frequencies).

Example Frequency Response   Fig. 4.17 shows a typical frequency response given by (4.102) when $f_r(y) = \kappa(y - y_0)$, i.e., the point mass is attached to a linear spring having static equilibrium at $y = y_0$. The linear response for the case of a mass fixed at $y = y_0$ has been superimposed. Results obtained by numerically integrating the full, un-averaged equations (4.95)–(4.96) are also indicated, and appears to agree with those obtained by averaging. Hence, the averaged system adequately captures the dynamics of the full system. The left-bent response-curves reflect a softening nonlinearity, and display the typical picture of stable branches connecting to unstable branches through points of vertical tangency.

It appears from the figure that, if free to slide, the point mass reduces vibrations of the string near the linear resonance frequency $(1 + 2\alpha \sin^2(\pi y_0))^{-1/2} \approx 0.97$. When the excitation frequency is outside the resonant domain, the mass does not move and the response approaches that for the fixed mass. The nonlinearly bent response implies that large amplitudes and amplitude jumps may occur slightly below linear resonance. Increasing very slowly the frequency of excitation across resonance, the amplitude of the string will jump to the upper stable branch of the curve at the point where the lower branch becomes unstable. Even so, stationary amplitudes will still be less than for the fixed-mass case. With a slow frequency *decrease*, the amplitudes will jump down from a value slightly larger than the peak amplitude for the fixed-mass case.

4.7.4   Response to Slow Frequency-Sweeps

Next we consider responses to slow frequency-sweeps across the fundamental resonance. Keep in mind that frequency responses as the one in Fig. 4.17 are valid

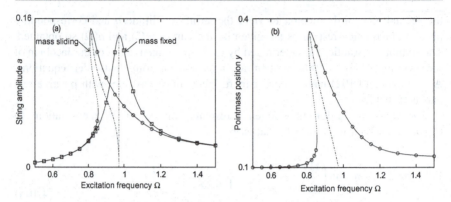

Fig. 4.17 Frequency responses $a(\Omega)$ and $y(\Omega)$ of base excited string with point mass sliding (spring–loaded) or fixed at $y = y_0$. Solid/dashed line: stable/unstable solutions of (4.102); ($\bigcirc, \square$): numerical integration of (4.95)–(4.96). Non-zero parameters: $\alpha = 0.3$, $w_A = 0.01$, $c_1 = 0.1$, $c_2 = 0.35$, $f_r(y) = \kappa(y - y_0)$, $\kappa = 0.03$, $y_0 = 0.1$

only for constant values of excitation frequency; they predict the stationary amplitudes obtained after initial transients (the homogeneous part of the solution) have decayed. They still remain valid if the frequency of excitation is changed very slowly. When this is not the case, such as with experimental frequency-sweeps or system startup/shutdowns at finite speed, the system will not have time to attain stationary amplitudes. One then needs to consider transient response during the passage of resonance. We thus reconsider the system (4.94) subjected to harmonic base excitation $w_0(\tau) = w_A \sin(\Phi(\tau))$ at a slowly varying frequency. The amplitude w_A is constant, whereas $\Phi(\tau) = \tau + v(\varepsilon\tau)$, $\varepsilon \ll 1$, so that v is a slowly varying function of time. The instantaneous frequency of excitation becomes $\Omega(\tau) = = 1 + \varepsilon \; \dot{v}(\varepsilon\tau)$, and we then consider a passage through the region of primary resonance $\Omega \approx 1$ for the following system:

$$\left(1 + 2\alpha \sin^2(\pi y)\right)\ddot{u} + \left(c_1 + 2\pi\alpha\dot{y}\sin(2\pi y)\right)\dot{u} + u$$
$$= w_A \left(\frac{4}{\pi} + 2\alpha\sin(\pi y)\right)\left(\Omega^2 \sin\Phi - \dot{\Omega}\cos\Phi\right),$$
$$\Phi(\tau) = \int \Omega(\tau)d\tau, \tag{4.104}$$
$$\ddot{y} + c_2\dot{y} + f_r(y) = -\frac{1}{2}\pi\sin(2\pi y)\ddot{u}u.$$

Methods for analyzing linear or nonlinear systems subjected to slowly varying excitations are described in, e.g., Nayfeh and Mook (1979) and Evan-Iwanowski (1976). For the present case we employ averaging.

With averaging the $\dot{\Omega}$-term in the first of Eq. (4.104) may be neglected when compared to Ω^2, since $\dot{\Omega} = \ddot{\Phi} = \varepsilon^2\ddot{v} \ll \Omega^2 \approx 1$. A Van der Pol transformation $(u, \dot{u})$ $\rightarrow (a \sin\psi, a\Omega(\tau)\cos\psi)$, where $a = a(\tau)$ and $\psi(\tau) = \Phi(\tau) + \theta(\tau)$, is then applied to

Eq. (4.104). Neglecting everywhere $\dot{\Omega}$-terms compared to Ω^2-terms, the transformed set of first-order equations becomes identical to that of the constant-frequency case, (4.97), with the only difference that now $\Omega = \Omega(\tau)$. Also, performing the averaging operations as described in Sect. 4.7.3, it is easily shown that the averaged equations become identical to those describing the constant-frequency case. Equations (4.99)–(4.100) are thus applicable too for the case of slow frequency-sweeps across resonance. But for sweeps, since Ω is a function of time, no stationary solutions of the averaged equations exist. To obtain $a(\tau)$ and $\theta(\tau)$, Eqs. (4.99)–(4.100) must be integrated numerically. This, however, is a much more stable and fast undertaking than to integrate the full un-averaged equations (4.104). The averaged equations (4.99)–(4.100) contain no fast-oscillating components, and this gives leave for employing much larger time-steps in the numerical procedure.

Only time-linear frequency-sweeps are considered. We assume the instantaneous frequency of excitation to be prescribed externally by:

$$\Omega(\tau) = \begin{cases} \Omega_0, & \tau < \tau_0, \\ \Omega_0 + r_s(\tau - \tau_0), & \tau_0 \leq \tau \leq \tau_1, r_s = (\Omega_1 - \Omega_0)/(\tau_1 - \tau_0). \end{cases} \quad (4.105)$$

As appears, the system is driven at constant frequency Ω_0 during an initial span of time $\tau \in [0; \tau_0[$, sufficiently long for transients to decay and stationary conditions to settle. Then, during $\tau \in [\tau_0; \tau_1]$, the frequency is either increased or decreased at a constant rate r_s towards the terminal frequency Ω_1. An evaluation of the function $v(\tau) = \int (1 - \Omega(\tau))d\tau$ will show that if the sweep rate r_s is small and if $\Omega_0 \approx 1$, then $v(\tau)$ is indeed a slowly varying function of time, as assumed.

Fig. 4.18 depicts time histories for a sweep through $\Omega \in [0.5; 1.5]$ at two different sweep rates r_s. Solid lines indicate results of numerically integrating the un-averaged equations (4.104). Dashed lines indicate results of numerically integrating the averaged equations (4.99)–(4.100) with $\Omega(\tau)$ given by (4.105). At the sweep rate $r_s = 0.002$ the averaged response $a(\tau)$ of the string accurately describes the envelope of the fast-oscillating full response $u(\tau)$ (Fig. 4.18(a), top). Also, the averaged response of the point mass closely follows the un-averaged response (Fig. 4.18(a), bottom). On raising the sweep rate to $r_s = 0.005$ small discrepancies between the averaged and the un-averaged responses become apparent (Fig. 4.18 (b)). Hence, for the particular parameter values chosen, values of sweep rates $r_s \leq 0.002$ will be considered sufficiently slow for averaged results to apply.

Fig. 4.19 compares frequency responses obtained at four different sweep rates when the point mass is respectively fixed (Fig. 4.19(a)) and spring-loaded (Fig. 4.19(b)). The stationary frequency responses of Fig. 4.17(a) (corresponding to $r_s = 0$) have been superimposed. With fixed point mass (Fig. 4.19(a)) the swept response for the slowest sweep rate $(r_s = \pm 2 \cdot 10^{-4})$ follows the stationary response $(r_s = 0)$ rather closely. When the point mass is free to slide (Fig. 4.19(b)), a jump-up to the stationary curve occurs for the up-sweep $(r_s = +2 \cdot 10^{-4})$, whereas a jump-down to the stationary curve occurs for the down-sweep

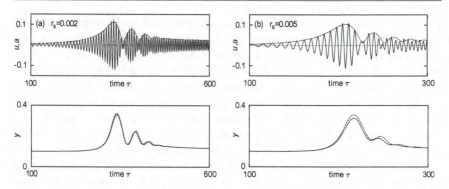

Fig. 4.18 Swept frequency responses of base excited string with sliding point mass. String motions $(u(\tau), a(\tau))$ and point mass position $y(\tau)$ for two values of sweep rate r_s. Solid line: $(u(\tau), y(\tau))$ by numerical integration of Eq. (4.104) with (4.105); dashed line: $(a(\tau), y(\tau))$ by numerical integration of averaged Eqs. (4.99)–(4.100) with $\Omega(\tau)$ given by (4.105). Non-zero parameters: $\alpha = 0.3$, $w_A = 0.01$, $c_1 = 0.1$, $c_2 = 0.35$, $f_r(y) = \kappa(y - y_0)$, $\kappa = 0.03$, $y_0 = 0.1$, $\Omega_0 = 0.5$, $\Omega_1 = 1.5$

$(r_s = -2 \cdot 10^{-4})$. Both jumps are accompanied by overshooting and damped beating. At a faster sweep rate $(r_s = \pm 2 \cdot 10^{-3})$ the right(left)-shift of peak amplitude associated with up(down)-sweep becomes more pronounced, as does the overshooting and beating for the case of sliding point mass (Fig. 4.19(b)). The swept excitation delays the onset of resonant vibrations, as compared to stationary excitation. Note that for up-sweeps $(r_s > 0)$ the maximum amplitudes are reduced if the mass is free to slide, though, at the cost of increased amplitudes during periods of beating. For down-sweeps $(r_s < 0)$ the maximum amplitudes are slightly increased if the mass is free to slide.

Thus, when the mass is free to slide the swept response is characterized by overshooting, beating, altered maximum response and delayed resonance. These features common to nonlinear systems driven by swept harmonic excitation (e.g., Evan-Iwanowski 1976).

4.7.5 Response to Near-Resonant Axial Excitation

We here consider obtaining frequency responses for the case of *axial* excitation of the string near its fundamental *parametric* resonance. Away from parametric resonance the string amplitudes are too small to excite sliding of the point mass, and the behavior of the system is essentially linear.

With harmonic mono-frequency excitation of the string axis, Eq. (4.94) apply with $w_0(\tau) = 0$ and $q(\tau) = q_A \sin(\Omega\tau)$. Here q_A is the constant amplitude of axial excitation and $\Omega \approx 2$ is the frequency of excitation, which is close to the fundamental parametric resonance of the string with no point mass. We thus consider the system:

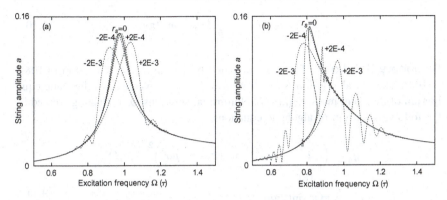

Fig. 4.19 Swept frequency responses $a(\Omega(\tau))$ of base excited string – effect of sweep rate r_s. (a) Point mass fixed at $y = y_0$; (b) Point mass sliding. Solid/dashed line: stationary amplitudes ($r_s = 0$) according to (4.102); dotted line: numerical integration of averaged system (4.99)–(4.100) with $\Omega(\tau)$ given by (4.105). Non-zero parameters: $\alpha = 0.3$, $w_A = 0.01$, $c_1 = 0.1$, $c_2 = 0.35$, $f_r(y) = \kappa(y - y_0)$, $\kappa = 0.03$, $y_0 = 0.1$, $\Omega_0 = 0.5$, $\Omega_1 = 1.5$, $\tau_0 = 100$

$$\left(1 + 2\alpha \sin^2(\pi y)\right)\ddot{u} + (c_1 + 2\pi\alpha\dot{y}\sin(2\pi y))\dot{u}$$
$$+ \left(1 + q_A \sin(\Omega\tau) + \mu u^2\right)u = 0, \tag{4.106}$$

$$\ddot{y} + c_2\dot{y} + f_r(y) = -\frac{1}{2}\pi \sin(2\pi y)\ddot{u}u. \tag{4.107}$$

Averaged System Following the approach of Sect. 4.7.3, a Van der Pol transformation $(u, \dot{u}) \rightarrow (a \sin\psi, \frac{1}{2} a\Omega \cos\psi)$ with $a = a(\tau)$, $\psi = \psi(\tau) = \frac{1}{2}\Omega\tau + \theta(\tau)$ with condition $\dot{a} \sin\psi + a\ \dot\theta \cos\psi \equiv 0$ turns (4.106) into a pair of first-order equations in $a(\tau)$ and $\theta(\tau)$. With $(\Omega^2-4, \alpha, c_1, q_A, a) \ll 1$, all right-hand terms of these equations will be small, implying that $a(\tau)$ and $\theta(\tau)$ are slowly varying as compared to the rapidly changing phase $\psi(\tau)$. Averaging the right-hand sides over $\psi \in [0, 2\pi]$ the averaged string equations become:

$$\left(1 + 2\alpha \sin^2(\pi y)\right)\Omega\dot{a} = -\frac{1}{2}(c_1 + 2\pi\alpha\dot{y}\sin(2\pi y))\Omega a$$
$$- \frac{1}{2}q_A \cos(2\theta)a,$$
$$\left(1 + 2\alpha \sin^2(\pi y)\right)\Omega a\dot\theta = -\frac{1}{4}\left(\Omega^2 - 4 + 2\alpha\Omega^2 \sin^2(\pi y)\right)a \tag{4.108}$$
$$+ \frac{1}{2}q_A \sin(2\theta)a + \frac{3}{4}\mu a^3.$$

As for the point mass equation (4.106), employing the Van der Pol transformation and subsequently averaging out rapidly varying terms ($c_2, f_r, a, \dot{a} \ll 1$) yields:

$$\ddot{y}+c_2\dot{y}+f_r(y) = \frac{1}{16}\pi\Omega^2 a^2 \sin(2\pi y).$$ (4.109)

Stationary Solutions These are obtained by letting $\dot{a}=\dot{\theta}=\dot{y}=\ddot{y}= 0$ in (4.108)–(4.109). The trivial solution is given by $a = 0$, $y = y_0$, where y_0 is the static equilibrium of the point mass, $f_r(y_0) = 0$. Non-trivial solutions (a, y, θ) are governed by the following system of algebraic equations:

$$q_A^2 = \left(\tfrac{1}{2}\Omega^2 - 2 + \alpha\Omega^2 \sin^2(\pi y) - \tfrac{3}{2}\mu a^2\right)^2 + (c_1\Omega)^2,$$

$$a^2 = \frac{16 f_r(y)}{\pi\Omega^2 \sin(2\pi y)},$$ (4.110)

$$\tan 2\theta = \frac{-\tfrac{1}{2}\left(\Omega^2 - 4 + 2\alpha\Omega^2 \sin^2(\pi y)\right) + \tfrac{3}{2}\mu a^2}{c_1\Omega}.$$

Frequency responses are then given by the curves $y(\Omega)$ and $a(\Omega)$. When $\mu = 0$ (nonlinear stretching ignored) the first equation readily yields the solution for the position y of the point mass, which upon substitution into the second equation provides the amplitude a of the string. When $\mu \neq 0$ the solution for y is obtained by numerically solving the first equation, substituting the second equation for a^2. Backbone curves are obtained by letting $q_A = c_1 = 0$, to yield:

$$\Omega^2 = \frac{4 + \left(48\mu f_r(y_{bbone})/\pi\Omega^2 \sin(2\pi y_{bbone})\right)}{1 + 2\alpha \sin^2(\pi y_{bbone})},$$

$$a_{bbone}^2 = \frac{16 f_r(y_{bbone})}{\pi\Omega^2 \sin(2\pi y_{bbone})}.$$ (4.111)

When $\mu = 0$ the first equation yields $y_{bbone} = \pi^{-1}\arcsin\left((2\Omega^{-2} - \tfrac{1}{2})/\alpha\right)^{1/2}$, $\Omega \leq 2$, which can be substituted into the second equation for obtaining the a-bone. When $\mu \neq 0$ the y-bone can be plotted with Ω as the dependent parameter, $\Omega = \Omega(y_{bbone})$.

Example Frequency Response Fig. 4.20 shows a typical frequency response as given by (4.110) with $f_r(y) = \kappa(y - y_0)$, i.e., the sliding point mass is restored by a linear spring. The effect of axial stretching is included by letting $\mu = 3.0$, a value chosen so as to limit the maximum resonant amplitude of the string with fixed mass to a realistic value of $a \approx 0.2$. Results obtained by numerically integrating the un-averaged equations (4.106)–(4.107) are superimposed, showing good agreement with the analytical predictions. As shown, the effect of the sliding point mass is to shift the response curve to the left through a nonlinear transition zone. It appears further from Fig. 4.20 that the response curve for the string with fixed mass is bent to the right. This indicates the stiffening effect of nonlinear stretching. When the mass is free to slide the response appears left-bent at low amplitudes (due to the softening effect of mass-sliding) and right-bent at higher amplitudes (due to the

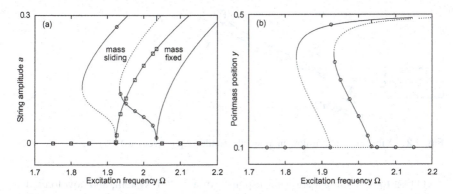

Fig. 4.20 Frequency responses $a(\Omega)$ and $y(\Omega)$ of axially excited string with a point mass sliding (spring-loaded) or fixed at $y = y_0$. Effect of axial stretching included ($\mu \neq 0$). Solid/dashed line: stable/unstable solutions of (4.110). ($\bigcirc, \square$): numerical integration of (4.106)–(4.107). Non-zero parameters: $\alpha = 0.1$, $q_A = 0.15$, $c_1 = 0.05$, $c_2 = 0.35$, $f_r(y) = \kappa(y - y_0)$, $\kappa = 0.03$, $y_0 = 0.1$, $\mu = 3.0$

stiffening effect of stretching). Except for a small range of excitation frequencies just below $\Omega \approx 1.92$, stable amplitudes of the string exist that are everywhere smaller than when the mass is fixed. Note that the folding of the sliding-mass response creates intervals of excitation frequencies having, respectively, one, two, three or four possible stationary amplitudes. In summary, when the mass is free to slide it effectively limits the amplitudes of the string.

4.7.6 Non-trivial Effects of Rotary Inertia

Some peculiar effects arise when one includes the effect of a finite rotary inertia J of the point mass (Thomsen 1996b). Small values of J imply a trivial increase of effective mass, whereas values larger than a critical threshold J_c cause nontrivial changes in system behavior. When $J > J_c$ most nonlinear effects of sliding reverse. Thus, the direction of sliding is reversed, equilibrium points interchange stability (stable/unstable becomes unstable/stable), and positions of the mass that maximizes/minimizes the effective mass of the system interchange. These effects may prove beneficial in contexts of vibration damping.

4.7.7 Summing Up

This section provided yet an example of vibration-induced movement of mass caused by nonlinear interaction. Also, it was demonstrated how to set up model equations using Hamilton's principle, how to introduce approximations and employ

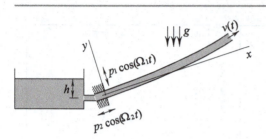

Fig. 4.21 Vibration-induced flow in a flexible cantilever pipe (Jensen 1997)

averaging for obtaining stationary responses of a 2-DOF system, and how to obtain transient responses to sweep excitation.

4.8 Vibration-Induced Fluid Flow in Pipes

Nonlinear vibrations may induce large-scale motion of solid bodies, as illustrated in Sects. 4.6–4.7. Can fluid particles be brought into organized movement as well?

Jensen (1997) examined this possibility by studying the flow of an incompressible fluid through a vibrating pipe. The system is shown in Fig. 4.21, where v (t) denotes the instantaneous speed of the fluid. In this section we summarize the results for a simplified system, neglecting gravity ($g = 0$) and considering lateral displacement of the base only ($p_2 = 0$).

The system is supposed to be resonantly excited at a frequency of excitation $\Omega \equiv \Omega_1$ near the i'th natural frequency ω_i of flexural pipe vibrations. Motions of the pipe and the fluid, respectively, are then approximately governed by the following nondimensional equations:

$$\ddot{q}_i + 2(\zeta_i\omega_i + \sqrt{\beta}k_{ii}U)\dot{q}_i + \left(\omega_i^2 + \sqrt{\beta}f_{ii}\dot{U} + g_{ii}U^2\right)q_i = p\vartheta_i\Omega^2 \cos\Omega\tau, \quad (4.112)$$

$$\dot{U} + \frac{1}{\sqrt{\beta}}\left(\alpha_j|U^{j-1}| + \frac{1}{2}U\right)U = -\sqrt{\beta}k_{ii}q_i\ddot{q}_i - \sqrt{\beta}h_{ii}(q_i\ddot{q}_i + \dot{q}_i\dot{q}_i), \quad (4.113)$$

where $q_i(t)$ denote the i'th modal amplitude of pipe vibrations, $U(t)$ is the fluid speed, and f_{ii}, g_{ii}, h_{ii} and ϑ_i are constants defined in terms of eigenfunctions for a cantilever beam. The parameters α, ζ_i and β represent, respectively, internal friction, external viscous damping, and the ratio of fluid mass to total mass. The integer parameter j indicates if the flow is predicted to be laminar ($j = 1$) or turbulent ($j = 2$).

Equations (4.112)–(4.113) are nonlinearly coupled in the variables q_i and U. As appears the forcing terms on the right-hand side of the fluid equation (4.113) are functions of the pipe amplitude q_i. Hence pipe vibrations cause the fluid to move axially. The effect is purely nonlinear, and thus negligible at low levels of pipe vibrations. The pipe equation (4.112) in turn shows that motions of the fluid change the stiffness and damping of the pipe, and thus affect pipe vibrations.

Approximate solutions to (4.112)–(4.113) can be obtained by multiple scales perturbation analysis. The requirement for eliminating secular terms results in the following set of frequency equations, governing stationary values of the modal beam amplitude b and mean fluid speed A:

$$\frac{1}{4}p^2\vartheta_i^2\Omega^4 = \left[\omega_i^2\left(\zeta_i\omega_i + \sqrt{\beta}k_{ii}A\right)^2 + \left(\frac{1}{2}g_{ii}A^2 - \omega_i(\Omega - \omega_i)\right)^2\right]b^2, \qquad (4.114)$$

$$\left(2\alpha_jA^{j-1} + A\right)A = \beta\omega_i^2k_{ii}b^2. \qquad (4.115)$$

It appears from (4.115) that pipe vibrations b may induce fluid motions with non-zero mean speed A. For turbulent flow ($j = 2$) (4.115) shows the mean fluid speed to be directly proportional to pipe amplitude, $A = b\sqrt{\beta\omega_i^2k_{ii}}/\sqrt{2\alpha_2 + 1}$.

Fig. 4.22 shows a typical first-mode frequency response as given by (4.114)–(4.115). The approximate analytical solutions are seen to agree with results obtained by numerical integration of (4.112)–(4.113). Thus the perturbation solution adequately captures the resonant behavior of the system. The upper curve in Fig. 4.22(b) shows the vibration amplitude to be expected when the fluid is fixed inside the pipe, corresponding to letting $U = \dot{U} = 0$ in (4.112). As shown in Fig. 4.22(a), near-resonant pipe excitations induce fluid motions at non-zero mean speed. Motions of the fluid in turn affect vibrations of the pipe, reducing its amplitude as compared to the fixed-fluid case.

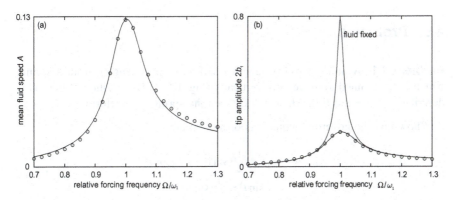

Fig. 4.22 Frequency responses showing (a) mean fluid speed A, and (b) pipe amplitude $2b_1$ versus excitation frequency Ω. Solid line: stable perturbation solution; ($\bigcirc$): numerical integration. ($i = 1$, $p = 0.01$, $\alpha_2 = 0.8$, $\beta = 0.2$, $\zeta = 0.01$.) (Jensen 1997; reprinted with permission from Elsevier)

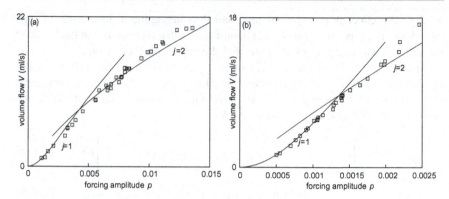

Fig. 4.23 Experimentally measured fluid flow V versus normalized pipe excitation amplitude p. Solid line: perturbation solution. (a) First-mode resonance ($\Omega = \omega_1$, $\alpha_1 = 0.1$, $\alpha_2 = 2.6$, $\zeta = 0.04$, $\beta = 0.39$; (b) Second-mode resonance ($\Omega = \omega_2$, $\alpha_1 = 0.22$, $\alpha_2 = 3.9$, $\zeta = 0.03$, $\beta = 0.39$.) (Jensen 1997; reprinted with permission from Elsevier)

Laboratory experiments were performed for a system similar to that of Fig. 4.21. Fig. 4.23 depicts results for the two lowest modes of vibration. Good agreement between experimental and theoretical results is noted.

Gravity was ignored for the simplified analysis above. However, theoretical and experimental results confirm the effect of vibration-induced fluid motion to be strong enough to overcome moderate levels of pipe inclinations in gravity.

The fluid could be replaced by a flexible string, which would then move through the pipe at non-zero mean speed. This possibility of transporting a flexible structure by means of vibrations was suggested and analyzed in Jensen (1996).

4.9 Problems

Problem 4.1 A uniform rod of length l and mass m is hanging from a spring, which is constrained to move only vertically (Fig. P4.1). The position of the rod is described by $x(t)$ and $\theta(t)$, where $x = 0$ when the spring is undeformed.

a) Show that the governing equations of motion are:

$$\ddot{u} + \omega_1^2 u = \frac{1}{2}\ddot{\theta}\sin\theta + \frac{1}{2}\dot{\theta}^2\cos\theta,$$
$$\ddot{\theta} + \omega_2^2\sin\theta = \frac{3}{2}\ddot{u}\sin\theta,$$

$$(4.116)$$

where $u = (x - x_e)/l$, $\omega_1^2 = k/m$, $\omega_2^2 = 3\ g/2\ l$, and x_e denote the statical equilibrium position of the upper end of the rod.

b) For the case of internal resonance $\omega_1 \approx 2\omega_2$, determine a zero order multiple scales expansion that is valid for small but finite amplitudes.

c) Can there be coupling between the x and the θ motions in the linear case? In the nonlinear?

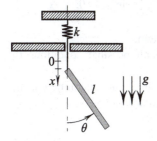

Fig. P4.1

Problem 4.2 A uniform rod of mass m and length l is hanging from a cart (Fig. P4.2). The cart has negligible mass, is restricted by a spring with stiffness k, and loaded by an external force $F(t)$.

a) Show that the governing equations of motion are:

$$\ddot{u} + \tilde{\omega}_1^2 u + \frac{1}{2}\ddot{\theta}\cos\theta - \frac{1}{2}\dot{\theta}^2\sin\theta = F(t),$$
$$\ddot{\theta} + \tilde{\omega}_2^2\sin\theta + \frac{3}{2}\ddot{u}\cos\theta = 0,$$

(4.117)

where $u = x/l$, $\tilde{\omega}_1^2 = k/m$ and $\tilde{\omega}_2^2 = 3\ g/2\ l$.

b) Expand the equations of motion to order three for small but finite values of u and θ, and determine the linear natural frequencies and mode shapes (ω_i, φ_i).

c) Decouple the linear part of the expanded equations of motion by using a modal transformation of coordinates $\mathbf{x} = \mathbf{\Phi}\mathbf{y}$, where $\mathbf{x} = \{u\ \theta\}^T$ and $\mathbf{\Phi} = [\varphi_1\ \varphi_2]$.

d) For the case $F(t) = Q\cos(\Omega t)$, perform a first-order multiple scales analysis to the stage where secular terms should be eliminated. List all possible external and internal resonances.

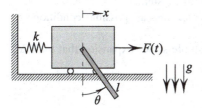

Fig. P4.2

Problem 4.3   A uniform rod of length l_2 and mass m is hanging from a massless chord of length l_1, as shown in Fig. P4.3.

a)   Show that the governing equations are

$$l_1\ddot{\theta}_1 + \frac{1}{2}l_2\ddot{\theta}_2 \cos(\theta_2 - \theta_1) - \frac{1}{2}l_2\dot{\theta}_2^2 \sin(\theta_2 - \theta_1) + g \sin\theta_1 = 0,$$

$$\frac{1}{3}l_2\ddot{\theta}_2 + \frac{1}{2}l_1\ddot{\theta}_1 \cos(\theta_2 - \theta_1) + \frac{1}{2}l_1\dot{\theta}_1^2 \sin(\theta_2 - \theta_1) + \frac{1}{2}g \sin\theta_2 = 0. \tag{4.118}$$

b)   Determine the linear natural frequencies of the system.

c)   Assuming small but finite amplitudes, determine all possible conditions of internal and external resonance.

d)   Choose l_2/l_1 to produce a three-to-one internal resonance, and determine for this case a uniformly valid first-order multiple scales expansion.

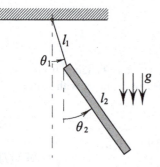

Fig. P4.3

Problem 4.4   A rigid beam with mass m and moment of inertia I is supported in a gravity field g by linear springs having stiffnesses k_1 and k_2 (Fig. P4.4). The center of gravity G of the system can move only vertically.

a)   Show that the equations of motion are:

$$m\ddot{x} + (k_1 + k_2)x + (k_1l_1 - k_2l_2) \sin\theta = -mg,$$

$$I\ddot{\theta} + (k_1l_1 - k_2l_2)x \cos\theta + \frac{1}{2}(k_1l_1^2 + k_2l_2^2) \sin 2\theta = 0. \tag{4.119}$$

b)   Determine the linear natural frequencies of the system.

c)   Assume small but finite amplitudes and determine all possible conditions of internal and external resonance.

d)   Determine a uniformly valid first-order multiple scales expansion that includes the case of modal coupling.

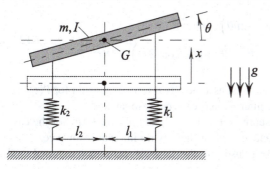

Fig. P4.4

Problem 4.5 The mass m in Fig. P4.5 slides in a gravity field g along a massless swinging rod, restricted by a linear spring having stiffness k and undeformed length L. The position of the rod is defined by the swing-angle $\theta(t)$ and the vertical position $Z(t)$ of the rod-hinge. The position of the mass m is defined by the spring-deflection $U(t)$. The hinge of the rod oscillates periodically, $Z(t) = Q\cos(\widetilde{\Omega}t)$, where Q and $\widetilde{\Omega}$ are externally controlled constants.

a) Using Lagrange's equations, show that the equations of motion are:

$$\ddot{\theta} + \frac{\left(g + Q\widetilde{\Omega}^2 \cos(\widetilde{\Omega}t)\right)\sin\theta + 2\dot{U}\dot{\theta}}{L+U} = 0,$$

$$\ddot{U} + \frac{k}{m}U - g\cos\theta - (L+U)\dot{\theta}^2 = Q\widetilde{\Omega}^2 \cos(\widetilde{\Omega}t)\cos\theta. \tag{4.120}$$

b) Nondimensionalize the equations of motion into the following form:

$$\ddot{\theta} + \frac{\left(1 + q\Omega^2 \cos(\Omega\tau)\right)\sin\theta + 2\dot{u}\dot{\theta}}{1+u} = 0,$$

$$\ddot{u} + \omega^2 u + (1 - \cos\theta) - (1+u)\dot{\theta}^2 = q\Omega^2 \cos(\Omega\tau)\cos\theta, \tag{4.121}$$

where:

$$\tau \equiv \omega_1 t, \quad \omega_1^2 \equiv \frac{g}{L+U_0}, \quad \omega_2^2 \equiv \frac{k}{m}, \quad U_0 \equiv \frac{mg}{k}, \quad q \equiv \frac{Q}{L+U_0},$$

$$u = u(\tau) \equiv \frac{U - U_0}{L+U_0}, \quad \theta = \theta(\tau), \quad \omega \equiv \frac{\omega_2}{\omega_1}, \quad \Omega \equiv \frac{\widetilde{\Omega}}{\omega_1}. \tag{4.122}$$

What are the physical interpretations of $\omega_{1,2}$, U_0, q, τ and u?

c) Add viscous damping terms $2\beta_1\dot{\theta}$ and $2\beta_2\dot{u}$ to the θ-equation and the u-equation of motion, respectively. Perform a Taylor expansion of the equations of motion for $(\theta, u) \approx (0, 0)$, retaining only leading order nonlinearities. Let $\varepsilon \ll 1$, and assume that $(u, \dot{u}, \ddot{u}, \theta, \dot{\theta}, \ddot{\theta}, \beta_{1,2}) = O(\varepsilon)$ and that $q = O(\varepsilon^2)$. Perform a substitutions of variables, $u \to \varepsilon u$, $\theta \to \varepsilon\theta$, etc., according to the assumptions, and show that small but finite vibrations are governed by

$$\ddot{\theta} + \theta = \varepsilon\left(-2\beta_1\dot{\theta} + u\theta - 2\dot{u}\dot{\theta}\right) + O(\varepsilon^2),$$

$$\ddot{u} + \omega^2 u = \varepsilon\left(-2\beta_2\dot{u} - \frac{1}{2}\dot{\theta}^2 + \dot{\theta}^2 + q\Omega^2\cos(\Omega\tau)\right) + O(\varepsilon^2).$$

(4.123)

d) Using the method of multiple scales, assume a two-term expansion for $\theta(\tau)$ and $u(\tau)$ and identify all possible internal and external resonances.

e) Calculate the approximate responses $\theta = \theta_0 + \varepsilon\theta_1$ and $u = u_0 + \varepsilon u_1$ for the case of primary external near-resonance, when there is no internal resonance.

f) Calculate the approximate response for the case of combined internal and external near-resonance.

g) Calculate the stationary amplitudes of the system for the case of combined external and internal resonance. Sketch typical force and frequency response curves. (For assessing stability of response branches use the results of a relevant example.)

h) Conclude your findings in brief form, using plain words. This part should be like the 'Summary and Conclusion' section of a journal paper, that is: Describe as clearly as possible the system you have considered, the (main) problem(s) you have solved, the most important results (in physical terms), the validity of results (assumptions and approximations), and a few suggestions for future work.

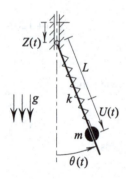

Fig. P4.5

Problem 4.6   For the pendulum with a sliding disk shown in Fig. 4.14 (p. 192):

a) Set up the equations of motion for the case where horizontal oscillations of the support of the form $Z_h(t) = Q_h\sin(\tilde{\Omega}t + \eta)$ appear in addition to the vertical oscillations $Z_v(t) = Q_v\sin(\tilde{\Omega}t)$, where η is a constant phase (so that the support describes an elliptical trajectory).

b) Consider for this system the existence and stability of the following configuration: The rod performs small amplitude vibrations near $\theta = 0$, and the disk performs small-amplitude vibrations near a fixed value of $U \in\,]0;\, l[$. (Use any method or combination of methods find solutions, including numerical integration of the equations of motion.)

Problem 4.7 Fig. P4.7 shows a section of a wind turbine wing, hung in an experimental test rig for investigating its dynamic behavior in airflow. For the present purpose the wing section can be considered a rigid solid body of mass m and mass moment of inertia J about the center of mass T_p. The vertical drag $P = K\varphi$ (small φ assumed) due to the horizontal airflow acts at distance a from T_p. The supports of the rig are characterized by linear stiffness $k_1 = k_2 = k > 0$, linear viscous damping $c_1 = c_2 = c > 0$, and distances L_1 and L_2 from T_p; here we let $L_1 = L_2 = L$.

a) Show that small motions of the wing section are governed by the linearized system

$$\mathbf{M\ddot{q}} + \mathbf{C\dot{q}} + \mathbf{Kq} = \mathbf{0}, \quad \mathbf{q} = \mathbf{q}(t), \tag{4.124}$$

where

$$\mathbf{q} = \begin{Bmatrix} y \\ \varphi \end{Bmatrix}; \quad \mathbf{M} = \begin{bmatrix} m & 0 \\ 0 & J \end{bmatrix}; \quad \mathbf{C} = \begin{bmatrix} 2c & 0 \\ 0 & 2cL^2 \end{bmatrix}; \quad \mathbf{K} = \begin{bmatrix} 2k & -K \\ 0 & 2kL^2 - aK \end{bmatrix}. \tag{4.125}$$

b) Classify the forces and the system (cf. Sect. 1.9).
c) Determine the values of rig support stiffness k that ensures the equilibrium $y = \varphi = 0$ to be stable at a given drag parameter K.
d) If the rig support is too flexible, will the instability of the static equilibrium be of the flutter or the divergence type?

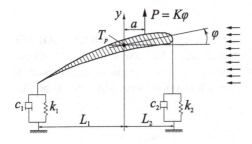

Fig. P4.7

Problem 4.8 In Fig. P4.8 a fluid of density ρ_f flows at uniform speed V through the inner cross-sectional area A_f of a simply supported elastic tube, which has bending stiffness EI, density ρ, material cross-sectional area A, viscous damping coefficient per unit length $c\rho A$, and length l.

a) Show that small transverse vibrations $u(x, t)$ are governed by the linear equation of motion:

$$(1+\alpha)\ddot{u} + c\dot{u} + \kappa u'''' + \alpha\left(V^2 u'' + 2V\dot{u}'\right) = 0,$$
$$u(0,t) = u''(0,t) = u(l,t) = u''(l,t) = 0,$$

$$(4.126)$$

where $u' = \partial u/\partial x$, $\dot{u} = \partial u/\partial t$, shear deformations and longitudinal and rotational inertia of the tube have been neglected, $\kappa = EI/\rho A$ is a measure of the transverse tube stiffness, and $\alpha = \rho_f A_f/\rho A$ the ratio of fluid mass to empty tube mass.

b) Perform a formal mode shape expansion (i.e. without specific choice of mode shapes) of the partial differential equation of motion.
c) Classify the forces and the discretized system (cf. Sect. 1.9).
d) Determine the critical value of the flow speed V beyond which the straight configuration of the tube becomes unstable.

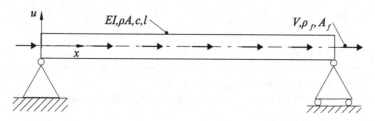

Fig. P4.8

Problem 4.9 In Fig. P4.9 fluid flows at speed $u > 0$ through two tubes AB and BC in a gravity field $g > 0$. The tubes can oscillate in the plane of the paper, and are identical, massless, rigid, have length l, and are linked at frictionless hinges at A and B. The linearized equations describing small oscillations ($\theta_1(t)$, $\theta_2(t)$) are, in nondimensional form:

$$p\left(8\ddot{\theta}_1 + 3\ddot{\theta}_2 + 6\dot{\theta}_1 + 12\dot{\theta}_2 - 6(\theta_1 - \theta_2)\right) + 9\theta_1 = 0,$$
$$p\left(3\ddot{\theta}_1 + 2\ddot{\theta}_2 + 6\dot{\theta}_2\right) + 3\theta_2 = 0,$$

$$(4.127)$$

where $p = u^2/(gl)$ is a fluid loading parameter, $\tau = ut/l$ is nondimensional time, and $\dot{\theta} = d\theta/d\tau$.

a) Classify the forces and the system (cf. Sect. 1.9).
b) For which flow speeds u is the equilibrium $\theta_1 = \theta_2 = 0$ stable? (Use the Routh-Hurwitz criterion; App. C.)
c) Can the model system be stabilized in the inverted position $\theta_1 = \theta_2 = \pi$? (Hint: Stability of $\theta_1 = \theta_2 = \pi$ for $g > 0$ corresponds to stability of $\theta_1 = \theta_2 = 0$ for $g < 0$.)

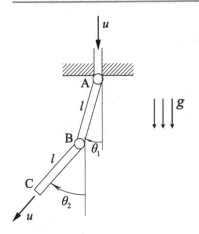

Fig. P4.9

Problem 4.10 Consider small vibrations $u = u(x, t)$ of the in-plane loaded thin plate in airflow shown in Fig. P4.10, governed by the linearized equation of motion:

$$\rho h \ddot{u} + \beta \dot{u} + D u'''' + q u'' + p u' = 0 \qquad (4.128)$$

where D is the bending stiffness, q the in-plane load, p the loading from the airflow, ρh mass per unit area, β the damping coefficient, all parameters are positive unless otherwise stated, $u' = \partial u / \partial x$, and $\dot{u} = \partial u / \partial t$. The plate is simply supported along $x = 0$ and $x = l$, and is considered infinitely long in the y-direction.

a) Perform a formal mode shape expansion (i.e. without specific choice of mode shapes) of the partial differential equation of motion.
b) Classify the forces and the system (cf. Sect. 1.9) when, respectively, $p = 0$ and $p > 0$... (see overleaf)
c) For a particular discretization, using expansion functions $\varphi_1 = \sin(\pi x/l)$ and $\varphi_2 = \sin(2\pi x/l)$, calculate the characteristic polynomial whose roots determines the stability of the equilibrium $u = 0$.
d) For $p = 0$, determine the critical value q^* of the in-plane load, so that the equilibrium $u = 0$ is stable if and only if $q < q^*$.
e) Now assume that $p > 0$, and that $q = \eta q^*$, where the parameter $\eta \in [0; 1[$ describes how near the in-plane load is to the critical value q^* at zero airflow. Calculate the critical value $p^* = p^*(\eta)$ of the airflow, so that the equilibrium $u = 0$ is stable if and only if $p < p^*$.

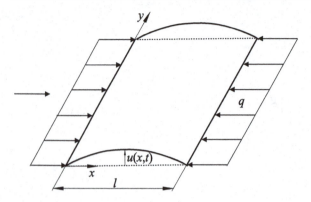

Fig. P4.10

Problem 4.11 Test engineers A and B have measured frequency response functions as shown in Fig. P4.11 for transverse vibrations of a beam stiffener which is critical to the safety of a larger structure. Fig. (a) shows the setup, and fig. (b) a typical time series when the input is applied at beam point #11 and the acceleration output is measured at point #2. Fig. (c) shows the corresponding frequency response function (FRF), measured with the beam hung in two rubber bands.

Their next task is to make a mathematical model of the vibrating beam, and their team manager wants a model that is *as simple as possible*. Test engineer A, to whom this seems straightforward, suggests using the well-known linear partial differential equation of motion describing flexural vibrations of a free-free uniform beam. But to engineer B the situation appears more complicated: Having just completed a course in advanced vibrations and stability, worries arise about overlooking several intricate phenomena (which A is happily unaware about). In particular, B wants to include the supporting rubber bands in the model, and to include material nonlinearities and geometrical nonlinearities.

So: *Is A's suggestion for a model adequate for reproducing the most important features of the experimental FRF in Figure (c)?*

Useful information

- Beam data: Length $l = 1$ m, cross-section $b \times h = 40 \times 10$ mm, mass $M = 3120$ grams, Young's modulus $E = 2.06 \times 10^{11}$ Pa.
- The FRF measurements are in the form of accelerance, i.e. $a(\omega)/f(\omega)$, where $a(\omega)$ and $f(\omega)$, respectively, is the Fourier transform of the measured acceleration $a(t)$ and force $f(t)$.
- The six smallest solutions of $\cos(\alpha) \cosh(\alpha) = 1$ is: 4.730, 7.853, 11.00, 14.13, 17.28, 20.42.].

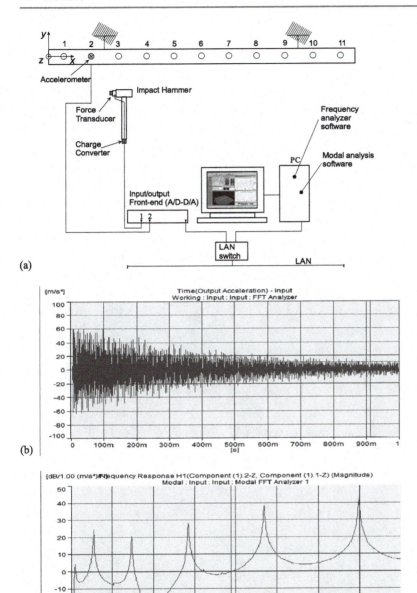

Fig. P4.11

5 Bifurcation Analysis

5.1 Introduction

Bifurcations mark the *qualitative* changes in system behavior that may occur when the parameters of a system are varied. For example, for a damped pendulum a bifurcation occurs when the damping parameter is changed from zero to a small positive value, because an undamped pendulum behaves in a qualitatively different way from a damped pendulum.

Bifurcation *analysis* is then concerned with why, when and how bifurcations occur. Bifurcation analysis can be cumbersome, though mostly worth the effort. The real strength of it comes in at an early stage of analysis, the one at which the *case* to analyze is decided upon. Bifurcation analysis reveals the critical cases and the kinds of qualitative shifts in behavior to be expected, and are thus particularly relevant where these are not immediately obvious. Further, even a basic knowledge on bifurcations helps in recognizing and understanding phenomena that occur universally across a variety of mathematical models. For example, the *Hopf bifurcation* (introduced already in Chap. 3) creates periodic oscillations out of equilibriums, and appears universally for physical, chemical, biological, economical and other systems. It is of course desirable to know the conditions under which such a dramatic change can take place.

The full story of bifurcation theory is long, mathematically involved, and far from complete. Initiated by Poincaré (1899), it remains an area of highly active research. This chapter is intended to provide only a brief introduction to notions and tools. We restrict ourselves to a rather simple and well-defined class of bifurcations, the so-called *codimension one bifurcations of equilibriums*. Besides being simple, they appear with so many systems that one should be able to recognize them when they occur (which they do for most examples of this book).

A rather pragmatic approach is chosen, and readers with a preference for mathematical rigor are referred to, e.g., Guckenheimer and Holmes (1983), Hale and Kocak (1991), Jackson (1991) or Troger and Steindl (1991). Nayfeh and

© The Author(s), under exclusive license to Springer Nature Switzerland AG 2021   271
J. J. Thomsen, *Vibrations and Stability*,
https://doi.org/10.1007/978-3-030-68045-9_5

Balachandran (1995) provide a readable and compact treatment with a focus on applications, as does Seydel (2010). There are algorithms for the numerical determination and tracking of bifurcations (Kaas-Petersen 1989; Nayfeh and Balachandran 1995; Dankowicz and Schilder 2013), which may be useful supplements to the mostly analytical methods described here, and ready-to-use software packages exist, e.g. AUTO (in Fortran, for UNIX/Linux), MatCont and COCO (MATLAB toolboxes), PyCont (Python toolbox).

After some introductory definitions of systems, bifurcations, and bifurcation conditions, we describe the possible bifurcations of codimension one. These are described in terms of one-dimensional generic systems. One-dimensional generic systems have relevance to higher-dimensional systems as well, as described in subsequent sections on methods for dimensional reduction. An introduction to continuation techniques is given, useful also for graphing bifurcation diagrams. We end the chapter by reconsidering, with an emphasis to bifurcations, some of the physical examples already encountered in Chaps. 3 and 4.

5.2 Systems, Bifurcations, and Bifurcation Conditions

5.2.1 Systems

We consider a general dynamic system described by a set of n autonomous first-order differential equations:

$$\dot{\mathbf{x}} = \mathbf{f}(\mathbf{x}, \boldsymbol{\mu}); \quad \mathbf{x} = \mathbf{x}(t) \in R^n, \ \boldsymbol{\mu} \in R^k, \tag{5.1}$$

where $\mathbf{x}$ is an n-vector of state-variables, $\mathbf{f}$ is an n-vector of generally nonlinear functions, and $\boldsymbol{\mu}$ is a k-vector of *control* or *bifurcation parameter s*. We consider as bifurcation parameters those among the system parameters for which a variation is concerned.

For example, the equation of motion (3.11) for the pendulum with an oscillating support, $\ddot{\theta} + 2\beta\omega_0\dot{\theta} + (\omega_0^2 - q\Omega^2 \cos(\Omega t))\sin\theta = 0$, can be recast in the form (5.1) with $n = 3$, $\mathbf{x} = \{\theta, v, z\}^T$, $\mathbf{f} = \{v, -2\beta\omega_0 v - (\omega_0^2 - q\Omega^2 \cos z)\sin\theta, \Omega\}^T$, and the vector $\boldsymbol{\mu}$ containing one or more of the parameters $(\beta, \omega_0, q, \Omega)$.

5.2.2 Bifurcations

As $t \to \pm\infty$ the state $\mathbf{x}(t)$ of the system (5.1) will approach a *limit set* in $\mathbf{x}$-space. The limit set can be a *singular point*, a *closed orbit*, or a *chaotic attractor*. For the underlying physical system these limit sets correspond to, respectively: equilibrium, periodic motion, or chaotic motion. If the parameters $\boldsymbol{\mu}$ of the system $\mathbf{f}$ are changed, then the limit sets may change as well. Typically a small change in μ produces only small *quantitative* changes in a limit set. For example, a slight change of μ could cause the location of a singular point to be displaced somewhat, or modify the shape of a periodic orbit.

However, for certain values of μ even a slight change in μ might set off more drastic changes in limit behavior. For example, the number and/or stability of singular points could change, a periodic orbit could appear, disappear or gain or lose stability – or a chaotic attractor could appear, disappear or change character. Such *qualitative* shifts in behavior are called *bifurcations*. A value of μ for which a bifurcation occurs is called a *bifurcation value* or *critical value*. If at some bifurcation value $\mu = \tilde{\mu}$ the stationary state is $x = \tilde{x}$, then $(\tilde{\mu}, \tilde{x})$ is a *bifurcation point*. A *bifurcation set* consists of the union of all bifurcation values in μ-space.

5.2.3 Bifurcation Conditions: Structural Instability

Bifurcations can occur only when a system is *structurally unstable*. A structurally *stable* system retains its qualitative properties even if its parameters are slightly perturbed. A structurally *unstable* system does not.

Structural stability has nothing to do with 'engineering stability', 'Lyapunov stability', or other common descriptors of the attracting or repelling properties of given limit sets. These notions of stability characterize properties of specific solutions $\tilde{x}$ in response to slight perturbations in x-space. *Structural stability*, by contrast, characterizes properties of the limit sets of a system $\mathbf{f}(\mathbf{x}, \mathbf{\mu})$ in response to slight perturbations in μ-space. Thus, a structurally stable system may well possess unstable limit sets, insofar as these do not change in number, character or stability when system parameters are slightly varied.

Consider, as an example, a lightly damped pendulum in gravity, but otherwise unforced. The *state* of upside-down equilibrium is unstable in the sense of Lyapunov, since a small perturbation causes the pendulum to escape from that state. The *system* is nevertheless structurally stable, since slight perturbations to the parameters of the system (natural frequency and damping) do not modify the qualitative behavior of the pendulum. However, with no damping the pendulum system becomes structurally unstable, since the dynamics of an undamped pendulum differ qualitatively from that of a pendulum subjected to even infinitesimal amounts of damping.

A system is structurally unstable if any of its singular points are *non-hyperbolic*. A singular point $\tilde{x}$ is *hyperbolic* (or *non-degenerate*) if the Jacobian evaluated at the point, $\mathbf{J}(\tilde{\mathbf{x}}, \mathbf{\mu}) = (\partial \mathbf{f}/\partial \mathbf{x})|_{\mathbf{x}=\tilde{\mathbf{x}}}$, has no eigenvalues with zero real part. So, any singular point for which the Jacobian has an eigenvalue with zero real part is *non-hyperbolic* (or *degenerate*).

Thus, structural stability breaks down and bifurcations can occur at those values of μ producing a system with one or more singular points having at least one Jacobian eigenvalue with a zero real part.

Note that structural instability is a *necessary* condition for bifurcations to occur. That is, a system must be structurally unstable for bifurcations to occur, but structural instability does not automatically imply bifurcation.

For example, the unforced pendulum system possesses Jacobian eigenvalues having zero eigenvalue when there is no damping, $\beta = 0$ (cf. Sect. 3.4.4). Hence for $\beta = 0$ the system is structurally unstable, meaning that the addition of slight damping (perturbation to $\beta = 0$) *may* change the qualitative properties of the system. Indeed, it was found in Sect. 3.4.4 that the addition of even infinitesimal damping changes the singular points at $\theta = p2\pi$ from centers to foci.

Assume now that we have located the bifurcation set of a given system $\mathbf{f}(\mathbf{x}, \boldsymbol{\mu})$. Thus, we know those values of system parameters $\boldsymbol{\mu}$ for which structural stability breaks down and bifurcations may occur that will change the qualitative behavior of the system. Then *how* does the behavior of the system change? To answer this we need to *unfold* the bifurcations, that is, to picture somehow the change in system behavior in response to changes in $\boldsymbol{\mu}$. We do this by drawing *bifurcation diagrams*. In most cases this can be accomplished only locally, that is, for each individual bifurcation value, and for *small* changes in $\boldsymbol{\mu}$. Connecting local pictures of bifurcations together into a global bifurcation diagram can be a painstaking procedure, if at all possible.

We next summarize the properties of the simplest of all local bifurcations, those of codimension one. Though simple, codimension one bifurcations find applications for real systems of high dimension as well. This holds true when the essential dynamics of a higher-dimensional system can be reduced to a one-dimensional *center manifold*, as described in Sects. 5.4 and 5.5.

5.3 Codimension One Bifurcations of Equilibriums

As the name suggests, codimension one bifurcations can be described on a one-dimensional manifold; it takes only a single parameter to unfold them. The *codimension* of a bifurcation is the smallest dimension of a parameter space that contains the bifurcation in a persistent way. The codimension is at least as large as the number of Jacobian eigenvalues having a zero real part (a purely imaginary pair counting as one). If there are no eigenvalues having a zero real part the codimension is zero, and no bifurcations occur. If there is a simple zero eigenvalue or a purely imaginary pair, the codimension is (usually) one, and bifurcations occur that can be unfolded through a single parameter.

If there is more than one Jacobian eigenvalue having a zero real part, then bifurcations of codimension two or higher can take place. These are more complicated, requiring at least two parameters for their unfolding. Hence, rather than the actual dimension n of a system $\mathbf{f}(\mathbf{x}, \boldsymbol{\mu})$, it is the number of eigenvalues having zero real parts (the *co*dimension) that determines the complexity of the bifurcations. This is not as strange as it may seem, since motions associated with non-zero eigenvalues ride on either contracting or expanding state spaces for which the dynamics is simple and well understood: Motions simply decay or grow exponentially with time.

We explain the codimension one bifurcations in terms of simple one-dimensional study cases. You may perceive these *generic* cases as reductions of full-blown higher dimensional systems possessing a single eigenvalue with a zero real part (in Sect. 5.5 we provide tools for this reduction).

5.3.1 The Pitchfork Bifurcation

The *pitchfork bifurcation* can be represented by the generic differential equation

$$\dot{x} = \mu x - x^3, \tag{5.2}$$

with singular points (found by letting $\dot{x} = 0$ and solving for x):

$$\tilde{x} = 0, \quad \tilde{x} = \pm\sqrt{\mu}. \tag{5.3}$$

If $\mu < 0$ there is one singular point, whereas if $\mu > 0$ there are three. The Jacobian of the system is:

$$J(x, \mu) = \frac{\partial \dot{x}}{\partial x} = \mu - 3x^2. \tag{5.4}$$

At the singular point $\tilde{x}$ the eigenvalue λ of the Jacobian becomes:

$$\lambda = \mu - 3\tilde{x}^2, \tag{5.5}$$

or

$$\begin{aligned} \lambda &= \mu \quad \text{for } \tilde{x} = 0, \\ \lambda &= -2\mu \quad \text{for } \tilde{x} = \pm\sqrt{\mu}. \end{aligned} \tag{5.6}$$

A bifurcation can occur when the Jacobian eigenvalue has a zero real part, that is, when $\mu = 0$. Hence, $\mu = 0$ is the (one and only) bifurcation value for this system.

When $\mu < 0$ the only singular point is $\tilde{x} = 0$. This point is stable since $\text{Re}(\lambda) = \mu < 0$. When $\mu > 0$ the singular point $\tilde{x} = 0$ becomes unstable ($\text{Re}(\lambda) > 0$), and two new singular points $\tilde{x} = \pm\sqrt{\mu}$ emerge, both stable since $\text{Re}(\lambda) = -2\mu < 0$.

We are now able to draw the *bifurcation diagram* of the generic pitchfork bifurcation (Fig. 5.1). A bifurcation diagram simply shows how singular points branch out and change stability at a bifurcation point. Indeed, the picture of the pitchfork bifurcation looks like a pitchfork.

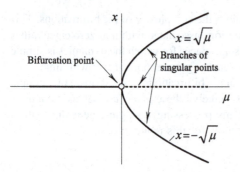

Fig. 5.1 The supercritical pitchfork bifurcation associated with $\dot{x} = \mu x - x^3$. (———) stable, (----) unstable

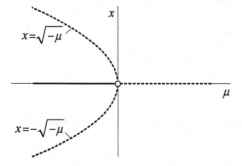

Fig. 5.2 The subcritical pitchfork bifurcation of $\dot{x} = \mu x + x^3$. (———) stable, (----) unstable

The pitchfork bifurcation described by (5.2) and Fig. 5.1 is called *supercritical*, because it occurs *beyond* the critical value of μ. Inverting the sign of the cubical term in (5.2), a *subcritical* pitchfork bifurcation appears instead. The diagram of the subcritical pitchfork is like that of the supercritical, though rotated 180 degrees and with inverted stability (Fig. 5.2). The distinction between subcritical and supercritical bifurcations, however, is not tied to the orientation of branches, but to the *stability* of these. During a supercritical bifurcation a stable solution turns unstable, and two stable branches emerge on each side of the unstable solution. During a subcritical bifurcation an unstable solution gains stability, and two branches of unstable solutions emerge.

We have come across several pitchfork bifurcations in this book already. For example, static buckling of a pinned-pinned column occurs through a supercritical pitchfork bifurcation. At pre-critical loadings the undeformed equilibrium position of the column is stable, whereas at a critical load the zero solution loses stability and bifurcates into two symmetrical and stable buckling modes. More examples will be considered in Sect. 5.14.

5.3.2 The Saddle-node Bifurcation

For describing the *saddle-node bifurcation* we consider the generic system

$$\dot{x} = \mu - x^2. \tag{5.7}$$

For $\mu < 0$ there are no singular points, whereas for $\mu > 0$ there are two, $\tilde{x} = \pm\sqrt{\mu}$. At the singular points the Jacobian becomes $J(\tilde{x}, \mu) = -2\tilde{x}$ with eigenvalue $\lambda = -2\tilde{x}$. The bifurcation value is $\mu = 0$, since at this value $\text{Re}(\lambda) = 0$. Examining the stability of singular points we find that $\text{Re}(\lambda) = -2\sqrt{\mu} < 0$ for $\tilde{x} = \sqrt{\mu}$ (a stable branch), whereas $\text{Re}(\lambda) = 2\sqrt{\mu} > 0$ for $\tilde{x} = -\sqrt{\mu}$ (an unstable branch). Fig. 5.3 shows the bifurcation diagram. Comparing to the pitchfork bifurcation, one sees that the saddle-node bifurcation does not involve a trivial (zero) solution, and that it lacks symmetry about the μ-axis.

The saddle-node bifurcation takes its name from the character of the two emerging singular points: Typically, in higher dimensional systems, the stable branch represents nodal points, whereas the unstable branch represents saddle points.

Saddle-node bifurcations are also called *tangent bifurcations*, due to the vertical tangency present at the bifurcation point. Other names are *turning point* or *fold bifurcations*, notions that will appear more obvious from the examples in Sect. 5.14.

5.3.3 The Transcritical Bifurcation

To describe the *transcritical bifurcation* we consider the system

$$\dot{x} = \mu x - x^2, \tag{5.8}$$

with two singular points $\tilde{x} = 0$ and $\tilde{x} = \mu$. The Jacobian is $J(\tilde{x}, \mu) = \mu - 2\tilde{x}$ with eigenvalue $\lambda = \mu - 2\tilde{x}$. The only bifurcation value is $\mu = 0$. It is found that the

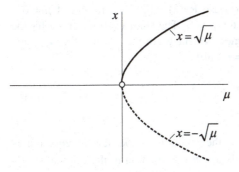

Fig. 5.3 The saddle-node bifurcation of $\tilde{x} = \mu - x^2$. (———) stable, (----) unstable

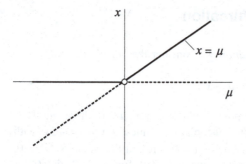

Fig. 5.4 The transcritical bifurcation of $\tilde{x} = \mu x - x^2$. (——) stable, (----) unstable

singular point $\tilde{x} = 0$ is stable for $\mu < 0$ and unstable for $\mu > 0$, and that the point $\tilde{x} = \mu$ is unstable for $\mu < 0$ and stable for $\mu > 0$. Thus, at a point of transcritical bifurcation two equilibrium solutions meet and exchange stability, as depicted in Fig. 5.4.

5.3.4 The Hopf Bifurcation

The *Hopf bifurcation* is characterized by the birth of a *limit cycle.*. That is, a statical equilibrium point bifurcates into periodic motion. This distinguishes the Hopf bifurcation from the pitchfork, saddle-node and transcritical bifurcations that involve only equilibrium points.

To unfold a Hopf bifurcation a *pair* of generic differential equations is required:

$$\dot{x} = -y + x\left(\mu - (x^2 + y^2)\right),$$
$$\dot{y} = x + y\left(\mu - (x^2 + y^2)\right). \tag{5.9}$$

The singular point $(\tilde{x}, \tilde{y}) = (0, 0)$ of the system has Jacobian eigenvalues $\lambda = \mu \pm i$. This point is stable (Re(λ) < 0) for $\mu < 0$ and unstable for $\mu > 0$. The bifurcation value is $\mu = 0$ since in this case Re(λ) = 0. Furthermore, at $\mu = 0$ a stable limit cycle emerges and exists for $\mu > 0$. The limit cycle is given by the solution set of $x^2 + y^2 = \mu$. This becomes evident when the system (5.9) is transformed into polar form ($x = r\cos\theta$, $y = r\sin\theta$):

$$\dot{r} = r(\mu - r^2),$$
$$\dot{\theta} = 1. \tag{5.10}$$

Clearly, $\dot{r} = 0$ when $r = \sqrt{\mu}$. Since the solution to the second equation is $\theta(t) = t$, the condition $\dot{r} = 0$ corresponds to periodic motion in the original (x, y)-coordinates: $x = r\cos\theta = \sqrt{\mu}\cos t$, and $y = r\sin\theta = \sqrt{\mu}\sin t$.

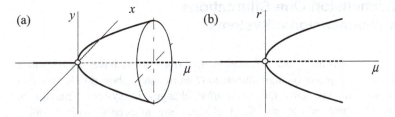

Fig. 5.5 Supercritical Hopf bifurcation of $\dot{x} = -y + x(\mu - (x^2 + y^2))$, $\dot{y} = x + y(\mu - (x^2 + y^2))$. (a) In (x, y, μ) space; (b) In (r, μ) space

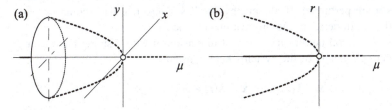

Fig. 5.6 Subcritical Hopf bifurcation of $\dot{x} = y + x(\mu + (x^2 + y^2))$, $\dot{y} = -x + y(\mu + (x^2 + y^2))$. (a) In (x, y, μ) space; (b) In (r, μ) space

Notice that the r-equation in (5.10) is identical to the generic equation (5.2) for the supercritical *pitchfork* bifurcation. Thus, the bifurcation diagram for the r-equation is a supercritical pitchfork (Fig. 5.5(b)). Since r is the amplitude of a periodic motion in the (x, y) plane we obtain the bifurcation diagram for the Hopf bifurcation shown in Fig. 5.5(a), where stable limit cycle motion occurs on the parabolic surface.

The Hopf bifurcation of (5.9) is *supercritical*, as is the pitchfork bifurcation associated with the r-equation in (5.10). The corresponding *subcritical* Hopf bifurcation (Fig. 5.6) is obtained for a sign-shifted variant of the system (5.9):

$$\dot{x} = y + x(\mu + (x^2 + y^2)),$$
$$\dot{y} = -x + y(\mu + (x^2 + y^2)) . \tag{5.11}$$

Note again, that whereas the pitchfork, saddle-node and transcritical bifurcations involve a simple zero eigenvalue at the bifurcation point, the Hopf bifurcation involves a *pair* of complex conjugated eigenvalues. During the bifurcation this pair of eigenvalues crosses the imaginary axis at the origin, that is, their real parts change sign.

5.4 Codimension One Bifurcations for N-dimensional Systems

Do the generic codimension one bifurcations described above have any relevance for real systems of possibly high dimension? Fortunately they do. Recall that bifurcation analysis concerns the *qualitative* changes in system behavior in response to parameter variations. Such changes are associated with Jacobian eigenvalues having a zero real part. Thus, the possible bifurcations depend on the number of eigenvalues having a zero real part, rather than on the dimension of the system. If for an n-dimensional system *all* Jacobian eigenvalues have non-zero real parts, then no bifurcations of the state $\mathbf{x}$ in question will occur when the system parameters $\boldsymbol{\mu}$ are perturbed. If *one* eigenvalue has a zero real part (counting multiplicity), then codimension one bifurcations can occur.

Consider a general nonlinear system of dimension n with a single parameter μ (i.e. all other parameters of the systems kept constant):

$$\dot{\mathbf{x}} = \mathbf{f}(\mathbf{x}, \mu); \quad \mathbf{x} = \mathbf{x}(t) \in R^n, \ \mu \in R. \tag{5.12}$$

Assume that a singular point $\mathbf{x} = \tilde{\mathbf{x}}$ exists, defined as a solution to the algebraic set of equations $\mathbf{f}(\mathbf{x}, \mu) = 0$. Near $\mathbf{x} = \tilde{\mathbf{x}}$ the behavior of the system is governed by the linearization:

$$\dot{\mathbf{x}} = \mathbf{J}(\tilde{\mathbf{x}}, \mu)(\mathbf{x} - \tilde{\mathbf{x}}); \quad \mathbf{J}(\mathbf{x}, \mu) = \frac{\partial \mathbf{f}}{\partial \mathbf{x}}. \tag{5.13}$$

At the singular point $\tilde{\mathbf{x}}$ the Jacobian $\mathbf{J}(\tilde{\mathbf{x}}, \mu)$ has n eigenvalues λ_j, $j = 1$, n. Bifurcations can occur if one or more of these eigenvalues has a zero real part. Let $\mu = \mu_0$ be a point for which $\mathrm{Re}(\lambda_j) = 0$ for one or more values of j. That is, $(\tilde{\mathbf{x}}, \mu_0)$ can be a bifurcation point. If there is only *one* eigenvalue with zero real part (an imaginary pair counting as one) the bifurcation will be of codimension one. Hence, it will qualitatively resemble one of the generic codimension one bifurcations described in Sect. 5.3. Generally, a simple zero eigenvalue indicates a saddle-node, a pitchfork or a transcritical bifurcation, whereas a purely imaginary pair indicates a Hopf bifurcation. However, without taking the nonlinear terms of (5.12) into account, we can make no useful statements regarding the post-bifurcational behavior.

Below we state two formal (and formidable) theorems establishing *necessary* conditions for, respectively, the saddle-node and the Hopf bifurcation to occur. Pitchfork and transcritical bifurcations are treated as variants of the saddle-node.

5.4.1 Saddle-Node Conditions

Here are the necessary conditions for saddle-node bifurcations to occur (adopted from Guckenheimer and Holmes 1983):

Theorem 5.1 (Saddle-node). Let $\dot{\mathbf{x}} = f(\mathbf{x}, \mu)$ be a system of differential equations in R^n depending on a single parameter μ. When $\mu = \mu_0$, assume that there is an equilibrium $\tilde{\mathbf{x}}$ for which the following conditions are satisfied:

(SN1): $\mathbf{J}(\tilde{\mathbf{x}}, \mu_0)$, where $\mathbf{J}(\mathbf{x}, \mu) = \partial f / \partial \mathbf{x}$, has a simple eigenvalue $\lambda = 0$ with right eigenvector $\mathbf{v}$ and left[1] eigenvector $\mathbf{w}$. Further, $\mathbf{J}(\tilde{\mathbf{x}}, \mu_0)$ has m eigenvalues with negative real part and $(n - m - 1)$ eigenvalues with positive real part (counting multiplicity).

(SN2): $\mathbf{w}^T \mathbf{g}(\tilde{\mathbf{x}}, \mu_0) \neq 0$, where $\mathbf{g}(\mathbf{x}, \mu) \equiv \partial \mathbf{f}(\mathbf{x}, \mu)/\partial \mu$.

(SN3): $\mathbf{w}^T \mathbf{h}(\tilde{\mathbf{x}}, \mu_0) \neq 0$, where $h_k(\tilde{\mathbf{x}}, \mu_0) \equiv \mathbf{v}^T \mathbf{H}_k(\tilde{\mathbf{x}}, \mu_0)\mathbf{v}$, $H_{k(ij)}(\mathbf{x}, \mu) \equiv \partial^2 f_k / \partial x_i \partial x_j$.

Then there is a smooth curve of equilibriums in $R^n \times R$ passing through the point $(\tilde{\mathbf{x}}, \mu_0)$, tangent to the hyperplane $R^n \times \{\mu_0\}$. Depending on the signs of the expressions in (SN2) and (SN3), there are no equilibriums near $(\tilde{\mathbf{x}}, \mu_0)$ when $\mu < \mu_0$ ($\mu > \mu_0$) and two equilibriums near $(\tilde{\mathbf{x}}, \mu_0)$ for each parameter value $\mu > \mu_0$ ($\mu < \mu_0$). The two equilibriums for $\dot{\mathbf{x}} = \mathbf{f}(\mathbf{x}, \mu)$ near $(\tilde{\mathbf{x}}, \mu_0)$ are hyperbolic and have stable and unstable manifolds of dimension, respectively, m and $n - m - 1$. The set of equations $\dot{\mathbf{x}} = \mathbf{f}(\mathbf{x}, \mu)$ satisfying (SN1)–(SN3) is open and dense in the space of C^∞ one-parameter families of vector fields with an equilibrium at $(\tilde{\mathbf{x}}, \mu_0)$ with a zero eigenvalue.

A bit of explanation may be in order:

Condition (SN1) ensures that $(\mu_0, \tilde{\mathbf{x}})$ is a bifurcation point with exactly one Jacobian eigenvalue having a zero real part, and that the imaginary part of this eigenvalue is zero (ruling out a Hopf bifurcation).

Conditions (SN2)–(SN3) are the so-called *transversality conditions*. If fulfilled, they ensure non-degenerate behavior with respect to the control parameter μ, and that the dominant nonlinear term is quadratic (*quadratic tangency*). For one-dimensional systems ($n = 1$), the expression in (SN3) reduces to $f_x''(\tilde{x}, \mu_0) \neq 0$, and this is seen to ensure a quadratic term.

To understand the notion of transversality we may consider the intersections of a function $f(x)$ with the *x*-axis. Intersections occur when $f(x) = 0$. They are *transverse* if $f'(x) \neq 0$ at the intersections – i.e. when the function really crosses the *x*-axis, rather than just touching it at a minimum or maximum. Transverse intersections imply that the number of intersections does not vary when f is slightly perturbed. *Non*-transverse intersections do not possess this property. For example, when

---

[1]The *left* eigenvector of a matrix $\mathbf{A}$ is the transpose of the usual (right) eigenvector of $\mathbf{A}^T$.

$f(x) = x^2$ the intersection at $x = 0$ is non-transverse, since $f'(0) = 0$. Perturbing the function to $f(x) = x^2 + \varepsilon$, we see that for $\varepsilon > 0$ there are no intersections, whereas for $\varepsilon < 0$ there are two.2

5.4.2 Transcritical and Pitchfork Conditions

Bifurcations with a simple zero eigenvalue will usually be saddle-nodes, unless something in the formulation of the problem prevents the saddle-node from occurring. This may be symmetry, or the presence of a trivial solution $x = 0$. In these cases a transcritical or a pitchfork bifurcation may replace the saddle-node.

The *transcritical* bifurcation is associated with the presence of a trivial solution from which bifurcations can occur. This prevents the condition (SN2) in Theorem 5.1 from being satisfied. The necessary conditions for the transcritical bifurcation then become those of the saddle-node, with (SN2) replaced by (Guckenheimer and Holmes 1983):

$$(SN2'): \mathbf{w}^T \mathbf{g}(\tilde{\mathbf{x}}, \mu_0) \neq 0, \text{ where } \mathbf{g}(\mathbf{x}, \mu) \equiv \mathbf{G}(\mathbf{x}, \mu)\mathbf{v}, \ G_{ij} \equiv \partial^2 f_i / \partial \mu \partial x_j$$

The *pitchfork* bifurcation is associated with systems obeying symmetry of some kind. For example, the Duffing equation $\dot{x} = y$, $\dot{y} = -y - x - x^3$, is invariant under the transformation $(x, y) \rightarrow (-x, -y)$. In general, systems obeying symmetry are described by functions $\mathbf{f}(\mathbf{x}, \mu)$ that are odd functions of $\mathbf{x}$. This prevents (SN3) in Theorem 5.1 from being satisfied. The necessary conditions for the pitchfork bifurcation becomes those of the saddle-node, with (SN3) replaced by an expression that contains *third*-order partial derivatives (see Guckenheimer and Holmes 1983). For one-dimensional systems it becomes $f_x'''(\tilde{x}, \mu_0) \neq 0$. For higher-dimensional systems there might be simpler ways to prove that the dominant nonlinear term is cubic.

5.4.3 Hopf Conditions

The importance of the Hopf bifurcation is perhaps best illustrated by the fact that at least one entire book has been devoted solely to the study of it (Marsden and McCracken 1976). We here come to the necessary conditions for this bifurcation to occur (adopted from Guckenheimer and Holmes 1983):

Theorem 5.2 (Hopf).   Suppose the system $\dot{\mathbf{x}} = \mathbf{f}(\mathbf{x}, \mu)$, $\mathbf{x}(t) \in R^n$, $\mu \in R$ has an equilibrium $(\tilde{\mathbf{x}}, \mu_0)$ at which the following conditions are satisfied:

(H1): $\mathbf{J}(\tilde{\mathbf{x}}, \mu_0)$, where $\mathbf{J}(\mathbf{x}, \mu) = \partial \mathbf{f}/\partial \mathbf{x}$, has a simple pair of pure imaginary eigenvalues and no other eigenvalues with zero real parts.

Then (H1) implies that there is a smooth curve of equilibriums $(\mu, \mathbf{x}(\mu))$ with $\mathbf{x}(\mu_0) = \tilde{\mathbf{x}}$. The eigenvalues $\lambda(\mu)$ and $\bar{\lambda}(\mu)$ of $\mathbf{J}(\mathbf{x}(\mu), \mu_0)$ which are imaginary at $\mu = \mu_0$ vary smoothly with μ. If, moreover:

(H2): $c_0 \equiv \frac{d}{d\mu} \mathrm{Re}(\lambda(\mu))|_{\mu=\mu_0} \neq 0$,

then there is a unique three-dimensional center manifold passing through the point $(\tilde{\mathbf{x}}, \mu_0)$ in $R^n \times R$ and a smooth system of coordinates (preserving the planes $\mu = $ constant) for which the Taylor expansion of degree three on the center manifold is given by:

$$\dot{x} = \left(c_0\mu + c_3(x^2 + y^2)\right)x - \left(\omega + c_1\mu + c_2(x^2 + y^2)\right)y,$$
$$\dot{y} = \left(\omega + c_1\mu + c_2(x^2 + y^2)\right)x + \left(c_0\mu + c_3(x^2 + y^2)\right)y, \tag{5.14}$$

which is expressed in polar coordinates ($x = r\cos\theta$, $y = r\sin\theta$) as

$$\dot{r} = (c_0\mu + c_3 r^2)r,$$
$$\dot{\theta} = \omega + c_1\mu + c_2 r^2. \tag{5.15}$$

If $c_3 \neq 0$, there is a surface of periodic solutions in the center manifold which has quadratic tangency with the eigenspace of $\lambda(\mu_0)$ and $\bar{\lambda}(\mu_0)$, agreeing to second order with the paraboloid $\mu = -c_3 r^2/c_0$. If $c_3 < 0$, then these periodic solutions are stable limit cycles, while if $c_3 > 0$, the periodic solutions are repelling.

Note that the generic system (5.9) (or (5.10)) appears as a special case of (5.14) (or (5.15)) with $(\omega, c_0, c_1, c_2, c_3) = (1, 1, 0, 0, -1)$.

Focusing on the essence of the Hopf theorem, you will appreciate that it is quite simple to employ for specific applications. Stripped from mathematical subtleties it states that:

- If (H1) for $\mu = \mu_0$ the linearized Jacobian at a singular point has a single pair of imaginary eigenvalues $(\lambda, \bar{\lambda})$, and no other eigenvalues with zero real part,
- and if (H2) the eigenvalues cross the imaginary axis transversely,
- and if the dominating nonlinear terms of the system are cubic,
- then limit cycles appear when μ is perturbed from μ_0 in the direction for which $\mathrm{Re}(\lambda) > 0$ (we call this birth of a limit cycle a *Hopf bifurcation*).
- If the cubic nonlinearity has a negative coefficient the Hopf bifurcation is *supercritical* (producing stable limit cycles), whereas with a positive coefficient the Hopf bifurcation is *subcritical* (producing unstable limit cycles).

The Hopf theorem includes the notion of a *center manifold*, to which we shall return in the next section. It appears that if a subsystem restricted to such a center manifold (defined by ω, c_0, ..., c_3) can be found, then we are able not only to prove the *existence* of limit cycles, but also to compute their *shape* as defined by the paraboloid equation $r^2 = -(c_0/c_3)\mu$.

However, the theorem provides no clues as to how this center manifold should be computed; it merely says that there *is* one. For many systems the determination of a suitable restriction to the center manifold can be a substantial undertaking, even though only the coefficient c_3 is needed (c_0 is quite easy to obtain). Partly due to this reason, the Hopf theorem is mostly used as a guide for *where* to look for periodic solutions for a given system. Having located a set of system parameters causing limit cycle behavior, a variety of other methods are at our disposal for determining their shape, for example, the perturbation methods described in Chaps. 3 and 4.

5.5 Center Manifold Reduction

The *center manifold theorem* provides a basis for systematically reducing the dimension of a system, while retaining its essential properties. For a nonlinear system of arbitrary dimension the linearized Jacobian eigenvalues may have positive, negative and zero real parts. As has already been demonstrated the 'interesting' dynamics – i.e. the *qualitative* change in system behavior – is associated with the subset of eigenvalues having zero real parts.

And that is what center manifold theory is all about: to throw away dimensions that just serve to describe 'uninteresting', hyperbolic dynamics. This has nothing to do with the common engineering practice of truncating discrete models to fit a given level of accuracy or computational power. For example, to capture with acceptable precision the behavior of a pinned-pinned column in response to a given load, you may need a Galerkin discretization retaining five modes, or a finite element model with 500 degrees of freedom. However, to capture the *qualitative* behavior, i.e. the bifurcations to be expected, a single autonomous differential equation may suffice (e.g., (5.2), (5.7) or (5.8)).

It *is* possible to proceed quite successfully through many kinds of nonlinear analysis without even knowing the existence of center manifolds. However, since they are so important in modern bifurcation analysis, and are entering into engineering literature, one should have at least some basic knowledge of their applicability.

This section, and the subsequent one on normal forms, is intended only to present the main ideas. Guckenheimer and Holmes (1983) provide a rather detailed discussion of center manifolds, as do Nayfeh (1993), Nayfeh and Balachandran (1995) and Troger and Steindl (1991). A pedagogical presentation with applications (for flow-induced oscillations) was provided by Holmes (1977), specifically addressing the engineering community. The exoticness of the subject in engineering contexts is reflected by a tendency, in scientific papers, to summarize and explain the most important concepts. A readable account of this kind is given in Païdoussis and Semler (1993) (see also Hsu 1983; Li and Païdoussis 1994).

5.5.1   The Center Manifold Theorem

We adopt from Guckenheimer and Holmes (1983) the following formulation for systems of differential equations having the form $\dot{\mathbf{x}} = \mathbf{f}(\mathbf{x})$, $\mathbf{x}(t) \in R^n$:

> **Theorem 5.3 (Center Manifold).**   Let $\mathbf{f}$ be a C^r vector field on R^n, vanishing at the origin ($\mathbf{f}(\mathbf{0}) = \mathbf{0}$), and let $\mathbf{A} = \mathbf{J}(\mathbf{0})$ where $\mathbf{J}(\mathbf{x}) = \partial\mathbf{f}/\partial\mathbf{x}$. Divide the eigenvalue spectrum of $\mathbf{A}$ into three parts, σ_s, σ_c and σ_u, such that
>
> $$\lambda \in \begin{cases} \sigma_s & \text{if } \operatorname{Re}(\lambda) < 0 \\ \sigma_c & \text{if } \operatorname{Re}(\lambda) = 0 \\ \sigma_u & \text{if } \operatorname{Re}(\lambda) > 0. \end{cases}$$
>
> Let the (generalized) eigenspaces of σ_s, σ_c and σ_u be $\mathbf{E}^s$, $\mathbf{E}^c$ and $\mathbf{E}^u$, respectively. Then there exists C^r stable and unstable invariant manifolds W^u and W^s tangent to $\mathbf{E}^u$ and $\mathbf{E}^s$ at $\mathbf{0}$, and a C^{r-1} center manifold W^c tangent to $\mathbf{E}^c$ at $\mathbf{0}$. The manifolds W^s, W^c and W^u are all invariant for the flow of $\mathbf{f}$. The stable and unstable manifolds are unique, but W^c needs not to be.

A *manifold* is just a smooth and continuous surface. And surfaces can locally be approximated by (generalized) tangents. For example, a circle is a manifold in R^2 (whereas a rectangle is not). The circle can locally be approximated by a tangent line. A spherical surface is an example of a manifold in R^3, which can locally be approximated by a tangent plane. Solutions to n-dimensional systems of differential equations ride on manifolds in R^m where $m \leq n$. These manifolds can locally be approximated by (generalized) tangents.

The *stable manifold* W^s of a singular point $\tilde{\mathbf{x}}$ of the system $\dot{\mathbf{x}} = \mathbf{f}(\mathbf{x})$ consists of all initial conditions $\mathbf{x}(0)$ for which $\mathbf{x}(t) \to \tilde{\mathbf{x}}$ for $t \to \infty$. This manifold has dimension n_s, where n_s is the number of Jacobian eigenvalues having a negative real part.

Similarly, the *unstable manifold* W^u has dimension n_u, and consists of all initial conditions for which $\mathbf{x}(t) \to \tilde{\mathbf{x}}$ for $t \to -\infty$.

The *center manifold* W^c has dimension n_c, and consists of all initial conditions for which $\mathbf{x}(t)$ neither grow nor decay with time.

Locally, the theorem says, these manifolds are tangents to the *eigenspaces* $\mathbf{E}^s$, $\mathbf{E}^c$ and $\mathbf{E}^u$, which are spaces spanned by the eigenvectors for the eigenvalues in, respectively, σ_s, σ_c and σ_u. Thus, near the singular point, the manifolds W^s, W^c and W^u can locally be approximated by the corresponding eigenspaces – just as a circle locally can be approximated by a line.

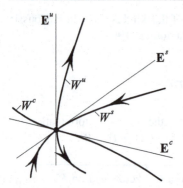

Fig. 5.7 Stable, unstable and center manifolds W^s, W^u and W^c, tangents to the eigenspaces $\mathbf{E}^s$, $\mathbf{E}^u$ and $\mathbf{E}^c$

The submanifolds W^s, W^c and W^u being *invariant* for the flow, means that a solution which is initiated on one of the manifolds stays on that particular manifold at all times.

Fig. 5.7 illustrates the concepts of stable, unstable and center manifolds and their tangency to eigenspaces for the case $n = 3$, $n_s = n_u = n_c = 1$. Note that without considering the nonlinear terms of $\mathbf{f}$, we cannot assign a flow direction to motions on the center manifold.

The center manifold theorem assumes a singular point at $\mathbf{x} = 0$. Hence, for a system $\dot{\mathbf{x}} = \mathbf{f}(\mathbf{x})$ with a singular point $\mathbf{x} = \tilde{\mathbf{x}} \neq 0$, one must shift the origin by a transform of state variables, $\mathbf{x} \to \mathbf{x} - \tilde{\mathbf{x}}$. So much for the theorem. Next we consider implications of it.

5.5.2   Implications of the Theorem

The center manifold theorem implies that the bifurcating system $\dot{\mathbf{x}} = \mathbf{f}(\mathbf{x})$ *locally* (near the bifurcation point) is topologically equivalent to:

$$
\begin{aligned}
\dot{\hat{\mathbf{x}}} &= \widehat{\mathbf{f}}(\widehat{\mathbf{x}}), & \widehat{\mathbf{x}} \in W^c \subset R^{n_c}, \\
\dot{\hat{\mathbf{y}}} &= -\widehat{\mathbf{y}}, & \widehat{\mathbf{y}} \in W^s \subset R^{n_s}, \\
\dot{\hat{\mathbf{z}}} &= \widehat{\mathbf{z}}, & \widehat{\mathbf{z}} \in W^u \subset R^{n_u}.
\end{aligned}
\tag{5.16}
$$

That is, one can split the full system into three independent subsystems that jointly describe the qualitative behavior of the full system near a bifurcation point. Here the second and the third subsystem govern motion on the stable and the unstable manifolds, respectively. The associated state-variables $\widehat{\mathbf{y}}(t)$ and $\widehat{\mathbf{z}}(t)$ are called *non-critical variables*. They describe the hyperbolic part of the system response (exponentially decaying or growing), which is not of concern here.

The first subsystem in (5.16) is the n_c-dimensional *center manifold reduction* to the original n-dimensional system $\dot{\mathbf{x}} = \mathbf{f}(\mathbf{x})$, near a particular point of bifurcation. It describes the non-hyperbolic part of the system response in terms of the n_c *critical variables* $\widehat{\mathbf{x}}(t)$. The center manifold reduction retains the bifurcational behavior of the full system, even though its dimension n_c may be far less than n.

The center manifold theorem merely states that subsystems of the form (5.16) exist, giving no hints as how to compute them. We next consider a systematic procedure for computing the center manifold reduction.

5.5.3 Computing the Center Manifold Reduction

In applications we usually study bifurcations of some *stable* state of a system. Thus, one can safely assume that there are no Jacobian eigenvalues with a positive real part ($n_u = 0$). We further assume that the system has been Taylor-expanded around the bifurcation point in question. Finally we assume that a modal transformation has been applied to the original system, so that its linear part is in block diagonal form (that is, in Jordan canonical form, cf. Sect. 3.4.5). By these assumptions the system under consideration can readily be split into the following subsystems:

$$
\begin{aligned}
\dot{\mathbf{x}} &= \mathbf{B}\mathbf{x} + \mathbf{f}(\mathbf{x}, \mathbf{y}), \quad \mathbf{x}(t) \in R^{n_c}, \\
\dot{\mathbf{y}} &= \mathbf{C}\mathbf{y} + \mathbf{g}(\mathbf{x}, \mathbf{y}), \quad \mathbf{y}(t) \in R^{n_s},
\end{aligned}
\tag{5.17}
$$

where $\mathbf{B}$ and $\mathbf{C}$ are $n_c \times n_c$ and $n_s \times n_s$ matrices whose eigenvalues have, respectively, zero and negative real parts, and where the nonlinear functions $\mathbf{f}$ and $\mathbf{g}$ vanish along with their first derivatives at the origin.

The first subsystem is the interesting one, describing motions on the center manifold in terms of the critical variables $\mathbf{x}(t)$. However, the non-critical variables $\mathbf{y}(t)$ also appear in this subsystem, through the nonlinear function $\mathbf{f}(\mathbf{x},\mathbf{y})$. To compute the center manifold reduction we need to eliminate $\mathbf{y}$ from $\mathbf{f}$.

Considering the second subsystem, one could be tempted to believe that $\mathbf{y}(t)$ would tend to zero exponentially fast, since all eigenvalues of the matrix $\mathbf{C}$ have negative real parts. The center manifold reduction would then readily be given as $\dot{\mathbf{x}} = \mathbf{B}\mathbf{x} + \mathbf{f}(\mathbf{x}, \mathbf{0})$. However, the argument that $\mathbf{y} \to \mathbf{0}$ due to $\mathbf{C}$ having negative real parts assumes $\mathbf{g}$ to be independent of $\mathbf{x}$. *If* $\mathbf{g}$ is independent of $\mathbf{x}$, then one can let $\mathbf{y} = \mathbf{0}$ to obtain the center manifold reduction. On the other hand, when $\mathbf{g}$ depends on $\mathbf{x}$ one cannot be sure that $\mathbf{y}$ will approach zero, and a more systematic procedure for eliminating $\mathbf{y}$ is in need.

For this purpose we note that, at the origin, the center manifold W^c is tangential to $\mathbf{E}^c$ (the $\mathbf{y} = \mathbf{0}$ space). The center manifold can therefore be represented by a local graph $\mathbf{h}(\mathbf{x})$:

$$W^c = \{(\mathbf{x}, \mathbf{y}) | \mathbf{y} = \mathbf{h}(\mathbf{x})\}, \quad \mathbf{h}(\mathbf{0}) = \mathbf{0}, \quad \mathbf{h}'(\mathbf{0}) = \mathbf{0}, \tag{5.18}$$

where $\mathbf{h}$ maps R^{n_c} on R^{n_s} and $\mathbf{h}'(\mathbf{x}) \equiv \partial\mathbf{h}/\partial\mathbf{x}$. Substituting $\mathbf{y} = \mathbf{h}(\mathbf{x})$ into (5.17) we obtain:

$$\begin{aligned}\dot{\mathbf{x}} &= \mathbf{B}\mathbf{x} + \mathbf{f}(\mathbf{x}, \mathbf{h}(\mathbf{x})), \\ \mathbf{h}'(\mathbf{x})\dot{\mathbf{x}} &= \mathbf{C}\mathbf{h}(\mathbf{x}) + \mathbf{g}(\mathbf{x}, \mathbf{h}(\mathbf{x})).\end{aligned} \tag{5.19}$$

Since at the origin $\mathbf{h}(\mathbf{x})$ is tangent to $\mathbf{y} = \mathbf{0}$, the first equation in (5.19) provides a good local approximation to solutions on the center manifold. To obtain an equation determining $\mathbf{h}(\mathbf{x})$ we can eliminate $\dot{\mathbf{x}}$ from the two equations in (5.19), giving:

$$\mathbf{h}'(\mathbf{x})[\mathbf{B}\mathbf{x} + \mathbf{f}(\mathbf{x}, \mathbf{h}(\mathbf{x}))] - \mathbf{C}\mathbf{h}(\mathbf{x}) - \mathbf{g}(\mathbf{x}, \mathbf{h}(\mathbf{x})) = \mathbf{0}, \tag{5.20}$$

with boundary conditions $\mathbf{h}(\mathbf{0}) = \mathbf{h}'(\mathbf{0}) = \mathbf{0}$.

In general this is a nonlinear partial differential equation, which does not allow solutions for $\mathbf{h}(\mathbf{x})$ to be obtained in closed form. However, approximate solutions that are valid near $\mathbf{x} = \mathbf{0}$ will suffice, and for this one can assume $\mathbf{h}$ to be a polynomial in $\mathbf{x}$. The coefficients of the polynomial can be determined so that (5.20) is fulfilled to any desired level of accuracy. The approximation to $\mathbf{h}(\mathbf{x})$ so obtained is then substituted back into the first equation in (5.19) to yield the center manifold reduction. The next section illustrates the procedure in terms of a simple example.

For bifurcation problems we need to slightly extend the center manifold procedure. This is because $\mathbf{h}$ will be a function of $\mathbf{x}$ *and* of the bifurcation parameters μ. We take that into account by simply augmenting to (5.17) a 'system' that expresses that the bifurcation parameters are time-independent:

$$\dot{\mu} = \mathbf{0}, \quad \mu \in R^k. \tag{5.21}$$

This is called the *suspension trick* (Carr 1981). The eigenvalues of the augmented system are all zero, and so application of the suspension trick increases the dimension of the center manifold from n_c to $n_c + k$.

Sometimes when the center manifold is of low dimension (one or two), the reduced system may resemble one of the generic bifurcating system already discussed in Sect. 5.3. The bifurcations of the original system are then known. If this is not the case, a further reduction into *normal form* may be required (cf. Sect. 5.6), by systematically identifying and removing 'inessential' nonlinear terms from the center manifold reduction.

5.5.4 An Example

As a simple example we consider a quadratic Duffing's equation with negative linear stiffness (Guckenheimer and Holmes 1983):

$$\ddot{u} + \dot{u} - ku + u^2 = 0, \tag{5.22}$$

for which we seek possible bifurcations as k is varied near $k = 0$. First rewrite the system as a set of first-order equations:

$$\begin{aligned} \dot{u} &= v, \\ \dot{v} &= -v + ku - u^2. \end{aligned} \tag{5.23}$$

This system has a singular point at $(u, v) = (0, 0)$, with associated Jacobian

$$\mathbf{J}(0, 0, k) = \begin{bmatrix} 0 & 1 \\ k & -1 \end{bmatrix}, \tag{5.24}$$

with eigenvalues $\lambda_{1,2} = -\frac{1}{2}(1 \pm \sqrt{1 + 4k})$. When $k = 0$ one eigenvalue has a zero real part, and $k = 0$ is thus a bifurcation value. At this value the two eigenvalues and their associated eigenvectors are:

$$\lambda_1 = 0, \quad \lambda_2 = -1, \quad \varphi_1 = \begin{Bmatrix} 1 \\ 0 \end{Bmatrix}, \quad \varphi_2 = \begin{Bmatrix} 1 \\ -1 \end{Bmatrix}. \tag{5.25}$$

Hence, according to the center manifold theorem, the essential dynamics of the system can be restricted to a one-dimensional center manifold associated with the single eigenvalue having a zero real part. However, since we attempt solving a bifurcation problem, the suspension trick is employed by augmenting to (5.23) the 'system' $\dot{k} = 0$. This system also has a zero eigenvalue, and so the center manifold will be *two*-dimensional.

The linear part of the combined system must be decoupled as far as possible, aiming at block diagonal subsystems of the form (5.17). This is easily accomplished through the invertible modal transformation:

$$\begin{Bmatrix} u \\ v \end{Bmatrix} = [\varphi_1 \quad \varphi_2] \begin{Bmatrix} x \\ y \end{Bmatrix} = \begin{bmatrix} 1 & 1 \\ 0 & -1 \end{bmatrix} \begin{Bmatrix} x \\ y \end{Bmatrix} = \begin{Bmatrix} x+y \\ -y \end{Bmatrix}. \tag{5.26}$$

Substituting into (5.23) and pre-multiplying by $[\varphi_1 \quad \varphi_2]^{-1}$ we obtain:

$$\begin{aligned} \dot{x} &= k(x+y) - (x+y)^2, \\ \dot{k} &= 0, \\ \dot{y} &= -y - k(x+y) + (x+y)^2, \end{aligned} \tag{5.27}$$

where the equation for the bifurcation parameter has been augmented. The two first equations correspond to the first subsystem of (5.17), the one possessing eigenvalues with a zero real part. The third system in (5.27) corresponds to the second subsystem of (5.17), describing the stable manifold. Note that the equations are decoupled in their linear parts (the term $k(x + y)$ is *non*linear, since k is considered a state variable), and that the nonlinear terms vanish along with their derivatives at the origin.

We then seek a Taylor expansion approximation to the center manifold:

$$y = h(x,k) = c_1x^2 + c_2kx + c_3k^2 + O(3), \qquad (5.28)$$

where $O(3)$ denote terms of order three (x^3, kx^2, k^2x, k^3) and higher. The expansion is chosen so as to satisfy the required boundary conditions, $h = \partial h/\partial x = \partial h/\partial k = 0$ at $(x, k) = (0, 0)$. Constant and linear terms have been omitted because they cannot satisfy these conditions. Equation (5.20) for the determination of $h(x, k)$ becomes:

$$\left[\begin{smallmatrix}\frac{\partial h}{\partial x} & \frac{\partial h}{\partial k}\end{smallmatrix}\right]\left\{\begin{smallmatrix}k(x+h) - (x+h)^2 \\ 0\end{smallmatrix}\right\} + h + k(x+h) - (x+h)^2 = 0. \qquad (5.29)$$

Substituting the expansion (5.28) for h it is found, equating to zero powers of x^2, xk, and k^2, that $c_1 = 1$, $c_2 = -1$ and $c_3 = 0$. Hence,

$$h(x,k) = x^2 - kx + O(3). \qquad (5.30)$$

The center manifold reduction is then obtained by substituting $y = h(x,k)$ into the first equation in (5.27):

$$\dot{x} = x(k - x) + O(3), \quad (\dot{k} = 0). \qquad (5.31)$$

Neglecting the $O(3)$ terms we can easily obtain the bifurcation diagram of the reduced system for *small* values of x and k. At the two singular points $x = 0$ and $x = k$ of (5.31) the Jacobian eigenvalues are, respectively, $\lambda = k$ and $\lambda = -k$. Thus the singular points along $x = 0$ are stable for $k < 0$ and unstable for $k > 0$, whereas singular points along $x = k$ are unstable for $k < 0$ and stable for $k > 0$. Fig. 5.8 shows the bifurcation diagram. Comparing it with Fig. 5.4 you should recognize a *transcritical* bifurcation. This is just what should be expected by mere inspection of

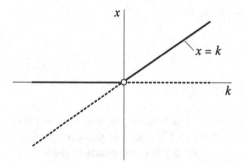

Fig. 5.8 Bifurcation near $k = 0$ of the system (5.22) (or (5.23)), as approximated by the center manifold reduction (5.31)

the center manifold reduction (5.31): To second order this system is identical to the generic system (5.8) for the transcritical bifurcation.

Seemingly, for this example we could just as well let $y = 0$ in the first of Eqs. (5.27) for obtaining the center manifold reduction (5.31). However, since x appears in the third equation of (5.27), one *cannot* deduce that y will tend to zero, even though the eigenvalues have all negative real parts.

5.5.5 Summing Up on Center Manifold Reduction

The center manifold theorem constitutes a basis for systematically reducing the dimension of a system of differential equations. The reduced system describes the qualitative dynamics on a submanifold that is associated with linearized eigenvalues having zero real parts. This can be used to examine local bifurcations of higher-dimensional nonlinear systems. An example was given for illustrative purposes. More realistic application examples can be found in, e.g., Guckenheimer and Holmes (1983), Holmes (1977), Païdoussis and Semler (1993), Li and Païdoussis (1994), Troger and Steindl (1991), Nayfeh (1993), and Nayfeh and Balachandran (1995).

The reduction process was described in terms of bifurcating *equilibriums*. It can be extended to study also bifurcations of *periodic solutions* (e.g., Nayfeh and Balachandran 1995). Though, for this purpose, one can instead employ a perturbation approach to obtain the modulation equations describing time evolution of oscillation amplitudes (cf. Chaps. 3 and 4), and then examine the bifurcating equilibriums of the modulation equations.

Center manifold reductions can alternatively be obtained by using perturbation analysis. For example, the double pendulum treated in Sect. 4.5 obeys the highly complicated four-dimensional set of equations (4.524.5). A multiple scales analysis (Sect. 4.5.3) performed near the Hopf bifurcation set reduces the system to (4.73)–(4.73), which is two-dimensional. This is the minimal dimension required for capturing Hopf bifurcations. The reduced system is a center manifold reduction, as indicated by the comment 'disregarding damped terms' just above (4.66), which corresponds to ignoring dimensions needed only to describe motions on the stable manifold.

5.6 Normal Form Reduction

Having reduced the essential dynamics of a system to a low-dimensional center manifold, the reduced system may still be very complicated. Typically a high number of nonlinear terms appear, of which only a few are essential for describing bifurcations of the system.

Normal form reduction is a technique for systematically removing in-essential nonlinear terms from a center manifold reduction. The technique involves a near-identity nonlinear transformation of the critical variables, which is so chosen as to make the transformed system as simple as possible. The result is a representation of the original system in so-called *normal form*. Normal forms exist and are complete for codimension one bifurcations, and fairly complete for codimension two problems.

The procedure for obtaining normal forms is considered outside the scope of this introduction to bifurcation analysis. We refer instead to Guckenheimer and Holmes (1983), Nayfeh (1993), Nayfeh and Balachandran (1995), and Troger and Steindl (1991).

Except for the simplest cases, the computational burden associated with obtaining normal forms may be overwhelming. Considering that the purpose of normal form reduction is to *simplify* analysis, it could be worthwhile considering alternatives, e.g., various perturbation methods. For example, the method of multiple scales (cf. Chaps. 3 and 4) also simplifies a system by eliminating 'in-essential' (non-secular) terms. Here, the identification of resonant terms can be viewed as a search for nonlinear terms that are 'essential', in the context given. Again, the double pendulum treated in Sect. 4.5 may serve to illustrate this: The original system (4.52) is four-dimensional and highly complicated. However, a multiple scales analysis (Sect. 4.5.3) reduces the system to the simple pair of Eqs. (4.73)–(4.73), which in turn is similar to the generic system (5.10) describing a Hopf bifurcation.

5.7 Bifurcating Periodic Solutions

Periodic solutions may bifurcate in response to changes in control parameters, just as equilibrium solutions may do. Again, linearized eigenvalues govern the nature of the bifurcations, and center manifold theory and normal forms can be employed for studying particular unfoldings (e.g., Guckenheimer and Holmes 1983; Troger and Steindl 1991; Nayfeh and Balachandran 1995). We now briefly mention some frequently encountered bifurcations of this type.

Symmetry-breaking Bifurcation Plotting the amplitude of periodic oscillations versus the control parameter, this bifurcation is similar to the pitchfork bifurcation of an equilibrium (Fig. 5.1). Though, in this case the zero solution of the pitchfork is replaced by a smooth curve in the upper half-plane. This curve corresponds to symmetric oscillations, whereas the branches of the bifurcation correspond to asymmetric oscillations. As with the pitchfork, the symmetry-breaking bifurcation can be supercritical or subcritical.

Cyclic-fold Bifurcation Here the amplitude of periodic oscillations bifurcates in a manner similar to the saddle-node bifurcation of equilibriums (Fig. 5.3), though with the bifurcation point located in the upper half-plane. Thus, the cyclic-fold is also referred to as a *tangent* or *turning point bifurcation*. The lack of solutions on the one side of this bifurcation does not imply that no solutions exist for the corresponding values of the control parameter, but only that such solutions cannot be described *locally*. To describe the post-bifurcation state one must usually rely on numerical solutions. This may reveal the existence of a remote disconnected attractor (point, periodic, chaotic), an unbounded solution, or a 'large' attractor corresponding to state variables much larger in magnitude than the pre-bifurcational values.

Transcritical Bifurcation The amplitude of periodic oscillations here bifurcates as for the transcritical bifurcation of equilibriums (Fig. 5.4), though, with the bifurcation point located in the upper half-plane. At transcritical bifurcations two solutions meet and 'exchange stability'.

Period-doubling (or flip) Bifurcation As with the symmetry-breaking bifurcation, a branch corresponding to periodic oscillations gains or loses stability upon continuously passing through the point of bifurcation. In addition the bifurcation creates a new state of periodic oscillations in which the period of oscillations is *doubled*. The period-doubling bifurcation can be *supercritical* (with a stable period-doubled solution) or *subcritical* (with an unstable period-doubled solution). Period-doublings frequently occur as precursors to chaos (cf. Chap. 6).

Secondary Hopf (or Neimark) Bifurcation Recall from Sect. 5.3.4 that a Hopf bifurcation corresponds to the creation of a periodic solution from an equilibrium (Fig. 5.5(a)). The *amplitudes* of oscillations can then be perceived as new equilibriums (for the equations governing modulation of amplitudes), as illustrated in Fig. 5.5(b). Now, in response to further variations of the control parameter, such an amplitude equilibrium can also experience a Hopf bifurcations. Since the equilibrium itself was created by a Hopf bifurcation, the new bifurcation is naturally called *secondary*. The outcome of a secondary Hopf bifurcation take the form of oscillations at a single frequency ω_1 (determined by the first Hopf bifurcation), with amplitudes that vary periodically with another frequency ω_2 (as determined by the secondary bifurcation). Typically the two frequencies ω_1 and ω_2 are *incommensurate* (i.e. ω_1/ω_2 is an irrational number) and the oscillations are then *quasiperiodic*. Quasiperiodic oscillations caused by secondary Hopf bifurcations often occur as precursors to chaos (cf. Chap. 6).

5.8 Grouping Bifurcations According to their Effect

At this stage we have encountered all possible local codimension one bifurcations of *equilibriums*: *pitchfork, saddle-node, transcritical* and *Hopf*. The pitchfork and Hopf bifurcations come in two variants: *supercritical* and *subcritical*. Further, we have mentioned some local bifurcations of *periodic* solutions: *Symmetry-breaking, cyclic-fold, transcritical, period-doubling* (or *flip*) and *secondary Hopf* (or

Neimark), with supercritical and subcritical variants of the symmetry-breaking and cyclic-fold bifurcations.

These bifurcations account for many of the qualitative changes in behavior experienced with physical systems, but not for all. There are codimension two and higher bifurcations, bifurcations of periodic solutions other than those already mentioned, and *global* bifurcations. Below we summarize (from Nayfeh and Balachandran 1995) some commonly encountered notions of bifurcations, describing their general *effects* rather than their particular *unfolding*.

Static bifurcations involve only singular points, i.e. equilibrium solutions, whereas *dynamic bifurcations* involve at least one branch representing motion of the system. The pitchfork, saddle-node and transcritical bifurcations are static bifurcations, whereas the Hopf bifurcation is dynamic.

Some bifurcations are *continuous* whereas other are *discontinuous* (or *catastrophic*), dependent on whether the states of the system vary continuously or discontinuously with the control parameters. The pitchfork bifurcation is continuous whereas the saddle-node is discontinuous.

During a *dangerous bifurcation* the current attractor (point, periodic or chaotic) of a system suddenly disappears, and the state jumps to a remote disconnected attractor. Dangerous bifurcations, also called *blue sky catastrophes*, are always discontinuous. The remote attractor can be bounded (point, periodic or chaotic) or unbounded. Whatever the case, such non-smooth changes in behavior may represent dangers to the life of real physical systems and their environment. Reversing the change in control parameters, the state may remain on the remote attractor well below the critical value, thus giving rise to *hysteresis* (different outcome dependent on the direction of parameter change). Saddle-node bifurcations and all kinds of subcritical bifurcations are usually considered to be dangerous.

During an *explosive bifurcation*, which is too discontinuous, the current attractor (point, periodic or chaotic) 'explodes' into a larger attractor. The new attractor includes the old one in form of a *ghost* or *phantom* attractor. Reversing the change in control parameters, the large attractor implodes into the original one.

5.9   On the Stability of Bifurcations to Perturbations

A bifurcation is said to be *stable* to a specified perturbation if this perturbation does not cause qualitative changes to the bifurcation diagram. We consider two examples.

5.9.1   Stability of a Saddle-node Bifurcation

We first consider the stability of the saddle-node bifurcation for the generic system (5.7). By adding to (5.7) a finite perturbation εx one obtains:

$$\dot{x} = \mu - x^2 + \varepsilon x. \tag{5.32}$$

The perturbation εx is chosen instead of ε, which would just add to μ and translate the bifurcation value. When $\mu \geq -\varepsilon^2/4$ the perturbed system has a pair of singular points at $x = \frac{1}{2}(\varepsilon \pm (\varepsilon^2 + 4\mu)^{1/2})$, but no singular points when $\mu < -\varepsilon^2/4$.

The Jacobian eigenvalues for the singular points are $\lambda = \pm(\varepsilon^2 + 4\mu)^{1/2}$. One eigenvalue has a zero real part when $\mu = -\varepsilon^2/4$, which is thus a bifurcation value. The corresponding bifurcation point is $(\mu, x) = (-\varepsilon^2/4, \varepsilon/2)$.

Fig. 5.9(a)–(c) shows the bifurcation diagrams for $\varepsilon < 0$, $\varepsilon = 0$, and $\varepsilon > 0$, respectively. The bifurcation diagrams are qualitatively alike, and so the saddle-node bifurcation associated with (5.7) is *stable* to the perturbation εx.

5.9.2   Stability of a Supercritical Pitchfork Bifurcation

Next we consider applying a perturbation ε to the generic system (5.2) for the supercritical pitchfork bifurcation. The perturbed system becomes:

$$\dot{x} = \mu x - x^3 + \varepsilon. \tag{5.33}$$

The bifurcation analysis is performed as above for the perturbed saddle-node bifurcation. Here, to avoid having to cope with a cubic polynomial for the determination of singular points, one can evaluate $\mu = \mu(x)$ instead of $x = x(\mu)$.

Three bifurcation diagrams are sketched in Fig. 5.10(a)–(c) corresponding to $\varepsilon < 0$, $\varepsilon = 0$, and $\varepsilon > 0$, respectively. The separation of one branch from the other two when $\varepsilon \neq 0$ implies that the perturbed behaves markedly different from the

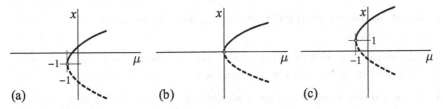

(a)   (b)   (c)

Fig. 5.9 Perturbed saddle-node bifurcations of $\dot{x} = \mu - x^2 + \varepsilon x$. (a) $\varepsilon = -2$, (b) $\varepsilon = 0$, (c) $\varepsilon = 2$

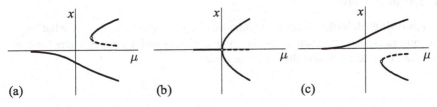

(a)   (b)   (c)

Fig. 5.10 Perturbed pitchfork bifurcations of $\dot{x} = \mu x - x^3 + \varepsilon$. (a) $\varepsilon < 0$, (b) $\varepsilon = 0$, (c) $\varepsilon > 0$

unperturbed system. Thus, the supercritical pitchfork bifurcation associated with (5.2) is *unstable* to the perturbation ε.

Real physical systems typically possess imperfections of some kind, which can be modeled as perturbations to the corresponding ideal system. For example, in the system (5.33), x could describe the deviation of the centerline of a column from the straight configuration, in response to a load μ. The perturbation ε would then describe an *imperfection* of the column, say, an initial curvature or localized buckle. An initially straight column ($\varepsilon = 0$) responds to increased loading as depicted in Fig. 5.10(b). That is, beyond the critical loading ($\mu = 0$ in this case) the straight configuration loses stability in favor a symmetric pair of buckled configurations. However, with an initial deviation from the straight configuration ($\varepsilon \neq 0$) the column responds as in Fig. 5.10(c) (or (a)), that is: Increased compressive loading just causes the column to deform further in the direction already determined by the initial imperfection, without the sudden loss of stability characterizing the ideal case $\varepsilon = 0$. Beyond the bifurcation value there are two possible stable configurations which, however, are not symmetric as for the case $\varepsilon = 0$. With small imperfections ($\varepsilon \ll 1$) the response is close to that of the perfect case, with the difference that one of stable branches is favored over the other.

5.10 Summing Up on Different Notions of Stability

By now we have encountered the term 'stability' in so many contexts, and with so many different meanings, that clearing up the concepts seems appropriate. The below list summarizes the different notions and their interpretation for continuous systems of differential equations, $\dot{x} = \mathbf{f}(t,\mathbf{x},\boldsymbol{\mu})$, where $\mathbf{x}(t)$ is a vector of state variables and $\boldsymbol{\mu}$ a vector of control parameters. Rigorous mathematical definitions can be found elsewhere, e.g., in the references given in the chapter introduction.

Bounded (or Lagrange) stability Requires a solution just to remain within finite limits, however large.

Lyapunov (or marginal or neutral or meta) stability Any solution $\mathbf{x}(t)$ coming near a Lyapunov stable solution $\tilde{\mathbf{x}}(t)$ stays near at all (future) times.

Uniform stability Same as Lyapunov stability, though only used with autonomous systems.

Asymptotic (or strict) stability Requires $\mathbf{x}(t) \rightarrow \tilde{\mathbf{x}}(t)$ for $t \rightarrow \infty$ (and implies Lyapunov stability).

Engineering stability Same as asymptotic stability (because that is what engineering safety is very much about: keeping structures and machines where they are put, in the state for which they were designed).

Orbital (or Poincaré) stability Any solution $\mathbf{x}(t)$ coming near *in state space* to an orbital stable solution stays near at all (future) times. Poincaré introduced this notion of stability to remedy the rather awkward fact that, for *nonlinear systems*, even steady periodic oscillations can be unstable in the sense of Lyapunov.[2]

Structural (or system) stability Structurally stable systems (of differential equations) retain their qualitative properties as the control parameters μ are slightly perturbed (cf. Sect. 5.2.3). Bifurcations can occur only for structurally *unstable* systems. Be careful to note that structural stability refers to perturbations of *control parameters* μ, whereas most other notions of stability refers to perturbations of *solutions* $\mathbf{x}(t)$.

Bifurcational stability A bifurcation is considered stable to a specified perturbation if this perturbation does not qualitatively change the bifurcation diagram (cf. Sect. 5.9).

Some confusion may arise when comparing with certain texts dealing exclusively with *linear* physical systems and differential equations. There one may find notions of stability similar to those given above, though with a quite different meaning, as well as notions that have meaning only for linear systems. For linear systems many features are implicitly assumed that do not hold for nonlinear systems. For example, linear systems cannot possess multiple solutions and finite post-bifurcation behavior. There is at most *one* solution or state to be concerned about (e.g., an equilibrium position), and thus concepts such as 'system', 'equilibrium' and 'structure' can be mixed or left undefined without causing confusion.

For example, some authors (e.g., Ziegler 1968) use the notion of structural stability to characterize properties of real engineering structures, rather than the structural properties of mathematical models. In this sense a column can be 'structurally stable', meaning that its design configuration is robust to small perturbations.

Also, some use the term 'system' to mean a real *physical* system, rather than a system of differential equations. In this sense 'system stability' implies robustness to perturbations of some particular state associated with the physical system concerned.

Further, for linear systems, concepts such as *static* and *dynamic stability* are used to describe what we would call static and dynamic *bifurcations* respectively.

The relevant interpretation of terms usually appears from the context, and should cause only little confusion. It *is*, however, desirable to employ concepts of stability in a consistent manner. In this respect the notions given above are recommended. These seem to be widely accepted across engineering and scientific disciplines, and hold generally for linear and nonlinear systems.

---

[2]Example: For an undamped pendulum the frequency of oscillation depends on the amplitude, if this is finite. Hence two solutions initiated at slightly different amplitudes will oscillate at slightly different frequencies. With time the two solutions will be running increasingly out of step, and are thus unstable in the sense of Lyapunov, even though the corresponding phase plane orbits stay close at all times.

5.11 Graphing Bifurcations: Numerical Continuation Techniques

We consider fixed point solutions (i.e. static equilibriums) of the dynamic system (5.1), but with only a single bifurcation parameter, i.e. $k = 1$ and $\mu = \{\alpha\}$, so that α is the only bifurcation parameter. The task is then to trace a curve $\mathbf{x} = \mathbf{x}(\alpha)$ (or some norm $\|\mathbf{x}\|$) with points $(\alpha, \mathbf{x})$, which are solutions to the *zero problem*:

$$\mathbf{f}(\mathbf{x}, \alpha) = \mathbf{0}, \tag{5.34}$$

where $(\mathbf{f}, \mathbf{x}, \alpha) \in R^n \times R^n \times R$. Since (5.34) only implicitly defines $\mathbf{x}$ as a function of α, this is typically not a trivial task, and will typically require numerical techniques. But it can be made less difficult by using *continuation* techniques, i.e. by proceeding from one calculated curve point to the next in small steps, rather than starting anew for every curve point. Here we provide a brief introduction to numerical continuation, and refer to the specialized literature for more methods and detail (e.g., Dankowicz and Schilder 2013; Keller 1986; Krauskopf et al. 2007; Nayfeh and Balachandran 1995; Seydel 2010).

There are several techniques in use for numerical continuation of fixed points or periodic solutions. These include sequential or natural parameter continuation, simplicial or piecewise linear continuation, (Davidenko-)Gauss-Newton continuation, and (pseudo-)arclength continuation. If a good approximation to a point at the solution curve is available, it is usually not difficult to quickly improve the accuracy of this approximation using, e.g., Newton-Raphson iteration. So the main problem is typically to provide initial approximations close enough for that to work, in particular near turning points, branches, or other singularities. The main difference among numerical continuation methods is how to accomplish coming close enough. In this introductory presentation we focus only on *single-parameter sequential continuation* and *pseudo-arclength continuation of fixed points*; this will highlight the main ideas, and prepare for other methods, if necessary.

5.11.1 Sequential Continuation

Sequential continuation, or *natural parameter* continuation, is usually straightforward to apply. As illustrated by Fig. 5.11(a) it just involves proceeding from a solution point $(\alpha_0, \mathbf{x}_0)$ at the solution curve to the next, by incrementing α a reasonably small step $\Delta\alpha$, i.e. $\alpha_1 = \alpha_0 + \Delta\alpha$, and assuming as a first approximation that $\mathbf{x}$ is unchanged, i.e. the prediction is $(\alpha_1, \mathbf{x}_1) = (\alpha_1, \mathbf{x}_0)$. The prediction is then corrected by solving $\mathbf{f}(\mathbf{x}_2, \alpha_1) = \mathbf{0}$ for $\mathbf{x}_2$ (using e.g. Newton-Raphson iteration or bisection or brute force) to find the next solution point $(\alpha_2, \mathbf{x}_2) = (\alpha_1, \mathbf{x}_2)$. This procedure is iterated, with the solution point $(\alpha_2, \mathbf{x}_2)$ now serving as a new starting point $(\alpha_0, \mathbf{x}_0)$.

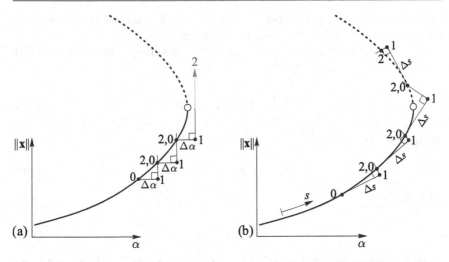

Fig. 5.11 Numerical continuation for determining the solution curve $(\alpha, \|\mathbf{x}\|)$ defined by $\mathbf{f}(\mathbf{x}, \alpha) = \mathbf{0}$. (a) Sequential continuation with prediction stepsize $\Delta\alpha$; (b) Pseudo-arclength continuation with prediction stepsize Δs. $\bigcirc$: turning point; ●: iteration point; 0/1/2: Starting/predicted/corrected point

As is evident from Fig. 5.11(a) sequential continuation will *not* work with turning points or branch points, no matter how small the increment in α; there is no way the solution points can be continuated round the turning point. Also, at regions with steep gradients (large $|d\mathbf{x}/d\alpha|$ or small $\|\partial\mathbf{f}/\partial\mathbf{x}\|$) very small α-increments will be needed, and performance inefficient.

5.11.2 Pseudo-arclength Continuation

Pseudo-arclength[3] continuation works by first making a linear *prediction step* along the current curve tangent, and then apply a *correction step* along a perpendicular line so as to catch up with the solution curve. As illustrated in Fig. 5.11(b) this will work also with turning points.

The essence in this method is to consider a *parametric representation* of the solution curve $(\alpha, \mathbf{x})$: By introducing a curve following parameter s we consider

---

[3]Literature on numerical continuation seems not to agree on what is 'pseudo' about this technique: (a) The linear tangent prediction is not along the true (curved) arclength and thus pseudo; (b) The Euclidian length restriction of the tangent step is imposed solely to supplement an equation to an otherwise underdetermined system, and thus pseudo; (c) The normal-to-tangent correction step is pseudo-, as compared to correction along a line of constant α. Here we just note that all these ingredients are indeed present, and contributes to make pseudo-arclength continuation workable, also with turning points.

both α and $\mathbf{x}$ functions of s, so that the solution curve is described by a set of points $(\alpha(s),\ \mathbf{x}(s))$, with s belonging to some relevant interval.

To calculate the relation between changes in α and changes in $\mathbf{x}$ on the solution curve, we differentiate (5.34) wrt s, which gives:

$$\frac{d\mathbf{f}}{ds} = \mathbf{f}_\mathbf{x}\mathbf{x}' + \mathbf{f}_\alpha\alpha' = \mathbf{0}, \tag{5.35}$$

or equivalently:

$$\mathbf{f}_\mathbf{x}\mathbf{z} = -\mathbf{f}_\alpha, \tag{5.36}$$

where

$$\mathbf{z} = \mathbf{x}'/\alpha'. \tag{5.37}$$

Here primes denote differentiation wrt the curve variable s, and subscript α and $\mathbf{x}$ denote partial differentiation wrt to the subscripted variable. Thus $\mathbf{f}_\alpha = \partial\mathbf{f}/\partial\alpha$, while $\mathbf{f}_\mathbf{x} = \partial\mathbf{f}/\partial\mathbf{x} = [\partial\mathbf{f}/\partial x_1\ \partial\mathbf{f}/\partial x_2\ \cdots\ \partial\mathbf{f}/\partial x_n]$ is the Jacobian of $\mathbf{f}$. This assumes that $\mathbf{f}_\mathbf{x}$ and $\mathbf{f}_\alpha$. exists and are continuous in the relevant range of $\mathbf{x}$ and α.

The system (5.35) of algebraic equations can also be written

$$[\mathbf{f}_\mathbf{x}\quad \mathbf{f}_\alpha]\left\{\begin{matrix}\mathbf{x}'\\\alpha'\end{matrix}\right\} = [\mathbf{f}_\mathbf{x}\quad \mathbf{f}_\alpha]\mathbf{t} = \mathbf{0}, \tag{5.38}$$

where $\mathbf{t} = \{\mathbf{x}'\ \alpha'\}^T$ is the tangent vector of the solution curve at the point $(\alpha, \mathbf{x})$. With $\mathbf{t}$ known we are able to take a step along the tangent to an already known solution point of the curve. However, the vector $\mathbf{t}$ contains $n + 1$ unknowns, while in (5.35) there are only n equations, so a constraint is needed. For that, as part of the technique, the Euclidian arclength normalization is chosen (also called the pseudo-arclength condition/constraint/normalization; it is just Pythagoras's theorem in $n+1$ space):

$$\sum_{j=1}^{n}(dx_j)^2 + (d\alpha)^2 = ds^2, \tag{5.39}$$

which after division with ds^2 can be written:

$$(\mathbf{x}')^T\mathbf{x}' + (\alpha')^2 = 1. \tag{5.40}$$

The system (5.38) of n linear equations and the scalar (nonlinear) equation (5.40) now defines what is necessary to determine the $n+1$ unknown values in the tangent vector $\mathbf{t}$.

With $\mathbf{t}$ known at a point of the solution curve (point 0 in Fig. 5.11(b)), a step of chosen length Δs can be taken in that direction, to a prediction point (point 1 in the figure). This point will generally not be a solution point, but with small Δs it will be close to. It can then be corrected back to the solution curve along a normal to the tangent vector, e.g. by Newton-Raphson iteration, to a new solution point (point 2 in the figure). The process then repeats, with the new solution point (2) serving as a starting point (0) for yet a tangent prediction.

To employ pseudo-arclength continuation to trace solutions $(\alpha, \mathbf{x})$ to the zero problem (5.34), the algorithm can be summarized as follows:

1. **Starting point, $(\alpha_0, \mathbf{x}_0)$:**

   1.1. *Initially:* Use any method (e.g. Newton-Raphson iteration from an approximate guess or bisection or brute force) to determine a solution point, preferably in a region where the system is expected to behave trivially (e.g. close to linearly).

   1.2. *Subsequently:* Use the last corrected point $(\alpha_2, \mathbf{x}_2)$ obtained during the continuation process.

2. **Direction for the continuation, i.e. determine the sign of α':**

$$\text{sgn}(\alpha') = \text{sgn}\left(\frac{d\alpha}{ds}\right) = \begin{cases} +1 & \text{for } \alpha \text{ increasing with } s \\ -1 & \text{for } \alpha \text{ decreasing with } s \end{cases}. \tag{5.41}$$

   2.1. *Initially:* Just choose arbitrarily.

   2.2. *Subsequently:* Keep sign unchanged, until a turning point has been passed. At a turning point the Jacobian of $\mathbf{f}$ is singular,[4] i.e. $|\mathbf{f_x}| = 0$. Thus the passage of a turning point can be detected a shift of sign in $|\mathbf{f_x}|$, i.e. by the fulfillment of the condition $|\mathbf{f_x}(\mathbf{x}_2, \alpha_2)\,||\mathbf{f_x}(\mathbf{x}_0, \alpha_0)| < 0$.

3. **Prediction step, to $(\alpha_1, \mathbf{x}_1)$:**

   3.1. Solve the linear algebraic equations (5.36), evaluated at the starting point, for $\mathbf{z}_0$:

$$\mathbf{f_x}(\mathbf{x}_0, \alpha_0)\mathbf{z}_0 = -\mathbf{f}_\alpha(\mathbf{x}_0, \alpha_0). \tag{5.42}$$

   3.2. Compute α'_0 from (5.40) with (5.37) inserted, and using (5.41) to choose among the positive/negative solution:

$$\alpha'_0 = \text{sgn}(\alpha'_0)\Big/\sqrt{1 + \mathbf{z}_0^T\mathbf{z}_0}. \tag{5.43}$$

---

[4]At a turning point α does not change with s (cf. Fig. 5.11(b)), i.e. $d\alpha/ds = \alpha' = 0$. Inserting this into (5.35) gives $\mathbf{f_x}\mathbf{x}' = \mathbf{0}$, which since $\mathbf{x}' \neq \mathbf{0}$ implies that $\mathbf{f_x}$ must be singular at this point.

3.3. Compute $\mathbf{x}'_0$ from (5.37):

$$\mathbf{x}'_0 = \alpha'_0 \mathbf{z}_0. \tag{5.44}$$

3.4. Compute the prediction point $(\alpha_1, \mathbf{x}_1)$, by extrapolating a stepsize Δs along the tangent at $(\alpha_0, \mathbf{x}_0)$, as defined by the tangent vector $\mathbf{t} = \{\mathbf{x_0}'\, \alpha_0'\}^T$ (choose Δs to give sufficiently close curve points, and reduce if problems near turning points):

$$\begin{aligned} \alpha_1 &= \alpha_0 + \alpha'_0 \Delta s, \\ \mathbf{x}_1 &= \mathbf{x}_0 + \mathbf{x}'_0 \Delta s. \end{aligned} \tag{5.45}$$

4. **Correction step, to** $(\alpha_2, \mathbf{x}_2)$, using Newton-Raphson iteration along a normal to the tangent, from $(\alpha_1, \mathbf{x}_1)$, so as to solve (5.34) (i.e. $\mathbf{f}(\alpha_2, \mathbf{x}_2) = 0$):

   4.1. Let $k = 0$, $\alpha^k = \alpha_1$, $\mathbf{x}^k = \mathbf{x}_1$, where superscript indicates iteration number.

   4.2. Increment $k \rightarrow k + 1$, and solve the following linear algebraic systems for $\mathbf{z}_1$ and $\mathbf{z}_2$ (Nayfeh and Balachandran 1995):

$$\begin{aligned} \mathbf{f}_\mathbf{x}(\mathbf{x}^k, \alpha^k)\mathbf{z}_1 &= -\mathbf{f}(\mathbf{x}^k, \alpha^k), \\ \mathbf{f}_\mathbf{x}(\mathbf{x}^k, \alpha^k)\mathbf{z}_2 &= -\mathbf{f}_\alpha(\mathbf{x}^k, \alpha^k). \end{aligned} \tag{5.46}$$

   4.3. Compute the parameter increments $\Delta\alpha^{k+1}$ and $\Delta\mathbf{x}^{k+1}$ from (Nayfeh and Balachandran 1995):

$$\begin{aligned} \Delta\alpha^{k+1} &= -\frac{g(\mathbf{x}^k, \alpha^k) + \mathbf{z}_1^T \mathbf{x}'_0}{\alpha'_0 + \mathbf{z}_2^T \mathbf{x}'_0}, \\ \Delta\mathbf{x}^{k+1} &= \mathbf{z}_1 + \mathbf{z}_2 \Delta\alpha^{k+1}, \end{aligned} \tag{5.47}$$

   where

$$g(\mathbf{x}, \alpha) = (\mathbf{x} - \mathbf{x}_0)^T \mathbf{x}'_0 + (\alpha - \alpha_0)^T \alpha'_0 - \Delta s. \tag{5.48}$$

   4.4. Compute α^{k+1} and $\mathbf{x}^{k+1}$ by linear extrapolation (with r as a relaxation parameter; $r = 1$ as long as $\|\mathbf{f}\|$ is reduced in each step, if not let $r \rightarrow r/2$ until $\|\mathbf{f}\|$ starts reducing):

$$\begin{aligned} \alpha^{k+1} &= \alpha^k + r\Delta\alpha^{k+1}, \\ \mathbf{x}^{k+1} &= \mathbf{x}^k + r\Delta\mathbf{x}^{k+1}. \end{aligned} \tag{5.49}$$

> 4.5. Check convergence: Is $\|\mathbf{f}(\mathbf{x}^{k+1}, \alpha^{k+1})\| < \varepsilon$?
>
> 4.5.1. If *no*, repeat from step 4.2.
> 4.5.2. If *yes*, the solution $(\alpha_2, \mathbf{x}_2) = (\alpha^{k+1}, \mathbf{x}^{k+1})$ is accepted, and the next curve point is calculated by repeating from step 1.2, with $(\alpha_2, \mathbf{x}_2)$ as the new starting point $(\alpha_0, \mathbf{x}_0)$.

To experiment with pseudo-arclength continuation consider Problem 5.6. It starts out simply, by tracing out the unit circle ($\alpha^2 + x^2 - 1 = 0$), and proceeds to the generic perturbed saddle-node bifurcation ($\dot{x} = \alpha - x^2 + 2x$) and perturbed pitchfork bifurcation ($\dot{x} = \alpha x - x^3 + 2$, with disconnected branches), and further to frequency responses for Duffing's equation ($\ddot{u} + 2\beta\omega_0\dot{u} + \omega_0^2 u + \gamma u^3 = q\cos\Omega t$) and the parametrically excited pendulum ($\ddot{\theta} + 2\varepsilon\beta\dot{\theta} + (1 - \varepsilon q\omega^2\cos(\omega\tau))\theta + \varepsilon\gamma\theta^3 = 0$).

While pseudo-arclength continuation works also with turning points, *branch points*, such as with the pitchfork bifurcation (Fig. 5.10(b)), will pose difficulties, which may call for more advanced techniques (Nayfeh and Balachandran 1995). But a simplistic approach may also work: As described in the next section the presence of a branch point can be detected by monitoring the rank of the matrix $[\mathbf{f}_\mathbf{x} \ \mathbf{f}_\alpha]$. Near the branch a brute force solution can be attempted (at least for lower values of n), by selecting a value $\alpha = \alpha^*$ just after the branch, and solving (5.34) for this value of α on a reasonably close grid in $\mathbf{x}$-space. This will give a number of approximate solutions $\mathbf{x}^*$, one for each branch; each point $(\alpha^*, \mathbf{x}^*)$ can then be used as a starting point for a separate arclength continuation on each branch.

Another way of tackling branch points is to temporarily remove them, by a small perturbation. As an example consider again Fig. 5.10, with Fig. (b) for the generic pitchfork bifurcation showing a branch point at (0,0). As illustrated in Figs. (a) and (c), by adding a small perturbation the branch is removed, or rather: the bifurcation diagram splits into two separate, disconnected curves – one with a turning point, and another one which is just trivial (with neither branches or turning points). So, effectively we could produce the bifurcation diagram in Fig. (b), with the branch, using pseudo-arclength continuation for a slightly perturbed system (small ε), in two continuation runs, one started on each of the separated branches. Generally this means to (1) detect the presence of a branch point (see next section), (2) Shift to solving a perturbed system, $\mathbf{f}(\mathbf{x}, \alpha) \pm \varepsilon = \mathbf{0}$, $|\varepsilon| \ll 1$, and (3) continuating solutions for the perturbed system.

5.11.3 Locating Bifurcation Points

Bifurcation points can be branch points or just turns/folds. As already illustrated (see again Fig. 5.11(b)) pseudo-arclength continuation is capable of continuating a solution beyond a turning point. Branches can also be followed, though unless more sophisticated algorithms are used, initial starting points needs to be located on each branch. But how do we locate the bifurcation points? Due to the finite stepsize Δs, the pseudo-arclength algorithm is unlikely to end up exactly at the turning or branch point; in fact that would make the algorithm fail, since the Jacobian $\mathbf{f_x}$ is singular at the turning point so that (5.42) is unsolvable.

However, the singularity of $\mathbf{f_x}$ at turning and branch points can be used to locate bifurcation points, e.g. by monitoring $|\mathbf{f_x}|$, or the rank or eigenvalues of $\mathbf{f_x}$ versus the parameter α or s. As a static bifurcation point is approached, one of the eigenvalues of $\mathbf{f_x}$ will approach zero (and the rank of $\mathbf{f_x}$ will decrease to $n - 1$). As a Hopf bifurcation is approached, the eigenvalues of $\mathbf{f_x}$ will approach a purely imaginary pair $\pm i\omega$, with all other eigenvalues having nonzero real parts.

Furthermore, the rank of the extended matrix $[\mathbf{f_x}\ \mathbf{f_\alpha}]$ (i.e. the coefficient matrix of the homogeneous system (5.38) for determining the tangent vector) can be used to determine if the bifurcation is a turning or a branch point: $[\mathbf{f_x}\ \mathbf{f_\alpha}]$ has rank n at saddle-node bifurcations, but rank $n - 1$ at other static bifurcation points (Nayfeh and Balachandran 1995). This reflects that saddle-node bifurcations have well-defined and continuously changing tangents ($\mathbf{t}$ in (5.38)), while this is not so for the other static bifurcations; the pitchfork and the transcritical bifurcations, with their branches, have no well-defined tangent. So: $\mathbf{f} = \mathbf{0}$ with rank$(\mathbf{f_x}) = n - 1$ marks a bifurcation; this bifurcation is of the turning point type if rank$([\mathbf{f_x}\ \mathbf{f_\alpha}]) = n$, while of the branching type if rank$([\mathbf{f_x}\ \mathbf{f_\alpha}]) = n - 1$.

The bifurcation detection methods described in this section are called *indirect*; they do not imply systematic search for bifurcation points, but are based just on monitoring properties of quantities that are anyway calculated during the continuation process. *Direct* methods, by contrast, relies on systematically searching for bifurcation points, e.g. by setting up equations that have bifurcation points as solutions. For example, at a bifurcation point $(\mathbf{x}, \alpha)$ must satisfy:

$$
\begin{aligned}
\mathbf{f}(\mathbf{x}, \alpha) &= \mathbf{0}, \\
\mathbf{f_x}(\mathbf{x}, \alpha)\mathbf{u} &= \mathbf{0}, \\
\mathbf{u}^T\mathbf{u} &= 1,
\end{aligned}
\tag{5.50}
$$

where the first equation is just (5.34), while the second and third equation jointly express the singularity of $\mathbf{f_x}$ at the bifurcation point (the unit-normalization ensures $\mathbf{u}$ is nontrivial). This system can be solved by using a Newton-Raphson procedure, as is further detailed in Nayfeh and Balachandran (1995).

5.12 Bifurcation Analysis and Continuation in Lab Experiments

Working in the lab or field with vibration problems often involves measuring frequency responses, using controlled or natural dynamic input, and measured output (accelerations, velocities, or displacement). When system response regimes are close to linear, today's preferred excitation is typically either impulsive (e.g. by modal impact hammers) or broadband pseudo-random noise (e.g. by vibration shakers) (Brandt 2011; Ewins 2000). Frequency sweeps can also be used, though less efficiently. But whatever the excitation, as long as the systems behaves approximately linearly there should be no difference (except for the noise) in the resulting measured frequency responses. And since linear systems cannot bifurcate, there will certainly be no turning points or branching bifurcations.

With nonlinear systems this is all different. In particular the frequency response will look differently for different response levels. This appears clearly from, e.g., Fig. 3.18, depicting for a cubically hardening nonlinearity the straight-up trivial resonance peak for low levels of excitation, and a strongly bent-towards-higher-frequencies peak for higher excitation levels. If lab measurements were to be made for such a system (could be a slender beam clamped between unmovable supports), only sweep testing could give a meaningful picture of the frequency response. But even this would be misleading, or at least only partially informing:

First, for excitation levels provoking nonlinearity, the frequency response obtained from sweeping *up* in frequency would be quite different from that obtained from sweeping *down*, i.e. there would be *hysteresis*: The resonance peak for the upsweep would have a vertical drop on the right side, and would be wider than the downsweep peak – which would have a vertical *increase* on the right side, but be less wide that the upsweep peak. To a trained analysist this would be a clear sign of nonlinearity (along with the presence of higher harmonics in the spectrum).

Secondly, the top of the peak would probably not be observed. Certainly not so during downsweeps, since the stationary state only jumps up to the high-level response branch at a some frequency below the peak top frequency. But also not during upsweeps; dependent on the sweep rate, and the presence of other disturbances, the response in such experiments typically drops down to the low-level response branch well before the peak top frequency.[5]

And thirdly, the unstable branch of the frequency response will not be observed at all. This is more important that one might think: A theoretically predicted frequency response such as, e.g., the one Fig. 3.18 needs experimental testing to be trusted. As for the way we have been doing science for hundreds of years, since Galileo, a physical theory can only be considered scientific if (1) it can be *falsified* (physical theories cannot be 'proved'), and (2) it can be tested experimentally. So, the prediction of the unstable branches of nonlinear frequency responses are not really 'scientific' – unless they can be experimentally observed.

---

[5]The same happens with numerical simulation, where a frequency response for a nonlinear system is obtained by sweeping up/incrementing frequency (rather than by numerical continuation; Sect. 5.11.2): The response typically drops down on the low-level branch well before the peak top frequency. To come closer to the peak top, one needs to use very small frequency increments, wait until all transients have surely decayed, and start simulation with initial conditions corresponding to the solution obtained for the previous frequency.

These troublesome features are due to the presence of bifurcations. For the example of Fig. 3.18, which is quite typical for applications, there are two saddle-node bifurcations, one at each of the two curve points with vertical tangent. Performing a frequency sweep corresponds to using sequential continuation (Sect. 5.11.1). And as described in Sect. 5.11.1, and illustrated in Fig. 5.11(a), this breaks down at turning points, and thus at saddle-node bifurcations. In purely numerical analysis this can be overcome by using pseudo-arclength continuation (Sect. 5.11.2 and Fig. 5.11(b)). But how should something similar be accomplished in a lab experiment? And in particular – how should it be possible to continuate the response curve at the *unstable* branches? Even if the system somehow could be pushed onto a point on the unstable branch, by the definition of stability it would only stay there very shortly, so there would be no chance of continuating to a neighbor point.

Only rather recently has it become possible to tackle the abovementioned serious obstacles for proper lab measurement of nonlinear frequency responses, so that these can be observed in full, i.e. with the peak top included, without hysteresis, and including even the unstable branches. The techniques go by names such as *control-based continuation (in experiments), experimental bifurcation analysis,* and *experimental tracking of bifurcations.* The actual implementation of these involves an advanced mix of experimental, numerical, mathematical, statistical, and signal processing techniques. Here we just summarize the main ideas, and otherwise refer to the specialized literature, e.g., Sieber and Krauskopf (2008), Sieber et al. (2011), Barton et al. (2012), Bureau et al. (2013, 2014), Barton (2017), Renson et al. (2019).

Basically control-based continuation, in lab experiments, works by proceeding from one equilibrium state to a neighboring one by using the same kind of tangent-predictor and corrector technique as with pseudo-arclength numerical continuation (Sect. 5.11.2). That is, except for some important differences: There is no mathematical model of the experimental system involved, so the gradient information needed for the Jacobian has to be estimated from measured data. Also, though at stable branches the correction step is not difficult (the system automatically ends up at the equilibrium state, if within the attraction zone), at *unstable* branches automatic feedback control is needed. The condition for this control to be effective is that the control force vanishes at the equilibrium (stable or unstable), i.e. it should be *non-invasive,* when at the equilibrium.

With control-based continuation, since there is no mathematical model, there is no model zero-problem such as (5.34). Instead an equivalent zero-problem is formulated based on the control force, specifying an equilibrium as a state where the control force is zero, i.e. is non-invasive. A feedback control is then devised which will change the system state so as to reduce the applied control force to zero; thus the control force acts as the "reference signal", as a proxy for the (unknown) system equilibrium state. The control target for the system is then determined in terms of the Fourier coefficients of the response that will reduce the control force to zero. The control turns both stable and unstable equilibrium states into asymptotically stable ones. The controller does not change the equilibrium solution itself, but only its linearization, in a way so as to stabilize otherwise unstable orbits.

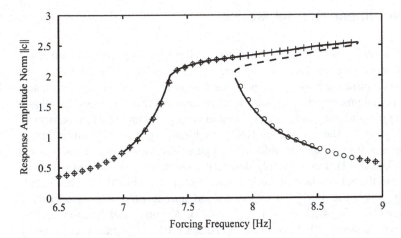

Fig. 5.12 Experimental frequency response for a harmonically forced nonlinear impact oscillator, obtained by both traditional frequency sweeps and control-based continuation. +/O: Measured response with increasing/decreasing frequency sweep; solid/dashed line: measured stable/unstable response by control-based continuation (Bureau et al. 2014)

Fig. 5.12 shows an example result from Bureau et al. (2014) (see also Rebouças et al. 2017), where a frequency response for an experimental cantilever beam with a two-sided stop has been obtained both by traditional up/down frequency sweeps (+/O), and by control-based continuation (solid/dashed line for stable/unstable solutions), with control forces delivered by electromagnetic actuation. As appears the two approaches produce closely agreeing results, though only control-based continuation can reveal the unstable branch of the frequency response, and also reaches the top of the resonance peak (at least as far as the stable and unstable branch appears to meet).

For the same system, and using the same control-based scheme, an *isola* was also detected and tracked (not shown) – that is, a family of stable and unstable equilibrium branches that are detached from the primary resonance curve in the bifurcation diagram. This isola, created by a 1:3 subharmonic resonance, was first predicted using a simple mathematical model of the impacting beam (Elmegård et al. 2014), and then in Bureau et al. (2014) detected experimentally using a traditional sweep, and traced out with both stable and unstable branches using control-based continuation.

Not only forced frequency responses can be tracked using control-based actuation, but also the *backbones* of such responses (Renson et al. 2016). As an alternative to the traditional rather simple (but also troublesome) way of extracting frequency response backbones from resonant decay measurements, control-based continuation is then applied with a control law specified so as to maintain phase quadrature (phase offset $\pi/2$ between excitation and response signals), which is what characterizes states on the frequency response backbone.

5.13 Nonlinear Normal Modes

Nonlinear normal modes (NNMs) offer yet another means of studying and illustrating important characteristics of vibrating systems, including their bifurcations.
Just like phase plane diagrams, frequency response functions, backbone curves, and
bifurcation diagrams, which are all data transforming and reducing tools that allows
for better communicating and understanding the complex motions of, in particular,
nonlinear systems. The science of NNMs is highly specialized, still in active
development, and important and difficult enough to warrant an entire book in itself
(like Kerschen 2014). Here we only devote it a short introduction and overview
(partly due to the authors lack of working expertise in the subfield), maybe enough
to stir further interest. For good overviews and introductions see, e.g., Vakakis
(1997), Kerschen (2014), Kerschen et al. (2009), Avramov and Mikhlin (2013),
Mikhlin and Avramov (2010), Noël and Kerschen (2017). For examples of application and more specialized subtopics see, e.g., Hill et al. (2017), Lacarbonara et al.
(2016), Renson, Kerschen and Cochelin (2016), Gendelman (2014), Peeters,
Kerschen and Golinval (2011), Lacarbonara, Rega and Nayfeh (2003), Shaw and
Pierre (1993).

The contemporary definition of NNMs is rather spacious and inclusive: *A
nonlinear normal mode is any periodic (not necessarily synchronous) motion of a
system* (Kerschen et al. 2009; Hill et al. 2017), though there is also a definition
based on invariant manifolds (Shaw and Pierre 1993). Here 'system' is mostly
understood to be a *conservative* system, or its conservative *part*, though the NNM
concept can also be extended to include damping and external forcing (e.g.,
Gendelman 2014). NNMs have a long history, back to the pioneering work of
Rosenberg (1960), but have gained increased interest since the 1990s, apparently
beginning with the works of, e.g. Shaw and Pierre (1993) and Vakakis (1997).

The various definitions of NNMs also covers the well-known (cf. Sect. 1.3.2)
linear normal modes (LNMs) – i.e. the linear mode shapes, if the system in
question is a physical structure where the concept of 'shape' makes sense. NNMs
can be seen as a generalization of linear normal modes. However, there are some
substantial differences:

First, LNMs are *invariant*, that is, motion started on any LNM remain in that
mode at all times. This generally breaks down with nonlinearity, so NNMs are not
necessarily invariant in that sense; energy imparted into on an NNM may go into
other NNMs, as we have already seen in Chap. 4 with nonlinear interaction.

Second, LNMs can be used in *superposition*: Owing to the orthogonality (i.e.
linear independency) of the LNMs, free and forced vibrations of a system can be
expressed as linear combinations of LNMs. NNMs cannot be used this way; generally they are not orthogonal, and with nonlinear systems the response to a sum of
influences does not necessarily equal the sum of responses to each separate
influence.

Third, with LNMs there is little debate on how to picture them: An LNM usually
comes in the form of a discrete set of numbers, or a continuous function, that

represents the physical configuration of the system in a straightforward manner; we perceive an LNM simply as the system configuration 'frozen' in time, and then visualize (mentally or by computer animation) the system vibrating in this shape, with all material points moving time-harmonically and synchronously in either phase or anti-phase. For NNMs the definition is much more general – an NNM is just '*a periodic motion*', and there are many ways we can picture periodic motions. So, while we think of LNMs as simply the vibrating system frozen at an instance of time during trivial time-harmonic oscillation, with NNMs this is not possible – the motion can rarely be represented in such a simple fashion. NNMs can therefore be thought of rather as just 'the periodic motion' of the system, and then we have to decide how to picture that motion. That means choosing which motion-describing *variables* to work with, and which *projection* of these to graph.

A typical way of depicting NNMs is by *configuration plots*, where (at least) two state variables are graphed versus each other. That could be two displacement coordinates, or two modal coefficients for a (linearly) mode shape expanded non-linear system. With linear systems such graphs appear trivial: Since LNMs can be multiplied by any constant, and still be the same LNM, the graphs will be straight lines in configuration space, with generally positive/negative slope for in/anti-phase motion. For NMM's the graphs in configuration plots will generally be curved lines.

Frequency–energy plots (FEPs) is another, and very common and convenient, way to picture NMMs, typically in combination with configuration plots. In a FEP, NNM motions are represented by the oscillation frequency corresponding to the minimal period of the periodic motion, versus the total mechanical energy associated with that motion. Several NNMs can be represented in the same plot. Again, with LNMs such graphs are trivial, just straight horizontal lines, since with linear system the free oscillation frequency is independent on amplitude and thus on energy level. However, as we have seen repeatedly in Chaps. 3 and 4, with non-linear systems the free oscillation frequencies typically depend on amplitude, so NNM typically display as curved lines in FEPs – at least when the energy becomes large enough for nonlinearity to be of significance. So typical FEPs of NNMs start out for low energies with a number of straight horizontal lines, one for each NNM, and then as energy increases the lines deform into curves. At certain critical levels of energy *bifurcations* may occur, and thus FEP curves can appear rather complicated, maybe splitting out and even self-intersecting. FEPs are typically accompanied by miniature inserts of configuration plots along the FEP-curves, giving hints of how the system vibrates at a given energy level and frequency.

Of what use are NMMs? As already mentioned they are *not* useful for modal expansion the way LNMs are. But they can aid in understanding and communicating how the behavior of a nonlinear system changes with, e.g., the level of energy, in particular when it comes to identify qualitative changes such as bifurcations and amplitude-frequency dependencies. To some extent NNM plots can be seen as extension or supplements to the well-known nonlinear frequency responses, and (in particular) frequency response *backbones* (cf. Sect. 3.6.6).

Though NMMs are mostly only defined and used for the conservative part of a system, they can be useful also for predicting or understanding the damped and *forced* response. This parallels the usefulness of frequency response backbone curves (Sect. 3.6.6), which are also derived solely for the unforced and undamped system and thus conservative system, but nevertheless provide valuable cues to also the behavior of the forced, damped system: The backbone curves tell at which frequencies resonance may occur, in dependency of vibration amplitude, and whether the forced resonance peak is left/right-slanted (softening/stiffening non-linearity), whether isolas (disconnected resonance curves) can be expected, and whether and for which frequencies sub/superharmonic, internal, or combination resonances may occur.

NMMs can also be useful in the decision of whether nonlinearity needs to be accounted for at all, in a given application: If a FEP shows (almost) straight horizontal lines in the frequency and energy range of relevance, then nonlinearity can maybe just be ignored (at least stiffness type nonlinearity; damping nonlinearity may not necessarily introduce frequency-energy dependency). And with attempts of *model reduction* NMMs can be used to identify the critical vibration modes of interest, and help including consideration to features defining these, while ignoring less significant features.

To see specific examples of how to calculate and analyze NNMs the interested reader is referred to literature cited in the beginning of this section.

5.14 Examples of Bifurcating Systems

We return in this section to some of the examples dealt with in Chaps. 3 and 4, focusing now on the bifurcations occurring with these examples.

5.14.1 Midplane Stretching (Duffing's Equation)

Reconsider Duffing's equation:

$$\ddot{u} + \omega_0^2 u = \varepsilon \left(q \cos \Omega t - 2\beta \omega_0 \dot{u} - \gamma u^3 \right), \tag{5.51}$$

describing the dynamics of numerous physical systems, e.g., the transverse modal deformations $u(t)$ of a beam in the presence of midplane stretching and time-harmonic loading (Fig. 5.13(b) and Sect. 3.7.1).

A multiple scales perturbation analysis was performed for this system in Sect. 3.7.2. The near-resonant response ($\Omega \approx \omega_0$) was obtained in the form:

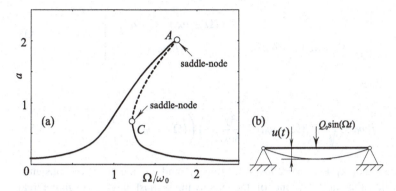

Fig. 5.13 (a) Typical near-resonant frequency response $a(\Omega/\omega_0)$ for the Duffing equation (5.51), showing saddle-node bifurcations. ($\gamma = 0.5$, $q = 0.2$, $\beta = 0.05$). (b) Example of a physical system displaying this response: a hinged-hinged beam subjected to time-harmonic loading and midplane stretching, with transverse beam deformations being described by $u(t) = a\cos(\Omega t - \psi) + O(\varepsilon)$

$$u(t) = a\cos(\Omega t - \psi) + O(\varepsilon), \tag{5.52}$$

where the slowly varying amplitude $a(t)$ and phase $\psi(t)$ are solutions of

$$
\begin{aligned}
a' &= -\beta\omega_0 a + \frac{q}{2\omega_0}\sin\psi, \\
a\psi' &= (\Omega - \omega_0)a - \frac{3\gamma}{8\omega_0}a^3 + \frac{q}{2\omega_0}\cos\psi.
\end{aligned}
\tag{5.53}
$$

The frequency response, describing stationary values of a as a function of Ω, was obtained by letting $a' = \psi' = 0$, resulting in (3.189). A typical response is shown in Fig. 5.13(a) (adopted from Fig. 3.17). Qualitative changes in the response occur at points A and C. We can now identify these as *saddle-node* bifurcations (cf. Fig. 5.3).

From Fig. 5.13(a) it should appear why saddle-nodes are also called *fold* or *turning point* or *tangent* bifurcations. The figure further illustrates why saddle-nodes belong to the class of *dangerous* bifurcations (Sect. 5.8): For example, beyond point A, and below C, the response escapes to 'distant attractors'. For the present case these attractors happen to be equilibriums (for the amplitude), though this would not be revealed by a local bifurcation analysis. (The 'local eye' cannot see what happens on the empty side of a saddle-node bifurcation.)

Forgetting for a moment that Fig. 5.13(a) has already been computed, we can rather easily check (5.53) for the presence of bifurcating equilibriums. At the equilibriums (defined by $a' = \psi' = 0$ in (5.53)) the Jacobian of the system becomes:

$$
\mathbf{J} = \begin{bmatrix} -\beta\omega_0 & -\left((\Omega - \omega_0) - \frac{3\gamma}{8\omega_0}a^2\right)a \\ \frac{1}{a}\left((\Omega - \omega_0) - \frac{9\gamma}{8\omega_0}a^2\right) & -\beta\omega_0 \end{bmatrix}, \tag{5.54}
$$

with eigenvalues:

$$
\lambda_{1,2} = -\beta\omega_0 \pm \sqrt{-\left((\Omega - \omega_0) - \frac{3\gamma}{8\omega_0}a^2\right)\left((\Omega - \omega_0) - \frac{9\gamma}{8\omega_0}a^2\right)}. \tag{5.55}
$$

It appears that $\lambda_1 + \lambda_2 = -2\beta\omega_0 < 0$, since β and ω_0 are positive constants, which implies that at least one of the eigenvalues must have a negative real part. Then there can be no Hopf bifurcations, since this would require a pure imaginary pair of eigenvalues. Further, at most one eigenvalue can have a zero part, since the other is always negative – which implies that *if* bifurcations exist, these will be saddle-node, pitchfork or transcritical bifurcations. As appears from (5.55) a zero eigenvalue occurs when:

$$
(\beta\omega_0)^2 + \left((\Omega - \omega_0) - \frac{3\gamma}{8\omega_0}a^2\right)\left((\Omega - \omega_0) - \frac{9\gamma}{8\omega_0}a^2\right) = 0. \tag{5.56}
$$

Combining this with the equation for the equilibrium values of a, defined by $a' = \psi' = 0$ in (5.53)), one obtains the bifurcation *set* in the parameter space (β, ω_0, Ω, γ) whereon bifurcations occur. Bifurcation *values* are then obtained by fixing all but a single parameter, say Ω. And bifurcation *points* (such as A and C in Fig. 5.13 (a)) are obtained by calculating the equilibrium value of a corresponding to the bifurcation value. The question still remains whether the bifurcation are saddle-nodes, pitchforks or transcritical. Here one could employ the theorems given in Sects. 5.4.1–5.4.2. However, the bifurcations are likely to be saddle-nodes, since (5.53) does not obey a zero solution and has no apparent symmetry (cf. Sect. 5.4.2). Alternatively, one could attempt reducing (5.53) to the one-dimensional center manifold defined by the zero eigenvalue. A subsequent reduction to normal form would be required if the center manifold did not immediately reduce to one of the generic forms. And then, finally, one could *sketch* the bifurcations *locally* near points A and C in Fig. 5.13(a). The interconnecting branch between A and C would not be revealed, and neither would the off-resonant 'tails' of the frequency response.

Do the results of the bifurcation approach justify the efforts? For this particular example one can argue that nothing was gained as compared to the approach of Sect. 3.7.2 (just plotting the branches of the frequency response and examining their stability). For more complicated systems, however, the situation might be different. Further, as stated in the chapter introduction, the real strength of bifurcation analysis comes in earlier in the analysis, where one has to decide upon the case to analyze.

5.14.2 Pendulum with a Moving Support (Parametric Excitation)

In Sect. 3.6 we considered a pendulum with a harmonically oscillating support (reshown in Fig. 5.14(b)). Small but finite pendulum rotations $\theta(\tau)$ are governed by a Duffing-type equation with parametric excitation:

$$\ddot{\theta} + 2\varepsilon\beta\dot{\theta} + \left(1 - \varepsilon q\omega^2 \cos \omega\tau\right)\theta + \varepsilon\gamma\theta^3 = 0. \qquad (5.57)$$

In Sect. 3.6.4 a multiple scales analysis was performed for the near-resonant case $\omega \approx 2$, with the following approximate result:

$$\theta(\tau) = a \cos\left(\tfrac{1}{2}(\omega\tau - \psi)\right) + O(\varepsilon), \qquad (5.58)$$

where the slowly varying amplitudes $a(\tau)$ and phases $\psi(\tau)$ are solutions of:

$$\begin{aligned}
a' &= -\beta a + \tfrac{1}{4}q\omega^2 a \sin \psi, \\
\psi' &= -\tfrac{3}{4}\gamma a^2 + \tfrac{1}{2}q\omega^2 \cos \psi + (\omega - 2).
\end{aligned} \qquad (5.59)$$

Stationary states are obtained by letting $a' = \psi' = 0$, which yields:

$$a^2 = 0 \quad \text{or} \quad a^2 = \frac{4}{3\gamma}\left((\omega - 2) \pm \sqrt{\left(\tfrac{1}{2}q\omega^2\right)^2 - (2\beta)^2}\right). \qquad (5.60)$$

Fig. 5.14(a) shows a typical frequency response $a(\omega)$ (adopted from Fig. 3.10). Qualitative changes in response occur at points A, C and D. We can now identify

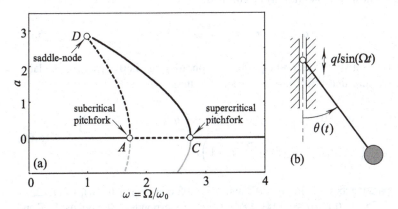

Fig. 5.14 (a) Frequency response $a(\omega)$ for the pendulum in (b) subjected to near-resonant parametric excitation. Bifurcations appear at A, C and D ($\gamma = -1/6$, $q = 0.2$, $\beta = 0.05$)

these changes as a *subcritical pitchfork* bifurcation at A, a *supercritical pitchfork* at C, and a *saddle-node* at D. Note that these bifurcations occur for the *equilibriums* of the system (5.59). However, these equilibriums describe amplitudes of *periodic solutions* for the original system (5.57). Thus, the subcritical pitchfork occurring at A for the system (5.59) corresponds to a *subcritical Hopf* bifurcation for the system (5.57) (cf. Sect. 5.4). Similarly, the supercritical pitchfork at C corresponds to a *supercritical Hopf bifurcation* for (5.57), and the saddle-node at D to a *cyclic-fold bifurcation* for (5.57).

To check that bifurcations indeed do occur as depicted we evaluate the Jacobian of (5.59):

$$\mathbf{J} = \begin{bmatrix} -\beta + \frac{1}{4}q\omega^2 \sin\psi & \frac{1}{4}q\omega^2 a \cos\psi \\ -\frac{3}{2}\gamma a & -\frac{1}{2}q\omega^2 \sin\psi \end{bmatrix}. \tag{5.61}$$

For the equilibrium points we can use $a' = \psi' = 0$ in (5.59) to eliminate $\sin\psi$ and $\cos\psi$ from (5.61), thus

$$\mathbf{J} = \begin{bmatrix} 0 & \frac{1}{2}a(\frac{3}{4}\gamma a^2 - (\omega - 2)) \\ -\frac{3}{2}\gamma a & -2\beta \end{bmatrix}, \tag{5.62}$$

for which the eigenvalues are governed by

$$\lambda^2 + 2\beta\lambda + \frac{3}{4}\gamma a^2 \left(\frac{3}{4}\gamma a^2 - (\omega - 2) \right) = 0. \tag{5.63}$$

The sum of solutions of this quadratic polynomial is $\lambda_1 + \lambda_2 = -2\beta < 0$, which rules out the possibility of a pure imaginary pair of eigenvalues. Hence, Hopf bifurcations cannot occur for the system (5.59). Further, since at most a simple zero eigenvalue can occur, the possible bifurcations include saddle-node, pitchfork and transcritical bifurcations. Letting $\lambda = 0$ in (5.63) it is found that such bifurcations require the following condition to be fulfilled:

$$a = 0 \quad \text{or} \quad a^2 = \frac{4}{3}(\omega - 2)/\gamma, \tag{5.64}$$

where a has to satisfy the second equilibrium condition in (5.60). Inserting the latter condition we find that bifurcations occur at the three points:

$$\begin{aligned} (\omega, a)_{1,2} &= (\hat{\omega}_{1,2}, 0), \quad \left[\hat{\omega}_{1,2} : \ (\hat{\omega} - 2)^2 - (\tfrac{1}{2}q\hat{\omega}^2)^2 + (2\beta)^2 = 0 \right], \\ (\omega, a)_3 &= \left(2\sqrt{\beta/q}, \ \sqrt{(8/3\gamma)}\sqrt{\beta/q - 1} \right). \end{aligned} \tag{5.65}$$

For the parameters $\gamma = -1/6$, $q = 0.2$ and $\beta = 0.05$ one obtains $(\omega, a)_1 \approx (1.72, 0)$, $(\omega, a)_2 \approx (2.75, 0)$, $(\omega, a)_3 \approx (1.00, 2.83)$, which corresponds to the points A, C and D in Fig. 5.14(a).

For examining the possibility only of saddle-node and pitchfork bifurcations, one can alternatively check for vertical tangencies of the equilibrium curves. For the non-trivial equilibriums we obtain by rearranging and squaring the second equation in (5.60) that

$$\left(\frac{3}{4}\gamma a^2 - (\omega - 2)\right)^2 = \left(\frac{1}{2}q\omega^2\right)^2 - (2\beta)^2, \tag{5.66}$$

or, differentiating both sides with respect to a:

$$2\left(\frac{3}{4}\gamma a^2 - (\omega - 2)\right)\left(\frac{3}{2}\gamma a - \frac{d\omega}{da}\right) = q\omega^2 2\omega \frac{d\omega}{da}. \tag{5.67}$$

Vertical tangency of the curve $a = a(\omega)$ requires $da/d\omega \to \infty$, that is, $d\omega/da \to 0$. Inserting $d\omega/da = 0$ in (5.67) it is found that vertical tangency occurs when

$$a = 0 \quad \text{or} \quad a^2 = \frac{4}{3}(\omega - 2)/\gamma, \tag{5.68}$$

where a has to satisfy the second equilibrium condition in (5.60). Since (5.68) is identical to (5.64) we obtain again the bifurcation points A, C and D of Fig. 5.14(a). By this approach the calculation and examination of eigenvalues are completely bypassed, though, only bifurcations characterized by vertical tangency will be revealed.

5.14.3 The Autoparametric Vibration Absorber

In Sect. 4.2 we studied the dynamics of the autoparametric vibration absorber (Fig. 4.1), a 2-DOF system for which the dynamics is governed by nonlinear modal interaction. A multiple scales analysis was performed for the case of combined near-external and near-internal resonance. That is, $\Omega \approx \omega_1 \approx 2\omega_2$, where Ω is the frequency of harmonic excitation, ω_1 the natural frequency of the system to be damped, and ω_2 the linearized natural frequency of the pendulum absorber.

Fig. 5.15(a) (adopted from Fig. 4.2) shows the force response of the absorber pendulum for the case of perfectly tuned external and internal resonance, $\Omega = \omega_1 = 2\omega_2$. The response curve depicts $b(q)$, that is, the stationary amplitude of pendulum rotations $\theta(t)$ as a function of the amplitude q of harmonic excitation, cf. (4.29). The symmetric lower part of the response, $-b(q)$, is not shown. Clearly, a *supercritical pitchfork bifurcation* appears at $q = q_1$. Since b determines the amplitude of pendulum rotations $\theta(t)$, a pitchfork bifurcation of b corresponds to a *Hopf bifurcation* of $\theta(t)$.

Fig. 5.15(b) (adopted from Fig. 4.3) shows the force response for the case of slightly detuned external resonance, that is, $\Omega \approx \omega_1 = 2\omega_2$. Comparing it to Fig. 5.15(a), the pitchfork bifurcation branching out from the trivial solution $b = 0$

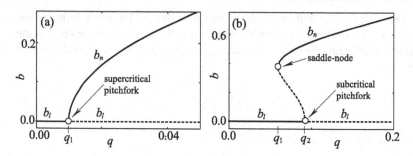

Fig. 5.15 Bifurcations of vibration amplitudes b for the pendulum part of the autoparametric vibration absorber. The bifurcation parameter q denotes the magnitude of the external loading. (a) Perfectly tuned internal and external resonance; (b) slightly detuned external resonance

is seen to change from *supercritical* to *subcritical*. We recall from Sect. 5.8 that subcritical bifurcations generally are considered *dangerous*, because they allow the state of a system to escape to a distant attractor. For this particular case, however, the distant attractor (b_n in the figure) is the one achieved at by design: The absorber pendulum starts moving and thus soaks up energy from the system to be damped. The distant attractor (b_n) is disconnected from the local one (b_l) only as seen by the 'local eye' of bifurcation analysis. Globally, as appears from Fig. 5.15(b), the two attractors are smoothly connected through a *saddle-node bifurcation* at q_1. We finally note that the saddle-node and the subcritical pitchfork node bifurcations occurring for the amplitude b corresponds to, respectively, a *subcritical Hopf* and *cyclic-fold* for the rotations $\theta(t)$ of the absorber pendulum.

The bifurcation diagram in Fig. 5.15(a) differs qualitatively from that in Fig. 5.15(b). This implies the supercritical pitchfork bifurcation in Fig. 5.15(a) to be *unstable* with respect to external detuning of the vibration absorber. Possibly this bifurcation is unstable to other relevant perturbations as well, and one should then expect *jumps* to occur for the corresponding physical system, rather than the smooth behavior depicted in Fig. 5.15(a).

The full set of equations describing motions of the vibration absorber is *five-dimensional* (corresponding to (4.9), which is a pair of second-order, non-autonomous ODEs). By using multiple scales analysis the dimension was reduced to *four* (corresponding to the four modulation equations (4.20)–(4.21)). The qualitative features of the four-dimensional system was shown to be governed by codimension one bifurcations. Thus, to describe only qualitative behavior, a one-dimensional system will suffice. This system can be obtained by reducing the modulation equations (4.20)–(4.21) to a one-dimensional center manifold, as described in Sect. 5.5.

5.14.4 The Partially Follower-loaded Double Pendulum

In Sect. 4.5 we analyzed the dynamics of the system reshown in Fig. 5.16(b), the follower-loaded double pendulum. The loading has a conservative component $(1 - \alpha p)$, always 'vertical', and a non-conservative component αp, always tangent to the upper pendulum bar. Here p quantifies the *level* of the loading, whereas α quantifies the *conservativeness* of loading. The limit cases are $\alpha = 0$ (purely conservative load) and $\alpha = 1$ (purely non-conservative). Physically, e.g., a tube inclined in gravity with fluid flowing inside will experience this type of loading, with α and p depending on inclination and flow velocity.

In Sect. 4.5.2 we examined the stability of the upright position of the pendulum (the *zero* solution $\theta_1 = \theta_2 = 0$), in dependence of the loading parameters α and p. The zero solution is stable for loading parameters corresponding to the white areas of Fig. 5.16(a) (adopted from Fig. 4.12), and unstable in grayed areas.

To find out what happens when the zero solution becomes unstable one can perform a perturbation analysis of the nonlinear system. In Sect. 4.5.3 a multiple scales analysis was performed for (α, p) near the curve $D_3 = 0$, $\alpha \in [\alpha_a; \alpha_b]$ in Fig. 5.16(a). It was found that for such loadings the pendulum may start oscillating at finite amplitude, that is, *supercritical Hopf bifurcations* occur. This is indicated in Fig. 5.16(a) by a small schematized Hopf bifurcation diagram across the stability boundary $D_3 = 0$. Similarly one can analyze the bifurcations occurring on the other stability boundaries, those determined by $H_4 = 0$ (Thomsen 1995). It appears from Fig. 5.16(a) that supercritical pitchfork bifurcations occur across the left and lower segment of $H_4 = 0$, whereas subcritical pitchfork bifurcations occur across the upper segment of $H_4 = 0$. The presence of a subcritical bifurcation across the upper

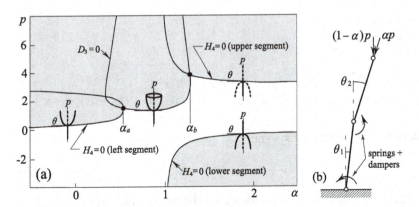

Fig. 5.16 (a) Stability and local bifurcations of the zero solution $\theta_1 = \theta_2 = 0$, for the double pendulum shown in (b). The zero solution is stable for loading parameters (α, p) corresponding to white areas, and unstable in grayed areas. Schematized bifurcation diagrams show what happens qualitatively upon crossing stability boundaries (Thomsen 1995)

segment of $H_4 = 0$ implies that, for loadings just below this segment, the zero solution is stable only to *small* disturbances. A strong disturbance may throw the state of the system beyond the unstable branches of the bifurcation. Indeed, numerical analysis show that the stable zero solution may explode into a large and complicated limit cycle, which is approached only after long periods of transient chaos (we shall return to this phenomenon in Chap. 6).

The *qualitative* aspects of the above results could have been obtained by using bifurcation analysis. Sect. 4.5.2 here provides the analysis of Jacobian eigenvalues, which is necessary for a subsequent bifurcation analysis. One-dimensional center manifold reductions could then be employed for qualitatively describing the codimension one bifurcations shown in Fig. 5.16(a).

Recall that center manifold reductions can alternatively be obtained by using perturbation analysis, as illustrated also by the present example: The full set of equations governing motions of the double pendulum is given by (4.52), which is *four*-dimensional, and rather complicated. The multiple scales analysis of Sect. 4.5.3, performed near the Hopf bifurcation set, reduces the system to (4.73)–(4.74), which is *two*-dimensional, and very simple. This is the minimal dimension required for capturing Hopf bifurcations. And the reduced system *is* a center manifold reduction, as indicated by the comment 'disregarding damped terms' (just above (4.66)), which corresponds to ignoring dimensions needed only to describe motions on the stable manifold.

5.15 Problems

Problem 5.1 For each of the systems below, examine and sketch the equilibriums as μ varies near $\mu = 0$. Identify the bifurcations, in case there are any.

(a) $\dot{x} = \mu x + x^5$
(b) $\dot{x} = \mu - x^5$
(c) $\dot{x} = \mu^2 x - x^3$

(d) $\dot{x} = -y + x(\mu + (x^2 + y^2)); \quad \dot{y} = x + y(\mu + (x^2 + y^2))$
(e) $\dot{x} = \mu x - \omega y + (\alpha x - \beta y)(x^2 + y^2); \quad \dot{y} = \omega x + \mu y + (\beta x + \alpha y)(x^2 + y^2)$

Problem 5.2 Consider the system:

$$\ddot{u} + \beta \dot{u} - u + u^2 = 0. \tag{5.69}$$

(a) Discuss possible bifurcations for the equilibriums of this system.
(b) Sketch the phase plane orbits for $\beta = 0$, $\beta < 0$, and $\beta > 0$, respectively.

Problem 5.3 Consider a beam with hinged-hinged supports that are fixed a given distance apart (Fig. P5.3). A single-mode approximation describing freely damped transverse oscillations takes the form (cf. (3.169)):

$$\ddot{u} + \dot{u} + ku + u^3 = 0. \qquad (5.70)$$

The stiffness k is positive, zero, or negative when the initial separation of supports corresponds to, respectively, pre-critical, critical or post-critical loading.

(a) Recast the system into first-order form, determine the singular points, and show that $k = 0$ is a bifurcation value. Examine the stability of singular points as k is varied near zero and sketch locally the bifurcation diagram.

(b) Show that the bifurcating system can be described on a two-dimensional center manifold. Compute an approximation to the center manifold, reduce the system to it, and examine the bifurcations near $k = 0$ for the reduced system.

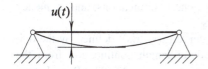

$u(t)$

Fig. P5.3

Problem 5.4 Consider a system subjected to nonlinear dissipative forces:

$$\ddot{u} + (2\beta + u^2)\dot{u} + \omega^2 u + u^3 = 0. \qquad (5.71)$$

(a) Show that Hopf bifurcations occur on the lines L_1 and L_2 that are defined by L_1: $\{\beta = 0, \omega^2 > 0\}$ and L_2: $\{2\beta = \omega^2, \omega^2 < 0\}$.

(b) Show that the bifurcations on L_1 occur at the singular point $(u, \dot{u}) = (0, 0)$ and are supercritical, whereas the bifurcations on L_2 occur at the singular points $(u, \dot{u}) = (\pm\sqrt{-\omega^2}, 0)$ and are subcritical.

Problem 5.5 Consider the system (Nayfeh and Balachandran 1995):

$$\ddot{\theta} + 2\beta\dot{\theta} + (\omega^2 - \Omega^2 \cos \theta) \sin \theta = 0 \qquad (5.72)$$

describing the position $\theta(t)$ of a mass sliding on a rotating hoop (Fig. P5.5). Here β is the coefficient of viscous damping, Ω the angular velocity of the hoop, and $\omega^2 = g/R$ where g is the acceleration due to gravity and R the hoop radius.

(a) For $\beta = 0$, determine the equilibrium positions of the system and sketch the phase plane orbits when $\Omega < \omega$, $\Omega = \omega$ and $\Omega > \omega$, respectively.

(b) For $\beta > 0$, examine and sketch the local bifurcations of equilibriums as Ω is increased from zero and beyond ω.

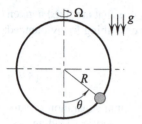

Fig. P5.5

Problem 5.6 This problem exercise trains using numerical pseudo-arclength continuation to trace out curves defined by zero-level sets of algebraic functions. This is relevant for calculating and plotting frequency responses from implicitly given frequency response equations, as well as other forms of bifurcation diagrams. The first problem is to trace out the unit circle, just to get ideas and basic coding fixed. Then you will be plotting some simple, generic bifurcations, and frequency response curves for some nonlinear oscillators.

(a) Set up an algorithm for computing the curve $(\alpha(s), x(s))$ defined by the unit circle, $F(x, \alpha) = \alpha^2 + x^2 - 1 = 0$, using pseudo-arclength continuation from the point $(\alpha(0), x(0)) = (\frac{1}{2}, \frac{\sqrt{3}}{2})$. Implement the algorithm in, e.g., MATLAB, and show the *predicted* as well as *corrected* points, along with the known solution (i.e. the unit circle), for a stepsize $\Delta s = 2\pi/20$ and $s \in [0; 2\pi]$.

(b) Check to see how your code works if the starting point $(\alpha(0), x(0))$ is not a solution point (i.e. part of the curve), but just a very approximate solution (e.g. a point in the vicinity of the curve).

(c) Modify the code to trace out the (α, x)-bifurcation diagram for the perturbed saddle-node bifurcation, corresponding to the generic system $\dot{x} = \alpha - x^2 + 2x$ (cf. Sect. 5.9.1 and Fig. 5.9(c)). This requires supplying an additional function for calculating *stability* corresponding to curve points, and that you indicate stability (use solid/dashed line for stable/unstable parts).

(d) Modify the code to trace out the (α, x)-bifurcation diagram for the perturbed pitchfork bifurcation, corresponding to the system $\dot{x} = \alpha x - x^3 + 2$ (cf. Sect. 5.9.2 and Fig. 5.9(c)). As compared to (c), you need to provide a feature for tracing out disconnected branches, corresponding to different starting points $(\tilde{\alpha}(0), \tilde{x}(0))$ provided by the user.

(e) Modify the code to calculate and plot the frequency response for Duffing's equation $(\ddot{u} + 2\beta\omega_0\dot{u} + \omega_0^2 u + \gamma u^3 = q\cos\Omega t$, cf. Sect. 3.7) near primary external resonance, using the multiple scales approximation for the frequency response equation and Jacobian matrix given in Sect. 3.7.2; this should give you something like Fig 3.17.

(f) Modify the code to calculate and plot the frequency response for the parametrically excited pendulum equation $(\ddot{\theta} + 2\varepsilon\beta\dot{\theta} + (1 - \varepsilon q\omega^2 \cos(\omega\tau))\theta + \varepsilon\gamma\theta^3 = 0$, cf. Sect. 3.6) near primary parametric resonance for small but finite oscillations, using the multiple scales approximation in Sect. 3.6.4 for the modulation equations (let $a' = \psi' = 0$ and square and add to find the frequency response equation); this should give you something like Fig. 3.10. And requires you to address the problems that come with *branching*.

6 Chaotic Vibrations

6.1 Introduction

Mathematical models sometimes behave unpredictably, as do many systems of the real world. If there are no stochastic components, or if these are considered to be inessential, we then speak about *deterministic chaos, chaotic dynamics* or, in case of vibrating structures, of *chaotic vibrations*.

This chapter provides an introduction to chaotic vibrations. What do they look like? How can they be described, where do they appear, and what are their origins? How can they be detected, predicted, suppressed, or perhaps utilized?

We focus on *practical* implications and applications. There is a lot of new and sophisticated mathematics associated with the study of chaos, in particular when it comes to the search for origins and causes. Without being totally ignorant, we shall leave most of this to the mathematicians.

Francis C. Moon has made great efforts to turn some of the mathematical conquests into useful engineering practice and laboratory experiments. His book (Moon 1987) on chaotic vibrations is intended for applied scientists and engineers, mainly within the field of mechanics; it is greatly recommended. Other relevant books are Acheson (1997), Guckenheimer and Holmes (1983), Jackson (1991), Ott (1993), Parker and Chua (1989), Schuster (1989), Thompson and Bishop (1994), Thompson and Stewart (1986), Tsonis (1992), Tufillaro et al. (1992), Skiadas (2016), Skiadas and Skiadas (2016), Awrejcewicz et al. (2016), and Amabili (2008). Gleick (1987) authored a popular introduction to the subject, including the exciting history of the birth of chaos science (recommended for the bedside table). Most material on chaotic vibrations is to be found in scientific papers, some of which will be referenced when appropriate. Rega, Settimi, and Lenci S (2020) provides a review of chaotic vibrations of one-dimensional elastic structures like beams, arches, and cables.

To some extent chaotic vibrations can be viewed simply as yet another phenomenon of which the vibration analyst should be aware. Just like resonance,

© The Author(s), under exclusive license to Springer Nature Switzerland AG 2021
J. J. Thomsen, *Vibrations and Stability*,
https://doi.org/10.1007/978-3-030-68045-9_6

subharmonics and amplitude jumps. To deal competently with resonances in engineering structures, you do not need to know about the philosophical discussions that were triggered by their discovery several hundred years ago (which was about the possible presence of heavenly harmonic phenomena). However, the discovery of chaotic phenomena has changed the way we look at nature and on mathematical models and computer output, as did the discovery of relativistic and quantum phenomena early in the twentieth century. So, before we embark on the more practical aspects of chaos, we briefly touch upon its general role in science.

In the beginning, there was Newton... Why do we set up mathematical models? Usually we do so to understand, explain, predict or control particular aspects of physical environment. Sometimes we are completely satisfied with a description on the average, in which case we rely on *stochastic* models. For example, temperature is a quite useful measure for describing the average kinetic energy of a huge number of molecules in a gas; Few will care about the individual molecules. At other times an averaged description is inadequate for the purpose. Then *deterministic* models are needed, devoid of stochastic components. For example, when constructing a grandfather clock we rely on the deterministic pendulum equation.

With deterministic systems all future and past states are uniquely defined by equations of motions and sets of initial conditions. That is, if you know the state of a deterministic system at a given instance of time, then you are able to predict its behavior at all times. This is at the very core of Newtonian mechanics, laid down in the *Principia* some 300 years ago (Newton 1686). During sequels of successful modeling and experiments the belief did almost take root in our minds that we could predict the future of *any* (non-atomic) system, though this might require an extremely detailed mathematical model. Complicated phenomena simply called for more complicated models, and bigger computers to simulate them. Ironically, it took only a most primitive computer to definitely demolish the belief.

...Then came a meteorologist, named Lorenz ... For many mathematical models the behavior is insensitive to the specific choice of initial conditions. For example, simulating on a computer the nonlinear pendulum equation $\ddot{\theta} + (g/l)\sin\theta = 0$ with initial conditions $\theta(t_0) = \dot{\theta} + (t_0) = k$, you will get virtually identical results whether $k = 1.00$ or $k = 1.01$. Checking the predictions in the laboratory you may even succeed by starting the pendulum with $k = 0.9$, and still conclude that you are able to predict the states of the system.

However, in 1961 Edward Lorenz, research meteorologist at MIT simulating weather-models on a computer, discovered that a simple third-order model displayed extreme sensitivity to initial conditions. Starting with initial conditions that were arbitrarily close in phase space, Lorenz observed the computer simulations quickly running out of step. To predict the future of the system its initial conditions were to be specified with infinite accuracy. This is impossible, with computers and in laboratory experiments, and so the true behavior of a simple and purely deterministic system was seemingly unpredictable.

The discovery had a significant impact on the scientific community, calling for a shift in the way we look at computer models and experiments. For some systems (e.g., the weather system) even the most refined mathematical modeling and the most powerful computers will not help in producing accurate long-term predictions. On the other hand, the discovery also implied that complicated behavior does not necessarily call for a complicated mathematical model.

Following the discovery of Lorenz, a huge number of systems displaying extreme sensitivity to initial conditions and chaotic behavior have been observed. They were always there, of course, but no one had looked at them with the right eyes[1]. You may observe them in nature, in mathematical models, in laboratory experiments, in human behavior, chemical reactions, fluid flow, history, economy, biology, in a dripping tap, the beat of a chicken's heart, the changing weather, a boxing match, a buckled beam, the firing of nerve cells, waterfalls, mechanical valves, etc. Opening your eyes it is hard to avoid the impression that non-chaotic regular behavior is associated only with a minor subset of systems in the real world.

... And now everything is chaos? No, it is not. It takes some specific conditions to produce chaos. Linear systems cannot behave chaotically. Neither can less-than-three autonomous first-order nonlinear differential equations. On the other hand, to put it a little sharply most real-world systems are chaotic in the sense that it is impossible to predict their state in a very distant future. For many of these systems we are able to predict their state within an accuracy and scale of time that serve our needs; they are *weakly chaotic*. Other systems are *strongly chaotic*, that is, we cannot predict their state with any reasonable accuracy within the time-scale of interest. Let us have a look at a simple example.

6.2 First Example of a Chaotic System

Consider a harmonically forced Duffing equation with negative linear stiffness:

$$\ddot{x} + \beta \dot{x} - \frac{1}{2}x + \frac{1}{2}x^3 = p \cos(\Omega t). \tag{6.1}$$

This equation represents a single-mode approximation to the magnetically buckled beam in Fig. 6.1, the now classical *Moon's beam* (e.g., Moon 1980). The equation governs the deflection $x(t)$ of the beam-tip in response to the harmonic loading of amplitude p and frequency Ω, in the presence of linear damping with coefficient β $(0 < \beta \ll 1)$. All variables and parameters of the equation are nondimensional.

---

[1]Poincaré seems to have been aware of what we now call chaos, while in the 1880s studying the three-body problem. But he was probably too far ahead of his contemporaries for anyone to appreciate the importance of this, and there were no computers to perform long-range integration of nonlinear differential equations.

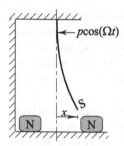

Fig. 6.1 Moon's beam

When $p = 0$ there are three static equilibrium positions: $x = 0$ (unstable) and $x = \pm 1$ (stable). Substituting $y = x \mp 1$, you will find that the linearized natural frequency for small oscillations around each of the buckled positions $x = \pm 1$ is $\tilde{\omega} = 1$. Thus, no matter how the beam is started, when $p = 0$ it will end up near one of the buckled positions, performing oscillations that decay at unit frequency.

When $p \neq 0$ the beam may oscillate around one or both of the buckled positions. Fig. 6.2(a) shows a result of a computer simulation for the first 100 s of response when $p = 0.15$, $\Omega = 1.2$, and $\beta = 0.1$. The solid and dashed curves reflect two slightly different sets of initial conditions: $x(0) = 1.10$ and $x(0) = 1.11$, respectively, with $\dot{x}(0) = 0$ for both responses. After a short period of transient motion the two solutions quickly catch up with each other, marching on forever in harmonic synchrony.

Fig. 6.2(b) shows the situation when the forcing level is raised to $p = 0.30$, keeping other parameters unchanged. After about 20 s the small difference in initial conditions has been magnified to a visible level, and from here on there is seemingly no correspondence between the two solutions. Both oscillate chaotically, swapping back and forth between the buckled solutions $x = \pm 1$. This will go on for hours of computer time, although, strictly speaking one cannot rule out the possibility that the solutions will finally settle down into some orderly and predictable kind of motion.

You may feel a slight suspicion creep into your mind, that what you see is merely computer garbage, say, signs of numerical instability. This is not the case, as we shall see, though of course the extreme sensitivity to initial conditions implies that changing the numerical solution procedure will also change details of the chaotic time-series. Experimental models of the buckled beam have been examined by several authors (this author copied them, just to convince himself); see, e.g., Holmes and Moon (1983), Moon (1987), and Tang and Dowell (1988). The experimental models behave as chaotically as the numerical simulations.

Plotting time-series is inconvenient for the study of chaotic motion, so we now present more appropriate tools.

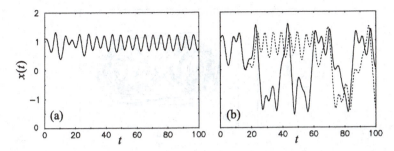

Fig. 6.2 First 100 s of a solution to (6.1) with $\Omega = 1.2$, $\beta = 0.1$ and two sets of initial conditions: $\dot{x}(0) = 0, x(0) = 1.10$ (solid line), and $x(0) = 1.11$ (dashed line). (**a**) $p = 0.15$, regular motion, insensitive to initial conditions. (**b**) $p = 0.30$, chaotic motion, extreme sensitivity to initial conditions

6.3 Tools for Detecting Chaotic Vibrations

Here we present a number of tools for detecting the presence of chaotic vibrations. Each tool is applied to the system (6.1) for two sets of system parameters corresponding to, respectively, regular and chaotic motion. Moon (1987) and Parker and Chua (1989) may be consulted for further information on each tool.

6.3.1 Phase Planes

Drawing phase plane orbits provides a simple method for distinguishing periodic from non-periodic and quasiperiodic motion. Chaotic motion looks complicated in the phase plane. However, so do certain other types of motion.

Rewriting the equations of motion as a number of autonomous first-order equations, an equal number of state variables span the phase space. The phase *plane* is a two-dimensional projection of the phase space, spanning two arbitrary state variables (usually a position and a velocity variable). Whereas the time variable runs on forever, it is usually possible to choose the variables of a phase plane so that motion in this plane becomes *bounded*.

The Duffing equation (6.1) can be rewritten as three first-order equations in the three state variables (x, y, z), where $y = \dot{x}$; and $z = \Omega t$. Hence, the $(x, \dot{x})$-plane is a phase plane. Sample phase plane orbits for (6.1) are shown in Fig. 6.3. The initial part of the solution was cut off, so the figures show only stationary, *post-transient* motion.

In Fig. 6.3(a) the solution traces out a closed orbit. This is a sign (but not a proof[2]) of regular periodic motion. A closed orbit crossing itself, as the double loop

---

[2]Closed orbits do not always represent periodic motion. For example, the system $(\dot{x} = 2ty$; $\dot{y} = -2tx)$ has solutions $x(t) = A\sin(t^2 + y), y(t) = A\cos(t^2 + y)$. These solutions are non-periodic, but nevertheless trace out the closed orbits $x^2 + y^2 = A^2$ in the (x, y) plane.

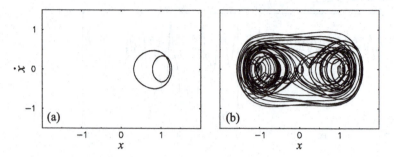

Fig. 6.3 Phase plane orbits of solutions to (6.1) with $\Omega = 1.2$ and $\beta = 0.1$. (a) $p = 0.27$, regular period-2 motion; (b) $p = 0.30$, chaotic motion

in the figure, is an indicator of *subharmonic* motion. That is, the fundamental frequency of the oscillation is lower than the frequency of the driving force. We term the motion represented by this figure a *period-2* orbit, since it takes two oscillations to return to any point on the curve. Similarly, when changing the parameters of a system, one may encounter period-1, period-3 and even higher order orbits.

In Fig. 6.3(b) the orbit tends to fill out a section of the phase plane in a rather complicated manner. Had the simulation been allowed to continue, the central part of the plane would be even more densely filled by orbits. A 'complicated'-looking phase plot is *one* indicator of chaotic motion. However, motion that rides on a complicated-looking orbit may very well be fully predictable, and thus non-chaotic. For example, a phase plot for a system with 1000 degrees of freedom may look complicated, even if the system is linear and thus certainly non-chaotic. Though you may find it difficult to predict the motion, your computer probably will not, since there is no sensitive dependence on initial conditions.

Another cause of complicated phase plots may be the presence of *quasiperiodic* motion. Quasiperiodic motion is characterized by oscillations at two or more frequencies that are *incommensurate*, bearing an irrational relationship to each other (such as $\sqrt{2}$ and π). Since the period of such a motion is infinite the phase plane orbit never repeats itself. Hence, quasiperiodic orbits fill up the phase plane, as do chaotic orbits, but they do so in a fully predictable manner since there is no sensitive dependence on initial conditions.

Phase plots are usually quite easy to obtain, with computer simulations and laboratory experiments. If, in the laboratory, only a single variable is accessible (say, position, velocity or acceleration), another variable can be obtained by integration or differentiation (analogue or digital). To avoid high-frequency-noise a high-quality low-pass filter should be used for smoothening prior to any differentiation of time-sampled signals.

The *embedding space* method is another technique for calculating phase plots when only a single variable is available. Also called the *pseudo-phase-space* method or the *delayed-coordinate technique*, it consists of plotting the value of the accessible signal $x(t)$ versus the value $x(t + T)$ of the same signal at a later time. The underlying principle is that $x(t + T)$ is related to $\dot{x}(t)$, since $\dot{x}(t) \approx (x(t + T) - x(t))/T$ for T small compared to the natural period of the system. Consequently, the qualitative features of a $(x(t), x(t + T))$-plot will resemble those of a $(x(t), \dot{x}(t))$-plot: Closed orbits will be closed orbits in both, and chaotic and quasiperiodic orbits fill up areas of both planes.

In summing up, we conclude that periodic motion reveals itself as a closed orbit in the phase plane, whereas chaotic and quasiperiodic motion both fills up areas of the plane.

6.3.2   Frequency Spectra

Most numerical software packages and laboratory signal analyzers include tools for computing the *Fourier spectrum* of a signal (cf. App. C.1.5). The spectrum associated with an arbitrary state variable is a useful tool for distinguishing between periodic and non-periodic motion, and between quasiperiodic and chaotic motion.

Fig. 6.4 shows the (log-magnitude) Fourier spectrum of $x(t)$ for two particular solutions to (6.1). In Fig. 6.4(a), a subharmonic period-2 motion manifests itself as discrete spikes at the driving frequency $\omega = \Omega = 1.2$, at the subharmonic $\omega = k\frac{1}{2}\Omega = 0.6$, and at the higher harmonics $\omega = k\frac{1}{2}\Omega$ of the subharmonic. In addition there is a (hardly visible) spike at $\omega = 0$, the *dc-component*, which shows that the average of $x(t)$ is displaced from zero; the beam oscillates around a buckled position.

In Fig. 6.4(b) the chaotic motion of the beam appears as broadband noise in the spectrum. Typical of chaotic spectra, one observes the dominating frequency $\omega = \Omega = 1.2$ as a spike lifting off from the noise floor.

Periodic motion always shows up as a discrete frequency spectrum. So does quasiperiodic motion, displaying the two or more incommensurate frequencies involved and possibly subharmonics, higher harmonics, and linear combinations of these. Chaotic motion produces a continuous broadband spectrum with off-lifting spikes at the dominating frequencies.

However, for systems with many degrees of freedom the spectrum sometimes appears continuous simply because so many frequencies are involved in the response. In laboratory experiments a large amount of measurement noise may also be responsible for a continuously looking spectrum with a few dominating spikes.

6.3.3   Poincaré Maps

Representing motion in a Poincaré map, it is usually easy to distinguish between periodic and non-periodic motion, between different kinds of periodic motion, and between chaotic and truly random motion.

As explained in Sect. 6.3.1 a phase plane representation of motion is obtained by plotting two components of a state vector $\mathbf{x}(t)$ versus each other. For (6.1) we may

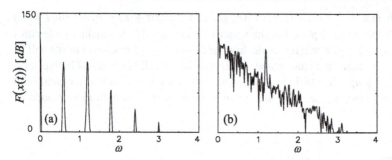

Fig. 6.4 Frequency spectra of solutions to (6.1) with $\Omega = 1.2$ and $\beta = 0.1$. (a) $p = 0.27$, regular period-2 motion; (b) $p = 0.30$, chaotic motion

plot $\dot{x}(t)$ versus $x(t)$. This plot traces out a continuous curve in the phase plane. Considering instead the values of the two state variables $\dot{x}$ and x only at discrete times t_k, $k = 0, 1, 2, \ldots$, the motion will appear as a sequence of dots in the phase plane. The continuous *flow* of states becomes replaced by a two-dimensional *map*, mapping states at time t_k on states at a later time t_{k+1}. Choosing the time instances t_k according to certain rules, the map will be a *Poincaré map*.

For systems subjected to external periodic forcing the Poincaré map is usually obtained by choosing $t_k = t_0 + kT$, $k = 0, 1, 2, \ldots$, where T is the period of the driving force. Thus for (6.1) we plot the sequence of points $(x(t_k), \dot{x}(t_k))$ with $t_k = t_0 + k2\pi/\Omega$. This is very much like observing the phase plane motion using a stroboscope that flashes at a particular phase Ωt_0 of the driving force, recording only the 'flash'-values of $(x(t), \dot{x}(t))$.

Fig. 6.5(a) shows a Poincaré map for (6.1) obtained with $t_0 = 0$. It corresponds to a sampling of the phase plot in Fig. 6.3(a) at times $t_k = k2\pi/\Omega$. The map consists of *a finite number of points*, implying periodic motion. In fact, there are *two* points, indicating period-2 motion. Period-n motion generally shows up as n points in the Poincaré map, whereas *quasiperiodic* motion reveals itself as infinitely many points filling up a closed curve.

Fig. 6.5(b) shows a Poincaré map similarly obtained, corresponding to the chaotic phase-plane orbit of Fig. 6.3(b). As simulation time marches on, more and more points will be added to the map. However, they will continue to do so in an orderly manner, filling out details of the strange creature you see in the figure. Indeed, what the figure shows is a two-dimensional cross-section of a so-called *strange attractor* on which the chaotic motion rides. We shall return to the 'strangeness' of strange attractors, which turns out to be associated with a non-integer topological dimension. Suffice here to say that a post-transient Poincaré map that consists neither of a finite number of points, nor of points filling up a closed curve, but nevertheless appears ordered – is a strong indicator of deterministic chaos. For some chaotic systems the Poincaré points will spread all over the map. This kind of behavior, known as *Hamiltonian chaos*, is typically found when there is no or very little dissipation in the system.

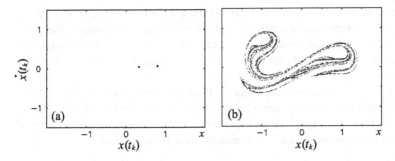

Fig. 6.5 Poincaré maps of solutions to (6.1) with $\Omega = 1.2$ and $\beta = 0.1$. (a) $p = 0.27$, regular period-2 motion; (b) $p = 0.30$, chaotic motion

Changing the *phase* of the Poincaré map, e.g. by letting $t_0 = \varphi_0/\Omega$ with $\varphi_0 \in [0; 2\pi]$, the map rotates and becomes somewhat deformed compared to Fig. 6.5 (b), though the topological features of the map remain unchanged. Plotting a sequence of Poincaré maps with φ_0 varied in steps from 0 through 2π, one sees the sheet-like structure in Fig. 6.5(b) stretching, folding and rotating, until it maps onto itself at $\varphi_0 = 2\pi$. Stretching and folding, as we shall see, is an important ingredient for chaos.

For truly autonomous systems there is no natural period for the driving force on which to base the Poincaré map. For example, forces of the follower-type (commonly encountered with flow-induced vibrations) do not depend explicitly on time. Then how should one choose the sampling times for a Poincaré map?

To give a clue to what can be done we rewrite (6.1) in autonomous form, so that the explicit dependence on time becomes hidden. Introducing a new variable $z = \Omega t - k2\pi$, (6.1) becomes:

$$\dot{x} = y$$
$$\dot{y} = -\beta\dot{x} + \frac{1}{2}x - \frac{1}{2}x^3 + p\cos z \qquad (6.2)$$
$$\dot{z} = \Omega.$$

As time proceeds, the solution $(x(t), y(t), z(t))$ to this set of equations traces out a curve in the (x, y, z) state space. To construct a Poincaré map we may define a two-dimensional surface in this space, and record all intersections of the $(x(t), y(t), z(t))$-curve with this surface. The surface should be transverse to the orbit everywhere. A particularly simple choice is to define the intersecting surface as a plane, $c_1 x + c_2 y + c_3 z = c_4$. Choosing the plane $z = 0$, as illustrated in Fig. 6.6, we obtain, since $z = \Omega t - k2\pi$, a Poincaré map sampled at $t_k = k2\pi/\Omega$. This is exactly the one shown in Fig. 6.5(b). The idea can be carried further to autonomous systems of dimension n by choosing, as the intersecting surface, a hyperplane of dimension $(n - 1)$. However, since orbits may never intersect an arbitrarily chosen plane, such a Poincaré map is not guaranteed to be well defined.

Experimental Poincaré maps can be obtained by using the external trigger facility present at most signal analyzers and oscilloscopes. For example, using the *XY*-ports of a storage oscilloscope to display the phase plane of a periodically driven system, one can usually set up an external triggering that intensifies the light beam at a particular phase of the driving force.

Summing up the possible limit sets of Poincaré maps, we find that a finite number of points correspond to periodic motion, an infinite number of points filling up a closed curve corresponds to quasiperiodic motion, and an infinite number of orderly distributed points (usually) corresponds to chaotic motion.

6.3.4 Lyapunov Exponents

Lyapunov exponents essentially measure the average rates of convergence or divergence of nearby orbits in phase space. A positive Lyapunov exponent indicates exponential separation of nearby orbits – a 'stretching' of phase space. Since for real systems the phase space is always bounded this separation cannot go on forever; Sooner or later orbits have to fold back towards the center of the phase space. This process of repeated stretching and folding is what characterizes chaos. Thus, a positive Lyapunov exponent, properly computed, is among the strongest indicators of chaotic motion.

Consider a general continuous system, written as a set of n autonomous first-order equations:

$$\dot{\mathbf{x}} = \mathbf{f}(\mathbf{x}); \quad \mathbf{x} = \mathbf{x}(t) \in R^n. \tag{6.3}$$

Start the system, by applying some initial conditions, and allow sufficient time for the system to reach an attractor (that is, cut off transients; cf. e.g. Table 1.4 in Sect. 1.7.5). The attractor can be an equilibrium point, a limit cycle or a strange attractor. Now, when the system appears to be on the attractor, define a new time zero. Record $\mathbf{x}(0)$, and consider the solution from thereon $\tilde{\mathbf{x}}(t), t > 0$. To this particular solution, moving on the particular attractor, we assign exactly n Lyapunov exponents $\hat{\lambda}_j, j = 1, n$. The set of Lyapunov exponents is defined as follows (Wolf et al. 1985):

$$\hat{\lambda}_j = \lim_{t \to \infty} \left[\frac{1}{t} \log_2 \left(\frac{r_j(t)}{r_j(0)} \right) \right], \quad j = 1, n,$$
$$\hat{\lambda}_1 \geq \hat{\lambda}_2 \geq \cdots \geq \hat{\lambda}_n, \tag{6.4}$$

where $r_j(t)$ measures the growth of an infinitesimal n-sphere of initial conditions at $t = 0$ in terms of the j'th ellipsoidal axis.

To understand the notion of Lyapunov exponents, consider a particular post-transient solution $\tilde{\mathbf{x}}(t)$ for a three-dimensional system (Fig. 6.7). At some instance of time t_0, compute another solution $\tilde{\mathbf{x}}(t) + \delta\tilde{\mathbf{x}}(t)$, with initial conditions

Fig. 6.6 A Poincaré section
with an intersecting orbit

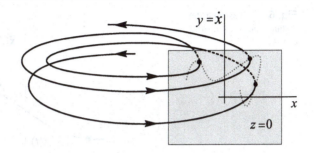

displaced a small distance ρ_0 from $\tilde{\mathbf{x}}(t_0)$. Locally, near t_0, the distance between the
original and the new solution will grow or shrink as $\rho(t) = \rho_0 2^{\lambda(t-t_0)}$, that is,
exponentially in time (the base-2 exponential turns out to be more convenient here
than the usual base-e). Imagine now that instead of following just *one* perturbed
solution, *all* solutions that start within a tiny ball of radius $r(t_0)$ are traced a small
instance of time ahead. The states $(x(t), y(t), z(t))$ of this bundle of trajectories,
initially within the ball of initial conditions, now evolve within an ellipsoid with
principal axes $r_j(t)$, $j = 1, 3$. The three Lyapunov exponents $\hat{\lambda}_{1,2,3}$ of this system
simply quantify the growth of each of the principal axes, on the average. Suppose,
e.g., that $\hat{\lambda}_1 > 0$ while $\hat{\lambda}_3 < 0\hat{\lambda}_2 < 0$. According to (6.4) this implies that one axis of
the ellipsoid is growing while the two others are shrinking. The exponential growth
can be observed only locally, since for real systems the states are bounded so that
escalating orbits sooner or later fold back. Hence, to estimate the Lyapunov
exponents of a solution, one takes the average of many local exponents spread over
the attractor.

As already indicated, a positive Lyapunov exponent implies extreme sensitivity
to initial conditions and thus chaotic motion on a strange attractor. Hence, to detect
whether an observed post-transient motion is chaotic or regular, we merely need to
compute the largest exponent $\hat{\lambda}_1$.

However, given the full *spectrum* of Lyapunov exponents $\hat{\lambda}_j, j = 1, n$, ordered as
a sequence $\hat{\lambda}_1 \geq \hat{\lambda}_2 \geq \ldots \geq \hat{\lambda}_n$, it is possible to give a more precise account of the
attractor involved (e.g., Parker and Chua 1989):

Theorem 6.1 Type of attractor in dependency of Lyapunov exponents $\hat{\lambda}_j$:

Stable equilibrium: $\hat{\lambda}_j < 0$ for $j = 1, n$

Stable limit cycle: $\hat{\lambda}_1 = 0$ and $\hat{\lambda}_j < 0$ for $j = 2, n$

Stable two-torus: $\hat{\lambda}_1 = \hat{\lambda}_2 = 0$ and $\hat{\lambda}_j < 0$ for $j = 3, n$

Stable K-torus: $\hat{\lambda}_1 = \cdots = \hat{\lambda}_K = 0$ and $\hat{\lambda}_j < 0$ for $j = K+1, n$

Chaotic: $\hat{\lambda}_1 > 0$

Fig. 6.7 Measuring the distance between nearby orbits

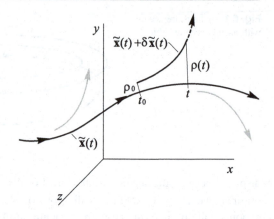

A K-torus is a higher-order limit cycle. For example, a two-torus can be visualized as the inner tube of a car tire, and motion on a two-torus as a trajectory that runs periodically on its surface. A *chaotic* attractor reveals itself as a *strange* attractor in the Poincaré map, consisting neither of a finite number of points, nor of points on a smooth curve[3]. We further have:

Theorem 6.2 On the sum of Lyapunov exponents::

$$\sum_{j=1}^{n} \hat{\lambda}_j < 0 \text{ for any dissipative system}$$

This means that the total volume of the ellipsoid of states will always shrink, even if the size of the ellipsoid may grow in certain directions. This is sometimes referred to as the *contracting nature of dissipative phase spaces*. Finally we have:

Theorem 6.3 On the presence of a zero-valued Lyapunov exponent:

$\hat{\lambda}_j = 0$ for at least one j for any limit set that is not an equilibrium point.

This zero exponent, present for all attractors other than equilibrium points, corresponds to the slowly changing magnitude of the principal axis that is tangential to the trajectory.

Theorems 6.1–6.3 can be used to exclude the possibility of chaos for one- and two-dimensional systems by the following arguments: Assume a chaotic attractor to

---

[3]In rare cases one finds strange *non-chaotic* attractors with $\hat{\lambda}_1 < 0$ (Romeiras et al. 1987).

be present. Then by Theorem 6.1 there is at least one positive exponent, and by Theorem 6.3 there is at least one zero exponent. Hence the dimension of the system must be at least two. However, by Theorem 6.2 the sum of all exponents must be negative, so a third, negative exponent must be also present. Consequently, the dimension of the system must be at least three[4].

As appeared, for *chaotic* solutions of three-dimensional systems the only possible Lyapunov spectrum is (+, 0, −). Furthermore, it must hold that $\hat{\lambda}_3 < -\hat{\lambda}_1$, to satisfy Theorem 6.2.

Similar arguments can be used to show that for four-dimensional systems there are three possible chaotic Lyapunov spectra: (+, 0, −, −), (+, +, 0, −), and (+, 0, 0, −). The second spectrum is referred to as *hyper-chaos*, whereas the third characterizes a chaotic two-torus (never observed, it seems).

How does one compute the Lyapunov exponents? It depends on whether the model under study is mathematical/numerical or experimental, and whether the entire Lyapunov spectrum is needed or just the largest exponent $\hat{\lambda}_1$ (the 'chaos-indicator') will suffice. In any case, an excellent account of these matters is given in Wolf et al. (1985) (see also Moon 1987; Parker and Chua 1989; Rosenstein et al. 1993; Kantz 1994). Here we only provide a clue to the numerical procedure for estimating the largest exponent $\hat{\lambda}_1$, for the case when the differential equations (6.3) are known[5].

The procedure works by integrating (6.3) numerically to obtain a post-transient reference trajectory $\tilde{\mathbf{x}}(t)$, while simultaneously integrating a set of variational equations:

$$\dot{\boldsymbol{\varepsilon}} = \mathbf{J}(\tilde{\mathbf{x}}(t))\boldsymbol{\varepsilon}, \quad \mathbf{J}(\mathbf{x}) = \frac{\partial \mathbf{f}}{\partial \mathbf{x}}, \qquad (6.5)$$

with unit-normalized initial conditions, $|\boldsymbol{\varepsilon}(t_0)| = 1$. Here $\mathbf{J}(\mathbf{x})$ defines the time-dependent Jacobian of $\mathbf{f}$. By integrating the variational equation we trace in time the distance $\boldsymbol{\varepsilon} = \mathbf{x} - \tilde{\mathbf{x}}$ between a perturbed trajectory and the reference trajectory. To capture the local 'stretches' while avoiding the global 'folds', the integrations are carried out only small amounts of time ahead. At the end of time-step t_k the local stretch is estimated by the local largest Lyapunov exponent:

$$\hat{\lambda}_1^{[k+1]} = \frac{1}{t_{k+1} - t_k} \log_2 \frac{|\boldsymbol{\varepsilon}(t_{k+1})|}{|\boldsymbol{\varepsilon}(t_k)|}. \qquad (6.6)$$

---

[4]This also follows from the Poincaré–Bendixon theorem: Forbidding intersection of trajectories in the plane, it excludes the possibility of chaos in two-dimensional systems.

[5]For calculating chaos indicators from measured data see Wolf et al. (1985) or Kantz (1994) for Lyapunov exponents, and Gottwald and Melbourne (2004) for an alternative indicator.

After N time-steps the estimate of the global largest exponent is computed as the average:

$$\hat{\lambda}_1^{[N]} = \frac{1}{N} \sum_{k=1}^{N} \lambda_1^{[k]}. \tag{6.7}$$

For chaotic attractors the reference trajectory will continue to visit new regions of phase space, possibly with different local Lyapunov exponents. Estimates of *positive* Lyapunov exponents therefore require some time to converge, and a running average estimate may continue to fluctuate. For non-chaotic attractors, i.e. equilibrium points and limit cycles, the trajectory repeatedly visits the same localized regions of phase space, giving a much faster rate of convergence for estimates of negative exponents.

Figure 6.8 depicts the converging estimate of the largest Lyapunov exponent $\hat{\lambda}_1$ for two solutions of the Duffing equation (6.1). One usually avoids estimating the unavoidable zero-valued exponent, which is anyway always present for limit sets that are not equilibrium points, cf. Theorem 6.3.

In Fig. 6.8(a), obtained for system parameters that yield stable period-2 motion, the estimate of the largest exponent converges towards the value -0.06. Since the zero-valued exponent has been ignored, the Lyapunov spectrum must have the form $(0, -, -)$. Then, by Theorem 6.1 we conclude that the motion is stable periodic, and thus fully predictable.

In Fig. 6.8(b) the system parameters have been chosen to yield chaotic motion. The estimate of the largest Lyapunov exponent fluctuates around a positive value, $\hat{\lambda}_1 \approx 0.19$. Hence, nearby trajectories diverge exponentially in time, motion takes place on a strange attractor, and the system behaves unpredictably in time.

Lyapunov exponents also find applicability for measuring the *predictability* of dynamical systems, and for characterizing the topological *dimension* of attractors; This is discussed in the two following sections.

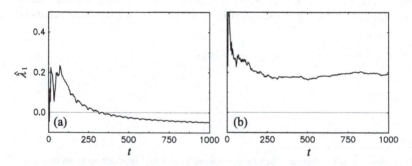

Fig. 6.8 Convergence of the largest Lyapunov exponent $\hat{\lambda}_1$ for (6.1) with $\Omega = 1.2$ and $\beta = 0.1$. (a) $p = 0.27$, regular period-2 motion; (b) $p = 0.30$, chaotic motion

6.3.5 Horizons of Predictability

Chaotic processes are not entirely unpredictable. Let us consider, as an example, the dynamics of the changing atmosphere as reflected in the ever-changing weather. These changes are commonly believed to behave chaotically, displaying extreme sensitivity to initial conditions. Still, we are able to predict states such as temperature, pressure and wind speed a few seconds ahead; they are unlikely to change much. Skilled meteorologists with access to computer models and 'initial conditions' sampled worldwide, can predict rather accurately details of the weather a few days ahead. However, they cannot tell exactly what the temperature will be at a certain geographic spot in ten days. And better computers, better models and more accurate initial conditions will improve only marginally on weather forecasts. We can make predictions on the average, such that winters will be colder than summers, and possibly forecast the global mean seasonal temperature for the next ten-year period. Though, it is impossible to predict the particular mean winter temperature 20 year ahead. Similarly, when watching the chaotic Moon's beam – or a boxing match, a dripping tap, a baby exploring your computer keyboard – it is possible to make accurate predictions only small instances of time ahead. Long-term predictions are obviously impossible in these and many other cases.

As we shall see chaotic processes destroy information. As information is destroyed predictions get worse.

Initial conditions represent information. For example, specifying a particular initial condition within an accuracy of one part per million amounts to $(\log 10^6/\log 2)$ ≈ 20 bits of information. For non-chaotic processes this information is preserved in time, so that from any particular state you may trace your way back to the initial conditions, or predict the state of the system ad infinitum (or *absurdum*). Non-chaotic processes are *information preserving*.

Chaotic processes *destroy* information. Any small inaccuracy in initial conditions is repeatedly stretched and folded. Since each stretch and fold represents a loss of information, sooner or later all information laid down in the initial conditions is lost.

To illustrate this degrading of information we imagine a baker who wants to color a clod of pie dough. Onto the dough he slips a drop of red color; we assume the drop contains 10^5 molecules of red. Initially the baker knows the position of every molecule to within an accuracy that corresponds to the size of the droplet. Then he stretches the dough. Knowing the details of the stretch his information is preserved, since nearby molecules stay near. He is able, at least theoretically, to trace each molecule back to where it came from. However, when the baker subsequently *folds* the dough he loses a tiny bit of information, since some nearby molecules may end up a far distance apart (those at the edge of the fold). Repeating this process a number of times the baker finds himself with a red clod of dough. That is, the 10^5 molecules have spread all over. All initial information is lost, and as for the position of each molecule the baker merely knows they are somewhere in the dough.

Lyapunov exponents quantify the average rate at which information degrades. Defined in terms of the base-2 logarithm, cf. (6.4), Lyapunov exponents are measured in bit/second.

Reconsidering the chaotic Duffing equation, we obtained at the end of Sect. 6.3.4 an estimate of the largest Lyapunov exponent, $\hat{\lambda}_1 \approx 0.19$. Hence, with initial conditions specified to within an accuracy of one part per million (20 bits), all bits of information will be lost after (20 bits)/(0.19 bits/second) ≈ 105 s, corresponding to approximately 20 periods of the driving force when $\Omega = 1.2$. Beyond this time – the *horizon of predictability* – nothing will be known about the state of the system, except than it would be somewhere on the chaotic attractor.

Generally, if the numerical or experimental accuracy is of the order 10^{-m}, then the time horizon for reasonable predictions is of the order

$$\tau_\infty = \frac{m}{\hat{\lambda}_1 \log 2}. \tag{6.8}$$

Back in Sect. 6.2 we showed two chaotic trajectories obtained for two sets of initial conditions, initially separated a relative distance 10^{-2} apart (Fig. 6.2(b)). The system parameters for this figure are identical to those of Fig. 6.8(b), so we know that $\hat{\lambda}_1 \approx 0.19$. With $m = 2$ we obtain from (6.8) that $\tau_\infty \approx 35$ s. Re-inspecting Fig. 6.2(b) you will find that this value roughly corresponds to the time at which the two solutions de-correlate.

6.3.6 Attractor Dimension

We now consider the *topological dimension* of different kinds of attractors and Poincaré maps. As you will see, the 'strangeness' of a strange attractor can be associated with its topological dimension.

Consider first a Poincaré map representing *periodic* motion. As we have seen it consists of a finite number of points (Fig. 6.5(a)). A point is a *zero-dimensional* geometrical object (since when you are at a point, no coordinates are required to specify where on the point you are).

Then think of a Poincaré map for *quasiperiodic* motion. It consists of an infinite number of points on a closed curve. Hence, the attractor reflects itself in the Poincaré map as a *one-dimensional* geometrical object.

If the motion under study was truly random – e.g., the forcing could be a white noise stochastic process – then dots would appear everywhere on the Poincaré map, filling out an area of the phase plane. An area is a *two-dimensional* object.

Now, which kind of object is the chaos-revealing Poincaré map in Fig. 6.5(b)? As mentioned in Sect. 6.3.3 new points will continue to fill out details of the strange object in an orderly manner. Hence, the map is not a collection of a finite number of points, nor is it a closed curve or an area, but rather something in between a curve and an area. It has a topological dimension of non-integer value, between one and two, that is: a *fractal* dimension.

There are far too few Poincaré points in Fig. 6.5(b) to reveal the finer details of its structure. The emergence of a fractal strange attractor on a high-resolution computer screen can be a stunning and pleasant experience, the object slowly taking form with new points adding still more detail to the image. As a characteristic of fractal objects, details appear at any level of magnification. Blowing up a tiny portion of Fig. 6.5(b) you will see the sheet-like structure repeated on a smaller scale. Subsequent magnifications can be carried on for as long as numerical resolution allows, with fine details appearing at any scale.

There are several measures quantifying the topological dimension of chaotic attractors, e.g., the *pointwise dimension, correlation dimension, information dimension, capacity, Hausdorff dimension* and *Lyapunov* dimension. A fine account of these and several other measures was given by Farmer et al. (1983). In many cases these different measures take on just about the same value.

The *Lyapunov dimension* d_L is a useful measure based on all but the most negative Lyapunov exponents. We adopt the following definition (Wolf et al. 1985):

$$d_L = K + \frac{\sum_{i=1}^{K} \hat{\lambda}_i}{\left| \hat{\lambda}_{K+1} \right|}, \qquad (6.9)$$

where the integer K is defined by the condition that

$$\sum_{i=1}^{K} \hat{\lambda}_i > 0 \quad \text{and} \quad \sum_{i=1}^{K+1} \hat{\lambda}_i < 0. \qquad (6.10)$$

As an example of calculating Lyapunov dimension we consider the Duffing system (6.1), with system parameters corresponding to, respectively, stable period-2 and chaotic motion (i.e. parameters corresponding to, respectively, the (a) and the (b) parts of Figs. 6.3, 6.4 and 6.5).

The full Lyapunov spectra were computed using the numerical algorithm described in Wolf et al. (1985). Equation (6.9) was then employed to yield results as shown in Table 6.1.

The attractors of the forced Duffing system reside in three-space, since there are three autonomous equations. What we observe in Poincaré maps are two-dimensional cross-sections of these attractors. Hence, to characterize the

Table 6.1 Lyapunov spectrum and Lyapunov dimension for Eq. (6.1) with parameters as in the caption for Fig. 6.3

	$\hat{\lambda}_1$	$\hat{\lambda}_2$	$\hat{\lambda}_3$	K	d_L
Period-2 motion	0	−0.05	−0.09	1	1
Chaotic motion	0.19	0	−0.33	2	2.58

dimensions of Poincaré sections we need in this case to subtract one dimension from d_L. We then infer from Table 6.1 that the Poincaré section for the periodic-2 case is zero-dimensional, i.e. a finite collection of points, whereas the chaotic Poincaré section has dimension 1.58, i.e. something in between a curve and an area.

Admittedly, the computation of attractor dimension did not reveal interesting news about the attractors of the system. A mere inspection of the Lyapunov spectrum would tell it all, with a little help from Theorems 6.1–6.3.

However, the concept of topological dimension has a more practical implication: it indicates *the smallest number of variables required to describe the dynamics on the attractor*. For example, in the regular period-2 case the dimension of the attractor is one. This implies that a single variable (governed by a first-order autonomous ODE) will suffice for qualitatively describing the motion. In the chaotic case the dimension of the attractor is larger than two, and thus the dynamics on the attractor cannot be captured with less than three variables. The more interesting observation comes in when you consider models, numerical or experimental, with *many* degrees of freedom. Imagine, for example, a space truss with 1,000 degrees of freedom and obviously intolerable chaotic behavior. The dimension of the chaotic attractor is unlikely to be 1,000, but rather 2.6 or 4.5, say. In this case you might reduce the model for the truss to order three or five without losing essential features of its dynamic behavior.

6.3.7 Basins of Attraction

A nonlinear system may have multiple stable solutions. Then the stationary motion actually observed depends on the initial conditions. Several examples of this phenomenon appeared in Chaps. 3–5. Consequently, for a fixed set of system parameters, chaos may appear or disappear by changing the initial conditions. To examine this dependency one may compute *basins of attraction*.

For example, a basin of attraction for the Duffing system (6.1) consists of a plane spanned by the initial conditions $x(t_0)$ and $\dot{x}(t_0)$. For each pair $(x(t_0), \dot{x}(t_0))$, in a relevant range with relevant resolution, the system is then integrated numerically and the stationary behavior is observed, classified and marked in the plane. Regular and chaotic motion could be assigned dots of different colors, or several colors or levels of gray could be used to further sub-classify the motion into period-2, quasiperiodic, etc. The largest Lyapunov exponent is a convenient tool for classifying different types of motion.

When chaos is involved the basins of attraction typically have fractal-like boundaries. This reflects the extreme sensitivity to initial conditions. The popular visualizations of the *Mandelbrot set* (Mandelbrot 1982; Dewdney 1985) are known for their highly fractal and beautiful structure. The Mandelbrot set is a basin of attraction for a system given by the one-dimensional map $z_{k+1} = z_k^2 + c$, where z and c are complex-valued and c is the initial condition.

Basins of attraction are rarely computed for systems having more than two state variables. This is because basins of dimension three and higher are very hard to interpret, should one have the computational power (and patience) to compute them. Therefore we are usually forced to circumvent the obstacle by employing 'typical' initial conditions, well aware that the system might behave otherwise for other sets of initial conditions. Or to postulate, rather loosely, that the system is chaotic for 'almost all' initial conditions.

6.3.8 Summary on Detection Tools

For characterizing, visualizing and quantifying chaotic phenomena a rather well developed set of tools is at our disposal: phase plane plots, frequency spectra, Poincaré maps, Lyapunov exponents, topological dimension and time-horizon estimates. A positive largest Lyapunov exponent is the strongest indicator of chaos, perhaps supported by a fractal Poincaré map and a broadband spectrum.

6.4 Universal Routes to Chaos

Chaos is typically observed only when system parameters are within certain ranges; Otherwise the behavior is regular. Transitions between regular and chaotic motions occur through *bifurcations*. Usually these bifurcations are *global*, that is, they cannot be deduced from local information on the flow of orbits near singular points and limit cycles.

A sequence of bifurcations taking a system from a regular state to a chaotic state is termed a *route to chaos*. Certain routes to chaos have been observed so often, and across so many systems and disciplines, that they seem to be somehow universal in character. This concerns the *period-doubling route*, the *quasiperiodic route*, the *transient route* and the *intermittency route*.

For applications the value of knowing these universal routes comes in when a single step of one of them is observed. One is then warned (or excited by the chance) that slight changes in system parameters may push the system into a chaotic regime. This is especially valuable if there is no or very limited control on system parameters – as for a traffic bridge, once it has been built – or if parameters cannot be changed just to satisfy the curiosity of the investigator – as with nuclear reactors and cardiac patients.

6.4.1 The Period-Doubling Route

This route involves a cascade of bifurcations, each characterized by a doubling of the period of motion. The process accumulates at a critical value of a system parameter beyond which the motion turns chaotic.

The period-doubling route to chaos has been observed for a large number of experimental systems, and for numerical models associated with differential equations (continuous flows) and difference equations (discrete maps). A classical example of the latter kind is the *logistic* or *quadratic map*:

$$x_{k+1} = \mu x_k(1 - x_k), \quad k = 0, 1, 2, \ldots . \tag{6.11}$$

This one-dimensional map originates from population biology. Nevertheless, it turns out that many systems associated with differential equations can locally be reduced to a quadratic map through the use of Poincaré sections.

Choosing an initial condition x_0, the iteration of the map (6.11) first produces some transient changes in x_k before the solution settles down into some limit behavior. Just as for continuous flows there are three possible limit sets: equilibrium points (for maps they are called *fixed points*), periodic motion (including quasiperiodic motion) and chaotic motion. The limit set depends on the system parameter μ, which is in this case considered a bifurcation parameter.

Fig. 6.9 shows a bifurcation diagram for the quadratic map. The map was iterated for 300 different values of $\mu \in [2.5; 4.0]$ with $x_0 = 0.1$. For each value of μ the first 100 iterates were considered transients, whereupon the next 25 iterates were plotted against μ. It appears that for $\mu \in [2.5; \mu_1]$ the iterates settle down onto a fixed point. We may consider this the 'period-1' motion. When $\mu = \mu_1$ the fixed point becomes unstable and the period doubles, i.e. the iterates visit two different values in turn. At a somewhat higher value of μ the period-2 solution becomes unstable in favor of a stable period-4 solution, i.e. the period has doubled once again. This process of period-doubling continues with still smaller changes in μ, accumulating at a critical value beyond which the motion turns chaotic.

Feigenbaum, in his famous work (1978, 1980), discovered that the sequence of bifurcation parameters μ_n at which the period doubles satisfy the relation:

$$\lim_{n \to \infty} \left(\frac{\mu_{n+1} - \mu_n}{\mu_n - \mu_{n-1}} \right) = \frac{1}{\delta}, \quad \delta = 4.6692 \cdots , \tag{6.12}$$

where the constant δ is the *Feigenbaum number* or *Feigenbaum's constant*.

Differentiating the right-hand side of (6.11) with respect to x_k, or sketching the function $\mu x_k(1 - x_k)$, it is found that the map has a *zero tangent* at $x_k = \frac{1}{2}$. Hence, the map is *non-invertible*, since given a value for x_{k+1} there are two possible values for x_k. The importance of Feigenbaum's work is that he showed the phenomenon of cascading period-doublings to be typical of maps that resemble (6.11), that is, maps possessing a zero tangent. Hence the *universal* appearance of the period-doubling route.

Examining Fig. 6.9 you might suggest that the crowding of dots in the chaotic regime merely represents motion that has period-doubled a huge number of times. This is not the case, as a calculation of the largest Lyapunov exponent would reveal (actually, for maps they are called Lyapunov *numbers*, being larger than unity for

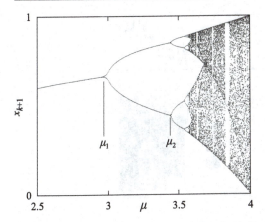

Fig. 6.9 Bifurcation diagram for the quadratic map (6.11), showing cascades of period-doublings and chaos

chaotic motion). Beyond the accumulation point the iterates behave unpredictably, with extreme sensitivity to initial conditions.

As mentioned above the period-doubling scenario can be observed for systems that are more complicated than the quadratic map. Usually, however, other scenarios are then involved as well. Slightly beyond a point of doubling accumulation the motion *is* chaotic. Though chaos may prevail beyond this point, it is more common to encounter 'windows' of regular periodic motion, after which chaos may re-enter the scene – perhaps through a quite different mechanism.

Fig. 6.10 shows a bifurcation diagram for the follower-loaded double pendulum dealt with analytically in Sect. 4.5 (*cf.* Fig. 4.11). For this four-dimensional system the local extremes of the up-bar positions $\theta_2(t)$ were plotted against α, the 'conservativeness' parameter of the problem ($\alpha = 1$ implies a perfectly following load) for a particular level of forcing, $p = 6$. As an aid for locating chaotic regimes the largest Lyapunov exponent $\hat{\lambda}_1$ has been calculated and plotted on top of the diagram; recall that $\hat{\lambda}_1 > 0$ implies chaos. It appears that as α is varied, then regular periodic motion alternates with chaos. Several universal routes appear in this figure, to be discussed in the following sections. It takes quite a significant magnification (not shown) of the diagram to reveal that period-doublings are responsible for the bifurcation into chaos at $\alpha = \alpha_2$, when α decreases from α_3.

6.4.2 The Quasiperiodic Route

This, also known as the *Ruelle-Takens-Newhouse* route, involves a transition into chaos through an interval of quasiperiodic motion.

Imagine a system with system parameters that yield a stable equilibrium point as the limit set. As we have seen, a change in system parameters may destabilize the

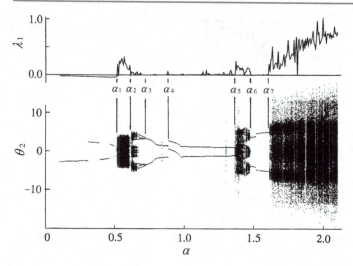

Fig. 6.10 Bifurcation diagram for the follower-loaded double pendulum. (Thomsen 1995)

equilibrium point in favor of a stable limit cycle, that is, a supercritical Hopf bifurcation may occur. Further parameter changes may cause the system to experience yet another Hopf bifurcation, inducing a second fundamental frequency into the motion. If these two frequencies are incommensurate the motion will be *quasiperiodic*, i.e. the solution rides on a two-torus. Now, should a *third* Hopf bifurcation occur, then motion on the corresponding three-torus will typically be highly unstable (Newhouse et al. 1978). With finite probability (as the mathematicians put it) the unstable three-torus will decay into a strange attractor. Thus, as the precursor to this type of chaos, one first observes periodic and then quasiperiodic motion as a system parameter is varied.

As an example of a system displaying the quasiperiodic route to chaos we reconsider the periodically forced non-shallow arch (dealt with analytically in Sect. 4.3). Fig. 6.11 shows a numerically obtained bifurcation map for the system. Values of the antisymmetric vibration amplitude f (cf. Fig. 4.6) at a particular phasing of the driving force have been plotted against the nondimensional excitation-frequency Ω for a constant level of forcing, $q = 0.2$. For each value of the bifurcation parameter Ω you may perceive the dots as a projection of an ordinary $(f_n, \dot{f}_n)$ Poincaré map onto the $\dot{f}_n$-axis. In the Poincaré map, as we have seen, quasiperiodic motion reveals itself as points on a closed curve. Projecting the map onto one axis, a line filled with points appears, just as for chaotic motion. In Fig. 6.11 the largest Lyapunov exponent has been plotted on top of the diagram to help distinguish quasiperiodic from chaotic motion.

Fig. 6.11(a) reveals that regular period-2 motion exists for $\Omega \in [\Omega_1; \Omega_3]$. Dots fill out the Poincaré lines between Ω_3 and Ω_5. Though, since $\hat{\lambda}_1 \approx 0$ the motion is not chaotic, but quasiperiodic. Beyond Ω_5 dots continue to fill out the lines, but since $\hat{\lambda}_1$ is now positive the motion is chaotic.

Fig. 6.11(b) shows the boxed region in Fig. 6.11(a), corresponding to quasiperiodic motion, enlarged 200 times. Narrow periodic and chaotic windows are seen to intervene in the quasiperiodic regime.

6.4.3 The Transient Route

This route to chaos involves chaotic transients of gradually increasing length.

It is quite common with nonlinear systems to observe chaotic motion at the *initial* parts of a time-series. During this time there is a positive largest Lyapunov exponent, and long-term predictions seem impossible. Often, however, there is a globally attracting equilibrium point or limit cycle to which the solution quickly converges. The motion starts chaotic but ends up regular.

Sometimes such chaotic transients increase in length as a system parameter is varied. They may last for so long that, by any reasonable means, the chaos observed must be considered a stationary feature. This is the transient route to chaos. It does not involve a sharp transition from regular to chaotic motion. Or rather, the transition value that appears in a bifurcation map depends on the time chosen for transient cut-off. If the cut-off time is raised, then some solutions that were formerly doomed chaotic now have time to settle down into regular motion.

It has been conjectured that the transient route is related to the collision of a chaotic attractor with a coexisting unstable limit cycle (Grebogi et al. 1985).

Returning to Fig. 6.11(a), showing a bifurcation diagram for the non-shallow arch, the transient route is observed upon entering the chaotic regime at $\Omega = \Omega_4$ from the right. For this figure the first 4,000 periods of the driving force were considered as transients. Raising the cut-off limit for transients even further, the chaotic transition zone at Ω_4 will move to the left.

It follows from the above discussion that transient cut-off times should be chosen with care (see e.g. Table 1.4 in Sect. 1.7.5). Some use a general limit beyond which chaotic motion is considered stationary, e.g., Moon (1987) suggests 5,000 periods of a fundamental system frequency. However, when the system under study models a particular system of the real world, it is recommended to choose the transient limit relative to the total life or operating time of the system. For example, 5,000 fundamental periods may be a short time when studying dynamics of turbine machinery, but a very long time for the study of ship motions due to sea waves.

6.4.4 The Intermittency Route

Chaotic intermittency reveals itself in a *time-series* plot of the motion. Here periods of regular motion alternate with periods of chaos.

The route starts with regular periodic motion. Then, as a system parameter is varied, short bursts of chaos pop up in between long periods of regular motion.

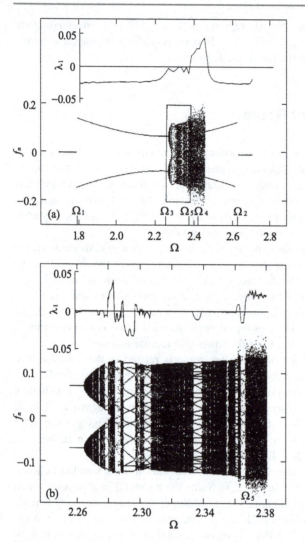

Fig. 6.11 Bifurcation diagrams for the non-shallow arch. The boxed region in (**a**) is shown enlarged in (**b**). ($q = 0.2$, $\omega = 2.44$, $\beta = 0.03$, $m = 3.32$, $\kappa = 2.63$) (Thomsen 1992)

Further increasing or decreasing the system parameter, the bursts become longer and more frequent, until at a critical point chaos is fully developed.

Intermittency is considered to be related to *inverse tangent bifurcations*, by which two fixed points of the Poincaré map (one stable, the other unstable) merge into one (e.g., Schuster 1989).

As for the so-called *type-I intermittency*, a universal scaling law has been proposed for the relation between a system parameter μ and the average length $\langle \tau \rangle$ of regular motion (e.g., Schuster 1989):

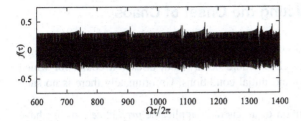

Fig. 6.12 Post-transient time record for the non-shallow arch, displaying chaotic intermittency. ($q = 0.3$, $\Omega = 1.7496$, $\omega = 2.44$, $\beta = 0.03$, $m = 3.32$, $\kappa = 2.63$) (Thomsen 1992)

$$\langle \tau \rangle \propto \frac{1}{\sqrt{\mu - \mu_c}}, \quad \mu > \mu_c, \tag{6.13}$$

where μ_c is a critical value beyond which chaotic bursts start to appear. As μ is increased beyond μ_c the average length of *laminar regions* (with regular motion) decreases, giving leave for increasingly long periods of chaotic motion.

Fig. 6.12 shows a post-transient time-series displaying chaotic intermittency for the non-shallow arch discussed in Sect. 4.3. Short chaotic bursts appear in between long periods of regular motion. The motion is unpredictable, even during periods of regular motion, since the time of the next chaotic burst is unpredictable.

For applications it is important to be aware of chaotic intermittency. For a real system, one cannot exclude *á priori* that it will display chaotic intermittency for certain values of the system parameters. Numerical simulations and laboratory experiments may show regular and seemingly predictable motion for as long as one is willing to wait. Still, the system may turn chaotic at the next instance of time.

Suppose a turbo-engine has been on the test bench for ten hours, displaying regular periodic motion all the time. Strictly speaking, you cannot rule out the possibly that wild chaotic dynamics may occur within the next test-hour. Should this occur, however, then according to (6.13) you can hope it will do so in short bursts that will not be harmful to the engine.

6.4.5   Summary on the Routes to Chaos

The *period-doubling, quasiperiodic, transient* and *intermittency* routes to chaos have been observed repeatedly for a variety of systems through many years. They are considered to be *universal*, though other routes presently unrevealed may exist. Observing indicators of any one route, we are warned that the system under study may be within reach of a chaotic attractor. Then small perturbations to the system parameters may render the system chaotic.

6.5 Tools for Predicting the Onset of Chaos

Given a set of model equations for a dynamical system, it would be nice having a general tool for predicting *if* chaos could occur. And, in case it could, for which values of system parameters and initial conditions. Unfortunately there is no such tool.

While awaiting the invention of a generally applicable *predictive* tool, we have to rely on less powerful methods. None of these methods are universally applicable, nor are they particularly simple to apply.

6.5.1 Criteria Related to the Universal Routes of Chaos

These criteria take as their starting point the postulated universality of certain routes to chaos: the period-doubling, quasiperiodic, transient and intermittency- routes described in Sect. 6.4. Observing characteristic events related to any particular route (e.g., period-doublings), one can attempt predicting critical values for chaos to occur by employing the corresponding universal scaling law.

Though in some cases predictions of acceptable quality have been obtained, one has to raise a flag of caution as to the uncritical employment of such procedures. It should be kept in mind that the term *universal* is to be associated with certain *classes* of systems. For example, Feigenbaum (1978, 1980) proved the period-doubling route to chaos to be a universal feature of systems that can be approximated by one-dimensional maps with a zero tangent. Many nonlinear systems cannot be approximated by such a map. Though such systems may display accumulating period-doublings, there is no guarantee that they will do so according to the Feigenbaum law (6.12).

Predictions Based on Observed Period-Doublings According to the Feigenbaum-scenario, period-doublings of the quadratic map accumulate according to (6.12), that is:

$$\lim_{n \to \infty} \left(\frac{\mu_{n+1} - \mu_n}{\mu_n - \mu_{n-1}} \right) = \frac{1}{\delta}, \quad \delta = 4.6692 \cdots, \tag{6.14}$$

where δ is the Feigenbaum number, and μ_n is the bifurcation value of the n'th period-doubling. In other words, at $\mu = \mu_n$ the fundamental period of motion T doubles from $2^{(n-1)}T$ to $2^n T$, cf. Fig. 6.9.

Assume that two subsequent period-doubling bifurcations have been observed at $\mu = \mu_{k-1}$ and $\mu = \mu_k$, respectively. One may then attempt computing the critical value μ_c for the onset of chaos by iterating (6.14):

$$\mu_{n+1} = \mu_n + \frac{1}{\delta}(\mu_n - \mu_{n-1}), \quad n = k, k+1, k+2, \ldots \tag{6.15}$$

$$\mu_{n+1} \to \mu_c \quad \text{for} \quad n \to \infty.$$

Of course, this assumes the system under study to obey the properties of a quadratic map. Also, the procedure ignores that Feigenbaum's constant δ only describe the process of accumulating bifurcations in the limit $n \to \infty$.

The two or more observed bifurcation values needed for this method may be obtained theoretically, experimentally or numerically.

If *numerically* obtained there is no need for the method described, since a more reliable approach for locating μ_c is to proceed simulating the system with other values of μ.

For *experimental* observations some justice is given to the approach, in particular when it is not possible to freely vary the system parameter in question. e.g., with systems in duty and many natural systems.

A more interesting application comes in when, in a *mathematical* model of a system, two subsequent period-doublings are predicted by purely analytical means. In this case the method of predicting a critical value for chaos becomes truly predictive, requiring only a set of model equations. If it is not known whether the system under study possesses the properties of a quadratic map, then careful numerical simulation is required to verify the predictive power of the criterion. Kapitaniak (1993), attempting to control chaos in a Duffing-type system, used a local perturbation method to locate a pair of period-doublings. Feigenbaum's law was then employed to predict the onset of chaos.

Predictions Based on Observed Transient or Intermittent Chaos With *type-I intermittency*, as described in Sect. 6.4.4, the average time of regular motion between chaotic bursts scales according to:

$$\langle \tau \rangle \propto \frac{1}{\sqrt{\mu - \mu_c}}, \quad \mu > \mu_c, \tag{6.16}$$

where μ_c is the critical value beyond which the first chaotic bursts appear. This scaling law applies too for certain systems displaying *transient* chaos. In this case $\langle \tau \rangle$ is the average length of transients, and μ_c is the limit value of μ that corresponds to fully developed chaos (e.g., Moon 1987). However, systems that exhibit *supertransient chaos* have been observed (Grebogi et al. 1985), in which case the length of transients scale according to

$$\langle \tau \rangle \propto k_1 \exp\left(\frac{k_2}{\sqrt{\mu - \mu_c}}\right), \quad \mu > \mu_c. \tag{6.17}$$

For systems exhibiting signs of intermittency, one could examine the average times of regular motion for a few values of μ. The critical value of μ_c could then be estimated by assuming (6.16) to be a valid scaling law for the system.

Similarly, when transient chaos is observed, an estimate for μ_c could be obtained by examining the variation of transient lengths, assuming (6.16) to be valid. As for the period-doubling criterion the observations could be obtained analytically, numerically or experimentally.

Predictions Based on Observed Quasiperiodic Motion The quasiperiodic route to chaos follows a path of subsequent Hopf bifurcations, as explained in Sect. 6.4.2. When a *third* Hopf bifurcation is just to appear, the system is likely to turn chaotic.

If, for a given system, one can set up analytical conditions for the occurrence of the first two bifurcations, then a predictive tool for quasiperiodic motion is at hand. For some systems the distance (in parameter space) from quasiperiodic motion to chaos is so short that a quasiperiodic criterion is effectively also a chaos criterion.

6.5.2 Searching for Homoclinic Tangles and Smale Horseshoes

Of the many efforts offered by mathematicians aiming to understand chaos, those relating to *homoclinic tangling* and *horseshoe maps* seem especially promising. It seems that most chaotic processes, if not all, are somehow related to the presence of horseshoe maps in the dynamics of the system. By the point of view to be discussed in this section, the presence of homoclinic tangling provides a fundamental mechanism behind the creation of horseshoe maps. Thus, we consider the following path of events:

Homoclinic intersections $\Rightarrow$ Homoclinic tangling $\Rightarrow$ Horseshoe maps
$\Rightarrow$ Extreme sensitivity to initial conditions $\Rightarrow$ Possibility of chaos

---

The first three terms will be defined below. For the moment we note that *if* the hypothesis regarding the path to chaos holds true, then a predictive, sufficient criterion may take as its starting point a search for homoclinic intersections.

In a section to follow we describe a particular method for locating homoclinic intersections for a certain class of systems, the so-called *Melnikov method*. To understand this method, and to grasp a probable mechanism behind chaos, we need to explain the concepts involved and how these relate to each other. Rigorous definitions and proofs can be found elsewhere (e.g., Guckenheimer and Holmes 1983).

Homoclinic Intersections By homoclinic intersections we refer to transverse crossings of the stable and unstable manifolds of a Poincaré map saddle-point.

Fig. 6.13 shows phase plane orbits for the Duffing equation (6.1), when there is no damping and no forcing. There is a saddle at (0, 0) and two centers at (±1, 0). A *homoclinic orbit* surrounds each center, connecting the stable and unstable branches of the saddle (recall from Sect. 3.4.3 that a homoclinic orbit connects a singular point with itself). When damping is added the centers become stable foci, and the homoclinic orbits break up. The saddle remains a saddle. Add then forcing to the system and the orbits will move around in a more complicated manner, though the presence of the saddle and the foci for the unforced problem will remain to affect the flow locally. The addition of forcing brings an extra state variable to the system, since the force depends explicitly on time. The system becomes three-dimensional, so that orbits may now intersect in the phase plane. This has nothing to do with homoclinic intersections.

We then consider instead the *Poincaré map* (PM) representation of the dynamics of a system. For the Duffing equation we sample the phase plane orbits at discrete times in phase with the driving force. In the Poincaré map, the continuous flow of phase plane orbits convert into sequences of points. These points may lie along certain curves in the PM. Such curves are called *orbits* (of the PM), just as for the phase plane. As we have seen, periodic motion reveals itself in the PM as one or more points. Even if there is only a single point, this point is called an orbit for the PM. Similarly, for quasiperiodic motion we obtain a closed orbit in the PM. If the system is in a transient or chaotic state the PM-points will move around as governed by the dynamics of the particular map.

The mapping of points in the PM is affected by the presence and the nature of *fixed points* – just as flows of phase plane orbits are locally affected by singular points. Near fixed points, new PM points will map according to the nature of the fixed point. For example, if a PM point is near a stable focus, then the next PM-point will map closer to this fixed point.

Though it is not obvious, the PM for the forced Duffing equation has a saddle-point near (0,0). Near this PM saddle, points are repelled from certain directions while attracted from other. Fig. 6.14 shows a sketch of a PM for the Duffing system (6.1) when the forcing is weak. The saddle near (0,0) has stable and unstable manifolds W^s and W^u, similar to the stable and unstable branches of a phase plane saddle. These manifolds are *invariant* for the map, meaning that any PM point on W^s will map on W^s, and similarly for W^u. Points on W^s will map closer and closer to the saddle, whereas point on W^u will map closer and closer to the two stable foci. The foci represent regular periodic motion in the phase plane (for the magnetically buckled beam each PM focus corresponds to periodic oscillations about one of the magnets). We note that W^s and W^u do not intersect.

However, in some cases the stable and unstable manifolds of a PM saddle may intersect. Fig. 6.15 shows a PM for the Duffing equation when the force has been raised to a level causing W^s (in solid line) and W^u (dashed) to intersect. This is called *homoclinic intersection*, and the points of intersection are *homoclinic points*. Let us consider a PM point falling at the intersection marked *I* in Fig. 6.15. Since this point

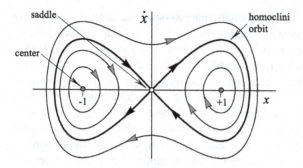

Fig. 6.13 Phase plane orbits for the unforced and undamped duffing equation

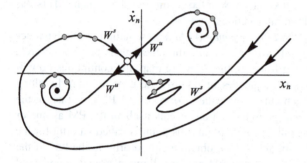

Fig. 6.14 Poincaré map for a weakly forced duffing system. The stable and unstable manifolds W^s and W^u of the saddle near $(0, 0)$ do not intersect

falls on W^s which is an invariant manifold, the next iterate of the map, $P^1(I)$, must also fall on W^s. By the same argument $P^1(I)$ must also fall on W^u. But then $P^1(I)$ must fall on an intersection between W^s and W^u, because this is the only way to fall on both manifolds. We can continue this reasoning and show that subsequent PM points $P^2(I)$, $P^3(I)$, ... will all fall on manifold intersections, as will the backward iterates $P^{-1}(I)$, $P^{-2}(I)$, ... Hence, if the stable and unstable manifolds intersect once, they will continue to do so an infinite number of times. Further, the forward iterates $P^2(I)$, $P^3(I)$, ... approach the saddle along the stable manifold, whereas the backward iterates $P^{-1}(I)$, $P^{-2}(I)$, ... approach the saddle along the unstable manifold. Hence, the homoclinic points fall on an orbit of the PM originating from and ending at the saddle. Naturally, this orbit is called the *homoclinic orbit* for the PM.

Homoclinic Tangling As originally observed by Poincaré, homoclinic intersections always cause wild oscillations near the saddle. Near the saddle, as appears from Fig. 6.15, the stable and unstable manifolds get bunched up, creating an image known as *homoclinic tangling*. These tangles beat at the heart of chaos, because in the region of homoclinic tangling initial conditions are subjected to a process of violent stretching and folding. We refer to this stretching and folding as *extreme sensitivity to initial conditions*.

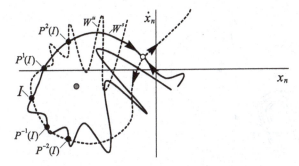

Fig. 6.15 Poincaré map for a strongly forced duffing system. The stable and unstable saddle manifolds W^s and W^u intersect, causing homoclinic tangling near the saddle

Horseshoes Homoclinic tangling leads to *Smale horseshoes* in the dynamics of the PM (e.g., Guckenheimer and Holmes 1983). *Horseshoes* are just the geometrical descriptors of the process of stretching and folding of initial conditions, and Stephen Smale (1967) was the one to uncover their importance for the chaotic dynamics of maps.

If a system is governed by a map of the horseshoe type, then a ball of initial conditions in phase space is mapped onto a new shape in which the original ball is stretched and folded (Fig. 6.16). For dissipative systems $\dot{\mathbf{x}} = \mathbf{f}(\mathbf{x})$, the divergence of the vector field $\mathbf{f}(\mathbf{x})$ is always *negative*, so that volumes get mapped onto smaller volumes. We say that dissipative systems are *volume contracting*. However, near the unstable manifolds of a Poincaré map saddle, volumes are *stretched* in the direction of the manifold. Since the total volume must decrease, the ball of initial conditions must contract more than it stretches. Near homoclinic points, if present, the volume is also *folded* – and a horseshoe appears (at least we visualize it this way).

After many iterations of the map this process of repeated stretching and folding produces a fractal-like structure, and the precise information as to which orbit originated where is lost. For each subsequent stretch and fold, more and more precision is required to relate initial conditions to later states of the system. With finite precision devices (computers, humans, rulers, watches, etc.) accurate prediction becomes impossible.

6.5.3 The Melnikov Criterion

As appears from the above discussion, homoclinic tangling is likely to produce chaos. Homoclinic tangling occurs when the stable and unstable manifolds of a Poincaré saddle intersect. A measure of the *distance* between two such manifolds provides a means for determining when intersections occur.

The *Melnikov function* provides this measure, as a function of the system parameters. The *Melnikov criterion* then states that chaos becomes possible when

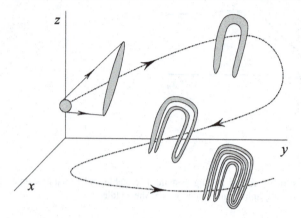

Fig. 6.16 Evolution of a ball of initial conditions for a system with homoclinic tangling and horseshoe mapping

the Melnikov function has a *simple zero*, that is, when the stable and unstable saddle manifolds of the Poincaré map intersect transversely.

We present here a Melnikov function for systems that can be expressed in the following form:

$$\dot{x} = \frac{\partial H}{\partial y} + \varepsilon f_1(x, y, t),$$
$$\dot{y} = -\frac{\partial H}{\partial x} + \varepsilon f_2(x, y, t),$$
(6.18)

where $H(x, y)$ is the Hamiltonian of the corresponding undamped and unforced system, ε is a small parameter, and the functions f_i are T-periodic in time, $f_i(t + T) = f_i(t)$. It is further assumed that a *saddle-point* exists for the unperturbed Hamiltonian problem, i.e. when $\varepsilon = 0$ in (6.18). A Poincaré map obtained in phase with the periodic terms f_i then too has a saddle-point, with stable and unstable manifolds W^s and W^u (Guckenheimer and Holmes 1983). The separation between W^s and W^u in the Poincaré map is given by the Melnikov function:

$$M(t_0) = \int_{-\infty}^{\infty} \mathbf{f}^T(\hat{x}, \hat{y}, t + t_0) \cdot \nabla H(\hat{x}, \hat{y}) dt,$$
(6.19)

where $\mathbf{f}^T = \{f_1, f_2\}$, ∇H is the gradient of H, $(\hat{x}(t), \hat{y}(t))$ describes the homoclinic (or heteroclinic) saddle orbit of the unperturbed Hamiltonian system, and t_0 measures the distance along the unperturbed homoclinic (or heteroclinic) orbit. Homoclinic intersections occur when $M(t_0) = 0$.

As an application example we consider a Duffing equation with negative linear stiffness, weak damping and small forcing:

$$\ddot{x} + \varepsilon\beta\dot{x} - x + x^3 = \varepsilon p\cos(\Omega t), \quad \varepsilon \ll 1, \tag{6.20}$$

which can be brought into the form (6.1) for the buckled beam by shifting the time variable, $\tau = \sqrt{2}t$. Written in first-order form (6.20) becomes

$$\begin{aligned}\dot{x} &= y, \\ \dot{y} &= x - x^3 + \varepsilon(p\cos(\Omega t) - \beta y).\end{aligned} \tag{6.21}$$

This system has the form (6.18), with Hamiltonian energy and perturbations, respectively:

$$\begin{aligned}H &= \tfrac{1}{2}y^2 - \tfrac{1}{2}x^2 + \tfrac{1}{4}x^4, \\ f_1 &= 0, \quad f_2 = p\cos(\Omega t) - \beta y.\end{aligned} \tag{6.22}$$

In the (x, y) phase plane the unperturbed system $(\varepsilon = 0)$ has centers at $(x, y) = (\pm 1, 0)$ and a saddle at $(x, y) = (0, 0)$, cf. Fig. 6.13. The two homoclinic orbits emanating from the saddle are given by the solution to $H(x, y) = 0$, that is:

$$\hat{y} = \pm\sqrt{\hat{x}^2 - \tfrac{1}{2}\hat{x}^4}. \tag{6.23}$$

By definition $\dot{x} = y$ implies that $dt = dx/y$, i.e. for the homoclinic orbit it holds that $dt = d\hat{x}/\hat{y}$. Substituting into the latter equation (6.23) for $\hat{y}$, we may integrate both sides to obtain t as a function of $\hat{x}$, invert to find $\hat{x}(t)$, and then insert this into (6.23) to give $\hat{y}(t)$. The result becomes:

$$\begin{aligned}\hat{x}(t) &= \sqrt{2}\operatorname{sech}(t), \\ \hat{y}(t) &= -\sqrt{2}\operatorname{sech}(t)\tanh(t).\end{aligned} \tag{6.24}$$

The Melnikov function (6.19) then becomes

$$\begin{aligned}M(t_0) &= \int_{-\infty}^{\infty} \{0, \quad p\cos\Omega(t + t_0) - \beta\hat{y}(t)\} \bullet \begin{Bmatrix} -\hat{x}(t) + \hat{x}(t)^3 \\ \hat{y}(t) \end{Bmatrix} dt \\ &= -\sqrt{2}p\int_{-\infty}^{\infty} \operatorname{sech} t \tanh t \cos\Omega(t + t_0)dt - 2\beta\int_{-\infty}^{\infty}\operatorname{sech}^2 t\tanh^2 t\,dt.\end{aligned} \tag{6.25}$$

Evaluating the integrals (the first by the method of residues) it is found that:

$$M(t_0) = -\tfrac{4}{3}\beta - \sqrt{2}p\pi\Omega\operatorname{sech}(\pi\Omega/2)\sin(\Omega t_0), \tag{6.26}$$

With β, π, $\Omega > 0$, quadratic tangency between the stable and unstable manifolds occurs when $M(t_0)$ just becomes zero, that is, when $\sin(\Omega t_0) = 1$ and

$$p = p_c = \frac{4}{3\sqrt{2}} \frac{\beta}{\pi\Omega} \cosh(\pi\Omega/2). \qquad (6.27)$$

When $p > p_c$ stable and unstable manifolds intersect transversely, homoclinic tangling occurs, and chaos becomes possible.

The result (6.27) holds for the system (6.20). It can be applied too for (6.1), by replacing the term $\cosh(\pi\Omega/2)$ in (6.27) with $\cosh(\pi\Omega/\sqrt{2})$.

Experiences with this criterion have shown it to predict rather accurately the threshold value for homoclinic tangling, but also that chaos typically sets in somewhat above this threshold. The Melnikov criterion seems to yield *lower* bounds, below which chaos cannot occur, rather than a trigger value for chaos. So, it appears not to be *sufficient*, as the link of events in the box on p. 326 would suggest. It even may not be *necessary*, since other basic mechanisms for chaos may exist.

Fig. 6.17 shows Poincaré map orbits for the Duffing equation $\ddot{u} + 0.25\dot{u} - u + u^3 = p\cos t$. The orbits should be perceived as closely spaced Poincaré points. Using (6.27) with $\beta = 0.25$ and $\Omega = 1$ we predict homoclinic intersections to occur when $p \geq p_c \approx 0.188$. Fig. 6.17(a) – where the unperturbed homoclinic loops have been superimposed in dashed line for reference – holds for $p = 0.11 < p_c$. There are no intersections between the stable and unstable manifolds of the Poincaré saddle near (0,0). In Fig. 6.17(b) the level of forcing has been raised to $p = 0.19 \approx p_c$. As an impressive verification of the theoretical results, quadratic tangency is noted at the point P_1. In Fig. 6.17(c) the forcing is $p = 0.30 > p_c$ and, correspondingly, homoclinic intersections occur.

Moon (e.g., 1987) has compared predictions given by the Melnikov criterion to experimental data for the magnetically buckled beam. The Melnikov criterion in this case provided a correct lower bound on chaos, actually much lower than the experimentally observed chaos. On this basis Moon suggested a heuristic criterion that would be *sufficient* for chaos to appear in this specific system. We shall return to this criterion in Sect. 6.5.4.

Melnikov criteria have been elaborated for a number of specific systems, most of them of the form (6.18). For example, Moon et al. (1985) presents a Melnikov criterion for the equation of motion associated with a pendulum in a magnetic field:

$$\ddot{\theta} + \beta\dot{\theta} + \sin\theta = p\cos\theta\cos(\Omega t), \qquad (6.28)$$

while Yagasaki et al. (1990) provide a criterion for the Duffing equation with combined parametrical and external periodic forcing:

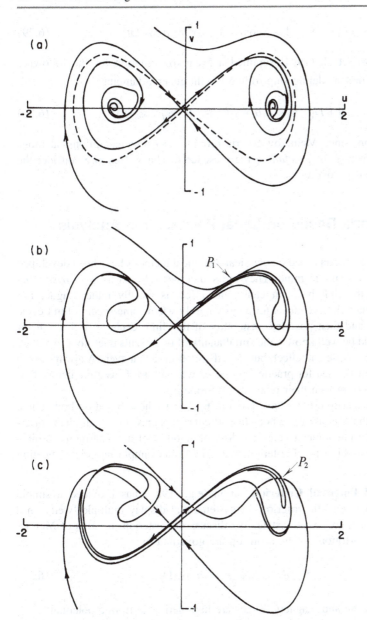

Fig. 6.17 Poincaré maps for $\ddot{u} + 0.25\dot{u} - u + u^3 = p\cos t$ showing: (**a**) no homoclinic intersections for $p = 0.11$; (**b**) quadratic tangency for $p = 0.19$; (**c**) Transverse intersections for $p = 0.30$. (computed by Yoshisuke Ueda as a personal communication to Guckenheimer and Holmes 1983)

$$\ddot{x} + \beta\dot{x} + (1 + b\cos\omega t)x + ax^3 = p\cos(\Omega t). \tag{6.29}$$

Also, Bikdash et al. (1994) supplied a Melnikov criterion for the following equation of motion modeling ship-rolling θ with general damping $D(\theta, \dot{\theta})$:

$$\ddot{\theta} + D(\theta, \dot{\theta}) + c_1\theta + c_2\theta^3 + c_3\theta^5 = p\cos(\Omega t) \tag{6.30}$$

In conclusion, since Melnikov criteria provide *necessary* conditions, at most, they can be effective for *precluding* the presence of chaos. Next we consider the search for *sufficient* conditions.

6.5.4   Criteria Based on Local Perturbation Analysis

A few predictive criteria based on classical analytical methods have been developed with particular systems in mind. They are sometimes referred to as *heuristic criteria*, because the link between cause and effect is usually rather vague. For example, one may observe chaos for a system only when some specific limit cycle turns unstable, and then postulate instability of this limit cycle as a criterion for chaos. As should be well known, the simultaneity of two events does not prove they are connected by cause and effect, but still, if we observe that event A always occur along with event B, then for practical purposes we may use B as a predictor of A, even if we cannot proven their relationship formally.

In many cases such criteria have proven superior to those based on homoclinic tangling, e.g., the Melnikov method. In particular, systems for which chaos originates from *local* phenomena –e.g., the loss of stability of a subharmonic orbit – seem amenable to this kind of criterion. We provide an example and briefly mention a few more.

The Multiwell Potential Criterion Moon (e.g., 1987) has provided a simple though workable heuristic criterion for systems that display multiple 'wells' in a plot of potential energy versus state. We discuss the criterion in terms of Moon's beam (Fig. 6.1). Motions of the beam-tip are governed by:

$$\ddot{x} + \beta\dot{x} - \frac{1}{2}x + \frac{1}{2}x^3 = p\cos(\Omega t). \tag{6.31}$$

Since the elastic and magnetic forces are here derivable from a potential:

$$V(x) = -\frac{1}{4}x^2 + \frac{1}{8}x^4, \tag{6.32}$$

we may as well write (6.31) in a form to which the multiwell potential criterion applies more generally:

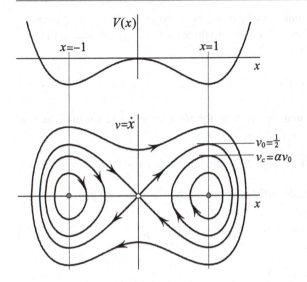

Fig. 6.18 Potential function $V(x)$ and phase plane orbits $(x(t),v(t))$ for the unforced and undamped duffing's equation (6.31)

$$\ddot{x} + \beta\dot{x} + \frac{\partial V}{\partial x} = p\cos(\Omega t). \tag{6.33}$$

Fig. 6.18 shows the potential (6.32), along with phase plane orbits for the unforced and undamped system. The phase plane orbits are obtained by letting $p = \beta = 0$ in (6.31), and using the same approach as in Sect. 3.4.2 for a similar system to give $v^2 = 2(C - V(x))$, where $v = \dot{x}$ and each value of the arbitrary constant C gives a particular orbit. The homoclinic orbit, through $(x, v) = (0, 0)$, is obtained for $C = 0$, i.e. $v_{\text{hom}}^2 = -2V(x)$.

For small values of the forcing p, as described in Sect. 6.2, the beam is observed to oscillate in one of the potential wells at $x = \pm 1$. For higher values of p the beam will jump chaotically between the two wells. Moon's idea was to estimate the value of p causing oscillation amplitudes just large enough to overcome the potential barrier at $x = 0$, and to link that event to the onset of chaos. Inspired by numerical simulations he postulated the existence of a *critical velocity* v_c for escaping a potential well. This velocity was proposed to be near the maximum velocity v_0 on the saddle separatrix of the unforced, undamped phase plane (see Fig. 6.18), that is:

$$v_c = \alpha v_0, \quad \alpha \approx 1, \tag{6.34}$$

where the constant α is to be determined empirically. To find v_0 we use the expression for the homoclinic orbit, $v_{\text{hom}}^2 = -2V(x)$, and find that there is a maximum for $x = \pm 1$ (where $dV/dx = 0$) with value $v_0 = v_{\text{hom}}(\pm) = \frac{1}{2}$.

Now, what is the maximum velocity of the beam when oscillating in a potential well? Rewriting (6.31) in terms of a new coordinate $u = x - 1$, centered at the well at $x = 1$, one obtains:

$$\ddot{u} + \beta \dot{u} + u + \frac{3}{2}u^2 + \frac{1}{2}u^3 = p \cos \Omega t; \quad x = u + 1. \tag{6.35}$$

Performing a local perturbation analysis for small u and p Moon finds that, to first order:

$$u(t) = A(\Omega) \cos(\Omega t) + \text{(higher order terms)}, \tag{6.36}$$

where the stationary amplitude $A(\Omega)$ is given by the frequency response equation

$$A^2 \left[\left(1 - \Omega^2 - \frac{3}{2}A^2 \right)^2 + \beta^2 \Omega^2 \right] = p^2. \tag{6.37}$$

By (6.36) the maximum velocity is

$$\dot{x}_{max} = |\dot{u}(t)|_{max} = |-\Omega A(\Omega) \sin \Omega t|_{max} = \Omega A(\Omega), \tag{6.38}$$

which takes on the critical value $v_c = a v_0 = \frac{1}{2}\alpha$ when

$$\Omega A(\Omega) = \frac{1}{2}\alpha. \tag{6.39}$$

Substituting this into (6.37), the critical velocity v_c is found to correspond to a critical amplitude of excitation p_c:

$$p_c = \frac{\alpha}{2\Omega} \left[\left(1 - \Omega^2 - \frac{3}{8}\frac{\alpha^2}{\Omega^2} \right)^2 + \beta^2 \Omega^2 \right]^{1/2}, \quad \alpha \approx 1. \tag{6.40}$$

Using $p > p_c$ as a criterion for chaos and $\alpha = 0.86$, Moon finds a fair agreement with experimental observations when $\beta \ll 1$ and Ω is within $\pm 50\%$ of the natural frequency of the beam.

Fig. 6.19 compares, for the system (6.31), the multiwell potential criterion (6.40) (solid line) to the Melnikov criterion (6.27) (dashed lined). Thresholds of chaos as observed by numerical simulation are also depicted (as circles). Seemingly, for this case the multiwell potential criterion provides a closer fit to the numerical results than does the Melnikov criterion.

The multiwell criterion has been successfully applied to other multiwell potential problems (Moon 1987).

Other Criteria based on Perturbation Analysis. In a number of papers Szemplinska-Stupnicka and co-workers employ local perturbation techniques for the study of the Duffing-system:

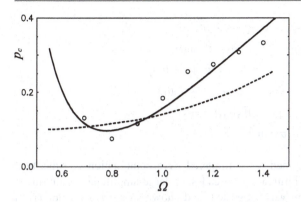

Fig. 6.19 (——) Multiwell potential estimates of critical forcing amplitude p_c for the system (6.31) when $\beta = 0.1$, as given by (6.40) with $\alpha = 0.93$. (- - - -) Melnikov estimates, as given by (6.27) with '2' replaced by '$\sqrt{2}$'. (o o o) Numerical simulation, threshold for chaos as determined by the presence of a positive Lyapunov exponent ($x(0) = \dot{x}(0) = 0.01$)

$$\ddot{x} + \beta\dot{x} - \frac{1}{2}x + \frac{1}{2}x^3 = p\cos(\Omega t), \tag{6.41}$$

and links the onset of chaos to the loss of stability of certain periodic orbits (e.g., Szemplinska-Stupnicka 1992; Szemplinska-Stupnicka and Rudowski 1992; Rudowski and Szemplinska-Stupnicka 1987). Their criterion agrees more closely with numerical simulation than does the Melnikov criterion.

A similar criterion was given for the following asymmetric and parametrically excited Duffing system (Szemplinska-Stupnicka et al. 1989):

$$\ddot{x} + 2\beta\dot{x} + (k_2 + k_3\cos(\Omega t))x + \alpha x^2 + \beta x^3 = 0. \tag{6.42}$$

Dowell and Pezeshki (1986, 1988) too considered criteria for the Duffing equation with negative linear stiffness. Numerical evidence was given in Dowell and Pezeshki (1988) that, for this particular system, chaos occurs when there is a near intersection of stable and unstable limit cycles in the phase space.

Thomsen (1992) provided a perturbation-based criterion for a non-shallow arch governed by the coupled equations:

$$\ddot{f} + 2\beta\dot{f} + (1 - m\omega^2 u)f = 0,$$
$$\ddot{u} + 2\beta\omega\dot{u} + \omega^2 u + \kappa(f\ddot{f} + \dot{f}^2) = (q/m)\cos(\Omega\tau). \tag{6.43}$$

The criterion, showing good agreement with numerical simulation, links the onset of chaos to loss of limit cycle stability near primary resonance $\Omega \approx \omega$.

In a study (Thomsen 1995) of chaos for a follower-loaded elastic double pendulum governed by:

$$(1+m)\ddot{\theta}_1 + \cos(\theta_2 - \theta_1)\ddot{\theta}_2 + (c_1 + c_2)\dot{\theta}_1$$
$$- c_2\dot{\theta}_2 + 2\theta_1 - \theta_2 + \dot{\theta}_2^2 \sin(\theta_1 - \theta_2)$$
$$= p((1-\alpha)\sin\theta_1 + \alpha\sin(\theta_1 - \theta_2)),$$
$$\cos(\theta_2 - \theta_1)\ddot{\theta}_1 + \ddot{\theta}_2 + c_2(\dot{\theta}_2 - \dot{\theta}_1)$$
$$- \theta_1 + \theta_2 - \dot{\theta}_1^2 \sin(\theta_1 - \theta_2)$$
$$= p(1-\alpha)\sin\theta_2,$$

(6.44)

the onset of chaos was assumed to be triggered by a combination of subcritical pitchfork bifurcations and bifurcating cascades of large-amplitude equilibriums. A simple analytical criterion was suggested, and shown to agree with numerical simulations.

As a final example we mention the study by Nayfeh and Khdeir (1986), employing perturbation techniques to predict the onset of period-doubling or tripling as a precursor to chaos for a model of a ship in regular sea waves.

6.5.5 Criteria for Conservative Chaos

Criteria exist for predicting chaos in non-dissipative conservative systems, e.g., *Chirikov's overlap criterion* (Schuster 1989). One might believe that criteria for zero dissipation would hold approximately for cases of *weak* dissipation as well. However, they do not. Systems with zero dissipation behave qualitatively different from systems with dissipation, however small. Thus, criteria for conservative chaos hold only for strictly non-dissipative systems, such as can be encountered in the field of celestial mechanics and plasma physics.

6.6 Mechanical Systems and Chaos

Chaos research has fabricated an impressively long list of mechanical systems displaying chaos. We here pinpoint only a few characteristic examples. Many of these systems obey equations of motion with a similar structure. We therefore group the examples according to mathematical rather than physical characteristic. We concentrate on equations of motion that occur repeatedly within the field of solid mechanics. These should be recognized as potentially chaotic to any vibration analyst. Encountering a nonlinear system that is *not* included below, it is wise to suspect it to behave chaotically anyway. Most physical systems may turn chaotic in response to strong nonlinearities and forcing. Starting with systems of the lowest possible dimension for chaos to appear, $D = 3$, we move on to higher order systems, $D = 4, 5, >5$. Most examples are given only a short mention and a bibliographic reference, whereas a few are described in more detail.

6.6.1 The Lorenz System (D = 3)

As a simple model for thermally induced fluid convection in the atmosphere, the Lorenz equations has probably limited relevance to problems of solid mechanics. However, since they are often referred to in the chaos literature you should know them anyway. Here, then, are the celebrated *Lorenz equations* that triggered chaos research back in 1961:

$$
\begin{aligned}
\dot{x} &= \sigma(y - x), \\
\dot{y} &= \rho x - y - xz, \\
\dot{z} &= xy - \beta z.
\end{aligned}
\tag{6.45}
$$

A favorite set of parameters for experts in the field is $\sigma = 10$, $\beta = 8/3$ and $\rho > 25$, for which there is a saddle at the origin and two unstable nodes. Motion then takes place on the chaotic *Lorenz attractor*, a beautiful creature which is pictured in many textbooks on chaos (e.g., Gleick 1987; Schuster 1989).

Note that the set of Lorenz equations is of the minimal dimension required for chaos to appear. Chaos requires $D \geq 3$, corresponding to at least three autonomous or two non-autonomous first-order ODEs.

6.6.2 Duffing-Type Systems (D = 3)

A basic form of a Duffing system with external forcing is:

$$
\ddot{x} + \beta\dot{x} + \omega^2 x + \gamma x^3 = p\cos(\Omega t).
\tag{6.46}
$$

Nondimensional forms are often encountered, such as (6.1). The parameter ω^2 can be negative, e.g. when buckling is involved, and the sign of γ describes whether the nonlinear restoring force is softening ($\gamma < 0$) or hardening ($\gamma > 0$). As should be clear from numerous examples in Chaps. 3 and 5, the basic Duffing system can display all kinds of intricate dynamic behavior. In particular when $\omega^2 < 0$ chaos may prevail for wide ranges of system parameters. Numerical simulations of the chaotic behavior have been verified experimentally, e.g., with the magnetically buckled Moon's beam.

Since we have dealt with it thoroughly no further examples seem necessary to clarify the immense importance of the basic Duffing system. One reason for (6.46) to occur so frequently in solid mechanics is that many structures possess symmetrical restoring forces, are only slightly deformed, and are periodically loaded. Taylor-expanding the equations of motion for such systems (perhaps after reduction to a single-DOF system), the quadratic term in the expansion will cancel due to symmetry, and the first nonlinear term will be cubic.

Variants of the basic Duffing system exist. For example, a beam subjected to transverse as well as axial loading obeys a Duffing-like equation with combined *parametric* and *external* excitation (e.g., Yagasaki et al. 1990):

$$\ddot{x} + \beta\dot{x} + \left(\omega^2 + q\cos(\Omega_2 t)\right)x + \gamma x^3 = p\cos(\Omega_1 t). \qquad (6.47)$$

For a slightly curved and axially loaded beam, an *asymmetric* (quadratic) term sneaks in along with parametric excitation (Szemplinska-Stupnicka et al. 1989):

$$\ddot{x} + \beta\dot{x} + \left(\omega^2 + q\cos(\Omega t)\right)x + \gamma_1 x^2 + \gamma_2 x^3 = 0. \qquad (6.48)$$

A similar quadratic term appears when Taylor-expanding solutions of beam, plate and shell problems about *buckled* states of static equilibrium. For example, if $\omega^2\gamma < 0$ then (6.46) has a nontrivial static equilibrium at $\tilde{x} = \sqrt{-\omega^2/\gamma}$, which is stable when $\omega^2 < 0$. This equilibrium corresponds to a buckled state when (6.46) governs transverse vibrations of an axially compressed beam. For describing small but finite vibrations near the buckled equilibrium one employs a shift of the dependent variable, $y(t) = x(t) - \tilde{x}$, and this causes a quadratic term to appear in the equation for y.

For a mass moving on a belt conveyor at speed v there could be *nonlinear energy dissipation*, in the form of velocity terms expressing damping and friction (Narayanan and Jayaraman 1991):

$$\ddot{x} + \beta_1(\dot{x} - v)^3 - \beta_1(\dot{x} - v) + \mu\mathrm{sgn}(\dot{x} - v) + \omega^2 x + \gamma x^3 = p\cos\Omega t. \qquad (6.49)$$

The above variants all display chaotic dynamics.

Midplane stretching, large rotations and numerous other features of solid structures may cause Duffing-type behavior. Even with otherwise linear structures, Duffing systems may arise whenever *feedback* control is added. For example, if the transverse stiffness of a linear structure is to be controlled dynamically by axial forcing, i.e. by tension-control, then ordinary quadratic control causes the *controlled* system to be of the Duffing-type.

6.6.3 Pendulum-Type Systems (D = 3)

Based on the equation of motion for the damped and unforced pendulum:

$$\ddot{\theta} + \beta\dot{\theta} + \omega_0^2\sin\theta = 0, \qquad (6.50)$$

a variety of pendulum-type systems exist which differ in the excitation term. Most of them display chaotic dynamics.

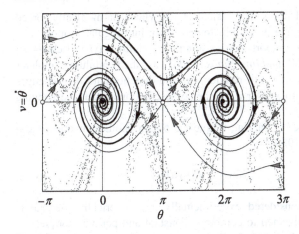

Fig. 6.20 Poincaré map for the parametrically excited pendulum system (6.51), overlaid by phase plane orbits for the corresponding unforced problem. The Poincaré points displayed correspond to 1750 periods of the driving force. ($\omega_0 = 1.0$, $\Omega = 2.2$, $q = 3.3$, $\beta = 0.1$)

With *parametric* excitation the pendulum equation may take the form studied in Chap. 3:

$$\ddot{\theta} + 2\beta\omega_0\dot{\theta} + \left(\omega_0^2 - q\Omega^2 \cos(\Omega t)\right)\sin\theta = 0, \qquad (6.51)$$

which models a pendulum with a periodically displaced support (Fig. 3.5).

In Chap. 3 a local perturbation analysis for (6.51) provided the frequency response plot shown in Fig. 3.10. Checking the analytical results, by numerical simulation or laboratory experiments, you will find fair agreement within the interval of primary parametric resonance ($\Omega \approx 2\omega_0$) *only* when the amplitude of rotation is not too large. For example, if in a laboratory experiment the excitation frequency Ω is swept through the range of Fig. 3.10, the observed response will follow the theoretical curve until the amplitude comes close to $\pi/2$. The motion then turns strongly chaotic[6].

The experimental chaos is confirmed by numerical simulations, for which Fig. 6.20 shows a Poincaré map. A ghost of a strange attractor is seen behind the analytical phase plane orbits for the corresponding unforced problem.

Leven and Koch (1981) seem to be the first considering chaotic dynamics for (6.51) (see also Leven et al. 1985). Moon (1987) reports that period-doublings have been observed in numerical solutions, and that a Feigenbaum number of $\delta = 4.74$ was calculated for the sixth subharmonic bifurcation.

---

[6]Thus one is reminded that the perturbation solution is only *locally* stable. Though still a stable solution, for large θ it is outperformed by a strange attractor which is globally stable.

The author has observed hours of transient chaos with a harmonically forced laboratory pendulum, before the pendulum settled down into regular motion.

Bishop and Clifford (1996) report on a seemingly intermittent type of chaos for the pendulum, which they call *tumbling chaos*. In this case the pendulum completes an apparently random number of full rotations in one direction and then changes direction of rotation, sometimes after a number of oscillations about the hanging position.

With an external periodically varying torque we obtain another chaotic pendulum system (e.g., Schuster 1989):

$$\ddot{\theta} + \beta\dot{\theta} + \omega_0^2 \sin\theta = Q\cos(\Omega t). \tag{6.52}$$

Blackburn et al. (1987) considered, experimentally and numerically, the chaotic dynamics of a pendulum subjected to combined constant and periodic torque:

$$\ddot{\theta} + \beta\dot{\theta} + \omega_0^2 \sin\theta = Q_0 + Q_1 \cos(\Omega t). \tag{6.53}$$

Finally we mention a chaotic pendulum system studied experimentally and numerically by Moon, Cusumano and Holmes (1987):

$$\ddot{\theta} + \beta\dot{\theta} + \omega_0^2 \sin\theta + q_1 \cos(\Omega t)\cos\theta = 0, \tag{6.54}$$

for which a Melnikov criterion was provided. This parametric equation models the rotation of a magnetic dipole in a time-periodic magnetic field.

6.6.4   Piecewise Linear Systems (D $\geq$ 3)

Piecewise linear systems behave linearly within certain intervals of the state variables. Many mechanical systems having free play or stops, i.e. *impact systems* belong to this class (Babitsky 1998; Burton 1968; Kobrinskii 1969). So do systems with bi-linear or multilinear elastic restoring forces, see, e.g., Shaw and Holmes (1983). Fig. 6.21 shows some examples.

The beam in Fig. 6.21(a) is similar to the Moon beam, though with the magnets replaced by a rigid stop. The chaotic dynamics of this system have been examined numerically and experimentally in a number of studies, in particular by Moon and Shaw (e.g., 1983). The partial differential equations of motion are linear with linear boundary conditions at the clamped end, $w(0,t) = w'(0,t) = 0$. However, at $x = l$ the beam is considered to be free $(w''(l,t) = w'''(l,t) = 0)$ when $w(l, t) < \Delta$, while simply supported $(w; (l,t) = w''(l,t) = 0)$ when $w(l, t) = \Delta$ and the beam moves to the right $(\dot{w}(l,t) > 0)$. Thus, the beam performs linear vibrations between the times of impacting with the stop. A single-mode approximation for the equation of motion takes the nondimensional form (Moon and Shaw 1983):

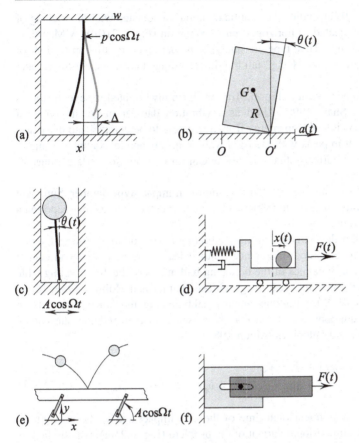

Fig. 6.21 Piecewise linear systems. (a) Stopped beam; (b) rocking object; (c) pendulum-type vibration damper; (d) another impact damper; (e) bouncing ball; (f) joint with play

$$\ddot{u} + \beta \dot{u} + \begin{pmatrix} 1 \\ c_1 \end{pmatrix} u = \begin{pmatrix} 1 \\ c_2 \end{pmatrix} q \cos(\Omega t) + \begin{pmatrix} 0 \\ c_3 \frac{\Delta}{c_4} \end{pmatrix}, \qquad (6.55)$$

where u is the modal amplitude and $c_{1,2,3,4}$ are constants. The upper values are used for $u < u_0 = c_4 \Delta$ and the lower ones for $u > u_0$. Note that the elastic restoring term corresponds to the action of the bi-linear spring in Fig. 3.4a. The system must be solved numerically, either directly or through iteration of a *map*. A map can be constructed because the system behaves as a linear single-DOF oscillator between the times of impacts. At the k'th impact one shifts the equation of motion, e.g., by using lower instead of upper values in (6.55), and solves the new equation using the current impact state $(u(t_k), \dot{u}(t_k))$ as the initial condition. The solution so obtained is valid until the next impact occurs at time t_{k+1}. Hence one can write down a *difference equation* that maps $(u(t_k), \dot{u}(t_k))$ onto $(u(t_{k+1}), \dot{u}(t_{k+1}))$. The dynamic behavior of the system is then traced by iterating the map.

Yim and Lin (1991) studied the chaotic dynamics associated with the rocking of slender objects subjected to horizontal base excitation (Fig. 6.21(b)). A Melnikov criterion was set up. This type of analysis is pertinent to the protection of earthquake-excited structures, e.g., tall buildings, storage tanks, ancient towers and nuclear reactors.

The impacting pendulum in Fig. 6.21(c) has been investigated in several studies (e.g., Moore and Shaw 1990). Used as a vibration damper in certain types of rotating machinery, it soaks up energy from a structure to be damped and dissipates the energy through impacts with the rigid wall. With the above examples in mind, you might presume correctly that this device can turn chaotic for certain values of the system parameters.

Fig. 6.21(d) shows another implementation of an impact-type damper. Sung and Yu (1992) examined the system numerically. Among other findings they reported a period-doubling route to chaos.

Putting a small roller-ball on a strongly vibrating table, you may observe the ball performing a strange random-like dance (until it leaves the plate and you cannot find it). The chaotic dynamics of the bouncing ball at a periodically vibrating table have been studied for many years (e.g., Guckenheimer and Holmes 1983; Moon 1987; Tufillaro et al. 1992). For one-dimensional bouncing, the ball moving only up and down, the equations of motion may be posed as a map relating subsequent post-impact times and velocities (Moon 1987):

$$
\begin{aligned}
T_{k+1} &= T_k + v_k, \\
v_{k+1} &= \alpha v_k - \gamma \cos(T_k + v_k), \quad k = 0, 1, \ldots .
\end{aligned}
\tag{6.56}
$$

Here T_k is the nondimensional time of the k'th impact and v_k the post-impact velocity. The two-dimensional variant of the problem (Fig. 6.21(e)) was studied by Kozol and Brach (1991).

As a final example we consider the simple device in Fig. 6.21(f). Two rigid bars, the one excited by a time-varying force, are joined by a smooth pin with *play*. This system was studied by Li et al. (1990), motivated by experimental observations of chaotic responses for a pinned truss structure. Assuming inelastic collisions and time-harmonic forcing, the motion of the pin $x(t)$ is governed by the nondimensional equation of motion:

$$
\begin{aligned}
\ddot{x} &= A \sin t \quad \text{for} |x| < 1, \\
\dot{x} &\rightarrow -r\dot{x} \quad \text{for} |x| = 1,
\end{aligned}
\tag{6.57}
$$

where A is the forcing amplitude and r the coefficient of restitution. Period-doublings and chaos were reported for large values of A. This system is called a *zero-stiffness impact oscillator*, since there is no elastic restoring term.

6.6.5 Coupled Autonomous Systems (D ≥ 4)

Coupled autonomous systems are governed by two or more *second order autonomous* ODEs, corresponding to four or more first-order ODEs. Most systems with follower-type loading belong to this class. With follower systems we often encounter the terms *divergence* and *flutter*, which relate to destabilizations of equilibriums by pitchfork and Hopf bifurcations, respectively. Since follower-loaded systems are *truly autonomous*, with no explicit time dependence even for the underlying second-order ODEs, there is no natural time period to choose for Poincaré maps. Instead one can choose a hyperplane in phase space and plot the points of intersections of orbits with that plane (cf. Sect. 6.3.3).

The Follower-loaded Double Pendulum This system, shown in Fig. 4.11, may serve to illustrate some fundamental properties of coupled autonomous systems. The equations of motion are:

$$
\begin{aligned}
(1+m)\ddot{\theta}_1 &+ \cos(\theta_2 - \theta_1)\ddot{\theta}_2 + (c_1 + c_2)\dot{\theta}_1 \\
&- c_2\dot{\theta}_2 + 2\theta_1 - \theta_2 + \dot{\theta}_2^2 \sin(\theta_1 - \theta_2) \\
&= p((1-\alpha)\sin\theta_1 + \alpha\sin(\theta_1 - \theta_2)), \\
\cos(\theta_2 - \theta_1)\ddot{\theta}_1 &+ \ddot{\theta}_2 + c_2(\dot{\theta}_2 - \dot{\theta}_1) \\
&- \theta_1 + \theta_2 - \dot{\theta}_1^2 \sin(\theta_1 - \theta_2) \\
&= p(1-\alpha)\sin\theta_2,
\end{aligned}
\tag{6.58}
$$

where θ_1 and θ_2 describe motions of the two pendulum arms, m is a mass ratio, $c_{1,2}$ the damping parameters, p the load magnitude and α the 'conservativeness'-parameter ($\alpha = 0$: conservative load; $\alpha = 1$: perfectly following load). In Chap. 4 we Taylor-expanded the system to order three and performed a *local* perturbation analysis near the Hopf bifurcation set of the linear stability diagram (Fig. 4.12). The perturbation results agreed with numerical simulations, at least for *small* values of p (Fig. 4.13). To study *global* bifurcations, perhaps leading to chaos, a numerical simulation of the un-approximated system (6.44) is required.

Fig. 6.22 shows some typical responses of the system as obtained by numerical integration of (6.44). The periodic phase plane orbit in Fig. 6.22(a) agrees approximately with the local perturbation solution.

The center of the periodic orbit in Fig. 6.22(b) is offset from zero, reflecting that the lower pendulum arm oscillates about a nontrivial equilibrium $\theta_1 \approx 2$ radians $\approx$ 115°. This is a result of a *global* bifurcation of the equilibrium into a large-amplitude limit cycle. It was not predicted by the perturbation analysis, which explicitly assumes small but finite values of $\theta_{1,2}$.

The response in Fig. 6.22(c) appears chaotic, with large-amplitude orbits filling the phase plane and a broadband frequency spectrum. The corresponding Poincaré section (not shown) and Lyapunov spectrum with $\hat{\lambda}_1 \approx 0.67$ confirms this response

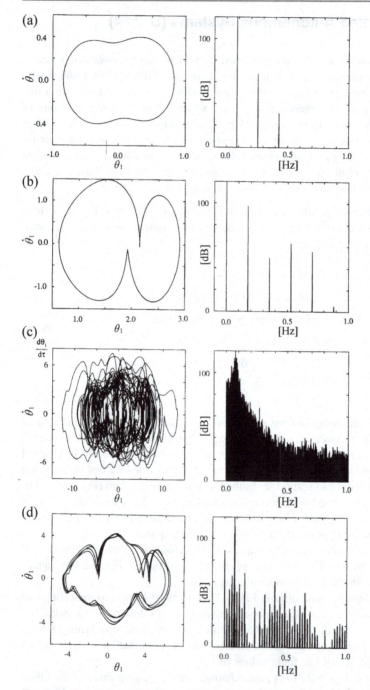

Fig. 6.22 Dynamics of the follower-loaded elastic double pendulum when $m = 2$ and $c = 0.1$. Post-transient phase plane orbits (left) and power spectra (right) for (**a**) $(\alpha, p) = (0.9, 2.0)$; (**b**) $(\alpha, p) = (0.9, 8.0)$; (**c**) $(\alpha, p) = (2.1, 5.0)$; (**d**) $(\alpha, p) = (2.1, 3.0)$. (Thomsen 1995)

to be chaotic. For this response the value of (α, p) is in a range where, according to perturbation analysis, the zero solution is destabilized both by linear and by non-linear terms. This causes an explosive rise in amplitude, allowing the orbit to explore a much larger region of phase space. In this larger region one can show that the dynamic behavior of the double pendulum is affected by the action of no less than twenty-seven states of equilibrium with mixed kinds of stability. Adding the like existence of limit cycles, one is not surprised that the flow of orbits displays extreme sensitivity to initial conditions.

In Fig. 6.22(d) the value of (α, p) corresponds to a situation where the zero-solution is stabilized by linear terms, but *de*stabilized by nonlinear terms. That is, the system is on the dangerous, some may say 'exciting', side of a subcritical pitchfork bifurcation for the oscillation amplitudes. Recall that in this case a stable zero solution can be destabilized by a strong disturbance. This is confirmed by the figure, which was computed for initial conditions corresponding to a strong velocity impulse. The large-amplitude periodic orbit in the figure was approached only after a very long chaotic transient (8,000 nondimensional seconds).

Figure 6.23 illustrates the different types of stationary motions of the pendulum encountered when varying the loading parameters (α, p). The analytical stability diagram obtained in Chap. 4 (Fig. 4.12) has been superimposed. The figure was obtained by computing the largest Lyapunov exponent $\hat{\lambda}_1$ on a 512×330 grid spanning the (α,p)-range, and then coloring the grid-points according to the value of $\hat{\lambda}_1$. Periodic motion (gray, $\hat{\lambda}_1 \approx 0$) prevails in regions where, according to pertur-bation analysis, the only solution is a stable limit cycle. This region is defined by $\alpha_a < \alpha < \alpha_b$ and p above the curve $D_3 = 0$ of supercritical Hopf bifurcations. Chaotic motion (black, $\hat{\lambda}_1 > 0$) prevails in regions where, according to perturbation analysis, subcritical pitchfork bifurcations have occurred, that is, where $\alpha > \alpha_b$ and p is above the upper branch of $H_4 = 0$. On these observations one may suggest a simple analytical and seemingly *sufficient* criterion for chaos to occur, though it does not hold for the periodic window (gray band) inside the main region of chaos.

Chaos-like phenomena for the follower-loaded double pendulum with *step-loading* have been examined in several studies by Kounadis (e.g., 1991 and references cited herein). Other variants of the double pendulum are likely to behave chaotically as well, e.g., the double pendulum with eccentric load or load-dependent stiffness treated in Guran and Plaut (1993).

Fluid-conveying Pipes A large number of studies consider the dynamics of *fluid-conveying pipes* (Fig. 6.24). M. P. Païdoussis, in particular, has contributed to this area (for chaos-related studies see, e.g., Païdoussis and Semler 1993; Païdoussis et al. 1989). Physically, a fluid-conveying pipe is a far more complicated system than the double pendulum discussed above. However, applying mode expansion for the PDE of the system in Fig. 6.24, one can reduce the dynamics of the system to the following set of approximating ODEs:

$$\ddot{q}_i + c_{ij}\dot{q}_j + k_{ij}q_j$$

$$= -\varepsilon \sum_{j,k,l=1}^{N} \left(\alpha_{ijkl}q_jq_kq_l + \beta_{ijkl}q_jq_k\dot{q}_l + \gamma_{ijkl}q_j\dot{q}_k\dot{q}_l \right), \quad i = 1, N, \quad (6.59)$$

where $q_i(t)$, $i = 1, N$ are the modal amplitudes, α,β,γ are constants and ε is a small parameter. Written as a set of $2N$ first-order equations, these turn out to be similar to (4.52)–(4.53), approximating motions of the follower-loaded double pendulum.

Experience seems to show that the fluid-conveying pipe, with physically realistic parameters, does not behave chaotically in its *fundamental* form, as in Fig. 6.24 but without the intermediate spring-support). With the spring-support chaos shows up (Païdoussis and Semler 1993). Also, chaos appears when an intermediate support with play is applied (Païdoussis et al. 1989; Païdoussis et al. 1991; Makrides and Edelstein 1992).

Langthjem (1995a) observed chaotic-like transients with a finite element model of a pipe in its fundamental form, but with system parameters chosen so as to maximize the lowest flutter-load. Langthjem shows how the optimizing system parameters bring the system close to a codimension-2 bifurcation point (corresponding to a double Hopf bifurcation); This might trigger complicated dynamics, including chaos.

Panel Flutter. Dowell (1982), in an early study of chaotic autonomous systems, considered airflow passing along the surface of a buckled elastic plate (Fig. 6.25). This aero-elastic problem involves *panel flutter*. Panel flutter occurred on the outer skin of the early Saturn rocket boosters, putting man on the moon in the early 1970s. Mathematically the problem resembles (6.59), implying that for some values of system parameters the flutter may turn chaotic.

Though without particular relevance to chaos, the early paper by Holmes (1977) provides a splendid mathematical background for the study of coupled autonomous systems.

6.6.6 Autoparametric Systems (D $\geq$ 5)

Recall from Chap. 4, that autoparametric phenomena can occur for nonlinearly coupled systems having near-integer relationships between two or more of the linearized natural frequencies. The integer-relationship causing *internal resonance* depends on the order of the nonlinearities present. When combined with *external resonance*, as we have seen, internal resonance may cause large-amplitude motions and nonlinear interaction of modes, even for weak forcing. Systems subjected to combined internal and external resonance may rather easily turn chaotic. We reconsider below two examples from Chap. 4, focusing here on chaotic motion.

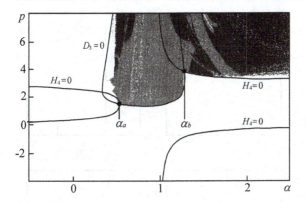

Fig. 6.23 Dynamics of the follower-loaded double pendulum with $m = 2$ and $c = 0.1$; regions of similar type of motion in the plane of loading parameters (α, p). Black: chaotic response $(\hat{\lambda}_1 > 0)$; gray: periodic response $(\hat{\lambda}_1 \approx 0)$; White: static response $(\hat{\lambda}_1 < 0)$. (Thomsen 1995)

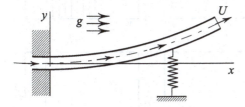

Fig. 6.24 Fluid-conveying pipe, clamped and spring-supported

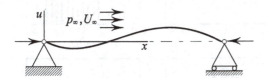

Fig. 6.25 The fluttering panel (Dowell 1982)

The Non-shallow Arch This inherently autoparametric system, shown in Fig. 4.6, obeys coupled equations of motion having quadratic nonlinearities:

$$\ddot{f} + 2\beta\dot{f} + (1 - m\omega^2 u)f = 0,$$
$$\ddot{u} + 2\beta\omega\dot{u} + \omega^2 u + \kappa(f\ddot{f} + \dot{f}^2) = \frac{q}{m}\cos(\Omega\tau), \quad (6.60)$$

where the variables f and u describe the amplitudes of antisymmetric and symmetric vibrations, respectively. In Chap. 4 a local perturbation analysis was performed, predicting large amplitudes of the autoparametrically excited antisymmetric mode when $\Omega \approx \omega$.

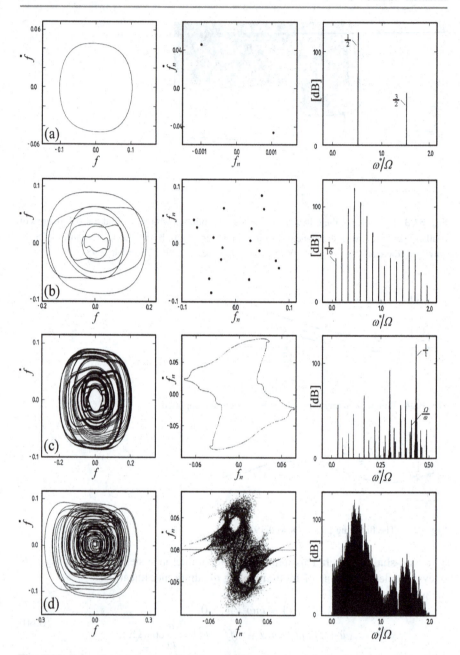

Fig. 6.26 Dynamics of the non-shallow arch. Post-transient phase plane trajectories (left), poincaré sections (middle) and power spectra (right). (**a**) $\Omega = 2.25$: period-2, $\hat{\lambda}^2 = -0.007$; (**b**) $\Omega = 2.295$: period-16, $\hat{\lambda}_1 = -0.03$; (**c**) $\Omega = 2.318$: quasi-periodic, $\hat{\lambda}_1 \approx 0$; (**d**) $\Omega = 2.4$: chaotic, $\hat{\lambda}_1 = -0.026$; (for all plots $q = 0.2$, $\omega = 2.44$, $\beta = 0.03$, $m = 3.32$ and $\kappa = 2.63$) (Thomsen 1992)

Fig. 6.26 displays typical solutions as obtained by numerical integration of (6.60). Phase planes, Poincaré maps and frequency spectra for the antisymmetric amplitude $f(t)$ are shown for four values of the excitation frequency Ω. Note how quasiperiodic motion (Fig. 6.26(c) resembles chaotic motion (Fig. 6.26(d)) in the *phase* plane, whereas in the Poincaré map they are easily distinguished. The routes to chaos for this system, as was discussed in Sect. 6.4, involve transient chaos, period-doublings, quasiperiodic oscillations, and intermittency.

Fig. 6.27(a) depicts regions of chaotic motions in the plane of the loading parameters Ω and q, as obtained by evaluating the largest Lyapunov exponents on a 100×100 grid. Grid-points corresponding to chaotic motion $(\hat{\lambda}_1 > 0)$ are shown in dark gray, whereas quasiperiodic motion $(\hat{\lambda}_1 \approx 0)$ is shown in lighter gray. A nonlinear stability diagram for the zero solution (Fig. 4.10, obtained by local perturbation analysis) has been superimposed. Chaos is seen to prevail in the knife-shaped region C. In this region, according to the perturbation analysis of Sect. 4.3.3, all small-amplitude solutions are unstable. Since a simple analytical expression exists for the boundary curve of region C we here have, seemingly, an approximate *sufficient* condition for chaos to occur.

Fig. 6.27(b) shows the *magnitudes* of the largest Lyapunov exponents on a grid similar to that of Fig. 6.27(a). To expound the numbers, we note that if $\hat{\lambda}_1 = 0.1$ and initial conditions are specified to within one part per million (20 bits), then after $20/0.1 = 200$ nondimensional seconds of time all information about the state of the system is lost. This period of time corresponds to merely 48–89 forcing periods when Ω is in the range 1.5–2.8. For chaotic motions, it appears from the figure, higher magnitudes of the loading q cause $\hat{\lambda}_1$ to increase, and thus lowers the time-horizon of predictability.

The Autoparametric Vibration Absorber This system, shown in Fig. 4.1, also obeys coupled equations having quadratic nonlinearities:

$$\ddot{x} + 2\beta_1\dot{x} + \omega_1^2 x + \gamma_1\theta^2 = q\cos(\Omega t),$$
$$\ddot{\theta} + 2\beta_2\dot{\theta} + \omega_2^2\theta + \gamma_2 x\theta = 0. \tag{6.61}$$

The vibration absorber is especially designed so as to utilize internal two-to-one resonance, and is thus inherently autoparametric. The equations of motion are similar in structure to (6.60) for the non-shallow arch. As we saw in Chap. 4 a local perturbation analysis yields similar results for the two systems (compare, e.g., Fig. 4.3 for the absorber to Fig. 4.8 for the arch). Hence, what has been said above on chaotic motions of the non-shallow arch is likely to hold as well for the autoparametric vibration absorber.

Other Autoparametric Systems Among the large number of nonlinear studies by Prof. Ali H. Nayfeh and co-workers at Virginia Polytechnic Institute and State University, many treat autoparametric systems, and some of these consider chaos. A system consisting of two orthogonally clamped beams with a two-to-one internal

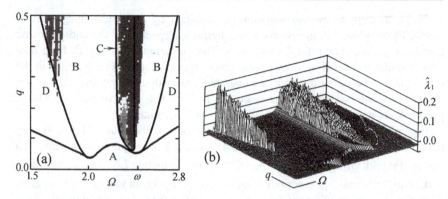

Fig. 6.27 The non-shallow arch. (**a**) Types of motion as a function of loading parameters (Ω, q): Chaos (dark gray); quasiperiodic motion (lighter gray); equilibrium or periodic motion (white). (**b**) largest Lyapunov exponent $\hat{\lambda}_1$. ($\omega = 2.44$, $\beta = 0.03$, $m = 3.32$, $\kappa = 2.63$) (Thomsen 1992)

resonance have been studied analytically and experimentally in a number of papers (e.g., Nayfeh 1989; Nayfeh and Balachandran 1989; Nayfeh et al. 1989; Balachandran and Nayfeh 1991). Chaotic vibrations of cylindrical shells, also possessing internal resonance, were described in Nayfeh and Raouf (1987). The rolling of ships in regular sea waves (Nayfeh and Khdeir 1986) is governed by coupled equations having quadratic nonlinearities, and thus chaos should be considered a possibility (though this would require an unlucky combination of a badly designed ship and resonant sea waves); see also Hatwal et al. (1983a, b).

6.6.7   High-Order Systems (D > 5)

There seems to be few studies dealing with chaotic vibrations of structures having many degrees of freedom. As an example, Moon and Li (1990) performed a numerical and experimental study of a pin-jointed truss structure with 153 degrees of freedom (Fig. 6.28). Broadband chaotic-like vibrations were observed, and conjectured to be associated with free play in the connecting joints (see also Li et al. 1990).

6.6.8   Other Systems

We here mention a few more cases of chaos for mechanical systems. See also Moon (1987), and the interview with Moon published in Goldstein (1990).

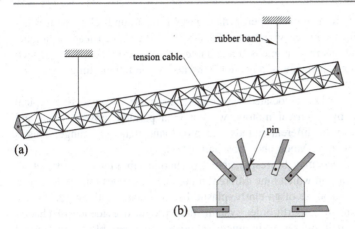

Fig. 6.28 Experimental truss structure investigated by Moon and Li (1990). (**a**) Truss supported by rubber bands; (**b**) nonlinear pin joint

As appears from Sects. 6.6.4 and 6.6.7, structures having *flexible joints* may perform chaotically, as may be the case when the joints are subjected to dry *friction* (e.g. Feeny et al. 1998; Feeny and Moon 2000; Ibrahim 1992a, b; Popp 1992). This may be relevant, e.g., for large space structures that are only lightly damped. For example, the large truss-structures planned to be put into space will contain test modules, inside which vibrations should be kept to a minimum. Since no rocket will carry a truss having a length comparable to several football grounds, these structures will have to be taken up in parts and assembled on location, probably with flexible joints. If welding was used the structure would essentially be linear; Vibrations would be predictable, and therefore controllable. Controlling chaotic structures is more difficult.

Pick-and-place robots typically have to be rather heavy, in order to minimize vibrations when their path includes sudden changes of motion. A *fast, lightweight robot* may display all kinds of unwanted vibrations, including chaos.

Impact printers may turn chaotic when driven at high speeds. Just beyond a critical speed the printer starts skipping single letters. At higher speeds it may print letters at random (Hendriks 1983).

If *dynamic control* is added to a system an unpleasant side effect may be chaotic vibrations of the *controlled* system, if the controller is not properly tuned. Keep in mind that even a linear structure controlled by a linear controller may combine into a *nonlinear* system. Then complicated motions may arise, including chaos. (An example: applying linear tension-control to a linear beam system causes the controlled system to be influenced by quadratic nonlinearities).

Vibration induced sliding can be accompanied by chaotic vibrations. For example, chaos was observed for the string with a sliding point mass dealt with in Sect. 4.7 (Thomsen 1996a).

Flow-induced vibrations may turn chaotic, as we have already seen. For example, in nuclear reactors there are fuel rods that are cooled by surrounding fluids. Also, there are heat exchangers where gases flow around tubes, with water flowing inside. Flow-induced vibrations of tubes or fuel rods can lead to fatigue. For *periodic* vibrations this fatigue is 'predictable', so that services can be planned in time. The author is unaware of any studies related to the fatigue-time associated with *chaotic* vibrations.

Machine noise is often associated with chaotic processes. Considering typical machinery driven by electrical motors, we put in a periodic power signal and observe a symphony of different sounds in a broadband frequency range. This is deterministic chaos (e.g., Moon and Broschart 1991).

Plasticity and *creep* may be responsible for chaotic vibrations. Poddar et al. (1988) describe a case in which nine different finite-element codes totally disagreed upon the transient response of an elastic-plastic beam. Reducing the model of the elastic-plastic beam to a fourth-order system, they provided evidence of chaotic vibrations. Thus, small differences in numerical procedures were blown up to yield highly different results after a number of time steps.

6.7 Elastostatical Chaos

Chaos is usually associated with *dynamics*, that is, with *initial value problems* defined on the infinite domain of a time variable. *Boundary value problems*, on the other hand, seem to preclude chaos, because boundaries are defined on the finite domain of a physical object.

Nevertheless, it has been demonstrated that certain complicated phenomena encountered in elastostatics become explainable when considered as *asymptotically* chaotic processes (El Naschie and Al Athel 1989; El Naschie 1990a, b). This is pertinent, e.g., to localized buckling of shells and the formation of soliton-like homoclinic loops at seemingly random locations along very long steel tapes. The term *elastostatical chaos* was introduced to describe the phenomenon.

As an example we consider the differential equation governing static deformations of a simply supported *Euler elastica* having length l and bending stiffness EI:

$$\varphi'' + \lambda^2 \sin \varphi = 0; \quad \lambda^2 \equiv \frac{P}{EI}, \varphi'(0) = \varphi'(l) = 0, \tag{6.62}$$

where $\varphi = \varphi(s)$ is the angle of cross-sectional rotation along an arch-measuring variable s, P is the compressive axial force and $\varphi' = d\varphi/ds$. Suppose the elastica possesses an initial imperfection in form of a sinusoidal crookedness $\eta(s) = a\sin(\omega s)$. Then, replacing φ by $\varphi + \eta$, (6.62) becomes:

$$(\varphi + a\sin(\omega s))'' + \lambda^2 \sin(\varphi + a\sin(\omega s)) = 0, \tag{6.63}$$

or, when for a small crookedness one can assume $a \ll 1$, to first order:

$$\varphi'' + \lambda^2 \sin \varphi = a\omega^2 \sin(\omega s),$$

$$\varphi'(0) = \varphi'(l) = 0. \tag{6.64}$$

Introducing a nondimensional length-measuring parameter $\tau = \omega s$, the equation transforms into

$$\ddot{\varphi} + \tilde{\lambda}^2 \sin \varphi = a \sin \tau, \quad \varphi = \varphi(\tau), \dot{\varphi}(0) = \dot{\varphi}(\omega l) = 0, \tag{6.65}$$

where $\dot{\varphi} = d\varphi/d\tau$ and $\hat{\lambda} = \lambda/\omega$. Considering τ as a time-like variable, then (6.65) is identical to the dynamic equation of motion for an undamped pendulum subjected to an external periodic torque – a system that is known to be chaotic in *time*. However, whereas the pendulum equation is part of an *initial value problem* defined on $\tau \in [\tau_0; \infty[$, the elastica Eq. (6.65) is part of a *boundary value problem* defined on $\tau \in [0; \omega l]$.

Still, when $\omega l \to \infty$ the boundary value problem approaches an initial value problem. Asymptotically, for structures extending semi-infinitely into space, there is mathematically no difference between the two problems. Hence, certain solutions of the elastica problem will change chaotically with the space coordinate, just as certain solutions of the corresponding initial value problem change chaotically in time.

Fig. 6.29 shows some possible configurations of the Euler elastica. The configurations was computed by solving (6.65) as an initial value problem with $\varphi(0) = 3.1 \approx \pi$, $\dot{\varphi}(0) = 0$, $\omega l = 200$ and $a = 0.01$. Having obtained a solution $\varphi(\tau)$ for $0 \leq \tau \leq \omega l$, the configuration of the elastica can be calculated by applying the geometrical relations $d\tau(\cos\varphi) = dx$ and $d\tau(\sin\varphi) = dy$ (see Fig. 6.29(a)), that is:

$$x(\tau) = \int_0^\tau \cos \varphi(\xi)\, d\xi, \quad y(\tau) = \int_0^\tau \sin \varphi(\xi)\, d\xi. \tag{6.66}$$

For all cases shown in the figure the nondimensional load $\tilde{\lambda}^2$ was sufficiently large that the sliding support could meet, and subsequently pass through, the fixed support (we ignore for the moment the physical impossibility of this).

Fig. 6.29(b), (c), (d) show the classical regular loops obtained already by Euler (1744). They are called *Euler loops*. However, in Fig. 6.29(e) loops appear seemingly at random. Increasing the nondimensional length ωl of the elastica, more and more loops will pop up, though one cannot predict where.

The Poincaré map shown in Fig. 6.30 was computed for a very long elastica ($\omega l = 50,000$), to check whether (6.65) performs chaotically for the parameters of Fig. 6.29(e). Indeed, the fractal appearance of the map indicates chaos, at least for the *initial value* problem.

The critical reader might doom this example a purely mathematical artifact. Very large deformations are involved, and magic boundary supports appear, which can pass through each other in space. However, this is a consequence of being confined to *plane* deformations. For a number of three-dimensional cases El Naschie (1990a,b)

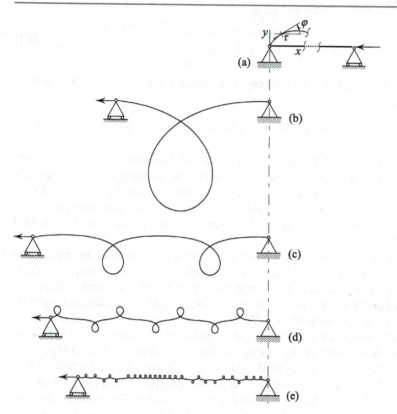

Fig. 6.29 Euler loops of a buckled elastica with a periodical imperfection. The figures show solutions to (6.65) for $a = 0.01$, $\omega l = 200$, $\varphi(0) = 3.1$, and $\dot{\varphi}(0) = 0$. (**a**) Undeformed elastica, definition of variables; (**b**) $\hat{\lambda}^2 = 0.001$; (**c**) $\hat{\lambda}^2 = 0.01$; (**d**) $\hat{\lambda}^2 = 0.1$; (**e**) $\hat{\lambda}^2 = 1$. In (**e**) loops appear seemingly at random (elastostatical chaos)

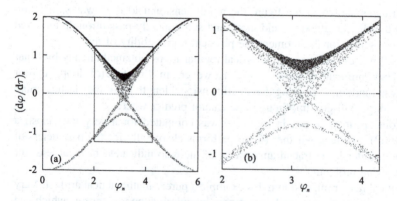

Fig. 6.30 (**a**) Poincaré map for (6.65). (**b**) Enlargement of the boxed region in (a). (Parameters as for Fig. 6.29(e), except that $\omega l = 50,000$)

provided strong experimental and numerical evidence of elastostatical chaos. For example, solving the 3D-elastica problem of a very long twisted steel tape as an *initial value* problem, the numerical solution qualitatively agrees with the random soliton loops actually observed in experiments.

An experiment you can try setting up for yourself requires only a long flat rubber band. Stretch it, twist it, and release it very slowly. Loops may form at random locations along the band, in agreement with the chaotic solutions of (6.65) which holds for this case too. Considering *only* the experimental observations, you might suspect the loops to occur at locations having some kind of material or geometrical imperfection. However, as a consequence of the statical-dynamical analogy, loops will form at random even if the structure was ideal with no imperfections at all.

6.8 Spatial and Spatiotemporal Chaos

For some systems the spatial variables may describe disordered patterns, fixed in time or slowly varying. The hallmark of this phenomenon is a rapid decay of spatial correlations, so that one cannot predict the formation of more distant patterns from local observations. This is called *spatial chaos*, to distinguish it from *temporal* chaos associated with disordered behavior in time.

Elastostatical chaos (Sect. 6.7) is one example of spatial chaos. Another type is related to granular materials or fluids. For example, spatial chaos has been observed in the form of disordered patterns of standing waves on the surface of a fluid enclosed in a vertically oscillating container (e.g., Tufillaro et al. 1989).

Certain coupled dynamical systems may display spatial as well as temporal disorder. Then disordered patterns move around unpredictably in time. This is called *spatiotemporal chaos*. See Cross and Hohenberg (1994) for an introduction to this new and rather unsettled area of research.

6.9 Controlling Chaos

What good is chaos? Research shows that it can be used to stabilize lasers, electronic circuits and animal hearts. Also, one may utilize chaotic dynamics in the design of systems that can rapidly switch between a large number of states in response to very small control forces.

Clever uses of chaos may be a promising area for future research, though, at present it is beyond the scope of this introduction to chaos. We briefly mention some basic ideas, and refer interested readers to, e.g., Bishop and Xu (1996), Chen (2000), Ditto and Pecora (1993), Kapitaniak (1993,1996), Lenci S, Rega G (2003), Myneni et al. (1999), Ott et al. (1990), Pecora and Caroll (1990), Shinbrot et al. (1990), and Thompson and Bishop (1994).

One basic idea for utilizing chaos explores the fact that many chaotic attractors have imbedded within them an *infinite number of unstable periodic orbits*. Though this can be proved mathematically, you may grasp it intuitively by reconsidering the period-doubling route to chaos (cf. Fig. 6.9): At each period-doubling a stable periodic orbit turns unstable in favor of a new stable periodic orbit, having twice the period of oscillation. The unstable periodic orbit remains a possible type of motion, though usually we do not observe it in numerical simulations or laboratory experiments—just like pencils are rarely observed standing in their upright position of unstable equilibrium. Right beyond the k'th period-doubling there are k unstable periodic orbits with periods 1, 2, 4, ..., 2^{k-1}, along with the newly emerged stable period-2^k orbit. Recall from Sect. 6.4.1 that the sequence of period-doublings accumulates at a critical value of the bifurcation parameter. Beyond that point the number of unstable periodic orbits has grown to become infinite, and chaos is the stable type of motion.

Now, assume we observe a system in a chaotic state of motion. The system is capable of displaying an unlimited number of different periodic motions, but we do not observe them because they are all unstable. *If* any particular unstable orbit could be stabilized by choice, then one would be able to produce and control a variety of observable periodic motions. Physically, the task of stabilizing an unstable orbit corresponds to making a pencil stand upright on the palm of your hand, that is, to stabilize its unstable equilibrium. What would you do? A workable approach is to move the hand quickly from side to side and back and forth. Mathematically this trick corresponds to moving a specified fixed point of a Poincaré map. So, to stabilize a particular orbit of a chaotically vibrating system one needs to: 1) obtain an experimental Poincaré map, 2) locate a particular fixed point of this map, and 3) move the fixed point systematically in time by applying small perturbations to a control parameter of the system.

Curiously, without even requiring a mathematical model of the system to be controlled, in a number of cases this technique has proved to be rather easy to apply. For example, several of the unstable periodic orbits for a chaotic magneto-elastic ribbon, clamped in a magnetic field, has been stabilized by using very small adjustments of the magnetic field strength (Ditto et al. 1990; Ditto and Pecora 1993).

6.10 Closing Comments

Summary Let us state the most troublesome feature of any nonlinear system: For a nonlinear system you may have more than one outcome, and you may not be able to predict which one it will be. Even if you can, you may not be able to predict details of it, because it may turn out to be chaotic.

However, we have tools at our disposal for *detecting* if the response is really chaotic and hence unpredictable, or just complicated. There are tools for *quantifying* and *visualizing* the chaotic response, so that it can be described in useful terms. It is sometimes possible to *predict* conditions for chaos, so that one can design around or perhaps *utilize* it. *Time-horizons of predictability* can be estimated, wherein at least some prediction is meaningful. We know a *basic mechanism* behind chaos: the repeated *stretching and folding* of phase space, in turn creating *extreme sensitivity to initial conditions*. A number of systems have been mentioned, mathematical and physical, for which chaos has been observed and at least partially explored. And we are aware that chaos is not a 'singular' phenomenon, i.e. something produced artificially by computers simulating degenerate systems with pathological parameters. Rather, chaos is such a common phenomenon within all areas of science, nature and human daily life, that one wonders why computers were ever required to open our eyes to it.

Bringing your Knowledge to the Kitchen Though chaos may cause a plenitude of problems one should not overlook the more joyful aspects. Hopefully you will find excitement in observing chaotic phenomena wherever you go.

A reasonably equipped kitchen is a good place to start. For example, in a typical refrigerator one may observe bottles or (worse) cooling elements that rattle periodically or sometimes chaotically in response to the purely deterministic 50 or 60 Hz AC-input. Also, depending on the volume flow, the kitchen tap may drip periodically, chaotically or not at all.

To study flow-induced vibrations you can pour a little milk into a bowl, which has a heavy glass lid, and cook it for a while in a microwave oven. Spectacular Hopf bifurcations may arise when vapor begins escaping at the lid in a periodic fashion. With full power, you may even witness bifurcations into chaos (then milk will spill all over the place and you realize that curiosity has its price).

There are even chaotic toys and art on the commercial market, e.g., a device termed 'space ball' which is basically a chaotic pendulum.

Life on the Edge of Chaos Our genes seem to appreciate chaos. Ears, eyes, brains, etc. are highly responsive to *changes* in stimuli. Watching a pendulum that starts swinging we lose alertness and become bored as soon as the motion settles down into orderly and predictable oscillations. The chaotic 'space ball' sells because it mixes up the orderly motion of an ordinary pendulum with a chaotic process. Order is appreciated because it allows for predicting events of the future – an important factor for evolutionary survival. Our nervous system is alert to disturbances of predictable order, because these force us to revise predictions and possibly take actions to prevent or achieve something.

Complete randomness is just as boring as complete order. For example, pure tones and random noise are interesting only to the moment when we find out they arise from the radio (and thus present no danger or chance of benefit). When this radio plays a piece of music we are more alert, since music represents a more proper mix of repeated and non-repeated stimuli. You may find that many of the interesting, exciting and delightful things in life are characterized by 'structured disorder', that is: By something in-between perfect order and pure randomness.

6.11 Problems

Problem 6.1 A large truss structure to be put into space is to carry test modules in which vibrations must be minimized. The designers experience vibration problems during ground tests of the structure. Your assistance is required.

a) An experimental vibration test is performed using broadband frequency excitation. The output frequency spectrum has a continuous appearance. Designer A claims this is due to the huge amount of excited degrees of freedom, whereas designer B suggests the possibility of chaotic vibrations due to loose joints. Does it matter whether A or B is right?

b) To examine the possibility of chaotic motions the designers employ a finite element model of the truss. Harmonic excitation at frequency Ω is assumed. At $\Omega = \Omega_1$ a Poincaré map of the numerical solution shows two dots. At Ω_2 the dots fill out a closed curve, and at Ω_3 the dots fill out some strange creature in the plane. Which kinds of motions are indicated at $\Omega_{1,2,3}$, respectively?

c) The designers decide to reduce the model to order five. With the reduced model they find no qualitative changes at $\Omega = \Omega_{1,2,3}$, as compared to the full model. At Ω_3 the full spectrum of Lyapunov-exponents turns out to be (0.30, 0.00, −0.05, −0.10, −0.20). Which kind of motion is indicated by this spectrum? Compute the Lyapunov dimension d_L, and assess whether the model of the truss can be further reduced, say, to order three.

d) The designers discuss whether the vibration problems at excitation frequency Ω_3 could be solved by adding adaptive control to the truss members. For this they intend to use a controller that requires positions of the truss to be predictable 1 s ahead. Feedback is provided by truss positions measured with 8 bits of accuracy ($\approx 0.4\%$). Do you see any obstacles for this kind of control?

e) One designer re-tests the real truss at a constant excitation frequency Ω_4. Within the first minute he notices a few chaotic bursts. During the second test minute the motion becomes regular. The designer decides that the first chaotic bursts were merely transients, and shuts off the experiment. Can he be right? Can he be wrong?

Problem 6.2 Consider a damped pendulum subjected to a periodically varying torque at the hinge. The swing angle $\theta(t)$ is governed by:

$$\ddot{\theta} + \sin\theta + \varepsilon 2\beta\dot{\theta} = \varepsilon p\cos(\Omega t), \quad \varepsilon \ll 1. \tag{6.67}$$

a) Show that for $\varepsilon = 0$ the system has a pair of heteroclinic phase plane orbits between $(\theta, \theta, = (\pm\pi, 0)$.

b) Show that the heteroclinic orbit in the positive half plane can be described by $(\theta(t), \theta, (t)) = (4\arctan(e^t) - \pi, 2\,\mathrm{sech}(t))$, where $\theta(0) = 0$ defines the origin of t.

c) Show that the Melnikov function for the system is:

$$M(t_0) = -16\beta + 2\pi p \operatorname{sech}(\pi\Omega/2)\cos(\Omega t_0). \qquad (6.68)$$

(Hint: $\int_{-\infty}^{\infty} \operatorname{sech}(t)\cos(\Omega t)\,dt = \pi\operatorname{sech}(\pi\Omega/2)$)

d) Set up a Melnikov criterion for predicting when the stable and unstable manifolds of the Poincaré map intersect.

e) Using numerical simulation, examine whether the Melnikov criterion is capable of predicting the occurrence of chaos for this system.

Problem 6.3 Consider the Duffing system with a double-well potential:

$$\ddot{y} + 2\beta\dot{y} - \frac{1}{2}y + \gamma y^3 = p\Omega^2\cos(\Omega t). \qquad (6.69)$$

a) Using the method of multiple scales, set up a frequency response equation for the determination of stationary amplitudes of small but finite oscillations in a potential well for the case of weak damping and excitation.

b) Use the frequency response equation obtained above for setting up a multiwell criterion for chaos.

c) Using numerical simulation, examine whether the criterion is applicable for the present system.

Problem 6.4 Consider a pendulum whose hinge moves periodically up and down at frequency Ω and amplitude A. The swing angle $\theta(t)$ is governed by:

$$\ddot{\theta} + 2\beta\omega_0\dot{\theta} + \left(\omega_0^2 - A\Omega^2\cos(\Omega t)\right)\sin\theta = 0, \qquad (6.70)$$

where the following parameters can be considered fixed:

$$\omega_0 = 1, \quad \beta = 0.05, \quad \Omega = 2.15, \quad \theta(0) = 0.01, \quad \dot{\theta}(0) = 0. \qquad (6.71)$$

a) Using numerical simulation, examine the phase plane orbits and determine the lowest value of A for which the pendulum seems to perform chaotic oscillations. Make sure the solution is chaotic by examining frequency spectrum, Poincaré map and largest Lyapunov exponent (remember to cut off transients).

b) Investigate the routes to chaos involved when A is varied across the value critical for chaos. Attempt sketching a bifurcation diagram for A. Do you observe period-doublings, intermittent chaos, transient chaos, or quasiperiodic responses?

c) Estimate the topological dimension of the chaotic attractor.

d) Estimate the time-horizon of predictability t_∞ when the uncertainty of initial conditions amounts to $\Delta = 10^{-3}$.

(e) Simulate the system for two slightly different sets of initial conditions corresponding to the value of Δ given above. Inspect the time-series plots, and estimate the time t_d at which the two solutions start to diverge. Compare t_d to t_∞.

7 Special Effects of High-Frequency Excitation

7.1 Introduction

What happens with a system being excited at a very high frequency, far beyond the highest underdamped natural frequency? You might think it simply responds at a similar high frequency in a trivial manner, since resonance effects are not at play. Indeed, vibration experts are typically trained to neglect possible high-frequency (HF) components of excitations. This often makes sense, since mechanical systems are lowpass filters, with a frequency response quickly vanishing beyond the highest subcritically damped natural frequency.

But probably you already know what happens with a simple pendulum, whose support is vibrated vertically at small amplitude and a frequency much higher than the fundamental natural frequency: The pendulum stabilizes in its upside-down position (a phenomenon we shall dwell further into below). This is a classic example of a 'slow' (in the sense 'average') effect of HF excitation, lasting as long as the HF excitation is sufficiently energetic. Other examples include changing dry friction into apparent viscous damping; transportation of mass (e.g. solid bodies, granular material, or fluids), and apparent changes in system stiffness, natural frequency, stability, and equilibriums.

Even a small-amplitude symmetric HF excitation may feed off significant asymmetric effects, and cause drastic changes to the low-frequency properties of a system. The effects may be interesting or strange, useful or disturbing, dangerous or disastrous – dependent on the circumstances. Or even be unimportant in many cases, but one cannot know without being capable of predicting and analyzing them.

Mechanical HF excitation provides the working principle behind many industrial applications, e.g., transport of material on vibration feeders, auto focusing of camera lenses, submersion of piles, or separation of solid materials according to size or density (e.g. Blekhman 2000; Blekhman and Sorokin 2016; Sorokin et al. 2010; Cabboi et al. 2020). In the context of *dynamic materials* (Lurie 2007), induced HF vibrations could be used to create materials whose effective properties change instantly, by changing properties of the HF signal (Blekhman 2008; Thomsen and Blekhman 2007, Abusoua and Daqaq 2018), and mechanical system parameters can be estimated using HF excitation (Abusoua and Daqaq 2017). Nature also utilizes slow effects of HF excitation, e.g., in enabling primitive organisms to drift and swim. And then there are also harmful effects, e.g., self-unscrewing bolts and nuts on vibrating machinery, and needle pointer instruments displaying bias error when operating in strongly vibrating environments. Some HF excitation phenomena appear well outside usual experience, e.g. fluid can be made to float above air by vibrating the air-fluid system strongly at high frequency, and even with objects floating upside down on the levitated air-fluid interface, as if gravity is reversed (Apffel et al. 2020, Sorokin and Blekhman 2020).

In this chapter we first present an outline of a method for conveniently analyzing systems with – and effects of – HF excitation. Hereafter the main ideas are illustrated in terms of three simple examples, representative of three main effects of HF excitation, which we shall call *stiffening, biasing*, and *smoothening*. Then various levels of generalizations are described, and finally some more involved examples are briefly presented to illustrate the variety and character of applications. The main reference in this area is the monograph by Blekhman (2000), while (Blekhman 2004), Fidlin (2002), Jensen (1999b), and Thomsen (2003a), can be considered as more specialized offspring.

7.2   The Method of Direct Separation of Motions (MDSM)

Here we describe a mathematical tool for dealing conveniently with linear and nonlinear systems subjected to high frequency excitation, the *Method of Direct Separation (or Partition) of Motions* (MDSM). The MDSM transforms a set of differential equations, generally nonlinear, into two subsets: one describing the 'fast' (rapidly oscillating) components of motion, and another one the 'slow' (in comparison) components. Typically, in applications, the subset describing slow motions is of primary interest, whereas the fast motions constitute a small and trivial overlay that is only interesting by its effect on the slow motions. The subset of 'slow' equations takes into account the fast forces through added terms, so-called *vibrational forces*, which describe the averaged influence of the fast forces.

Traditional perturbation approaches, such as the multiple scaling and averaging methods, also produce approximate equations describing slow solution components, known as *modulation equations* (e.g. Nayfeh and Mook 1979). These methods can be used as well for problems involving rapidly oscillating terms. For example, averaging is used for such problems in Chelomei (1981), Fidlin (1999, 2001, 2004a,b), Fidlin and Thomsen (2001), and Sethna (1967) – and the method of multiple scales in Fidlin (2000), Hansen (2000), Tcherniak (1999), and Tcherniak and Thomsen (1998). But the MDSM is particularly well suited, in terms of relatively easy application and interpretation, when the forces of excitation are rapidly changing as compared to the natural frequencies.

Originating from Kapitza's heuristic approach for a specific problem (Kapitza 1951, 1965; Landau and Lifshitz 1976), the MDSM was formalized, generalized, named, and applied to a wide variety of physical systems and phenomena by I. I. Blekhman (e.g. Blekhman 1976, 1994, 2000, 2004). Below we present an outline of the method; for rigorous derivations and theorems and extensive examples of application Blekhman (2000) should be consulted.

7.2.1 Outline of the MDSM

A typical candidate for the method is a dynamical system of this form:

$$\mathbf{M}(\mathbf{x}, \dot{\mathbf{x}}, t)\ddot{\mathbf{x}} = \tilde{\mathbf{s}}(\mathbf{x}, \dot{\mathbf{x}}, t) + \tilde{\mathbf{f}}(\mathbf{x}, \dot{\mathbf{x}}, t, \tau), \tag{7.1}$$

where $\mathbf{x} = \mathbf{x}(t) \in R^n$ is a time-dependent state vector, $\mathbf{M}$ is a mass matrix, $\tilde{\mathbf{f}}$ and $\tilde{\mathbf{s}}$ are generalized force vectors, and $\dot{\mathbf{x}} = dx/dt$. The forces contained in $\tilde{\mathbf{f}}$ are assumed to be 2π-periodic with respect to a *fast time scale* $\tau = \Omega t$, that is: $\tilde{\mathbf{f}}(\mathbf{x}, \dot{\mathbf{x}}, t, \tau) = \tilde{\mathbf{f}}(\mathbf{x}, \dot{\mathbf{x}}, t, \tau + j2\pi), j = 1, 2, \ldots$ The frequency Ω is supposed to be much higher than some characteristic frequency ω of the corresponding unforced, linearized system. Assuming $\mathbf{M}$ to be positively definite, we rearrange the system to be decoupled in the acceleration terms:

$$\begin{aligned}\ddot{\mathbf{x}} &= \mathbf{M}(\mathbf{x}, \dot{\mathbf{x}}, t)^{-1}\left(\tilde{\mathbf{s}}(\mathbf{x}, \dot{\mathbf{x}}, t) + \tilde{\mathbf{f}}(\mathbf{x}, \dot{\mathbf{x}}, t, \tau)\right) \\ &\equiv \mathbf{s}(\mathbf{x}, \dot{\mathbf{x}}, t) + \mathbf{f}(\mathbf{x}, \dot{\mathbf{x}}, t, \tau),\end{aligned} \tag{7.2}$$

where $\mathbf{s}$ holds the *slow forces* and $\mathbf{f}$ the *fast forces*.

The key step with the MDSM is to assume that there are slowly varying components of solutions to (7.2), which can be separated from the fast ones as follows:

$$\mathbf{x} = \mathbf{x}(t, \tau) = \mathbf{z}(t) + \boldsymbol{\varphi}(t, \tau), \tag{7.3}$$

where $\mathbf{z}(t)$ holds the *slow* components of $\mathbf{x}$, and $\varphi(t,\tau)$ the *fast* components being 2π-periodic in the fast time τ. (Mnemonic: $\mathbf{z}$ for 'zlow' and φ for 'phast'). The transformation $\mathbf{x} \rightarrow (\mathbf{z}, \varphi)$ increases the number of variables from n to $2n$. Hence, for the transform to be unique, we must impose n additional constraints. For this we require the average of φ over one period $T = 2\pi/\Omega$ of the rapidly oscillating component of $\mathbf{f}$ to vanish identically, i.e.:

$$\langle \boldsymbol{\varphi}(t, \tau) \rangle = \mathbf{0}, \tag{7.4}$$

where $\langle\rangle$ is the (fast time) *averaging operator*, defined as follows:

$$\langle \mathbf{h}(t,\tau)\rangle \equiv T^{-1}\int_0^T \mathbf{h}(t,\Omega t)\,dt = (2\pi)^{-1}\int_0^{2\pi}\mathbf{h}(t,\tau)\,d\tau. \tag{7.5}$$

Since Ω is assumed to be large, the period $T = 2\pi/\Omega$ will be small, and thus the *slow time* t can be considered 'frozen' at a constant value, while the fast time τ varies from 0 to 2π during the time interval of integration. The averaging operator is linear, i.e. for arbitrary constants c_1 and c_2 and functions $\mathbf{h}_1$ and $\mathbf{h}_2$ it holds that:

$$\langle c_1\mathbf{h}_1(t,\tau)+c_2\mathbf{h}_2(t,\tau)\rangle = c_1\langle\mathbf{h}_1(t,\tau)\rangle + c_2\langle\mathbf{h}_2(t,\tau)\rangle. \tag{7.6}$$

Also, functions that are independent of the fast time τ are unaffected by averaging:

$$\langle\mathbf{h}(t)\rangle = \mathbf{h}(t). \tag{7.7}$$

Consequently, by (7.3), (7.4), (7.6), and (7.7):

$$\langle\mathbf{x}(t,\tau)\rangle = \langle\mathbf{z}(t)\rangle + \langle\boldsymbol{\varphi}(t,\tau)\rangle = \mathbf{z}(t), \tag{7.8}$$

so that the slow motions $\mathbf{z}$ are just the (fast time) average of the full motions $\mathbf{x}$. Further, for functions $\mathbf{h}(t,\tau)$ that are 2π-periodic in τ it holds that:

$$\langle\dot{\mathbf{h}}(t,\tau)\rangle = (2\pi)^{-1}\int_0^{2\pi}\frac{d\mathbf{h}(t,\tau)}{\Omega^{-1}d\tau}d\tau = (2\pi/\Omega)^{-1}[\mathbf{h}(t,2\pi)-\mathbf{h}(t,0)] = 0,$$

$$\langle\ddot{\mathbf{h}}(t,\tau)\rangle = (2\pi)^{-1}\int_0^{2\pi}\frac{d\dot{\mathbf{h}}(t,\tau)}{\Omega^{-1}d\tau}d\tau = (2\pi/\Omega)^{-1}[\dot{\mathbf{h}}(t,2\pi)-\dot{\mathbf{h}}(t,0)] = 0. \tag{7.9}$$

We now perform the transformation of variables by inserting (7.3) into (7.2):

$$\ddot{\mathbf{z}}+\ddot{\boldsymbol{\varphi}} = \mathbf{s}(\mathbf{z}+\boldsymbol{\varphi},\dot{\mathbf{z}}+\dot{\boldsymbol{\varphi}},t)+\mathbf{f}(\mathbf{z}+\boldsymbol{\varphi},\dot{\mathbf{z}}+\dot{\boldsymbol{\varphi}},t,\tau). \tag{7.10}$$

Applying the averaging operator to both sides of this equation yields:

$$\langle\ddot{\mathbf{z}}+\ddot{\boldsymbol{\varphi}}\rangle = \langle\mathbf{s}(\mathbf{z}+\boldsymbol{\varphi},\dot{\mathbf{z}}+\dot{\boldsymbol{\varphi}},t)+\mathbf{f}(\mathbf{z}+\boldsymbol{\varphi},\dot{\mathbf{z}}+\dot{\boldsymbol{\varphi}},t,\tau)\rangle, \tag{7.11}$$

which, due to (7.6), (7.7), and (7.9) reduce to a set of equations governing the slow motions $\mathbf{z}(t)$:

$$\ddot{\mathbf{z}} = \mathbf{s}(\mathbf{z},\dot{\mathbf{z}},t)+\langle\mathbf{s}_1(\mathbf{z},\dot{\mathbf{z}},\boldsymbol{\varphi},\dot{\boldsymbol{\varphi}},t)\rangle + \langle\mathbf{f}(\mathbf{z}+\boldsymbol{\varphi},\dot{\mathbf{z}}+\dot{\boldsymbol{\varphi}},t,\tau)\rangle, \tag{7.12}$$

where s_1 defines a function that vanishes when there are no fast motions:

$$s_1(\mathbf{z}, \dot{\mathbf{z}}, \boldsymbol{\varphi}, \dot{\boldsymbol{\varphi}}, t) \equiv s(\mathbf{z} + \boldsymbol{\varphi}, \dot{\mathbf{z}} + \dot{\boldsymbol{\varphi}}, t) - s(\mathbf{z}, \dot{\mathbf{z}}, t). \tag{7.13}$$

A corresponding equation governing the fast motions $\varphi(t, \tau)$ is then obtained by subtracting (7.12) from (7.10):

$$\begin{aligned}
\ddot{\boldsymbol{\varphi}} &= s_1(\mathbf{z}, \dot{\mathbf{z}}, \boldsymbol{\varphi}, \dot{\boldsymbol{\varphi}}, t) - \langle s_1(\mathbf{z}, \dot{\mathbf{z}}, \boldsymbol{\varphi}, \dot{\boldsymbol{\varphi}}, t)\rangle \\
&\quad + f(\mathbf{z} + \boldsymbol{\varphi}, \dot{\mathbf{z}} + \dot{\boldsymbol{\varphi}}, t, \tau) - \langle f(\mathbf{z} + \boldsymbol{\varphi}, \dot{\mathbf{z}} + \dot{\boldsymbol{\varphi}}, t, \tau)\rangle.
\end{aligned} \tag{7.14}$$

Equations (7.12) and (7.14) just express the original system (7.2) in terms of the new variables $\mathbf{z}$ and $\boldsymbol{\varphi}$. They are not simpler to solve. Typically, however, one is mainly interested in the slow components of motion $\mathbf{z}(t)$. For determining $\mathbf{z}(t)$ approximately, only a first approximation for the fast motions $\varphi(t, \tau)$ is required, since in (7.12) φ occurs in averaged terms only. It is here the view on the MDSM as a perturbation method comes in, since the equations for the fast motions φ are typically solved in terms of a small perturbation parameter $\varepsilon = \Omega^{-1}$. If the approximate solution φ^* is correct to order ε^m, then $\varphi = \varphi^* + O(\varepsilon^{m+1})$, and the error in the approximation for the slow motions $\mathbf{z}$ can be estimated using magnitude-order symbols. (The error will depend on the functions s and f, cf. (7.12)–(7.13).)

Thus, in applying the MDSM one first transforms the original equations of motion (7.1) into the form (7.2), and further into (7.12) and (7.14). Next, a first approximation $\varphi = \varphi^*$ for the fast motions is determined from (7.14), and the approximate solution φ^* is substituted for φ in (7.12). Finally one performs the fast-time averaging in (7.12), and possibly attempts solving the averaged equation for the slow motions $\mathbf{z}$.

Often, however, important observations and statements can be made just by studying the terms of the averaged system, without actually solving the equations. For example, the effects of stiffening, biasing, and smoothening, to be described in subsequent sections, are all apparent directly on by inspecting the averaged equations governing the slow motions.

7.2.2 The Concept of Vibrational Force

Assume that a specific solution φ^* (or at least an approximation) is known for the fast motions $\varphi(t, \tau)$. Equation (7.12) for the slow motions z then becomes:

$$\ddot{\mathbf{z}} = s(\mathbf{z}, \dot{\mathbf{z}}, t) + \langle s_1(\mathbf{z}, \dot{\mathbf{z}}, \boldsymbol{\varphi}^*, \dot{\boldsymbol{\varphi}}^*, t)\rangle + \langle f(\mathbf{z} + \boldsymbol{\varphi}^*, \dot{\mathbf{z}} + \dot{\boldsymbol{\varphi}}^*, t, \tau)\rangle, \tag{7.15}$$

which can be rewritten as:

$$\ddot{\mathbf{z}} = \mathbf{s}(\mathbf{z}, \dot{\mathbf{z}}, t) + \mathbf{v}(\mathbf{z}, \dot{\mathbf{z}}, t), \tag{7.16}$$

where $\mathbf{v}$ defines the *vibrational forces*:

$$\mathbf{v}(\mathbf{z}, \dot{\mathbf{z}}, t) \equiv \langle \mathbf{s}_1(\mathbf{z}, \dot{\mathbf{z}}, \boldsymbol{\varphi}^*, \dot{\boldsymbol{\varphi}}^*, t) \rangle + \langle \mathbf{f}(\mathbf{z} + \boldsymbol{\varphi}^*, \dot{\mathbf{z}} + \dot{\boldsymbol{\varphi}}^*, t, \tau) \rangle. \tag{7.17}$$

Vibrational forces do not necessarily have the physical unit of force, but should be understood in a generalized sense to represent certain forcing terms in the equation of motion. The physical unit may correspond to acceleration of state variables such as translation, rotation, temperature, or electric charge.

Note that Eq. (7.16) for the slow motions $\mathbf{z}$ is quite similar to the original Eq. (7.2) for the full motions $\mathbf{x}$, though with the fast forces $\mathbf{f}$ replaced by (slow) vibrational forces $\mathbf{v}$, that have no explicit dependence on the fast time τ. Note also that the fast motions $\boldsymbol{\varphi}^*$ are determined as approximate solutions to (7.14); however, $\boldsymbol{\varphi}^*$ could also be calculated by purely numerical methods, or even measured experimentally.

To an observer filtering out the small HF components of the motions – as with measurement instruments, human senses, or dedicated lowpass filters – it will appear as if the system is influenced by the slow vibrational forces $\mathbf{v}$, while the fast forces $\mathbf{f}$ are not apparent. For example, one out of many ways to explain the well-known inversion of a pendulum on a vibrating support (Sect. 7.3.1) is to show that the effect of the HF excitation, on the average, corresponds to a slow vibrational force pulling the pendulum towards the upside-down equilibrium.

7.2.3   The MDSM Compared to Classic Perturbation Approaches

The MDSM is typically used with excitations well away from system resonances (Thomsen (2003b) represents an exception), and there should be components of the excitation that are rapidly oscillating.

Several perturbation approaches also work by splitting motions into slow and fast components. For example, as explained in Chap. 3, the method of multiple scales explicitly introduces independent time scales for capturing motions at different scales of time. Solutions are sought in the form of uniformly valid expansions, e.g., $u(t;\varepsilon) = u_0(T_0, T_1, \ldots, T_n) + \varepsilon u_1(T_0, T_1, \ldots, T_n) + \cdots + \varepsilon^n u_n(T_0, T_1, \ldots, T_n)$, where $\varepsilon \ll 1$ and $T_i = \varepsilon^i t$. Here $T_0 = t$ is the fast time, whereas $T_1, T_2, \ldots$ are slow times. Each term of the expansion usually becomes a *product* of slow and fast terms, e.g., $u = a(\varepsilon t)b(t) + O(\varepsilon)$.

With the MDSM, by contrast, solutions are obtained in the form of *sums* of slow and fast components, e.g., $u = z(\varepsilon t) + \varepsilon \varphi(t)$. Hence, the MDSM lends itself naturally at describing slow motions overlaid by small, fast oscillations (Fig. 7.1(a)),

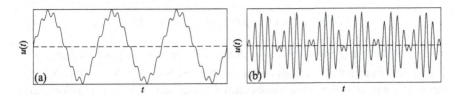

Fig. 7.1 (a) Large, slow motions o verlaid by small, fast oscillations, $u(t) = \sin(\varepsilon t) + \varepsilon \sin(\omega t)$, $\varepsilon \ll 1$; (b) Slowly modulated fast motion, $u(t) = \sin(\varepsilon \omega t)\sin(\omega t)$

whereas the method of multiple scales and similar are better suited for capturing slowly modulated fast motions (Fig. 7.1(b)).

7.3 Simple Examples

Next we use the MDSM to illustrate the stiffening, the biasing, and the smoothening effect of HF excitation in simple physical settings, which will also introduce much of the terminology to be used subsequently.

7.3.1 Pendulum on a Vibrating Support (Stiffening and Biasing)

Studies of the pendulum with a rapidly vibrating support (Fig. 7.2) have a long history (e.g., Stephenson 1908a,b, 1909; Hirsch 1930; Lowenstern 1932; Kapitza 1951; Panovko and Gubanova 1965; Bogdanoff and Citron 1965), which is still continuing (e.g., Acheson 1993, 1995; Acheson and Mullin 1993; Anderson and Tadjbakhsh 1989; Fenn et al. 1998; Guran and Plaut 1993; Jensen et al. 2004; Sah et al. 2013; Schmitt and Bayly 1998; Sudor and Bishop 1999; Weibel et al. 1997). Here it serves to illustrate the stiffening and biasing effects of HF excitation, which in turn influence the effective natural frequency and the existence and stability of equilibriums. The equation of motion is:

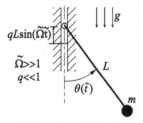

Fig. 7.2 Pendulum with a vibrating support

$$\ddot{\theta} + 2\beta\dot{\theta} + \left(1 - q\Omega^2 \sin(\Omega t)\right)\sin\theta = 0, \tag{7.18}$$

where θ is the swing angle, $t = \omega_0 \tilde{t}$ is nondimensional time, $\tilde{t}$ physical time, $\omega_0 = \sqrt{g/L}$ the linear natural frequency for oscillations near $\theta = 0$, $\dot{\theta} = d\theta/dt$, β the damping ratio, q the relative amplitude of prescribed support oscillations, and $\Omega = \tilde{\Omega}/\omega_0$ the nondimensional frequency of this excitation. We consider the case of small but rapid vibrations of the support, i.e. $q \ll 1$ and $\Omega \gg 1$, with $q\Omega = O$ (1). Approximate solutions can be obtained by a number of different perturbation methods, considering $\varepsilon = \Omega^{-1}$ as a small parameter. Using the MDSM (Sect. 7.2.1), the motions $\theta(t)$ are split into slow and fast components as follows:

$$\theta = \theta(t, \tau) = z(t) + \Omega^{-1}\varphi(t, \tau), \quad \tau = \Omega t, \tag{7.19}$$

where z describes the slow motions (at the time-scale of free pendulum oscillations), and $\Omega^{-1}\varphi$ is a small overlay of fast motions (at the rapid rate of support vibrations). We perceive t as a slow time scale and $\tau = \Omega t$ as a fast time scale. The slow motions z are those of primary concern, whereas the details of the fast overlay φ are interesting only by their effect on z.

Considering (7.19) as a transform of variables, $\theta \rightarrow (z, \varphi)$, a constraint is needed to make the transform unique. For this it is convenient to require a zero fast-time average of the fast motions:

$$\langle \varphi(t, \tau) \rangle \equiv \frac{1}{2\pi} \int_0^{2\pi} \varphi(t, \tau)\, d\tau = 0, \tag{7.20}$$

where $\langle \rangle$ defines time averaging over one period of the fast excitation, with the slow time t considered fixed.

Using (7.19) and the chain rule to calculate the time derivatives of θ in terms of the new variables z and φ, we find:

$$\begin{aligned}
\frac{d\theta}{dt} &= \frac{d(z + \Omega^{-1}\varphi)}{dt} = \frac{dz}{dt} + \Omega^{-1}\frac{d\varphi}{dt} = \frac{dz}{dt} + \Omega^{-1}\left(\frac{\partial\varphi}{\partial t} + \frac{\partial\varphi}{\partial \tau}\frac{d\tau}{dt}\right) \\
&= \dot{z} + \varphi' + \Omega^{-1}\dot{\varphi}, \\
\frac{d^2\theta}{dt^2} &= \frac{d}{dt}\left(\dot{z} + \varphi' + \Omega^{-1}\dot{\varphi}\right) = \ddot{z} + \left(\dot{\varphi}' + \Omega\varphi''\right) + \Omega^{-1}(\ddot{\varphi} + \Omega\dot{\varphi}') \\
&= \Omega\varphi'' + \ddot{z} + 2\dot{\varphi}' + \Omega^{-1}\ddot{\varphi},
\end{aligned} \tag{7.21}$$

where $\dot{\varphi} \equiv \partial\varphi/\partial t$, and $\varphi' \equiv \partial\varphi/\partial\tau$, and $d\tau/dt = \Omega$ has been employed. Then we are ready to substitute (7.19) and (7.21) into (7.18), obtaining:

$$\begin{aligned}
&(\Omega\varphi'' + \ddot{z} + 2\dot{\varphi}' + \Omega^{-1}\ddot{\varphi}) + 2\beta(\dot{z} + \varphi' + \Omega^{-1}\dot{\varphi}) \\
&+ (1 - q\Omega^2 \sin\tau)\sin(z + \Omega^{-1}\varphi) = 0,
\end{aligned} \tag{7.22}$$

Next insert the Taylor-expansion of $\sin(z + \Omega^{-1}\varphi) = \sin(z) + \Omega^{-1}\varphi\cos(z) + O(\Omega^{-2})$ for $\Omega^{-1} \ll 1$, rearrange according to magnitude order (recall that $q\Omega = O(1)$), and obtain the equation governing the fast motions φ:

$$\begin{aligned}
\varphi'' = {}&q\Omega \sin z \sin\tau \\
&- \Omega^{-1}(\ddot{z} + 2\dot{\varphi}' + 2\beta(\dot{z} + \varphi')) \\
&+ \sin z - q\Omega\,\varphi\cos z \sin\tau) + O(\Omega^{-2}),
\end{aligned} \tag{7.23}$$

A first order approximate solution to this equation is found simply by integrating the dominating terms twice with respect to τ, noting that z is independent of τ:

$$\varphi(t, \tau) = -q\Omega \sin z \sin\tau + O(\Omega^{-1}), \tag{7.24}$$

which satisfies the requirement $\langle\varphi\rangle = 0$. To determine the slow motions z we average (7.23) and obtain, using (7.20) and rearranging:

$$\ddot{z} + 2\beta\dot{z} + \sin z - q\Omega\cos z\langle\varphi\sin\tau\rangle = O(\Omega^{-1}), \tag{7.25}$$

or, substituting (7.24) for φ, noting that $\langle\sin^2\tau\rangle = 1/2$, and dropping small terms of the order Ω^{-1}:

$$\ddot{z} + 2\beta\dot{z} + \left(1 + \frac{1}{2}q^2\Omega^2\cos z\right)\sin z = 0. \tag{7.26}$$

Hence, by (7.19) and (7.24) the pendulum motions are given by:

$$\theta(t) = z(t) - q\,\sin(z)\sin(\Omega t) + O(\Omega^{-2}). \tag{7.27}$$

It appears, since $q \ll 1$, that the rapidly oscillating overlay is very small; it constitutes the trivial effect of the HF excitation. Nontrivial effects appear in the dynamics of the slow motions z. It appears from (7.26) that z is governed by a differential equation quite similar to the original one (7.18), though, with the non-autonomous excitation term $q\Omega^2\sin(\Omega t)$ replaced by the autonomous term $1/2$ $(q\Omega)^2\cos(z)$, which describes an average effect of the HF excitation. The effective restoring force in the presence of HF excitation thus changes from $f_r(z) = \sin(z)$ to:

$$f_r(z) = \left(1 + \frac{1}{2}(q\Omega)^2\cos z\right)\sin z, \tag{7.28}$$

which is depicted in Fig. 7.3(a) for three values of the excitation intensity $q\Omega$. Of concern here is the *effective stiffness*, which we define as the slope of the restoring force curve at the static equilibriums. As appears from Fig. 7.3(b), at the two equilibriums corresponding to the straight downward or upward pointing pendulum ($z = 0, \pm\pi$), this stiffness increases in the presence of HF excitation (i.e. when $q\Omega \neq 0$). Specifically, using (7.28) we find that the stiffness for both equilibriums increases by the same quantity $1/2(q\Omega)^2$:

$$\frac{df_r}{dz} = \begin{cases} 1 + \frac{1}{2}(q\Omega)^2 & \text{for } z = 0, \\ -1 + \frac{1}{2}(q\Omega)^2 & \text{for } z = \pm\pi. \end{cases} \tag{7.29}$$

For the equilibrium at $z = 0$ the change in effective stiffness corresponds to a change in effective natural frequency, from unity to ω_1,

$$\omega_1 = \sqrt{1 + \frac{1}{2}(q\Omega)^2}. \tag{7.30}$$

Thus, near the down-pointing equilibrium, the free pendulum oscillations occur at a higher frequency when the support is vibrating rapidly up and down. For example, with a pendulum clock mounted on a table vibrating at intensity $q\Omega = 1$, the clock will run $\sqrt{1 + \frac{1}{2}} - 1 \approx 22\%$ faster than when the table is at rest.

For the up-pointing equilibrium at $z = \pm\pi$ the situation is different, because in the absence of HF excitation this equilibrium is unstable – as reflected by a negative stiffness (cf. (7.29) and Fig. 7.3(b)). Thus an increase in effective stiffness, if sufficiently large, may stabilize this equilibrium. As appears from (7.29) this occurs when $(q\Omega)^2 > 2$, which is a well-known result (e.g., Blekhman 2000; Kapitza 1951; Nayfeh and Mook 1979; Panovko and Gubanova 1965; Stephenson 1908a). With this condition fulfilled, small disturbances to the upright equilibrium of the pendulum decay at an effective natural frequency ω_2 that increases with the level $q\Omega$ of excitation:

$$\omega_2 = \sqrt{\frac{1}{2}(q\Omega)^2 - 1}. \tag{7.31}$$

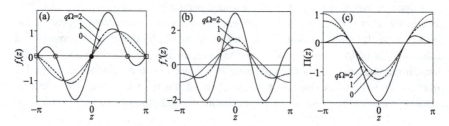

Fig. 7.3 (a) Effective restoring force $f_r(z)$ in the presence of different levels of intensity $q\Omega$ of rapid support vibrations. Equilibrium points: ● stable, ○ unstable, □ stable if $df_r/dz > 0$ (i.e. only for the $q\Omega = 2$ curve); (b) effective linear stiffness $f_r'(z) = df_r/dz$; (c) potential energy $\Pi(z) = \int f_r dz$

Also, under such conditions two new equilibriums emerge symmetrically about the vertical axis, $z = \pm\arccos(-2(q\Omega)^{-2})$. One can show that these equilibriums are always unstable, and thus act as 'potential barriers' that have to be overcome if the pendulum is to be moved between the two stable equilibriums at $z = 0$ and $z = \pm\pi$ (cf. the curve for $q\Omega = 2$ in Fig. 7.3(a) and (c)).

It appears that the change in effective stiffness, here considered the primary effect, has derived effects in the form of changes of natural frequencies, stabilization of equilibriums, and emergence of new equilibriums.

Another derived effect is a change in the effective nonlinearity of the system. This can be seen when Taylor-expanding (7.26) and including leading order nonlinearities. Here one finds the following approximate equations of motion, valid near the vertical equilibriums:

for $z \approx 0$:

$$\ddot{z} + 2\beta\dot{z} + \left(1 + \tfrac{1}{2}q^2\Omega^2\right)z - \tfrac{1}{6}(1 + 2q^2\Omega^2)z^3 + O(z^5) = 0,$$

for $z \approx \pm\pi$:

$$\ddot{z} + 2\beta\dot{z} - \left(1 - \tfrac{1}{2}q^2\Omega^2\right)(z - \pi) + \tfrac{1}{6}(1 - 2q^2\Omega^2)(z - \pi)^3 + O\left((z - \pi)^5\right) = 0.$$

(7.32)

Considering the leading nonlinear term z^3, it appears that for $z \approx 0$ its softening (negative coefficient) character becomes more pronounced with increased level $q\Omega$ of excitation, whereas for $z \approx \pm\pi$ the hardening character (positive coefficient) of the nonlinearity turns softening for sufficiently large $q\Omega$.

The present system may also serve to illustrate an example of the *biasing* effect of HF excitation. For this we imagine the pendulum to be excited and swinging in the *horizontal* plane, so that the gravity term in (7.18) and (7.26) (the number one) is replaced by zero. Then, in the absence of the HF excitation, all positions θ or z are equilibriums for the pendulum, i.e. the pendulum has no preference or bias for pointing into any particular direction. However, as appears from (7.26) with the gravity term zeroed, when $q\Omega \neq 0$ there are four equilibriums: $z = 0, \pi, \pm\pi/2$, and one can easily show that only $z = 0, \pi$ are stable. Hence, in the presence of rapid support excitation the pendulum is biased to line up with the direction of excitation. To an observer it will seem as there are 'gravity-like' forces pulling towards these directions, about which the pendulum may oscillate at the (slow) frequency $q\Omega/\sqrt{2}$.

The effects mentioned above are not just mathematical artifacts; they are quite easily demonstrated in the laboratory by using a small pendulum driven e.g. by an electric jigsaw (Michaelis 1985) or loudspeaker (Kalmus 1970). Also, the pendulum Eq. (7.18), in particular if Taylor-expanded to order three, is representative of a great many other systems and structures. For example, a single-mode approximation for a beam with pulsating axial excitation has this form, and thus the effects described above are pertinent for this case as well (Chelomei 1956, 1983; Jensen 2000a; Jensen, et al. 2000; Mailybaev and Seyranian 2009; Yabuno and Tsumoto 2007). Then the results for the upright pendulum equilibrium apply if the average axial beam load exceeds the buckling load, whereas the results for the down pointing pendulum equilibrium apply for sub-critical beam loads. The situation with a horizontal pendulum – having no restoring force in the absence of support excitation – corresponds to a beam that has no transverse stiffness at all, i.e. to an untaught string.

7.3.2 Mass on a Vibrating Plane (Smoothening and Biasing)

Fig. 7.4(a) shows the system: a mass m of characteristic dimension L, attached by a spring of stiffness K to a horizontal plane that vibrates at a small amplitude qL and high frequency $\tilde{\Omega}$. The friction between mass and plane is given by a simple asymmetric Coulomb form, i.e. the friction force is $mg\mu(\dot{X})$ where the coefficient μ depends on the direction of motion as shown in Fig. 7.4(b) (many materials possess this property, e.g. the skin or fur of animals, as do certain asymmetrical processes such as vibrational piling and penetration):

$$\mu(\dot{X}) = \begin{cases} \mu_+ & \text{for } \dot{X} > 0 \\ -\mu_- & \text{for } \dot{X} < 0 \end{cases}; \quad \mu(\dot{X}) \in [-\mu_-; \mu_+] \quad \text{for } \dot{X} = 0. \tag{7.33}$$

The nondimensional equation of motion is:

$$\ddot{x} + \omega^2 x + \bar{\mu}\,\text{sgn}(\dot{x}) + \mu_\Delta = q\Omega^2 \sin(\Omega t), \tag{7.34}$$

where $\bar{\mu}$ is the *average* coefficient of friction, and μ_Δ the *asymmetry* in friction:

$$\begin{aligned} \bar{\mu} &= \frac{1}{2}(\mu_+ + \mu_-), \\ \mu_\Delta &= \frac{1}{2}(\mu_+ - \mu_-), \end{aligned} \tag{7.35}$$

and where $x = X/L$ denotes position, $t = \omega_0 \tilde{t}$ nondimensional time ($\tilde{t}$ physical time), $\dot{x} = dx/dt$, $\omega_0 = \sqrt{g/L}$ a characteristic frequency, $\omega = \sqrt{K/m}/\omega_0$ the nondimensional natural frequency for small oscillations when there is no friction, q the nondimensional excitation amplitude, and $\Omega = \tilde{\Omega}/\omega_0$ the nondimensional excitation frequency. By using $\sqrt{g/L}$ rather than $\sqrt{K/m}$ as the characteristic frequency we allow for setting $\omega^2 = 0$, to study what happens when there is no spring. The vibrations from the plane are assumed to be sufficiently strong to cause the mass to slide in both directions, i.e. $q\Omega^2 > \max(\mu_+, \mu_-)$.

(a) (b)

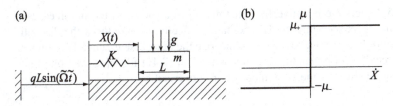

Fig. 7.4 (a) Mass on a vibrating plane; (b) Coefficients of dry friction

Again we determine approximate solutions for the case of high excitation frequency $\Omega \gg \omega$ and small excitation amplitude $q \ll 1$, by splitting motions into slow and fast components:

$$x = x(t, \tau) = z(t) + \Omega^{-1}\varphi(t, \tau), \tag{7.36}$$

where z holds the slow motions, $\Omega^{-1}\varphi$ is a small overlay of zero average fast motions, and $\tau = \Omega t$ is the fast time scale. Proceeding as for the pendulum example in the previous section we find that the fast motions are:

$$\varphi(t, \tau) = -q\Omega \sin \tau + O(\Omega^{-1}), \tag{7.37}$$

while the slow motions are governed by:

$$\ddot{z} + \omega^2 z + \bar{\mu}\langle \mathrm{sgn}(\dot{z} - q\Omega \cos \tau) \rangle + \mu_\Delta = 0, \tag{7.38}$$

where $\langle \rangle$ denotes the averaging operator defined by (7.20) or (7.5). Hence, by (7.36) and (7.37), motions of the mass are given by:

$$x(t) = z(t) - q \sin(\Omega t) + O(\Omega^{-2}). \tag{7.39}$$

The average term in (7.38) can be calculated by noting that the argument of the signum function is positive for $\tau \in [\tau_1; \tau_2]$ where $t_1 = \arccos(\dot{z}/q\Omega)$ and $\tau_2 = 2\pi - \tau_1$, and otherwise negative. As a result the equation of slow motions becomes:

$$\ddot{z} + \omega^2 z + \bar{f}(\dot{z}) = 0, \tag{7.40}$$

where $\bar{f}$ is the *effective friction force*:

$$\bar{f}(\dot{z}) = \begin{cases} \bar{\mu}\,\mathrm{sgn}(\dot{z}) + \mu_\Delta & \text{for } |\dot{z}| \geq q\Omega, \\ \bar{\mu}(1 - \frac{2}{\pi}\arccos(\dot{z}/q\Omega)) + \mu_\Delta & \text{for } |\dot{z}| < q\Omega. \end{cases} \tag{7.41}$$

Comparing to the original Eq. (7.34) for x, it appears that the non-autonomous excitation term has disappeared, and that the term representing friction force has changed for velocities $|\dot{z}| < q\Omega$.

Fig. 7.5 depicts the effective friction force $\bar{f}$. Comparing to the unforced case (dashed line), it is seen that the HF excitation effectively *smoothens* the discontinuity at $\dot{z} = 0$. In fact it makes sense to linearize the otherwise essentially nonlinear friction force: Taylor-expanding the arccos-term in (7.41) for $|\dot{z}| \ll q\Omega$, one finds that for small velocities the slow motions are governed by:

$$\ddot{z} + \omega^2 z + \bar{\beta}\dot{z} + \mu_\Delta = 0, \quad \bar{\beta} = \frac{2}{\pi} \frac{\bar{\mu}}{q\Omega}, \tag{7.42}$$

from which a well-known result appears (e.g. Blekhman 2000): HF excitation can make dry friction appear as linear viscous damping, with the equivalent damping coefficient $\bar{\beta}$ gradually vanishing as the intensity $q\Omega$ of excitation is increased. Indeed this is an everyday experience, e.g. with objects that seems to 'float' on a vibrating surface, which is used in many technological processes, e.g. vibrational piling (cf. Problem 7.3), and vibrational loosening of sticking parts.

As a second effect the HF excitation may introduce *bias* – a preference to certain states over others. First we note that in the absence of HF excitation the mass is in static equilibrium as long as the spring force is too small to overcome friction, i.e. as long as $-\mu_+ < \omega^2 x < \mu_-$ or, in terms of the average friction and asymmetry in friction: as long as $|\omega^2 x + \mu_\Delta| < \bar{\mu}$; Thus all $x \in [-\mu_+/\omega^2 ; \mu_-/\omega^2]$ are equilibrium positions for the mass. To find static equilibriums in the presence of HF excitation we let $\dot{z} = \ddot{z} = 0$ in (7.40), solve for z and find the single solution $z = \tilde{z}$,

$$\tilde{z} = -\frac{\mu_\Delta}{\omega^2}. \tag{7.43}$$

Thus the mass is biased to occupy a particular position, rather than a range of positions – an effect that increases with asymmetry in friction and with spring flexibility ($\propto \omega^{-2}$). This kind of bias may explain the incidence of misreading scale instruments in strongly vibrating environments, which makes the 'hidden' asymmetry μ_Δ apparent. For a further analysis of effects of HF excitation in the presence of friction or nonlinear damping see e.g. Chatterjee et al. (2003, 2004), Kapelke et al. (2017).

Another example of bias can occur when there is no restoring spring, i.e. when $\omega^2 = 0$. Then it is found, letting $\ddot{z} = 0$ in (7.40) and solving for $\dot{z}$, that a steady equilibrium solution exists, corresponding to a situation where the mass moves at

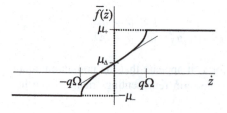

Fig. 7.5 Effective friction force $\bar{f}(\dot{z})$ in the presence of HF excitation of amplitude $q \ll 1$ and frequency $\Omega \gg 1$, as given by (7.41). (—) $q\Omega \neq 0$, (– – –) $q\Omega = 0$

constant speed $\dot{z} = \tilde{z}$,

$$\tilde{z} = -q\Omega \, \sin\left(\frac{\pi}{2}\mu_\Delta/\bar{\mu}\right)$$
$$\approx -\frac{\pi}{2}q\Omega\mu_\Delta/\bar{\mu} \quad \text{for} \quad \mu_\Delta/\bar{\mu} \ll 1. \tag{7.44}$$

This speed, the *drift velocity*, vanishes as the coefficients of friction become identical, i.e. as $\mu_\Delta \to 0$. Indeed, the asymmetry does not need to be in the friction coefficients: A quite similar effect is induced if the vibrating plane is excited along a skew direction (different from vertical and horizontal), or if the mass has an internal degree of freedom oriented at a skew angle with respect to gravity. This kind of bias accounts for many commonly observed phenomena in strongly vibrating environments, e.g. self-loosening of screws and nuts and vibrational transportation of objects. Blekhman and Dzhanelidze's (1964) have provided a major account on such effects; see also Chatterjee et al. (2005), and Hunnekens et al. (2011).

A strikingly simple way of creating travelling motions out of vibration is the *bristlebot* – a device to be made at the kitchen table in a few minutes and at very little cost[1]: (1) Cut the brush end of a toothbrush and bend its bristles a little back over; (2) Mount a small pager motor (e.g. a cell phone vibration motor) to the top using double sided tape; (3) Stick a 3-volt button cell battery at the tape, with one of the pager motor leads below; (4) Bend the other lead to touch the side of the battery; (5) Watch the bristlebot as crawls its way on a surface. This manifestation of the biasing effect is caused by the asymmetry in the orientation of the bristles, and thus the friction forces, which in the presence of vertically imposed vibrations implies that horizontal motions are a little larger in one direction than in the other, thus creating a net average motion in one direction. For a theoretical analysis of bristlebot motion see e.g. Cicconofri and DeSimone (2016).

7.3.3 Brumberg's Pipe (Smoothening and Biasing)

Fig. 7.6 shows Brumberg's Pipe (Brumberg 1970), which is shaken horizontally and vertically at high frequencies $\tilde{\Omega}$ and $2\tilde{\Omega}$, respectively, in a gravity field g. Inside the pipe slides a solid mass of characteristic length L and coefficient of friction μ. Under proper relative phasing of the two excitations, the longitudinal inertia forces acting on the mass are directed upwards at those intervals of time where the transverse normal forces, and thus the friction forces, are weakest – whereas at time intervals where the longitudinal inertia is directed downwards, the friction forces are strongest. Thus the mass moves a little longer upwards than downwards during each period of vertical oscillation, traveling up the pipe, against gravity, at a

---

[1]Instructions e.g. on YouTube. (The author clocked 7 min. for making a first bristle-bot.)

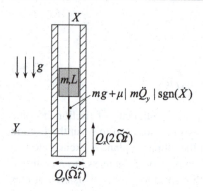

Fig. 7.6 Brumberg's pipe

constant average speed with an overlay of HF oscillations. For the present purpose we only consider harmonic excitations with a relative phasing that maximizes this effect (Blekhman 2000):

$$Q_x = -a_x L \sin(2\tilde{\Omega}\tilde{t}), \quad Q_y = a_y L \sin(\tilde{\Omega}\tilde{t}), \tag{7.45}$$

by which the nondimensional equation of motion becomes:

$$\ddot{x} + 1 + \mu a_y \Omega^2 |\sin(\Omega t)| \mathrm{sgn}(\dot{x}) + 4 a_x \Omega^2 \sin(2\Omega t) = 0, \tag{7.46}$$

where $x = x/L$ denotes the longitudinal position of m relative to the pipe, $t = \omega_0 \tilde{t}$ is nondimensional time, $\omega_0 = \sqrt{g/L}$ a characteristic frequency, $\dot{x} = dx/dt$, a_x and a_y are the horizontal and vertical excitation amplitudes, and $\Omega = \tilde{\Omega}/\omega_0$ the fundamental excitation frequency. For the magnitude of parameters we assume the excitation is high in frequency and small in amplitude, and that friction is weak, i.e. $\Omega^{-1} \ll 1$, $a_{x,y} = O(\Omega^{-1})$, $a_{x,y}\Omega = O(1)$, and $\mu = O(\Omega^{-1})$.

Brumberg obtained an exact but complicated solution to (7.46). Here we use the MDSM, splitting the motions into slow and fast components:

$$x = x(t, \tau) = z(t) + \Omega^{-1}\varphi(t, \tau), \tag{7.47}$$

where $\tau = \Omega t$ is the fast time, and φ is a small overlay of fast oscillations having zero average. Proceeding as for two previous examples one finds that the fast motions φ are given by:

$$\varphi(t, \tau) = a_x \Omega \sin(2\tau), \tag{7.48}$$

while the slow motions z are governed by:

$$\ddot{z} + 1 + \mu a_y \Omega^2 \langle \mathrm{sgn}(\dot{z} + 2a_x \Omega \cos(2\tau)) | \sin \tau | \rangle = 0, \tag{7.49}$$

where $\langle\rangle$ is the averaging operator defined by (7.20) or (7.5). Calculating the averaging term, the final equation for the slow motions becomes:

$$\ddot{z} + 1 + \frac{2}{\pi}\mu a_y \Omega^2 \bar{h}(\dot{z}) = 0, \tag{7.50}$$

where the function $\bar{h}$ is defined by:

$$\bar{h}(\dot{z}) = \begin{cases} \text{sgn}(\dot{z}) & \text{for } |\dot{z}| \geq 2a_x\Omega, \\ 1 - \sqrt{2(1 - \dot{z}/(2a_x\Omega))} & \text{for } |\dot{z}| < 2a_x\Omega. \end{cases} \tag{7.51}$$

To see if there are stationary solutions corresponding to a situation where the mass moves at constant average speed $\dot{z}(t) = \tilde{z}$ up the pipe, we let $(\dot{z}, \ddot{z}) = (\tilde{z}, 0)$ in (7.50)–(7.51), solve for $\tilde{z}$ and find:

$$\tilde{z} = a_x\Omega\left(2 - \left(1 + \frac{1}{\frac{2}{\pi}\mu a_y \Omega^2}\right)^2\right), \quad \text{for } \frac{2(\sqrt{2}-1)}{\pi}\mu a_y \Omega^2 \geq 1, \tag{7.52}$$

where the latter condition ensures existence and stability of the solution, and the main assumptions on parameters should be recalled: $a_{x,y}\Omega = O(1)$, $\mu\Omega = O(1)$, and $\Omega \gg 1$. Hence, in the presence of HF excitation of proper strength and phasing, the mass is biased to move at constant speed, against gravity.

Brumberg's pipe shares features with the mass on a vibrating plane in the previous example, as is revealed when Taylor-expanding (7.50)–(7.51) for small $\dot{z}$, re-arranging, and neglecting small terms of the order $\dot{z}^2$ and higher:

$$\ddot{z} + \frac{1}{\sqrt{2}\pi}\mu\Omega\left(\frac{a_y}{a_x}\right)\dot{z} = \frac{2(\sqrt{2}-1)}{\pi}\mu a_y \Omega^2 - 1. \tag{7.53}$$

Two features are apparent from this equation of slow motions: First there is a smoothening effect, where the essentially nonlinear dissipation term in the original Eq. (7.46) (representing dry friction) is replaced by a linearizable dissipative term (representing equivalent viscous damping). Secondly there is a biasing effect, represented by the first term of the right-hand side of (7.53), which may change the original bias of the mass, from moving downwards to moving upwards in gravity; this happens when $\mu a_y \Omega^2$ is large enough to make the right-hand side positive.

7.4 A Slight but Useful Generalization

The equations of motion (7.18) and (7.34) for, respectively, the pendulum example and the sliding mass have a similar form, which is quite common in applications:

$$\ddot{u} + s(u, \dot{u}) = \Omega f(u) \sin(\Omega t), \tag{7.54}$$

which also includes some of the most frequently occurring equations in vibration theory: the *Duffing equation*, the *Mathieu* equation, and the *van der Pol equation* (for a special treatment of the latter with HF excitation see e.g. Belhaq and Fahsi 2008; Belhaq and Sah 2008). Here $s = O(1)$ hold the 'slow' forces, $f = O(1)$ describe the position-dependent part of the 'fast' forces, and $\Omega \gg 1$. Using the MDSM we let

$$u = u(t, \tau) = z(t) + \Omega^{-1} \varphi(t, \tau), \tag{7.55}$$

where $\tau = \Omega t$. Performing the same operations as for the examples in Sect. 7.3, or the general procedure outlined in Sect. 7.2.1, it is found that the fast motions φ are given by:

$$\varphi = -f(z) \sin \tau + O(\Omega^{-1}), \tag{7.56}$$

while the slow motions z are governed by:

$$\ddot{z} + s(z, \dot{z}) = V_f(z) + V_s(z, \dot{z}), \tag{7.57}$$

where the *vibrational forces* V_s and V_f are given by

$$\begin{aligned} V_f(z) &= -\tfrac{1}{2} f(z) f'_u(z), \\ V_s(z, \dot{z}) &= \langle s(z, \dot{z}) - s(z, \dot{z} - f(z) \cos \tau) \rangle, \end{aligned} \tag{7.58}$$

where $f_u' = \partial f / \partial u$.

Again, the equation for the slow motions z is autonomous, with its left-hand side similar to the original equation of motion (7.54), while the vibrational forces V_s and V_f incorporate the average effect of the HF excitation term in (7.54). The vibrational force V_s vanishes if s is linear in the velocity terms, and thus contributes in cases with nonlinear dissipation, such as the example with the mass on a plane with dry friction (Sect. 7.3.2). Similarly, the vibrational force V_f vanishes if f is a constant, and thus contributes in cases with parametric excitation, such as the example with the pendulum on a vibrating support (Sect. 7.3.1).

Several general results regarding the stiffening, biasing, and smoothening effects of HF excitation may be deduced from (7.57)–(7.58). For example, the change in effective stiffness for an equilibrium $z = \tilde{z}$ of (7.57) is $\Delta k = \tfrac{1}{2} (f f'_u)'_u \big|_{u=\tilde{z}}$, which can be both positive and negative. For the pendulum Eq. (7.18), with $f(\theta) = q\Omega \sin\theta$, this gives an increase in stiffness, $\Delta k = 1/2 q^2 \Omega^2$, for both pendulum equilibriums ($z = 0$ and $z = \pi$). On the other hand, if the pendulum support is excited horizontally instead of vertically, then $f(\theta) = q\Omega \cos\theta$, which gives a *reduction* in stiffness, $\Delta k = -1/2 q^2 \Omega^2$, for the equilibriums $z = 0, \pi$.

The changes in stiffness in turn imply changes in natural frequencies and equilibrium stability, which can also be calculated in similar general terms. However, we shall postpone this to the following section dealing with a far more general class of systems.

7.5 A Fairly General Class of Discrete Systems

The general system (7.54) analyzed in the previous section covers many, but far from all relevant systems with HF excitation. For example, in the equation of motion (7.46) for Brumberg's pipe, a velocity $\dot{x}$ appears in the HF excitation term, thus excluding this system from the class (7.54). Also, to claim generality we need to include systems with several degrees of freedom, and HF excitations more general than a mono-frequency time harmonic. On the other hand, the system (7.1), used to outline the MDSM, is so general that it is very difficult to extract meaningful information regarding the HF effects of concern here (though see Kremer 2016).

So, as a useful compromise, we consider a general class of systems with HF excitation that is broad enough to cover a large variety of systems of scientific and industrial interest, but also sufficiently specific to provide results that are physically interpretable. The purpose of the analysis is to set up expressions governing the slow motions, and to quantify the stiffening, the biasing, and the smoothening effects for this general system. As should be evident from the examples in Sect. 7.3, these effects are understood to characterize the *averaged* behavior of the system, i.e. what can be observed when ignoring the details of a small overlay of HF oscillations. This corresponds to lowpass filtering the observed response – either digitally, if the response consists of computer simulated or real sampled data, or analogously, as with response perceived by band limited devices such as physical measurement instruments and human senses.

Of course there are many ways to study HF effects in a general sense, differing in the level of generalization, and the particular effects at focus; the following is based on Thomsen (2002, 2003a, 2005); for alternative attempts with a similar goal see e.g. Blekhman and Sorokin (2016), Fidlin (2006), Kremer (2016).

7.5.1 The System

We consider dynamical systems that can be modeled by a finite number of second order ordinary differential equations, generally nonlinear, and with time-explicit HF excitation (Thomsen 2002):

$$\mathbf{M}(\mathbf{u},t)\frac{d^2\mathbf{u}}{dt^2}+\mathbf{s}(\mathbf{u},\frac{d\mathbf{u}}{dt},t)+\sum_{j=1}^{m}\left(\mathbf{h}_j(\mathbf{u},\frac{d\mathbf{u}}{dt},t)+\Omega\mathbf{f}_j(\mathbf{u},t)\right)\frac{\partial^2}{\partial(\Omega t)^2}\xi_j(t,\Omega t)=\mathbf{0},$$

$$\mathbf{u}=\mathbf{u}(t);\quad \mathbf{u}(0)=\mathbf{u}_0;\quad (d\mathbf{u}/dt)_{t=0}=\dot{\mathbf{u}}_0;\quad \Omega\gg 1,$$

$$(7.59)$$

where $\mathbf{u}$ is an n-vector describing the positional state of the system at time t, $\mathbf{M}$ is a positive definite $n \times n$ matrix describing inertia forces, $\mathbf{s}$ is an n-vector of 'slow' forces, Ω is a large number representing a fundamental excitation frequency, $\mathbf{h}_j$ and $\mathbf{f}_j$ are n-vectors that, jointly with the scalar time functions ξ_j, describe 'fast' or rapidly oscillating excitations, and the functions ξ_j and their first and second derivatives with respect to Ωt are 2π-periodic in Ωt with zero average (i.e. any non-zero averages should be extracted and moved to the $\mathbf{s}$ functions). Thus the three main terms of the matrix equation describe a balance between inertia forces ($\mathbf{M}$), slowly changing internal and external forces ($\mathbf{s}$), and rapidly oscillating forces ($\mathbf{h}_j$, $\mathbf{f}_j$, ξ_j). The class of systems described by (7.59) is sufficiently broad to cover many applications involving finite-degree-of-freedom or discretized continuous mechanical systems, including those described in Sect. 7.3 and those shown in Fig. 7.7.

The notions of 'slow' and 'fast' refer to two distinct characteristic time scales or frequencies characterizing motions of the system: First there is a time scale t and a characteristic frequency ω describing motions of the system when all $\xi_j = 0$; For example, one can take as ω the largest natural frequency of the linearized unloaded system, i.e. the largest real root of the characteristic polynomial $|\mathbf{J}(\mathbf{0}, \mathbf{0}, t) - \omega^2\mathbf{M}(\mathbf{0}, t)| = 0$, where $\mathbf{J} \equiv \partial\mathbf{s}/\partial\mathbf{u}$. And then there is another time scale Ωt, much faster than t, which describes fluctuations imposed by external excitations of characteristic frequency $\Omega \gg \omega$, with time variations specified by ξ_j. It is further assumed that all functions in (7.59) are generally nonlinear functions of their arguments, that they are of magnitude order unity or lower (i.e. $O(\|\mathbf{M}\|) = O(\|\mathbf{s}\|) = O(\|\mathbf{h}_j\|) = O(\|\mathbf{f}_j\|) = O(\|\mathbf{M}\|) = O(\xi_j) \le 1$), that ξ_j are bounded on $[0;2\pi]$, and that $\mathbf{f}_j$ and $\mathbf{M}$ are bounded with continuous first derivatives with respect to $\mathbf{u}$ (whereas $\mathbf{h}_j$ and $\mathbf{s}$ are not necessarily continuous). The assumptions on magnitude order and boundedness are not required to hold in general (this would preclude even linear functions such as $\mathbf{s} = \mathbf{u}$), but only for the type of solutions under consideration; Typically any kind of unstable or strongly resonant solution causes the assumptions to be violated.

In (7.59) $\mathbf{f}_j$ is multiplied be the large parameter Ω while $\mathbf{h}_j$ is not, i.e. the part of the fast forces depending on velocity is assumed to be of lower order of magnitude than the part depending only on position. In some special applications this is not the case, i.e. there is a strong velocity-dependent HF excitation. The more complicated analysis of this case was given by Fidlin (1999), Blekhman and Sorokin (2010), and Kremer (2016).

The HF excitation is assumed to be provided explicitly by some external energy source, as specified by the functions ξ_j, i.e. (7.59) is non-autonomous in the fast time Ωt; this is the typical case in applications. However, the HF excitation could

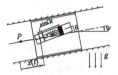

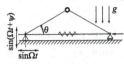

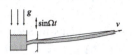

(a) Vibration-induced movement using friction layers and a HF-resonator (Thomsen 2000; Fidlin and Thomsen 2001).

(b) Change of equilibriums, stability, natural frequencies, and nonlinear response for a HF-excited two-bar link (Tcherniak and Thomsen 1998).

(c) Using resonant HF excitation to pump fluid (Jensen 1997) or continuous material (Jensen 1996) through pipes.

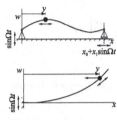

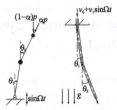

(d) Using vibration-induced sliding to damp resonant vibrations of strings and beams (Babitsky and Veprik 1993; Thomsen 1996a,b; Miranda and Thomsen 1998).

(e) Chelomei's pendulum: Has a freely sliding disk. With HF excitation both the disk and the pendulum stabilizes against gravity (Chelomei 1983; Blekhman and Malakhova 1986; Thomsen and Tcherniak 2001).

(f) Stabilization and change of nonlinear behavior for follower-loaded double-pendulums with HF support excitation (left) (Jensen 1998, 2000b), and for articulated pipes carrying HF pulsating fluid (right) (Jensen 1999a).

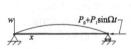

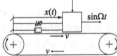

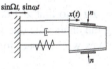

(g) Increasing buckling load and natural frequencies for a column by HF excitation (Chelomei 1956, 1983; Jensen 2000a; Jensen, et al. 2000; Mailybaev and Seyranian 2009).

(h) Quenching friction-induced stick-slip vibration by using HF excitation (Thomsen 1999; Chatterjee et al. 2004).

(i) Quenching chaotic oscillations for a stick-slip friction oscillator using HF excitation (Feeny and Moon 2000).

Fig. 7.7 HF-excited systems representing stiffening (**b,e,f,g**), biasing (**a,b,c,d,e,h**), and smoothening (**a,h,i**) effects

also come from some *internal* oscillator degree of freedom, corresponding to a system model *autonomous* in Ωt. The MDSM can be extended to analyzing this case, as demonstrated by Sheheitli and Rand (2011, 2012) for a system model with three coupled van der Pol oscillators having vastly different frequencies, and a mass-spring-pendulum system with vastly different frequencies.

7.5.2   Example Functions

For the example systems in Sect. 7.3 one has, for the pendulum with a vibrating support: $\mathbf{u} = \theta$, $\mathbf{M} = 1$, $\mathbf{s} = 2\beta\dot{\theta} + \sin\theta$, $m = 1$, $\mathbf{h}_1 = 0$, $\mathbf{f}_1 = -q\Omega\sin\theta$, $\xi_1 = -\sin\Omega t$; for the mass on a vibrating plane: $\mathbf{u} = u$, $\mathbf{M} = 1$, $\mathbf{s} = \omega^2 u + \bar{\mu}\,\mathrm{sgn}(\dot{u}) + \mu_\Delta$, $m = 1$, $\mathbf{h}_1 = 0$, $\mathbf{f}_1 = -q\Omega$, $\xi = -\sin\Omega t$; and for Brumberg's pipe: $\mathbf{u} = x$, $\mathbf{M} = 1$, $\mathbf{s} = 1$, $m = 2$, $\mathbf{h}_1 = 0$, $\mathbf{f}_1 = -a_x\Omega$, $\xi_1 = \sin(2\Omega t)$, $\mathbf{h}_2 = -\mu a_y\Omega^2\mathrm{sgn}(\dot{x})$, $\mathbf{f}_2 = 0$, $\xi_2 = |\sin(\Omega t)| - 2/\pi$.

7.5.3   The Averaged System Governing the 'Slow' Motions

Using the MDSM to analyze the general system (7.59), we split the motions into slow and fast components:

$$\mathbf{u} = \mathbf{u}(t, \tau) = \mathbf{z}(t) + \Omega^{-1}\boldsymbol{\varphi}(t, \tau), \qquad (7.60)$$

where $\tau = \Omega t$ is the fast time, $\mathbf{z}(t)$ the slow motions, and $\Omega^{-1}\boldsymbol{\varphi}$ the fast motions which has small amplitude, is 2π-periodic in τ, and has zero fast-time average:

$$\langle\boldsymbol{\varphi}(t, \tau)\rangle \equiv \frac{1}{2\pi}\int_0^{2\pi}\boldsymbol{\varphi}(t, \tau)\,d\tau = 0, \qquad (7.61)$$

Then $\langle\mathbf{u}\rangle = \mathbf{z}(t)$, so that $\mathbf{z}(t)$ describes the fast-time average of the total motion $\mathbf{u}$. Also, using the chain rule one finds that the time derivatives of $\mathbf{u}$ transform into partial derivatives with respect to the two time scales t and τ as follows:

$$\begin{aligned}
\frac{d\mathbf{u}}{dt} &= \dot{\mathbf{z}} + \boldsymbol{\varphi}' + \Omega^{-1}\dot{\boldsymbol{\varphi}}, \\
\frac{d^2\mathbf{u}}{dt^2} &= \Omega\boldsymbol{\varphi}'' + \ddot{\mathbf{z}} + 2\dot{\boldsymbol{\varphi}}' + \Omega^{-1}\ddot{\boldsymbol{\varphi}},
\end{aligned} \qquad (7.62)$$

where $\dot{\boldsymbol{\varphi}} \equiv \partial\varphi/\partial t$ and $\boldsymbol{\varphi}' \equiv \partial\varphi/\partial\tau$. Inserting this and (7.60) into (7.59) we find:

$$\begin{aligned}
&\mathbf{M}(\mathbf{z} + \Omega^{-1}\boldsymbol{\varphi}, t)\left(\Omega\boldsymbol{\varphi}'' + \ddot{\mathbf{z}} + 2\dot{\boldsymbol{\varphi}}' + \Omega^{-1}\ddot{\boldsymbol{\varphi}}\right) + \mathbf{s}(\mathbf{z} + \Omega^{-1}\boldsymbol{\varphi}, \dot{\mathbf{z}} + \boldsymbol{\varphi}' + \Omega^{-1}\dot{\boldsymbol{\varphi}}, t) \\
&+ \sum_j\left(\mathbf{h}_j(\mathbf{z} + \Omega^{-1}\boldsymbol{\varphi}, \dot{\mathbf{z}} + \boldsymbol{\varphi}' + \Omega^{-1}\dot{\boldsymbol{\varphi}}, t) + \Omega\mathbf{f}_j(\mathbf{z} + \Omega^{-1}\boldsymbol{\varphi}, t)\right)\xi_j''(t, \tau) = \mathbf{0},
\end{aligned}$$

$$(7.63)$$

or, Taylor-expanding for $\Omega^{-1} \ll 1$ and rearranging:

$$
\begin{aligned}
\boldsymbol{\varphi}'' = &-\mathbf{M}^{-1}(\mathbf{z}, t) \sum_j \mathbf{f}_j(\mathbf{z}, t) \xi_j''(t, \tau) - \Omega^{-1}(\ddot{\mathbf{z}} + 2\dot{\boldsymbol{\varphi}}') \\
&- \Omega^{-1}\mathbf{M}^{-1}(\mathbf{z}, t)\mathbf{s}(\mathbf{z}, \dot{\mathbf{z}} + \boldsymbol{\varphi}', t) \\
&- \Omega^{-1}\mathbf{M}^{-1}(\mathbf{z}, t)\Bigg\{ \sum_j \left(\mathbf{h}_j(\mathbf{z}, \dot{\mathbf{z}} + \boldsymbol{\varphi}', t) + \nabla\mathbf{f}_j(\mathbf{z}, t)\boldsymbol{\varphi} \right) \xi_j''(t, \tau) \\
&+ \nabla\mathbf{m}(\mathbf{z}, t, \boldsymbol{\varphi}) \Bigg\} + O(\Omega^{-2}),
\end{aligned}
\tag{7.64}
$$

Here the symbol ∇ indicates derivatives with respect to position variables:

$$
\begin{aligned}
\nabla\mathbf{f}_j(\mathbf{u}, t) &\equiv \frac{\partial \mathbf{f}_j}{\partial \mathbf{u}} = \left[\frac{\partial \mathbf{f}_j}{\partial \mathbf{u}_{(1)}} \quad \cdots \quad \frac{\partial \mathbf{f}_j}{\partial \mathbf{u}_{(n)}} \right], \\
\nabla\mathbf{m}_{(i)}(\mathbf{u}, t, \boldsymbol{\varphi}) &\equiv (\boldsymbol{\varphi}'')^{\mathrm{T}} \nabla\mathbf{M}_{(i)}\boldsymbol{\varphi} \\
\nabla\mathbf{M}_{(i)}(\mathbf{u}, t) &\equiv \frac{\partial \mathbf{M}_{(i)}}{\partial \mathbf{u}} = \left[\frac{\partial \mathbf{M}_{(i)}}{\partial \mathbf{u}_{(1)}} \quad \cdots \quad \frac{\partial \mathbf{M}_{(i)}}{\partial \mathbf{u}_{(n)}} \right],
\end{aligned}
\tag{7.65}
$$

where a subscript in parenthesis denotes a particular vector element or matrix column, and T denotes vector or matrix transpose. Solving (7.64) for $\boldsymbol{\varphi}$ we find:

$$
\boldsymbol{\varphi}(t, \tau) = \hat{\boldsymbol{\varphi}}(t, \tau) + O(\Omega^{-1}),
\tag{7.66}
$$

where $\hat{\boldsymbol{\varphi}}$ is the zero order approximate solution that will be used in place of $\boldsymbol{\varphi}$, satisfying $\langle \hat{\boldsymbol{\varphi}} \rangle = 0$ and $|\hat{\boldsymbol{\varphi}}| = O(1)$:

$$
\hat{\boldsymbol{\varphi}}(t, \tau) = -\mathbf{M}^{-1}(\mathbf{z}, t) \sum_j \mathbf{f}_j(\mathbf{z}, t) \xi_j(t, \tau).
\tag{7.67}
$$

To obtain an equation governing the slow motions $\mathbf{z}(t)$ we employ the averaging operator $\langle \rangle$ to (7.64), recalling that $\langle \xi_j \rangle = \langle \xi_j' \rangle = \langle \xi_j'' \rangle = 0$ by assumption, and then insert the solution (7.66) for $\boldsymbol{\varphi}$, multiply by $\Omega\mathbf{M}$, neglect terms of order Ω^{-1} and lower, rearrange, and obtain:

$$
\mathbf{M}(\mathbf{z}, t)\ddot{\mathbf{z}} + \mathbf{s}(\mathbf{z}, \dot{\mathbf{z}}, t) + \mathbf{v}(\mathbf{z}, \dot{\mathbf{z}}, t) = \mathbf{0},
\tag{7.68}
$$

where $\mathbf{v}(\mathbf{z}, \dot{\mathbf{z}}, t)$ is the *vibrational force* (with $\hat{\boldsymbol{\varphi}}$ given by (7.67)):

$$
\begin{aligned}
\mathbf{v}(\mathbf{z}, \dot{\mathbf{z}}, t) = &\langle \mathbf{s}(\mathbf{z}, \dot{\mathbf{z}} + \hat{\boldsymbol{\varphi}}', t) - \mathbf{s}(\mathbf{z}, \dot{\mathbf{z}}, t) \rangle + \sum_j \langle \mathbf{h}_j(\mathbf{z}, \dot{\mathbf{z}} + \hat{\boldsymbol{\varphi}}', t) \xi_j''(t, \tau) \rangle \\
&+ \sum_j \nabla\mathbf{f}_j(\mathbf{z}, t) \langle \hat{\boldsymbol{\varphi}} \xi_j''(t, \tau) \rangle + \langle \nabla\mathbf{m}(\mathbf{z}, t, \hat{\boldsymbol{\varphi}}) \rangle,
\end{aligned}
\tag{7.69}
$$

As appears, Eq. (7.68) for the slow motions z is similar in form to the original equation of motion (7.59) for the full motions $\mathbf{u}$, though, with the integrated influence of fast forces $(\mathbf{h}_j + \Omega \mathbf{f}_j)\xi_j''$ accounted for by the vibrational forces $\mathbf{v}$, depending only on the slowly changing variables z and t. Any explicit dependence on the fast time variable $\tau = \Omega t$ has been averaged out, which makes the equation for z far easier to solve than the original equation for $\mathbf{u}$. This holds for analytical solutions, not least because there are powerful mathematical tools applying only to autonomous systems. And it holds for numerical solutions, where much larger time steps can be used since there is no need to keep track of the rapid oscillations at frequency Ω, and the numerical ill-posedness or 'stiffness' is reduced that arise when solving systems of differential equations on vastly different time scales.

The initial conditions needed to solve (7.68) for z is obtained from the original initial conditions in (7.59), which by (7.60), (7.62), and (7.66)–(7.67) becomes:

$$\mathbf{z}(0) = \mathbf{u}_0 - \Omega^{-1} \hat{\boldsymbol{\varphi}}(0,0); \quad \dot{\mathbf{z}}(0) = \dot{u}_0 - \hat{\boldsymbol{\varphi}}'(0,0) - \Omega^{-1} \dot{\hat{\boldsymbol{\varphi}}}(0,0). \tag{7.70}$$

This completes the separation of the full motions $\mathbf{u}$ into slow and fast components z and $\boldsymbol{\varphi}$, as given by (7.60) and (7.67), and (7.68)–(7.69).

7.5.4 Interpretation of Averaged Forcing Terms

In Eq. (7.68) for the slow motions z, the vibrational forces v describe average effects. They correspond to real physical forces having the same effect as the HF excitation, on the average. To an observer or measuring instrument filtering out small-amplitude HF vibrations, the response u from (7.59) is similar to the response z from (7.68). It should be recalled that the average of the HF excitation itself is assumed to be zero, $\langle \xi_j \rangle = 0$. In many cases its average effect is also zero, i.e. $v = 0$. So, it is instructive to consider each of the terms making up v, to see when non-zero average effects can appear:

The first term in (7.69) expresses the average of the nonlinear velocity-dependent terms of the slow forces $\mathbf{s}$. This term vanishes if $\mathbf{s}$ is independent of velocity, or is linear in velocity. However, it may be non-zero when $\mathbf{s}$ is nonlinear in velocity terms, such as for systems with dry friction (cf. the examples in Sect. 7.3 and Fig. 7.7(h,i)) or nonlinear damping (*cf.* Problem 7.6, and Chatterjee et al. 2003).

The second term expresses the average effect of the velocity-dependent parts $\mathbf{h}_j$ of the fast forces. This term disappears when the fast forces are independent of velocity, but generally does not disappear even if the velocity-dependency is linear. The example systems of Brumberg's pipe (Sect. 7.3.3) and Fig. 7.7(a,f(right)) display non-zero average terms of this type.

The third and fourth terms describe the average effect of strong, velocity independent parametrical excitation terms. These terms disappear if $\mathbf{f}_j$ and $\mathbf{M}$ are independent on **u**. Non-zero averages of this kind of terms are quite common where nontrivial effects of HF excitation appear, e.g. they occur with the example systems of the pendulum on a vibrating support (Sect. 7.3.1) and Fig. 7.7(a–g).

It is important to note that the vibrational force is generally a nonlinear function of ξ_j, since $\hat{\boldsymbol{\phi}}$ depends on ξ_j (cf. (7.67)). Consequently, if $\mathbf{v}_1$ is the vibrational force corresponding to a specific time variation of the excitation $\xi_1(t, \Omega t)$, and $\mathbf{v}_2$ is the force corresponding to another excitation $\xi_2(t, \Omega t)$, then the force $\mathbf{v}$ corresponding to the simultaneous action of both excitations does *not*, in general, equal $\mathbf{v}_1 + \mathbf{v}_2$. Thus one cannot readily predict the effects of combined excitations based on knowledge of isolated effects (Thomsen 2008a considers effects of multifrequency HF excitation). However, consideration to (7.69) shows that when **M** and $\mathbf{f}_j$ are independent of **u**, and $\mathbf{h}_j$ is independent of $d\mathbf{u}/dt$, and **s** is linear in or independent of $d\mathbf{u}/dt$ – then one can indeed calculate the vibrational force for a sum of excitations by adding up the vibrational forces for each excitation. None of the examples in Sect. 7.3 or Fig. 7.7 fulfills this necessary condition for the validity of superposition.

7.5.5   The Effects

Next we make general predictions for the stiffening, biasing, and smoothening effects, based on the averaged system with vibrational forces as given by (7.69). Note that (7.68) is valid on the quite general assumptions stated below the original system (7.59), whereas various additional assumptions may be stated during the following analysis. However, it will not be explicitly stated whether in each case the system functions **M**, **s**, **h**, and **f** are allowed to depend on the slow time t, since this is obvious from the context; thus the t in e.g. $f(u, t)$ will be kept even when f is actually not allowed to vary explicitly with t.

We shall base several results on a linearization of the averaged system (7.68)–(7.69) near a possible static equilibrium $(\mathbf{z}, \dot{\mathbf{z}}) = (\tilde{\mathbf{z}}, 0)$, where $\tilde{\mathbf{z}}$ is defined by:

$$\mathbf{s}(\tilde{\mathbf{z}}, \mathbf{0}, t) + \mathbf{v}(\tilde{\mathbf{z}}, \mathbf{0}, t) = \mathbf{0}, \tag{7.71}$$

where for consistency in notation we keep t in the function arguments, as explained above, even though the existence of static equilibriums generally requires **s** and **v** to be independent of t. The linearization is readily obtained using Taylor-expansion at the equilibrium, and discarding nonlinear terms:

$$\mathbf{M}(\tilde{\mathbf{z}}, t)\ddot{\mathbf{z}} + (\mathbf{C}(\tilde{\mathbf{z}}) + \Delta\mathbf{C}(\tilde{\mathbf{z}}))\dot{\mathbf{z}} + (\mathbf{K}(\tilde{\mathbf{z}}) + \Delta\mathbf{K}(\tilde{\mathbf{z}}))(\mathbf{z} - \tilde{\mathbf{z}}) = \mathbf{0}, \tag{7.72}$$

where, using derivative operators $\nabla g(\mathbf{u}, \dot{\mathbf{u}}) \equiv \partial g / \partial \mathbf{u}$ and $\dot{\nabla} g(\mathbf{u}, \dot{\mathbf{u}}) \equiv \partial g / \partial \dot{\mathbf{u}}$:

$$
\begin{aligned}
\mathbf{K}(\tilde{\mathbf{z}}) &= \nabla \mathbf{s}(\tilde{\mathbf{z}}, 0, t), \quad \Delta \mathbf{K}(\tilde{\mathbf{z}}) = \nabla \mathbf{v}(\tilde{\mathbf{z}}, 0, t), \\
\mathbf{C}(\tilde{\mathbf{z}}) &= \dot{\nabla} \mathbf{s}(\tilde{\mathbf{z}}, 0, t), \quad \Delta \mathbf{C}(\tilde{\mathbf{z}}) = \dot{\nabla} \mathbf{v}(\tilde{\mathbf{z}}, 0, t).
\end{aligned}
\tag{7.73}
$$

7.5.6   Stiffening

By stiffness we refer to the change in resistive static force per unit deformation near a static equilibrium. Stiffening then refers to a situation where the stiffness of an equilibrium of the equation of slow motions differs from the stiffness of the same equilibrium of the original equation of motion. Examples of stiffening appeared in the vibrated pendulum example (Sect. 7.3.1), and in Fig. 7.7(b,e,f,g).

For convenience we assume that, in absence of HF excitation, the general system (7.59) has a static equilibrium at $\mathbf{u} = \mathbf{0}$, so that $\mathbf{s}(\mathbf{0}, 0, t) = \mathbf{0}$ (other equilibriums $\mathbf{u} = \tilde{\mathbf{u}} \neq \mathbf{0}$ are treated by first applying a transform of coordinates, $\mathbf{u} \rightarrow \mathbf{u} - \tilde{\mathbf{u}}$), and that $\mathbf{s}$ has continuous derivatives with respect to $\mathbf{u}$ at that equilibrium. It is also assumed that this equilibrium remains an equilibrium in the presence of HF excitation, i.e. $\mathbf{v}(\mathbf{0}, \mathbf{0}, 0) = \mathbf{0}$ (the case $\mathbf{v}(\mathbf{0}, \mathbf{0}, 0) \neq \mathbf{0}$ is treated in the section on *biasing* further below).

We define the stiffening effect of HF excitation as the additional infinitesimal change in effective (generalized) force, be it positive or negative, which appears in response to an infinitesimal change in position near the equilibrium. Using (7.72)–(7.73) with $\tilde{\mathbf{z}} = \mathbf{0}$ we find that the stiffness changes from $\mathbf{K}(\mathbf{0})$ to $\mathbf{K}(\mathbf{0}) + \Delta \mathbf{K}(\mathbf{0})$,

$$
\mathbf{K}(\mathbf{0}) = \nabla \mathbf{s}(\mathbf{0}, 0, t), \quad \Delta \mathbf{K}(\mathbf{0}) = \nabla \mathbf{v}(\mathbf{0}, 0, t).
\tag{7.74}
$$

Only in rare cases will the slow forces $\mathbf{s}$ contribute to the stiffening effect $\Delta \mathbf{K}(\mathbf{0})$. First, conferring with (7.69) one finds that only those components of $\mathbf{s}$ that are nonlinear in velocity *and* linear in position can contribute to $\Delta \mathbf{K}(\mathbf{0})$. For the examples of this chapter, only the one in Fig. 7.7(i) has such a component, and in that case the relevant average vanishes at the equilibrium. Similarly, the velocity-dependent part $\mathbf{h}_j$ of the fast forces may theoretically contribute to the stiffening effect, but typically it does not.

So, the main responsible terms for the stiffening effects are the two last terms in (7.69); we denote their contribution by $\Delta \mathbf{K}^{\mathbf{f},\mathbf{m}}(\mathbf{0})$, so that by (7.74) and (7.69):

$$
\Delta \mathbf{K}^{\mathbf{f},\mathbf{m}}(\mathbf{0}) = \left. \frac{\partial}{\partial \mathbf{z}} \right|_{\mathbf{z}=0} \left(\sum_j \nabla \mathbf{f}_j(\mathbf{z}, t) \left\langle \hat{\boldsymbol{\phi}} \xi_j''(t, \tau) \right\rangle + \langle \nabla \mathbf{m}(\mathbf{z}, t, \hat{\boldsymbol{\phi}}) \rangle \right).
\tag{7.75}
$$

Inserting (7.67) for $\hat{\varphi}$, we find that the first averaging term becomes:

$$
\begin{aligned}
\left\langle \hat{\varphi}\, \xi_j''(t,\tau) \right\rangle &= -\mathbf{M}^{-1}(\mathbf{z},t) \sum_k \left\langle \xi_k \xi_j'' \right\rangle \mathbf{f}_k(\mathbf{z},t) \\
&= \mathbf{M}^{-1}(\mathbf{z},t) \sum_k \left\langle \xi_k' \xi_j' \right\rangle \mathbf{f}_k(\mathbf{z},t),
\end{aligned}
\tag{7.76}
$$

where the last equality follows from the fact that $\left\langle \xi_k \xi_j'' \right\rangle = -\left\langle \xi_k' \xi_j' \right\rangle$ when ξ_k is 2π-periodic in τ and has zero average (this can be verified by representing ξ_k by its Fourier series and calculating the average of $\xi_k \xi_j''$). In particular, $-\left\langle \xi_k \xi_k'' \right\rangle = \left(\xi_k'\right)^2$, which is the squared RMS value of the velocity of the excitation.

Similarly, for the i'th component of the second averaging term in (7.75) it is found that:

$$
\begin{aligned}
\left\langle \nabla \mathbf{m}_{(i)}(\mathbf{z},t,\hat{\varphi}) \right\rangle &= \left\langle (\hat{\varphi}'')^\mathsf{T} \nabla \mathbf{M}_{(i)} \hat{\varphi} \right\rangle \\
&= \left\langle \left(\mathbf{M}^{-1}(\mathbf{z},t) \sum_j \mathbf{f}_j(\mathbf{z},t)\xi_j'' \right)^\mathsf{T} \nabla \mathbf{M}_{(i)}(\mathbf{z},t)\mathbf{M}(\mathbf{z},t)^{-1} \sum_k \mathbf{f}_k(\mathbf{z},t)\xi_k \right\rangle \\
&= \sum_{j,k} \left\langle \xi_j' \xi_k' \right\rangle \left(\mathbf{M}^{-1}(\mathbf{z},t)\mathbf{f}_j(\mathbf{z},t) \right)^\mathsf{T} \nabla \mathbf{M}_{(i)}(\mathbf{z},t)\mathbf{M}^{-1}(\mathbf{z},t)\mathbf{f}_k(\mathbf{z},t).
\end{aligned}
\tag{7.77}
$$

Inserting these results into (7.75) and rearranging, the result becomes:

$$
\Delta \mathbf{K}^{\mathbf{f},\mathbf{m}}(\mathbf{0}) = \sum_{j,k} \left\langle \xi_j' \xi_k' \right\rangle \frac{\partial}{\partial \mathbf{u}}\Bigg|_{\mathbf{u}=\mathbf{z}=\mathbf{0}} \left(\nabla \mathbf{f}_j \mathbf{M}^{-1}\mathbf{f}_k + \left\{ \begin{array}{c} (\mathbf{M}^{-1}\mathbf{f}_j)^\mathsf{T} \nabla \mathbf{M}_{(1)} \mathbf{M}^{-1}\mathbf{f}_k \\ \vdots \\ (\mathbf{M}^{-1}\mathbf{f}_j)^\mathsf{T} \nabla \mathbf{M}_{(n)} \mathbf{M}^{-1}\mathbf{f}_k \end{array} \right\} \right).
\tag{7.78}
$$

To examine the meaning of this, we first note that the second term vanishes if $\mathbf{M}$ is constant, since then $\nabla \mathbf{M}_{(i)} = \mathbf{0}$. But this is just what can always be achieved by multiplying (7.59) by $\mathbf{M}^{-1}$ (since $|\mathbf{M}| > 0$ is assumed), whereby an equivalent system is obtained with a new mass matrix $\mathbf{M} = \mathbf{I}$, the unit diagonal matrix. Hence this second term carries no essential information, which cannot be inferred from the first term. We term the contribution to stiffness from this important first term (with $\mathbf{M} = \mathbf{I}$) *parametrically induced stiffness*, and denote it $\Delta \mathbf{K}^{\mathbf{f}}(\mathbf{0})$:

$$\Delta \mathbf{K}^{\mathbf{f}}(\mathbf{0}) = \sum_{j,k} \left\langle \xi_j' \xi_k' \right\rangle \frac{\partial \big((\nabla \mathbf{f}_j) \mathbf{f}_k \big)}{\partial \mathbf{u}} \bigg|_{\mathbf{u}=\mathbf{z}=\mathbf{0}}$$

$$= \sum_{j,k} \left\langle \xi_j' \xi_k' \right\rangle \left(\nabla \mathbf{f}_j \nabla \mathbf{f}_k + \mathbf{f}_k^{\mathrm{T}} \otimes \nabla^2 \mathbf{f}_j \right) \Big|_{\mathbf{u}=\mathbf{z}=\mathbf{0}}, \tag{7.79}$$

where the last term is an $n \times n$ matrix whose i'th column is given by

$$\left(\mathbf{f}_k^{\mathrm{T}} \otimes \nabla^2 \mathbf{f}_j \right)_{(i)} \equiv \mathbf{f}_k^{\mathrm{T}} \left(\nabla^2 \mathbf{f}_j \right)_{(i)}, \tag{7.80}$$

and $(\nabla^2 \mathbf{f}_j)_{(i)}$ is the Hessian matrix corresponding to $\mathbf{f}_{j(i)}$:

$$\left(\nabla^2 \mathbf{f}_j \right)_{(i)} \equiv \begin{bmatrix} \dfrac{\partial^2 \mathbf{f}_{j(i)}}{\partial \mathbf{u}_{(1)}^2} & \cdots & \dfrac{\partial^2 \mathbf{f}_{j(i)}}{\partial \mathbf{u}_{(1)} \partial \mathbf{u}_{(n)}} \\ \vdots & \ddots & \vdots \\ \dfrac{\partial^2 \mathbf{f}_{j(i)}}{\partial \mathbf{u}_{(n)} \partial \mathbf{u}_{(1)}} & \cdots & \dfrac{\partial^2 \mathbf{f}_{j(i)}}{\partial \mathbf{u}_{(n)}^2} \end{bmatrix}. \tag{7.81}$$

Expression (7.79) allows some general statements to be made about the parametrically induced stiffening effects, induced by $\mathbf{f}_k$ (or $\mathbf{M}^{-1}\mathbf{f}_k$) and ξ:

1. At least one of the functions $\mathbf{f}_k$ (or $\mathbf{M}$) must depend on $\mathbf{u}$ in order for the effective stiffness to change. This is equivalent to saying that the equations of motion written in standard second-order form (i.e. multiplying (7.59) by $\mathbf{M}^{-1}$) must have HF excitation terms that are *parametric* in character, whereas a purely *external* excitation will not produce such effects.[2]
2. The first contributive term in (7.79) is nonzero only if $\mathbf{f}_k$ is linearizable with a nonzero gradient at zero. It disappears for even functions, $\mathbf{f}_k(-\mathbf{u}, t) = \mathbf{f}_k(\mathbf{u}, t)$, and for functions that are essentially nonlinear. We may thus term its contribution *linearly induced parametrical stiffness*. If $\left\langle \xi_j' \xi_k' \right\rangle = 0$ for $j \neq k$, as is often the case in applications, then the contribution will always be positive definite, since $|\nabla \mathbf{f}_j \nabla \mathbf{f}_k| = |\nabla \mathbf{f}_k| |\nabla \mathbf{f}_k| > 0$ for $j = k$.
3. The second term in (7.79) is nonzero only when $\mathbf{f}_k$, has both a constant and a quadratic part when Taylor-expanded near the equilibrium; we may term it *nonlinearly induced parametrical stiffness*, since only nonlinear functions $\mathbf{f}_k$ can contribute to it. The change in stiffness can be positive or negative, and it disappears for odd functions $\mathbf{f}_k$, i.e. for $\mathbf{f}_k(-\mathbf{u}, t) = -\mathbf{f}_k(\mathbf{u}, t)$.

---

[2]This holds only to the first level of approximation. In the presence of *strong nonlinearity* in the slow function $\mathbf{s}$, higher order contributions becomes significant, and a purely external HF excitation can indeed change the effective stiffness (linear and nonlinear) significantly. In Thomsen (2008b) this is termed *external stiffening*; see also Blekhman (2000, 2008), Fidlin (2000, 2006), and Lazarov et al. (2008, 2009, 2010).

4. Terms of order three and higher of the Taylor-expansion of $\mathbf{f}_k$, do not contribute to stiffness (to the level of approximation employed), since their gradients vanish at the equilibrium.

5. The stiffening effect is linear, by the definition employed above. But this does not imply that approximately correct results are always obtained by using linearized equations of motions. For example, if $n = 1$ and $\mathbf{f} = f(u) = \sin u$, then the sum of the two terms in the parenthesis of (7.79) becomes $(f')^2 + ff'' \equiv p(u) = \cos^2 u - \sin^2 u$, so that $p(0) = 1$, which is the same as when starting with the linearization $f(u) \approx u$. However, if instead $f(u) = \cos u - 2$, then one again finds $p(0) = 1$, while using the linearization $f(u) \approx -1$ one finds $p(0) = 0$, so that, by an otherwise appropriate linearization, the linear stiffening effect is in this case totally overlooked. In general, referring to 2–4. above, if $\mathbf{f}$ contains constant terms, then quadratic nonlinearities in $\mathbf{u}$ (when Taylor-expanded) should be retained, or stiffening effects could be inadvertently overlooked.

6. The magnitude or "strength" of the stiffening effect increases linearly with the squared RMS velocity of the excitation velocity, i.e. with the input level of energy. Hence the particular details of the excitation time signals are unimportant for the effect, as long as the signals are periodic and have high frequency and small amplitude. (As for the significance of the HF waveform, see also Michaux (2008).) Table 7.1 provides cross-averages $\left\langle \xi'_j \xi'_k \right\rangle$ for some typical input signal forms for reference.

Changes in effective linear stiffness may cause changes in derived linear quantities, such as natural frequencies and stability. Using (7.72)–(7.73) with $\tilde{\mathbf{z}} = \mathbf{0}$, the linearized dynamics near the equilibrium is seen to be governed by:

$$\mathbf{M}(\mathbf{0}, t)\ddot{\mathbf{z}} + (\mathbf{C}(0) + \Delta\mathbf{C}(0))\dot{\mathbf{z}} + (\mathbf{K}(0) + \Delta\mathbf{K}(0))\mathbf{z} = \mathbf{0}. \qquad (7.82)$$

Table 7.1 Example input signals ξ_j and their corresponding accelerations ξ''_j and cross-averages $\left\langle \xi'_j \xi'_k \right\rangle$. Here δ_{jk} is the Kronecker delta, $\delta(\tau)$ is Dirac's delta function, all ξ-functions have zero average and are 2π-periodic in $\tau = \Omega t$ (modulo 2π), j and k are integers, and siv(τ) is a sawtooth function having the same zeroes and extrema as $\sin(\tau)$

		ξ_j	ξ''_j	$\left\langle \xi'_j \xi'_k \right\rangle$
Sine		$\sin(j\tau)$	$-j^2\sin(j\tau)$	$\frac{1}{2}j^2\delta_{jk}$
Sawtooth		siv$(j\tau)$	$\frac{2j}{\pi}\sum_{p=1}^{2j}(-1)^p\delta\left(\tau - \frac{\pi}{2j}(2p-1)\right)$	$\frac{18}{2\pi^2}j^2\delta_{jk}$
Rectified sine		$\lvert\sin(j\tau)\rvert - 2/\pi$	$-j^2\lvert\sin(j\tau)\rvert + 2j\sum_{p=1}^{2j}\delta\left(\tau - \frac{\pi}{j}p\right)$	$\frac{1}{2}j^2\delta_{jk}$
Shifted sine		$\sin(\tau + \psi_j)$	$-\sin(\tau + \psi_j)$	$\frac{1}{2}\cos(\psi_j - \psi_k)$

Thus the natural frequencies and stability associated with the equilibrium are determined by the eigenvalues $\lambda = \lambda_r$, $r = 1, \ldots, 2n$, which are roots of the characteristic polynomial:

$$\left| \lambda^2 \mathbf{M}(\mathbf{0}, t) + \lambda (\mathbf{C}(\mathbf{0}) + \Delta \mathbf{C}(\mathbf{0})) + \mathbf{K}(\mathbf{0}) + \Delta \mathbf{K}(\mathbf{0}) \right| = 0. \qquad (7.83)$$

The natural frequencies are given by $|\mathrm{Im}(\lambda_r)|$, while the equilibrium in question is stable to small disturbances only if $\mathrm{Re}(\lambda_r) < 0$ for all r. Both quantities are likely to differ from the values obtained when there is no HF excitation, i.e. when $\Delta \mathbf{C}(\mathbf{0}) = \Delta \mathbf{K}(\mathbf{0}) = \mathbf{0}$ in (7.83). For assessing stability in specific cases, it might be convenient to use Ziegler's system of classification (Sect. 1.9), by which statements on stability are inferred from properties such as symmetry and definiteness of the system matrices. Also, since effective structural properties (stiffness, damping, stability etc.) can be controlled by changing the structure or the HF excitation or both, some interesting inverse problems could be posed on using HF excitation as a design variable in designing structures with specific or optimal low-frequency properties.

7.5.7   Biasing

We define biasing as a change in (average) static equilibrium position as a consequence of HF excitation. Two important kinds will be considered: positional bias and velocity bias:

Positional bias refers to a fixed translation of a static equilibrium. Examples of positional bias appeared in the examples of the vibrated pendulum and the mass on a vibrated plane (Sect. 7.3.2), and in Fig. 7.7(b,d,e,h). Assuming, as in the previous section, that the general system (7.59) has a static equilibrium at $u = 0$ when unexcited, then HF excitation may change this state of affairs, so that the equilibrium for the slow components of motion changes from $(\mathbf{z}, \dot{\mathbf{z}}) = (\mathbf{0}, \mathbf{0})$ to $(\tilde{\mathbf{z}}, \mathbf{0})$, where $\tilde{\mathbf{z}} \neq \mathbf{0}$ is constant-valued. We term $\tilde{\mathbf{z}}$ a quasi-equilibrium for the full motions $\mathbf{u}$, because these will actually oscillate at small amplitude and high frequency about $\tilde{\mathbf{z}}$, cf. (7.60). To calculate $\tilde{\mathbf{z}}$ we use (7.68), inserting $(\mathbf{z}, \dot{\mathbf{z}}, \ddot{\mathbf{z}}) = (\tilde{\mathbf{z}}, \mathbf{0}, \mathbf{0})$, and find that $\tilde{\mathbf{z}}$ is the solution of the generally nonlinear set of algebraic equations (7.71), i.e.

$$\mathbf{s}(\tilde{\mathbf{z}}, \mathbf{0}, t) + \mathbf{v}(\tilde{\mathbf{z}}, \mathbf{0}, t) = \mathbf{0}, \qquad (7.84)$$

with $\mathbf{v}$ given by (7.69). The equilibrium $\tilde{\mathbf{z}}$ is stable when the real parts of the roots λ of the following characteristic polynomial are all negative:

$$\left| \lambda^2 \mathbf{M}(\tilde{\mathbf{z}}, t) + \lambda (\mathbf{C}(\tilde{\mathbf{z}}) + \Delta \mathbf{C}(\tilde{\mathbf{z}})) + \mathbf{K}(\tilde{\mathbf{z}}) + \Delta \mathbf{K}(\tilde{\mathbf{z}}) \right| = 0, \qquad (7.85)$$

where the matrices $\mathbf{C}$ and $\mathbf{K}$ are given by (7.73). More than one solution for $\tilde{\mathbf{z}}$ may exist, since $\mathbf{s}$ and $\mathbf{v}$ are generally nonlinear functions of position, or there may be no solutions at all. Here we consider the typical case of a relatively small bias, $|\tilde{\mathbf{z}}| \ll 1$, for which it makes sense to Taylor-expand (7.84) and solve for $\tilde{\mathbf{z}}$; This yields, recalling that $\mathbf{s}(\mathbf{0}, \mathbf{0}, t) = \mathbf{0}$:

$$\tilde{\mathbf{z}} = -(\mathbf{K}(0) + \Delta\mathbf{K}(0))^{-1}\mathbf{v}(0, 0, t) + O(|\tilde{\mathbf{z}}|^2), \qquad (7.86)$$

where the last term denotes small higher order contributions. Thus, to first order the positional bias is proportional to the HF induced vibrational forces $\mathbf{v}$ at the original equilibrium, and inversely proportional to the effective linear stiffness. Using (7.69) and (7.67) we find:

$$\begin{aligned} \mathbf{v}(0, 0, t) = \left\langle \mathbf{s}(0, \hat{\boldsymbol{\varphi}}'_0, t) \right\rangle + \sum_j \left\langle \mathbf{h}_j(0, \hat{\boldsymbol{\varphi}}'_0, t)\xi''_j(t, \tau) \right\rangle \\ + \sum_j \nabla\mathbf{f}_j(0, t)\left\langle \hat{\boldsymbol{\varphi}}_0 \xi''_j(t, \tau) \right\rangle + \left\langle \nabla\mathbf{m}(0, t, \hat{\boldsymbol{\varphi}}_0) \right\rangle, \end{aligned} \qquad (7.87)$$

where

$$\hat{\boldsymbol{\varphi}}_0(t, \tau) = -\mathbf{M}^{-1}(0, t)\sum_j \mathbf{f}_j(0, t)\xi_j(t, \tau). \qquad (7.88)$$

Thus there are several sources to bring up positional bias, corresponding to each of the four terms in (7.87) having a non-zero average. The contribution of the $\mathbf{s}$ term, to be denoted $\mathbf{v}^s$, can be partly examined by splitting $\mathbf{s}$ into components:

$$\mathbf{s}(\mathbf{u}, \dot{\mathbf{u}}, t) = \mathbf{s}_0 + \mathbf{S}_{11}\mathbf{u} + \mathbf{S}_{12}\dot{\mathbf{u}} + \mathbf{r}(\mathbf{u}, \dot{\mathbf{u}}, t), \qquad (7.89)$$

where $\mathbf{s}_0$, and $\mathbf{S}_{11,12}$ are constant-valued vectors and matrices, and $\mathbf{r}$ is the remainder part, which is essentially nonlinear in $\mathbf{u}$ and $\dot{\mathbf{u}}$ and satisfies $\mathbf{r}(\mathbf{0}, \mathbf{0}, t) + \mathbf{s}_0 = \mathbf{0}$. Then $\mathbf{v}^s$ becomes, recalling that the fast motions $\hat{\boldsymbol{\varphi}}$ has zero average:

$$\mathbf{v}^s = \left\langle \mathbf{s}(0, \hat{\boldsymbol{\varphi}}'_0, t) \right\rangle = \mathbf{s}_0 + \left\langle \mathbf{r}(0, \hat{\boldsymbol{\varphi}}'_0, t) \right\rangle, \qquad (7.90)$$

from which it appears that no bias occurs from this source if $\mathbf{s}$ is a linear function of $\mathbf{u}$ and $\dot{\mathbf{u}}$, since then $\mathbf{r} = \mathbf{0}$ which implies $\mathbf{s}_0 = \mathbf{0}$. This also holds if $\mathbf{s}$ is nonlinear in $\mathbf{u}$, but linear in $\dot{\mathbf{u}}$. Thus $\mathbf{s}$ should be nonlinear in $\dot{\mathbf{u}}$, e.g. as for systems with dry friction, for bias to occur from this source.

The second term of (7.87) may or may not contribute to positional biasing effect, dependent on the details of the functions $\mathbf{h}_j$, though it does not occur for the examples in this chapter. Considering the two last terms of (7.87), denoting their contribution by $\mathbf{v}^{f,m}$, we use (7.76) and (7.77) to substitute the average terms and find that:

$$\mathbf{v}^{\mathbf{f,m}} = \sum_{k,j} \left\langle \xi_k' \xi_j' \right\rangle \left(\nabla \mathbf{f}_j \mathbf{M}^{-1} \mathbf{f}_k + \left\{ \begin{array}{c} \left(\mathbf{M}^{-1}\mathbf{f}_k\right)^{\mathrm{T}} \nabla \mathbf{M}_{(1)} \mathbf{M}^{-1} \mathbf{f}_l \\ \vdots \\ \left(\mathbf{M}^{-1}\mathbf{f}_k\right)^{\mathrm{T}} \nabla \mathbf{M}_{(n)} \mathbf{M}^{-1} \mathbf{f}_l \end{array} \right\} \right)_{\mathbf{u}=\mathbf{z}=0} . \tag{7.91}$$

Then, by the same arguments as used below (7.78), we note that the essential information is carried by the first term, which we denote $\mathbf{v}^{\mathbf{f}}$ when $\mathbf{M} = \mathbf{I}$, i.e.:

$$\mathbf{v}^{\mathbf{f}}(\mathbf{0},\mathbf{0},t) = \sum_{j,k} \left\langle \xi_j' \xi_k' \right\rangle \nabla \mathbf{f}_j(\mathbf{0},t) \mathbf{f}_k(\mathbf{0},t). \tag{7.92}$$

Inspecting this expression, it appears that $\mathbf{v}^{\mathbf{f}} \neq \mathbf{0}$ requires that at least one of the functions $\mathbf{f}_k$ depends on $\mathbf{u}$, i.e. for bias to occur from this source, the HF excitation should be parametric in character. Furthermore, at least one $\mathbf{f}_k$ should be linearizable with a nonzero gradient at zero, while essentially nonlinear functions $\mathbf{f}_k$ do not contribute to $\mathbf{v}^{\mathbf{f}}$. Likewise, functions $\mathbf{f}_k$ that possesses symmetry – be it odd, $\mathbf{f}_k(-\mathbf{u}, t) = -\mathbf{f}_k(\mathbf{u}, t)$, or even, $\mathbf{f}_k(-\mathbf{u}, t) = \mathbf{f}_k(\mathbf{u}, t)$ – do not contribute to $\mathbf{v}^{\mathbf{f}}$. The latter requirement on asymmetry is essential, and can be illustrated by the simple function $\mathbf{f} = f(u) = 1 + u$, where the first term is evenly and the other one oddly symmetrical, while their sum is asymmetric. In this case the contribution to $\mathbf{v}^{\mathbf{f}}$ is proportional to $\nabla f(0)f(0) = f'(0)f(0) = 1$, whereas this value is zero if calculated for each of the terms 1 or u separately. In physical terms, the function $1 + u$ is representative, e.g., of structures subjected to a combination of external and parametrical HF excitation (cf. Fig. 7.7(b)). Then bias may occur, even though the time average of the excitation is zero, whereas it disappears if either of the two excitations are switched off.

As special cases of positional bias we mention those where, in the absence of HF excitation, the system has (1) no static equilibriums at all, i.e. $\mathbf{s}(\mathbf{u}, \mathbf{0}, t) = \mathbf{0}$ has no real-valued solutions, or (2) has a continuous range of equilibriums, i.e. $\mathbf{s}(\mathbf{u}, \mathbf{0}, t) = \mathbf{0}$ has infinitely many solutions. Even in these cases the HF excitation may create well-defined equilibriums, as given by possible solutions $\tilde{\mathbf{z}}$ to (7.71), though (7.86) only holds when these happens to be close to zero. Examples are here the horizontal pendulum (last part of the vibrated pendulum example, Sect. 7.3.1), and the mass with no spring on a vibrating plane (last part of the mass on a vibrated plane example, Sect. 7.3.2).

Velocity bias refers to the possible emergence of states of steady drifting with constant velocity, i.e. with slow motions of the form $(\mathbf{z}, \dot{\mathbf{z}}) = (\dot{\tilde{\mathbf{z}}}t, \dot{\tilde{\mathbf{z}}})$, where $\dot{\tilde{\mathbf{z}}}$ is the drift velocity. Examples of velocity bias appear in the last part of the mass on a vibrated plane example (Sect. 7.3.2, mass with no spring), with Brumberg's pipe (Sect. 7.3.3), and Fig. 7.7(a,c). Using (7.68), we find that $\dot{\tilde{\mathbf{z}}}$ should be a solution of:

$$\mathbf{s}(\dot{\tilde{\mathbf{z}}}t, \dot{\tilde{\mathbf{z}}}, t) + \mathbf{v}(\dot{\tilde{\mathbf{z}}}t, \dot{\tilde{\mathbf{z}}}, t) = \mathbf{0}. \tag{7.93}$$

If such a solution exists, one can show that the condition for its stability is that the real parts of the roots λ of the following characteristic polynomial are all negative:

$$\left| \lambda \mathbf{M}(\tilde{\dot{z}}t, t) + \nabla \mathbf{s}(\tilde{\dot{z}}t, \tilde{\mathbf{z}}, t) + \dot{\nabla} \mathbf{v}(\tilde{\dot{z}}t, \tilde{\mathbf{z}}, t) \right| = 0. \qquad (7.94)$$

A necessary condition for velocity bias to exist is that $\mathbf{M}$ and the expression in (7.93) are independent of t, that is: $\mathbf{s}$ and $\mathbf{v}$ or their sum, and $\mathbf{M}$, should be independent of $\mathbf{u}$ and of t.

For typical cases where the velocity bias is relatively small, we may solve (7.93) approximately by Taylor-expanding near $\tilde{\mathbf{z}} = 0$ and solving for the drift velocity $\tilde{\dot{\mathbf{z}}}$, the result becoming:

$$\tilde{\dot{\mathbf{z}}} = -\left(\nabla \mathbf{s}(\tilde{\dot{z}}t, \mathbf{0}, t) + \dot{\nabla} \mathbf{v}(\tilde{\dot{z}}t, \mathbf{0}, t) \right)^{-1} \left(\mathbf{s}(\tilde{\dot{z}}t, \mathbf{0}, t) + \mathbf{v}(\tilde{\dot{z}}t, \mathbf{0}, t) \right), \qquad (7.95)$$

which again is only a proper solution if independent of t (in typical cases the terms with $\tilde{\dot{z}}t$, cancel each other, so that $\tilde{\dot{\mathbf{z}}}$ does not need to be known in order to compute the right-hand side.)

7.5.8 Smoothening

Sometimes HF excitation has the effect of apparently removing discontinuities of a system; we refer to this as the smoothening effect. A classic example is systems modeled with Coulomb friction, where HF excitation may cause the discontinuity at zero velocity to seemingly disappear, and the damping appears viscous in character (cf. the vibrated pendulum and the mass on a vibrated plane in Sect. 7.3.2) and Fig. 7.7(a,h,i). To study this effect we assume that the slow forces $\mathbf{s}$ of the system (7.59) has a single discontinuity at zero velocity when there is no HF-excitation, i.e.

$$\mathbf{s}_+(\mathbf{u}, t) \neq \mathbf{s}_-(\mathbf{u}, t), \quad \text{where:}$$
$$\mathbf{s}_+(\mathbf{u}, t) \equiv \lim_{\dot{u} \to 0_+} \mathbf{s}(\mathbf{u}, \dot{\mathbf{u}}, t), \quad \mathbf{s}_-(\mathbf{u}, t) \equiv \lim_{\dot{u} \to 0_-} \mathbf{s}(\mathbf{u}, \dot{\mathbf{u}}, t). \qquad (7.96)$$

As appears from (7.68) the slow forces for the averaged system is $\mathbf{s} + \mathbf{v}$, which by (7.69) evaluates to:

$$\mathbf{s}(\mathbf{z}, \dot{\mathbf{z}}, t) + \mathbf{v}(\mathbf{z}, \dot{\mathbf{z}}, t) = \left\langle \mathbf{s}(\mathbf{z}, \dot{\mathbf{z}} + \hat{\boldsymbol{\varphi}}', t) \right\rangle$$
$$+ \sum_j \left\langle \mathbf{h}_j(\mathbf{z}, \dot{\mathbf{z}} + \hat{\boldsymbol{\varphi}}', t) \xi_j''(t, \tau) \right\rangle + \bar{\mathbf{r}}(\mathbf{z}, \dot{\mathbf{z}}, t), \qquad (7.97)$$

where the fast motions $\hat{\boldsymbol{\varphi}}$ are given by (7.67), and $\bar{\mathbf{r}}$ is a remainder term holding continuous functions that are unrelated to $\mathbf{s}$. It seems intuitively plausible that

s + v can be continuous at $\dot{\mathbf{z}} = \mathbf{0}$, even though **s** (and perhaps $\mathbf{h}_j$) is discontinuous: The right-hand side is calculated by averaging **s** (and $\mathbf{h}_j$) while the velocity argument $\dot{\mathbf{z}} + \hat{\boldsymbol{\varphi}}$ is racing back and forth across the discontinuity – and this average may not change much in the vicinity of $\dot{\mathbf{z}} = \mathbf{0}$.

To set up conditions for smoothening to occur, we first note that the vector character of the functions and their dependence on **u** and t is inessential for this, since the averaging process is performed element-wise, and with **u** and t considered constant. It therefore suffices to consider a general scalar function $s(\dot{z})$, bounded on R, and having a single discontinuity at $\dot{z} = 0$, and then consider the *smoothed image* $\tilde{s}$ of s:

$$\tilde{s}(\dot{z}) \equiv \langle s(\dot{z} + \varphi'(\tau)) \rangle, \tag{7.98}$$

where the fast motions $\varphi(\tau)$) are 2π-periodic in τ with zero average. Using the definition (7.61) of fast-time averaging one finds, splitting up the period of integration, that:

$$\tilde{s}(\dot{z}) = \frac{1}{2\pi}\left[\int_{\tau_-} s(\dot{z} + \varphi'(\tau))\,d\tau + \int_{\tau_0} s(\dot{z} + \varphi'(\tau))\,d\tau \right. $$
$$\left. + \int_{\tau_+} s(\dot{z} + \varphi'(\tau))\,d\tau \right], \tag{7.99}$$

where τ_-, τ_0, and τ_+ are the intervals of time where $\dot{z} + \varphi'(\tau) < 0, = 0$, and > 0, respectively, and $\tau_- \cup \tau_0 \cup \tau_+ = [0;2\pi]$. The limits of these intervals are solutions to the algebraic equation $\dot{z} + \varphi'(\tau) = 0$; we assume there are $n_\tau = n_\tau(\dot{z})$ such solutions and denote them $\tau_j = \tau_j(\dot{z}), j = 1, 2, \ldots, n_\tau$. By these definitions the first and the third *integrands* are assured to be continuous in $\dot{z}$, while continuity of the corresponding *integrals* requires continuity also of the limits of the integration intervals. The following observations regarding the continuity of $\tilde{s}(\dot{z})$ can then be made, noting that for every value of $\dot{z}$, the number of solutions n_τ is either zero, finite, or infinite:

1. If $n_\tau = 0$ (for some range of $\dot{z}$) then $\tilde{s}$ is continuous (in that range). Namely, when $\dot{z} + \varphi'(\tau)$ never sweeps the discontinuity at zero, then $\tilde{s}$ is given by either the first or the third integral in (7.99), which is a continuous function of $\dot{z}$ because the integrand s is continuous during τ_- or τ_+, and the interval of integration $[0;2\pi]$ is independent of $\dot{z}$.
2. If n_τ is infinite, the discontinuity of s at zero is still present in $\tilde{s}$. Namely, $n_\tau \to \infty$ means that $\dot{z} + \varphi'(\tau) = 0$ over finite sub-intervals of $[0;2\pi]$, i.e. τ_0 is not a point set. Then the second integral in (7.99) is to be performed over finite intervals of time where $\dot{z} + \varphi'(\tau) = 0$, i.e. at values where s is discontinuous (undefined); this integral therefore is discontinuous, as is consequently $\tilde{s}$.
3. If n_τ is finite, and the zeroes τ_j of $\dot{z} + \varphi'(\tau)$ are all simple, then $\tilde{s}$ is continuous for the corresponding values of $\dot{z}$. This is because a finite value of n_τ means that $\dot{z} + \varphi'(\tau)$ sweeps the discontinuity a finite number of times during a period,

i.e. τ_0 is a point set. Then the second integral in (7.99) vanishes, because s is bounded and the integration intervals are infinitely short. Consequently, continuity is assured if the two other integrals are continuous. Their integrands are continuous, because they are to be evaluated for the intervals τ_- and τ_+, where $\dot{z} + \varphi'(\tau)$ has constant sign and thus only sweeps the continuous part of s. But the boundaries of the integration intervals should be continuous as well, that is, τ_j should be continuous functions of $\dot{z}$. By the implicit function theorem this is true when $\varphi''(\tau_j) \neq 0$, that is, if τ_j are simple zeroes of $\dot{z} + \varphi'(\tau)$.

4. In particular, it follows from the above results that $\tilde{s}$ is continuous at $\dot{z} = 0$ if $\varphi'(\tau)$ has a finite number of simple zeroes during one period of fast motions.

As described the smoothening effect on the discontinuity at zero occurs only if the fast motions actually sweeps the discontinuity, and does not 'rest' there. An example function that fails to satisfy this is $\varphi' = -\sin(2\tau)$ for $\tau \in [0;\pi]$; $\varphi' = 0$ for $\tau \in [\pi;2\pi]$, which has simple zeroes at $\tau = 0$, $\pi/2$, but infinitely many non-simple zeroes for $\tau \in [\pi;2\pi]$; hence this function does not smooth discontinuities.

An example function that *does* satisfy the requirements is a simple harmonic, e.g. $\varphi' = -a\cos(\tau)$, which has simple zeroes at $\pi/2$ and $3\pi/2$ and thus is smoothening. Its effect on the discontinuous signum function appears in Table 7.2(a); it could represent the smoothening of the dry friction characteristic for a physical system subjected to HF excitation (see also the mass on a vibrated plane example in Sect. 7.3.2, and Fig. 7.5).

Table 7.2(b) shows how fast motions smoothens the discontinuity of a function that may more realistically model dry friction. Here the function s has negative slope at the origin and thus may cause unstable oscillations of a corresponding physical system, whereas the smoothed image has positive slope everywhere (For applications of this see Thomsen 1999; Feeny and Moon 2000; Chatterjee et al. 2004).

Discontinuities in derivatives may also be smoothened, as seen when replacing s and $\tilde{s}$ with $ds/d\dot{u}$ and $d\tilde{s}/d\dot{z}$ (or higher order derivatives) in the above expressions and arguments. Table 7.2(c,d) shows two such examples, which might represent idealized physical processes involving saturation and 'barrier' behavior, respectively, in both cases with discontinuous changes in slope HF-smoothened.

Even if s has no discontinuity, the smoothed image $\tilde{s}$ may still differ from s. Table 7.2(e–h) shows smoothed images of functions that are typically encountered in applications. The essential properties of such functions are captured by a third order polynomial with zero constant term and coefficients $s_{1,2,3}$:

$$s(\dot{z}) = s_1\dot{z} + s_2\dot{z}^2 + s_3\dot{z}^3, \tag{7.100}$$

whose smoothed image is, by (7.98) and recalling that $\langle \varphi' \rangle = 0$:

$$\tilde{s}(\dot{z}) = \left(s_2 \left\langle (\varphi')^2 \right\rangle + s_3 \left\langle (\varphi')^3 \right\rangle \right) + \left(s_1 + 3s_3 \left\langle (\varphi')^2 \right\rangle \right) \dot{z}$$
$$+ s_2\dot{z}^2 + s_3\dot{z}^3. \tag{7.101}$$

Table 7.2 Examples of slow functions $s(\dot{u})$ and their corresponding smoothed images $\tilde{s}(\dot{z})$ for $\varphi(\tau) = -a\sin(\tau)$. Lengthy expressions for $\tilde{s}_{c1}$ and $\tilde{s}_{c2}$ in (c) and (d) are omitted; they are smooth functions with smooth transitions at the boundaries of their intervals of definition

$s(\dot{u})$	$\tilde{s}(\dot{z}) \equiv \langle s(\dot{z} + \varphi'(\tau))\rangle, \quad \varphi' = -a\cos\tau$	$s\,(-), \tilde{s}\,(---)$
(a) $\mathrm{sgn}(\dot{u})$	$\begin{cases} 1 - \frac{2}{\pi}\arccos(\dot{z}/a), & \|\dot{z}\| \leq a \\ \mathrm{sgn}(\dot{z}), & \|\dot{z}\| > a \end{cases}$	
(b) $\mathrm{sgn}(\dot{u}) - \dot{u} + \frac{1}{3}\dot{u}^3$	$\begin{cases} 1 - \frac{2}{\pi}\arccos(\dot{z}/a) + (\frac{1}{4}a^2 - 1)\dot{z} + \frac{1}{3}\dot{z}^3 & \text{for } \|\dot{z}\| \leq a \\ \mathrm{sgn}(\dot{z}) + (\frac{1}{4}a^2 - 1)\dot{z} + \frac{1}{3}\dot{z}^3 & \text{for } \|\dot{z}\| > a \end{cases}$	
(c) $\begin{cases} \dot{u}, & \|\dot{u}\| \leq 1 \\ \mathrm{sgn}(\dot{u}), & \|\dot{u}\| > 1 \end{cases}$	$\begin{cases} \dot{z}, & \|\dot{z}\| \leq 1 - a \\ \tilde{s}_{c1}(\dot{z}), & \|\dot{z}\| \in [1-a; 1+a] \\ \mathrm{sgn}(\dot{z}), & \|\dot{z}\| > 1 + a \end{cases}$	

(continued)

Table 7.2 (continued)

$s(\tilde{u})$	$\tilde{s}(\dot{z}) \equiv \langle s(\dot{z}+\varphi'(\tau))\rangle, \quad \varphi' = -a\cos\tau$	$s\ (—),\ \tilde{s}\ (---)$
(d) $\begin{cases} 0, & \lvert\tilde{u}\rvert \le 1 \\ \tilde{u} - \text{sgn}(\tilde{u}), & \lvert\tilde{u}\rvert > 1 \end{cases}$	$\begin{cases} 0, & \lvert\dot{z}\rvert \le 1-a \\ \tilde{s}_{c2}(\dot{z}), & \lvert\dot{z}\rvert \in [1-a; 1+a] \\ \dot{z} - \text{sgn}(\dot{z}), & \lvert\dot{z}\rvert > 1+a \end{cases}$	
(e) $\dot{u}$	$\dot{z}$	
(f) $\dot{u}^2$	$\dot{z}^2 + \tfrac{1}{2}a^2$	

(continued)

Table 7.2 (continued)

$s(\dot{u})$	$\tilde{s}(\dot{z}) \equiv \langle s(\dot{z}+\varphi'(\tau))\rangle, \quad \varphi' = -a\cos\tau$	$s\,(—),\ \tilde{s}\,(---)$
(g) $\dot{u}^3$	$\dot{z}^3 + \tfrac{3}{2}a^2\dot{z}$	
(h) $|\dot{u}|\dot{u}$	$\begin{cases} \left(1 - \tfrac{2}{\pi}\arccos(\dot{z}/a)\right)\left(\dot{z}^2 + \tfrac{1}{2}a^2\right) + \tfrac{3}{\pi}a\dot{z}\sqrt{1 - (\dot{z}/a)^2}, & \text{for } |\dot{z}| \le a \\ |\dot{z}|\dot{z} + \tfrac{1}{2}a^2\mathrm{sgn}(\dot{z}) & \text{for } |\dot{z}| > a \end{cases}$	

Interestingly, the nonlinear parts of s (with coefficients s_2 and s_3) change the linear properties of $\tilde{s}$ (i.e. the first two groups of terms), whereas the nonlinear properties themselves are unaffected. For example, the slope of $\tilde{s}$ at $\dot{z} = 0$ is seen to increase, as compared to s if the cubic nonlinearity of s is progressive ($s_3 > 0$), and to decrease if it is recessive ($s_3 < 0$). Thus, the effective linear damping of a physical system might be partially controlled by using HF excitation, or be created out of nothing, as when $s_1 = 0$, $s_3 \neq 0$. Also, a constant drifting term is seen to appear, so that $\tilde{s}(0) \neq 0$ when there are nonlinear velocity terms in s.

Finally we recall that the above results for s and φ hold as well for the corresponding vector functions $\mathbf{s}$ and $\hat{\boldsymbol{\varphi}}$ of the general system (7.59).

7.5.9   Effects of Multiple HF Excitation Frequencies

When $m > 1$ in (7.59) the HF excitation may have more than a single excitation frequency. The consequence of this can be calculated using the above general results, with the following main conclusions (Thomsen 2008a):

- The stiffening effect is similar to that of mono-frequency excitation, provided the excitation frequencies are well separated.
- For non-close and non-resonant excitation frequencies, the change in effective static stiffness is proportional to the sum of squared excitation velocities, i.e. to the input energy level.
- With two or more close excitation frequencies, there is an additional contribution of slowly oscillating stiffness, having magnitude order similar to the change in static stiffness, and frequencies equal to the differences in non-close frequencies.
- With close excitation frequencies, strong parametrical resonance can occur at conditions that might not appear obvious, that is: when the *difference* in any two excitation frequencies comes near $2\tilde{\omega}/k$, where k is an integer (with $k = 1$ corresponding to primary parametric resonance), and $\tilde{\omega}$ is an effective natural frequency of the system – which due to the HF-excitation is actually shifted away from the natural frequency ω without HF-excitation.
- Strong multifrequency HF-excitation can stabilize unstable quasistatic equilibriums, just as can monofrequency HF-excitation. Generally this holds only if the frequencies are well separated, i.e. with differences much larger than the lowest natural frequencies of the system. With two or more close high frequencies, the effect of HFE may be stabilizing or destabilizing, dependent on particular parameter values. Thus continuous broadband and random excitation does not have a uniquely stabilizing effect paralleling that of monofrequency HF-excitation, or multifrequency HF-excitation with non-close frequencies. (See also Kremer 2018.)

7.5.10   Effects of Strong Damping

With vibrating systems damping is typically rather low; otherwise the vibration level would be too low to be of concern, or useful. But in some cases damping is inherently strong, e.g. with controlled structures (velocity feedback effectively acts as damping), and for micro/nano systems (the relative significance of damping increase with downsizing). With additional HF excitation nontrivial effects show up, of which the most important seems to be a change of equilibrium state with the level of damping (Fidlin and Thomsen 2007). For example, a strongly damped pendulum, with a hinge vibrated at high frequency along an elliptical path, will line up along a line offset from the vertical; the offset vanishes for very light or very strong damping, attaining a maximum that can be substantial, depending on the strength of the HF excitation. Such an effect could provide a key for explaining several experimental observations of effects of microwave radiation on large molecules (like proteins) and nanostructures in solvents.

7.6   A General Class of Linear Continuous Systems

Next we consider a general class of *continuous* systems with HF excitation, described by linear partial differential equations of the following form (Thomsen 2003b):

$$u_{tt} + L_c[u_t] + L_0[u] + \Omega \sum_{k=1}^{N} L_{1k}[u] \sin(m_k\Omega t + \psi_k)$$

$$= f_0(\mathbf{x}, t) + \Omega f_1(\mathbf{x}, t) \sin(m\Omega t + \psi), \quad \Omega \gg 1, \tag{7.102}$$

which models a variety of HF-excited continuous elastic structures, such as strings (Babitsky and Veprik 1993; Tcherniak 2000; Thomsen 1996a, 2003b; Zak 1984), rods, pipes (Jensen 1996, 1997), beams (Blekhman and Malakhova 1986; Chelomei 1983; Jensen 2000a; Jensen, et al. 2000; Miranda and Thomsen 1998; Krylov and Sorokin 1997; Thomsen 1996b; Thomsen and Tcherniak 2001), membranes (Thomsen 2003b), plates, flexible rotating disks (Hansen 1999, 2000), and shells.

In the model (7.102), $u = u(\mathbf{x}, t)$ describes a scalar displacement component of a continuous structure at spatial position $\mathbf{x} \in D$ where D is a bounded region, $( )_t \equiv \partial/\partial t$, L_c, L_0, and L_{1k} are linear spatial differential operators describing damping, stiffness, and spatial component of the HF excitation, respectively, $\Omega \gg 1$ is the fundamental excitation frequency, m_k and m are integers, ψ_k and ψ are phasings of the HF excitation, f_0 and f_1 are external distributed forces, with f_0 changing slowly as compared to the base period $2\pi/\Omega$ of the HF excitation, and f_1 containing rapidly fluctuating components. In general subscript 1 denote functions and operators related to terms with explicit HF variation, while subscript 0 denotes quantities without such variations. The two HF excitation terms are written with Ω as a factor, to indicate they are assumed to be "strong". Specifying the excitation in terms of harmonic time functions implies that the following results will appear much more transparent than if generalized time functions were used. Still, quite general time functions can be described for the parametric excitation in (7.102), by using the sum to describe a Fourier expansion of the time function. The system description is completed by specifying initial conditions $u(\mathbf{x}, 0)$ and $u_t(\mathbf{x}, 0)$, and linear homogeneous boundary conditions $B_r[u] = 0$, $r = 1, \ldots, 2q$ for $\mathbf{x} \in \partial D_r$, where $\partial D_r \subseteq \partial D$ denote parts of the boundary of D, q is the order of (7.102), and B_r contain spatial derivatives of order 0 through $q - 1$. The extension to the multivariable case can be performed by replacing the scalar functions u, f_0, and f_1 with suitable vectors, and is thus trivial.

The linear eigenvalue problem associated with the (7.102) is:

$$
\begin{aligned}
L_0[\varphi] &= \omega^2 \varphi, \quad \mathbf{x} \in D, \\
B_r[\varphi] &= 0, \qquad \mathbf{x} \in \partial D_r,
\end{aligned}
\tag{7.103}
$$

which is assumed to be self-adjoint, with eigenvalues $\omega_i{}^2$ and corresponding normalized eigenvectors $\varphi_i(\mathbf{x})$ satisfying the following orthogonality relations:

$$
\int_D \varphi_i \varphi_j \, dD = \delta_{ij}, \quad \int_D \varphi_i L_0[\varphi_j] \, dD = \omega_i^2 \delta_{ij}, \quad i, j = 1, 2, \cdots.
\tag{7.104}
$$

The excitation of the system is assumed to be small in displacement amplitude and high in frequency, and the motion u to consist of small, rapidly oscillating components superimposed on slowly changing components.

Next we shall set up approximate equations governing the slow or average components of motion, from which apparent changes in stiffness and related quantities can be predicted. We first derive the *Generalized No-Resonance Prediction* (GNRP), which holds for cases where resonance effects can be ignored, e.g., for structures with a low modal density and far-from-resonant HF excitation. However, in many cases of practical relevance the modal density is so high that resonance effects become highly significant, even if the excitation is not sharply resonant. The predictions provided by the GNRP are then of only limited value. For that case we derive the *Generalized Analytical Resonance Prediction* (GARP), which holds also when resonance effects are significant, though still not dominating.

7.6.1   The Generalized No-Resonance Prediction (GNRP)

This analysis is relevant when Ω is large and far away from any resonances. Using the MDSM we split the motions into slow and fast components u_0 and u_1:

$$u(\mathbf{x}, t) = u_0(\mathbf{x}, t) + \Omega^{-1} u_1(\mathbf{x}, t, \tau), \tag{7.105}$$

where $\tau = \Omega t$ is the fast time, and u_1 is 2π-periodic in τ with a zero fast-time average, $\langle u_1 \rangle = 0$. Performing the same steps as for the discrete system in Sect. 7.5.3, one finds the following equation governing the slow motions u_0:

$$u_{0tt} + L_c[u_{0t}] + L_0[u_0] + \sum_{k=1}^{N} \langle L_{1k}[u_1] \sin(m_k \tau + \psi_k) \rangle = f_0(\mathbf{x}, t), \tag{7.106}$$

while the fast motions u_1 are governed by:

$$u_{1\tau\tau} = -\sum_{k=1}^{N} L_{1k}[u_0] \sin(m_k \tau + \psi_k) + f_1(\mathbf{x}, t) \sin(m\tau + \psi) + O(\Omega^{-1}). \tag{7.107}$$

A first order solution for u_1, valid for $\Omega \gg 1$ and neglecting resonance effects, is readily obtained by integrating (7.107) (recalling that u_0 does not depend on τ):

$$\begin{aligned} u_1 &= \sum_{k=1}^{N} L_{1k}[u_0] m_k^{-2} \sin(m_k \tau + \psi_k) \\ &\quad - f_1(\mathbf{x}, t) m^{-2} \sin(m\tau + \psi) + O(\Omega^{-1}). \end{aligned} \tag{7.108}$$

Inserting this into (7.106) and evaluating the averaging term yields the following averaged system:

$$u_{0tt} + L_c[u_{0t}] + \bar{L}_0[u_0] = \bar{f}_0(\mathbf{x}, t) + O(\Omega^{-1}), \tag{7.109}$$

where

$$\begin{aligned} \bar{L}_0 &\equiv L_0 + \frac{1}{2} \sum_{k,j=1}^{N} m_j^{-2} \delta_{m_k m_j} \cos(\psi_k - \psi_j) L_{1k} L_{1j}, \\ \bar{f}_0 &\equiv f_0 - \frac{1}{2} m^{-2} \sum_{k=1}^{N} L_{1k}[f_1] \delta_{m_k m} \cos(\psi - \psi_k), \end{aligned} \tag{7.110}$$

and where it has been used that $\langle \sin(i\tau + \psi_i) \sin(j\tau + \psi_j) \rangle = 1/2\delta_{ij} \cos(\psi_i - \psi_j)$ for $i, j = 1, 2, \ldots$, and it should be recalled that some of the m_k's can be equal.

Thus the effect of the HF excitation, on the average, is to change the stiffness operator L_0, and change the slow external forces f_0. To observers (or measuring instruments) that do not notice (or is lowpass filtering) the small overlay of HF displacements, it will appear as if the stiffness and the slow forcing has changed. As appears the change in stiffness is independent of the external HF excitation f_1, whereas the change in slow forcing occurs only when external and parametric HF excitation are both present ($f_1 \neq 0$ and $L_1 \neq 0$).

A change in effective stiffness, in turn, changes related quantities, such as natural frequencies. To find the apparent natural frequencies $\bar{\omega}_i$ in the presence of HF excitation one should solve the eigenvalue problem (7.103) using $\bar{L}_0$ in place of L_0. Knowing the eigenvalues ω_i^2 and eigenfunctions φ_i for the unexcited system, an estimate for the change in natural frequencies can be obtained by pre-multiplying the eigenvalue problem $\bar{L}_0[\bar{\varphi}_i] = \bar{\omega}_i^2 \bar{\varphi}_i$ by $\bar{\varphi}_i i$, integrate over the domain of the system, assume the change in eigenfunction is negligible, and find:

$$\bar{\omega}_i^2 = \int_D \bar{\varphi}_i \bar{L}_0[\bar{\varphi}_i] \, dD \approx \int_D \varphi_i \bar{L}_0[\varphi_i] \, dD$$

$$= \omega_i^2 + \frac{1}{2} \sum_{k,j=1}^{N} m_j^{-2} \delta_{m_k m_j} \cos(\psi_k - \psi_j) \int_D \varphi_i L_{1k} L_{1j}[\varphi_i] \, dD. \tag{7.111}$$

Many theorems and methods for continuous systems rely on the self-adjointness of the relevant eigenvalue problems. It appears from the assumed self-adjointness of L_0 and the definition of $\bar{L}_0$ in (7.110), that the eigenvalue problem for $\bar{L}_0$ is self-adjoint if $\int_D u L_{1i} L_{1j}[v] \, dD = \int_D v L_{1i} L_{1j}[u] \, dD$ for $i, j = 1, 2, \ldots$, and for any two test functions $u(\mathbf{x})$ and $v(\mathbf{x})$ satisfying the boundary conditions.

7.6.2 The Generalized Analytical Resonance Prediction (GARP)

This analysis is relevant when Ω is large and the response is influenced by resonances in the vicinity of Ω. It should be emphasized that we are concerned with small but rapid vibrations of a system, since only then is it meaningful to consider the average vibration amplitude, i.e. the 'slow' motions, instead of the full motion. This in turn implies that conditions of sharp resonance, with accompanying large amplitudes, need not to be covered. The predictions will be in terms of mode shapes and natural frequencies. For this we consider a mode shape expanded[3] model of the general system (7.102), i.e. let

$$u(\mathbf{x}, t) = \mathbf{v}^T(t) \boldsymbol{\varphi}(\mathbf{x}), \tag{7.112}$$

---

[3]Or a Galerkin discretization, or an assumed-modes model – with correspondingly relaxed requirements on the functions in $\boldsymbol{\varphi}$.

where the vector $\boldsymbol{\varphi}$ holds the eigenfunctions $\varphi_i(\mathbf{x})$, $i = 1, n$, that are solutions to the eigenvalue problem (7.103)–(7.104), and the vector $\mathbf{v} = \mathbf{v}(t)$ holds the modal coefficients v_i, $i = 1, n$, that are solutions to the following linear system of ordinary differential equations (obtained by standard mode shape expansion):

$$
\begin{aligned}
\mathbf{v}_{tt} + \mathbf{c}\mathbf{v}_t + \boldsymbol{\omega}^2 \mathbf{v} + \Omega \sum_{k=1}^{N} \boldsymbol{\gamma}_k \mathbf{v} \sin(m_k \Omega t + \psi_k) \\
= \mathbf{p}_0(t) + \Omega \mathbf{p}_1(t) \sin(m\Omega t + \psi),
\end{aligned}
\tag{7.113}
$$

where the components of the matrices $\mathbf{c}$, $\boldsymbol{\omega}^2$, $\boldsymbol{\gamma}_k$, and vectors $\mathbf{p}_0$, $\mathbf{p}_1$ are:

$$
\begin{aligned}
c_{ij} &= \int_D \varphi_j L_c[\varphi_i]\, dD, \quad \omega_{ij}^2 = \delta_{ij}\omega_i^2, \\
\gamma_{k(ij)} &= \int_D \varphi_j L_{1k}[\varphi_i]\, dD, \\
p_{0i}(t) &= \int_D f_0(\mathbf{x}, t)\varphi_i\, dD, \\
p_{1i}(t) &= \int_D f_1(\mathbf{x}, t)\varphi_i\, dD, \quad i,j = 1,n, \quad k = 1,N.
\end{aligned}
\tag{7.114}
$$

External resonances occur for this system in two cases: (1) when $\mathbf{p}_1 \neq \mathbf{0}$ and $\Omega \approx \omega_i/m$; and (2) when $\mathbf{p}_0$ has a nonzero (long-term) average and $\Omega \approx \omega_i/m_k$, $i = 1, n$, $k = 1, N$.

Parametric resonances also occur in two cases: (a) when $\gamma_{k(ij)}\gamma_{k(ji)} > 0$ and $\Omega \approx (\omega_i + \omega_j)/m_k$, $i, j = 1, n$, for at least one $k = 1, N$; and (b) when $\gamma_{k(ij)}\gamma_{k(ji)} < 0$ and $\Omega \approx (\omega_i - \omega_j)/m_k$ for at least one k. For vanishing damping the width of the resonant regions are determined by the condition $\gamma_{k(ij)}\gamma_{k(ji)} > \omega_i\omega_j(m_k\Omega - (\omega_i - \omega_j))^2$ in case (a), and by $\gamma_{k(ij)}\gamma_{k(ji)} < -\omega_i\omega_j(m_k\Omega - (\omega_i - \omega_j))^2$ in case (b); see Nayfeh and Mook (1979) for a detailed treatment of these resonances, and for higher order approximations and consideration to damping.

To solve (7.113) approximately for $\Omega \gg 1$ we use the MDSM, splitting the motions into slow and fast components:

$$
\mathbf{v} = \mathbf{v}_0(t) + \Omega^{-1}\mathbf{v}_1(t, \tau); \quad \langle \mathbf{v}_1 \rangle = \mathbf{0}, \quad \tau = \Omega t,
\tag{7.115}
$$

where the equations for the slow motions $\mathbf{v}_0$ and the fast motions $\mathbf{v}_1$ become, respectively:

$$
\mathbf{v}_{0tt} + \mathbf{c}\mathbf{v}_{0t} + \boldsymbol{\omega}^2 \mathbf{v}_0 + \sum_{k=1}^{N} \boldsymbol{\gamma}_k \langle \mathbf{v}_1 \sin(m_k\tau + \psi_k)\rangle = \mathbf{p}_0(t),
\tag{7.116}
$$

and

$$
\begin{aligned}
\mathbf{v}_{1\tau\tau} &+ \sum_{k=1}^{N} \gamma_k \mathbf{v}_0 \sin(m_k\tau + \psi_k) - \mathbf{p}_1(t)\sin(m\tau + \psi) \\
&+ \Omega^{-2}\left[\mathbf{v}_{1tt} + c\mathbf{v}_{1t} + \omega^2 \mathbf{v}_1\right] \\
&+ \Omega^{-1}\Big[2\mathbf{v}_{1t\tau} + c\mathbf{v}_{1\tau} \\
&+ \sum_{k=1}^{N} \gamma_k (\mathbf{v}_1 \sin(m_k\tau + \psi_k) - \langle \mathbf{v}_1 \sin(m_k\tau + \psi_k)\rangle)\Big] = 0,
\end{aligned}
\tag{7.117}
$$

For the case of interest, where resonant effects are significant but not dominating, the term $\Omega^{-2}\omega^2\mathbf{v}_1$ in (7.117) is comparable in magnitude to the dominating first terms, so the equation of fast motion can be written:

$$
\begin{aligned}
\mathbf{v}_{1\tau\tau} + \Omega^{-2}\omega^2\mathbf{v}_1 &= -\sum_{k=1}^{N} \gamma_k \mathbf{v}_0 \sin(m_k\tau + \psi_k) \\
&+ \mathbf{p}_1(t)\sin(m\tau + \psi) + O(\Omega^{-1}),
\end{aligned}
\tag{7.118}
$$

with stationary solution (valid when Ω is away from sharp resonance):

$$
\begin{aligned}
\mathbf{v}_1 &= \sum_{k=1}^{N} \left(m_k^2\mathbf{I} - \Omega^{-2}\omega^2\right)^{-1}\gamma_k \mathbf{v}_0 \sin(m_k\tau + \psi_k) \\
&- \left(m^2\mathbf{I} - \Omega^{-2}\omega^2\right)^{-1}\mathbf{p}_1(t)\sin(m\tau + \psi) + O(\Omega^{-1}).
\end{aligned}
\tag{7.119}
$$

Inserting into (7.116) and evaluating the average, the equation for the slow motions becomes:

$$
\mathbf{v}_{0tt} + c\mathbf{v}_{0t} + \left(\omega^2 + \Delta\omega^2\right)\mathbf{v}_0 = \mathbf{p}_0(t) + \Delta\mathbf{p}_0(t),
\tag{7.120}
$$

where $\Delta\omega^2$ denotes the apparent change in stiffness:

$$
\Delta\omega^2 \equiv \frac{1}{2}\sum_{k,j=1}^{N} \gamma_k \left(m_j^2\mathbf{I} - \Omega^{-2}\omega^2\right)^{-1}\gamma_j \delta_{m_k m_j}\cos(\psi_k - \psi_j),
\tag{7.121}
$$

and $\Delta\mathbf{p}_0(t)$ denotes an apparent change in static load:

$$
\Delta\mathbf{p}_0(t) \equiv \frac{1}{2}\sum_{k=1}^{N} \gamma_k \left(m^2\mathbf{I} - \Omega^{-2}\omega^2\right)^{-1}\mathbf{p}_1(t)\delta_{m_k m}\cos(\psi_k - \psi).
\tag{7.122}
$$

Expressions (7.120)–(7.122) provide straight-forward predictions of apparent changes in stiffness and static loading, and facilitate predictions of related quantities such as natural frequencies, buckling loads, and static equilibriums – provided the frequency of excitation Ω is high, and not in sharp external or parametric resonance with the system. Next we illustrate application of the GNRP and the GARP in the form of two simple examples.

7.6.3   Example 1: Clamped String with HF Base Excitation

Fig. 7.8 shows the system: a clamped-free string of mass per unit length ρA and finite bending rigidity EI, subjected to gravity g and time-harmonic horizontal base excitation of velocity amplitude a and high frequency Ω. The displacement amplitude $\Omega^{-1}a$ is small, while the acceleration Ωa is strong. The linearized partial differential equation governing small transverse motions $u(x, t)$, and the boundary conditions, are:

$$u_{tt} + cu_t + \omega_0^2 l^4 u_{xxxx} - ((l - x)u_x)_x \Omega a \sin(\Omega t) = -g, \qquad (7.123)$$

$$u(0, t) = u_x(0, t) = u_{xx}(l, t) = u_{xxx}(l, t) = 0, \qquad (7.124)$$

where subscripts x and t denote partial derivatives with respect to space and time, c is the viscous damping coefficient, l the string length, and $\omega_0 = \sqrt{EI/(\rho A l^4)}$ is a characteristic frequency.

Equation (7.123) has the form (7.102), with $N = 1$, $L_c = c$, $L_0 = \omega_0^2 l^4 (\ )_{xxxx}$, $L_{11} = -a((l - x)(\ )_x)_x$, $m_1 = 1$, $f_0 = -g$, $\psi_1 = f_1 = 0$, and boundary conditions of the required type. To predict the response under far-from-resonance conditions we employ the GNRP, and use (7.109)–(7.110) to readily produce an approximation for the equation of slow motions:

$$u_{0tt} + cu_{0t} + \omega_0^2 l^4 u_{0xxxx} + \frac{1}{2}a^2 \left((l - x)((l - x)u_{0x})_{xx}\right)_x = -g, \qquad (7.125)$$

which can be rearranged into the following form:

$$u_{0tt} + cu_{0t} + \left(\bar{\omega}_0^2(x)l^4 u_{0xx}\right)_{xx} = -g, \qquad (7.126)$$

where

$$\bar{\omega}_0^2(x) \equiv \omega_0^2 + \frac{1}{2}(a/l)^2(1 - (x/l))^2. \qquad (7.127)$$

Here $\bar{\omega}_0^2(x)$ is seen to represent a distribution of effective bending stiffness (per unit mass) along the string. It is composed of a first term, ω_0^2, describing the structural or "real" stiffness of the string, and another one equivalencing the average

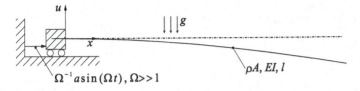

Fig. 7.8 Physical model of the base excited string

effect of the HF excitation; this part drops off quadratically with x towards the string tip. To an observer filtering out the small overlay of HF excitation, the additional *apparent* stiffness caused by the HF excitation cannot be distinguished from real structural stiffness with the equivalencing distribution.

Weakly resonant behavior is taken into account by using the GARP expressions (7.120)–(7.122) to compute equations governing the slow component $\mathbf{v}_0$ of the modal coefficients $\mathbf{v}(t)$; the results become (see Thomsen 2003b for details):

$$\mathbf{v}_{0tt} + c\mathbf{v}_{0t} + \left(\boldsymbol{\omega}^2 + \Delta\boldsymbol{\omega}^2\right)\mathbf{v}_0 = -g\boldsymbol{\beta}, \qquad (7.128)$$

where $\boldsymbol{\omega}$ is a diagonal matrix holding the squared natural frequencies of the string, $\{\boldsymbol{\beta}\}_i = \int_0^l \varphi_i\,dx \big/ \int_0^l \varphi_i^2\,dx$ where $\varphi_i(x)$ is the i'th mode shape, and the matrix $\Delta\boldsymbol{\omega}^2$ describes the change in effective stiffness due to the HF excitation:

$$\Delta\boldsymbol{\omega}^2 \equiv \tfrac{1}{2}a^2\boldsymbol{\gamma}\left(\mathbf{I} - \Omega^{-2}\boldsymbol{\omega}^2\right)^{-1}\boldsymbol{\gamma}, \qquad (7.129)$$

where $[\boldsymbol{\gamma}]_{ij} = \int_0^l (l-x)\varphi_{ix}\varphi_{jx}\,dx \big/ \int_0^l \varphi_i^2\,dx$. Such expressions can be used directly for predicting apparent changes in stiffness and related quantities, or along with (7.112) (with u_0 substituted for u) to predict the slow component $u_0(x, t)$ of the displacement field.

7.6.4   Example 2: Square Membrane with In-Plane HF Excitation

Next we consider predicting the free transverse vibrations of a square membrane, stretched in the (x, y) plane (Fig. 7.9). The membrane has density ρ and thickness h, and is clamped along the boundary lines $x = 0$ and $y = 0$, while held with tension N_0 per unit length along the two other boundary lines $x = 1$ and $y = 1$. The whole arrangement is shaken in the plane, so that points on the clamped boundary lines are translated at high angular frequency Ω along circular paths of small radius $\Omega^{-1}a$, $\Omega \gg 1$, $a = O(1)$. Assuming small out-of-plane vibrations $u(x, y, t)$, so that in-plane

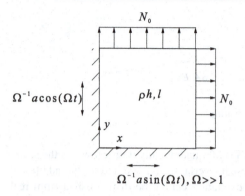

Fig. 7.9 Stretched membrane subjected to in-plane HF excitation

deformations can be ignored, the boundary tension N_0 can be held approximately constant and the equation of transverse motion becomes:

$$u_{tt} - \left(\left(c_0^2 + (l - x)\Omega a\,\sin(\Omega t)\right)u_x\right)_x - \left(\left(c_0^2 + (l - x)\Omega a\,\cos(\Omega t)\right)u_y\right)_y = 0, \quad (7.130)$$

where $c_0^2 \equiv N_0/\rho h$ is a squared wave speed, and the boundary conditions are $u(0, y, t) = u(l, y, t) = u(x, 0, t) = u(x, l, t) = 0$. This equation is of the form (7.102) with $N = 2$, $L_c = 0$, $L_0 = -c_0^2(()_{xx} + ()_{yy})$, $L_{11} = -a((l - x)()_x)_x$, $L_{12} = -a((l - y)()_y)_y$, $m_1 = m_2 = 1$, $\psi_1 = 0$, $\psi_2 = \pi/2$, and $f_0 = f_1 = 0$.

The slow component u_0 of the far-from-resonance response is given by the GNRP, for which application of (7.109)–(7.110) gives:

$$u_{0tt} + \bar{L}_0[u_0] = O(\Omega^{-1}), \quad \Omega \gg 1, \quad (7.131)$$

where the effective stiffness operator $\bar{L}_0$ can be written:

$$\begin{aligned}
\bar{L}_0 &= L_0 + \frac{1}{2}(L_{11}^2 + L_{12}^2) \\
&= -c_0^2\left(()_{xx} + ()_{yy}\right) \\
&\quad + \frac{1}{2}a^2\left[\left((l - x)^2()_{xx}\right)_{xx} + \left((l - y)^2()_{yy}\right)_{yy}\right].
\end{aligned} \quad (7.132)$$

As appears, in the presence of HF excitation ($a \neq 0$) the membrane exhibits bending-like stiffness (fourth order derivative terms) in addition to the stiffness already provided by stretching (second order derivative terms). This in turn implies, e.g., that more boundary conditions are required to solve for u_0, and that traveling waves will appear dispersive instead of non-dispersive.

The GARP is applicable to illustrate how the stiffness might be affected when resonances cannot be ignored, as will typically be the case. For brevity we assume here that $0 \ll \omega_1 < \Omega \ll \omega_2$, so that only the first resonance needs to be taken into account, and that only the change in ω_1 is to be predicted. Then (7.120)–(7.122) can be employed with $n = 1$, giving an equation for the slow component v_0 of the fundamental modal coefficient:

$$v_{0tt} + \left(\omega_1^2 + \Delta\omega_1^2\right)v_0 = 0; \quad \Delta\omega_1^2 \equiv \frac{1}{2}\left(1 - (\omega_1/\Omega)^2\right)^{-1}\left(\gamma_1^2 + \gamma_2^2\right), \quad (7.133)$$

where $\omega_1 = \sqrt{2}c_0\pi/l$ is the fundamental natural frequency for the unexcited membrane, and $\Delta\omega_1$ is the predicted change in this natural frequency due to HF excitation. Using (7.114) to calculate $\gamma_{1,2}$, and inserting the normalized fundamental eigenfunction (i.e. the first solution of $L_0[\varphi] = \omega^2\varphi$ with boundary conditions), $\varphi_1(x, y) = 2l^{-1}\sin(\pi x/l)\sin(\pi y/l)$, one finds $\gamma_1 = \gamma_2 = \pi^2 a/(2l)$ and thus

$$\Delta\omega_1 = \left(1 - (\omega_1/\Omega)^2\right)^{-1/2}\left(\frac{\pi^2 a}{2l}\right), \quad (7.134)$$

which is valid for $0 \ll \omega_1 < \Omega \ll \omega_2$ and Ω away from any parametric resonances. Whether this change in natural frequency is of any practical significance appears to depend on two factors: The relative velocity amplitude a of the HF excitation $\Omega^{-1}a/l$, and the closeness of Ω to ω_1.

7.7 Specific Systems and Results – Some Examples

7.7.1 Using HF Excitation to Quench Friction-Induced Vibrations

Analytical predictions have been derived, for how the smoothening effect of HF excitation can be used to eliminate unwanted friction-induced oscillations (Chatterjee et al. 2004; Feeny and Moon 2000; Nath and Chatterjee 2016; Thomsen 1999). Such self-excited oscillations may occur e.g. with sliding machine parts, squeaking train wheels, chattering machining tools, windscreen wiper blades, and certain types of vehicle brake squeal (e.g., Armstrong–Hélouvry et al. 1994; Feeny et al. 1998; Lancioni et al. 2016; Popp 1992). Indeed, the application of HF signals to smoothen nonlinearities is known in control science as dither control (Bellman et al. 1986; Feeny and Moon 2000; Lehman et al. 1996; Meerkov 1980; Shapiro and Zinn 1997; Zames and Shneydor 1976).

The analysis in Thomsen (1999) is given in terms of the classical mass-on-moving-belt model (Fig. 7.7(h)) with HF excitation added to the oscillator mass and nondimensional equation of motion:

$$\ddot{x} + 2\beta\dot{x} + x + \gamma^2\mu(\dot{x} - v_b) = a\Omega^2\sin(\Omega t), \quad (7.135)$$

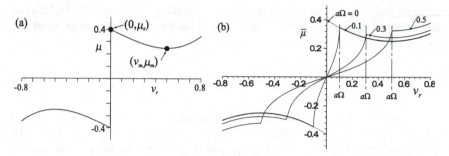

Fig. 7.10 (a) Friction function $\mu(v_r)$; (b) Effective friction characteristic $\bar{\mu}(v_r)$ for different intensities $a\Omega$ of HF excitation. Parameters: $\mu_s = 0.4$, $\mu_m = 0.25$, $v_m = 0.5$

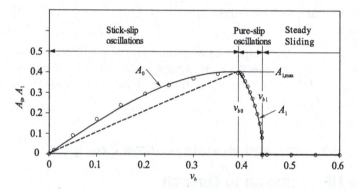

Fig. 7.11 Amplitude A of stable friction-induced oscillations as a function of excitation speed v_b. (—) analytical prediction; (O) Numerical simulation. Parameters: as in Fig. 7.10, and $\beta = 0.05$, $\gamma^2 = 1$

where x is the position of the mass, γ^2 a measure of the normal load, β the damping ratio, v_b the belt speed, $\Omega \gg 1$ the HF excitation frequency, and $a \ll 1$ its displacement amplitude. The friction coefficient μ versus sliding velocity v_r is:

$$\mu(v_r) = \mu_s \text{sgn}(v_r) - \frac{3}{2}(\mu_s - \mu_m)\left(v_r/v_m - \frac{1}{3}(v_r/v_m)^3\right), \qquad (7.136)$$

which has negative slope for small sliding velocities, and positive slope for higher velocities, as depicted in Fig. 7.10(a). (Such simple friction functions of just sliding velocity have been disputed, and demonstrated not to reproduce experimental tests satisfactorily; see, e.g., Cabboi et al. 2016).

Without HF excitation, stable self-excited oscillations occur for all belt speeds v_b lower than a critical value v_{b1}, as appears from Fig. 7.11 (Thomsen and Fidlin 2003). HF excitation may prevent such oscillations from building up, by effectively smoothening out the discontinuity of the friction characteristic and canceling its negative slope, as shown in Fig. 7.10(b). Here HF excitation has an effect similar to

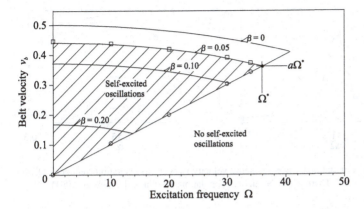

Fig. 7.12 Ranges of belt speeds v_b producing self-excited oscillations, as a function of excitation frequency Ω. (–,–) Analytical prediction; (O, □) numerical simulation. Parameters: as in Fig. 7.11, and $a = 0.01$

adding ordinary lubricant, effectively changing dry friction into viscous damping – with the important difference that HF excitation can easily be removed or changed. The consequence is depicted in the stability diagram of Fig. 7.12, showing that the interval of belt speeds causing self-excited oscillations shrinks and eventually vanishes as the frequency Ω of the HF excitation is increased.

In this example the HF excitation is applied tangentially to the surface. However, using a HF-modulated *normal force* also reduces the effective friction, and may significantly reduce or remove friction-induced oscillations (Cochard et al. 2003).

7.7.2 Displacement Due to HF Excitation and Asymmetric Friction

Thomsen (2000) and Fidlin and Thomsen (2001) analyzed a device where the biasing effect of HF excitation is used to make a slider propel itself controllably inside a stator (Fig. 7.7(a)). The slider is sandwiched between two friction layers of differing friction properties, and driven by an internal HF resonator. Inertia forces from the resonator make the slider oscillate back and forth at small amplitude and high frequency. With the resonator aligned askew to the friction layers, friction forces are less when the resonator presses against the layer of least friction, so the slider moves a bit longer in one direction than in the other during each vibration period – effectively in one direction with a small overlay of HF vibrations. The internal resonator might be realized using piezoelectric ceramics, as with traditional ultrasonic motors (Ueha and Tomikawa 1993). Compared to these, the driving principle considered here does not rely on the establishment of traveling waves, thus circumventing the need for two vibration exciters (source and absorber) and complicated electronics.

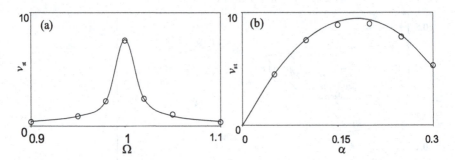

Fig. 7.13 Variation of stationary average slider velocity v_{st} with (**a**) excitation frequency Ω, and (**b**) resonator mass α. (–) Theory; (O) Numerical simulation. (Fidlin and Thomsen 2001)

Analytical expressions for the average sliding speed as a function of system parameters are derived, that generally agree well with numerical simulation results. As an example, Fig. 7.13(a) shows a maximum for the stationary average slider velocity v_{st} when the excitation frequency Ω is near the (unit-normalized) natural frequency of the internal resonator. Fig. 7.13(b) displays the influence of the (normalized) resonator mass α, showing a characteristic maximum at a specific value of resonator mass (where the increase in slider velocity due to increased inertia forces, are overridden by the effect of reduced oscillation amplitudes for the heavier resonator).

7.7.3 Chelomei's Pendulum – Resolving a Paradox

In 1982 V.N. Chelomei described an interesting experiment, involving a pendulum with a vertically vibrating support and a sliding disk at its rod (Fig. 7.14(a)): With small-amplitude HF excitation of the support, the pendulum was observed pointing upwards, with the disk floating almost at rest somewhere up the rod (Chelomei 1983). The phenomenon stirred significant scientific interest, but Chelomei's own explanation of it has been subjected to much controversy, as summarized in Thomsen and Tcherniak (2001).

The latter work provides what is believed to be a coherent theory explaining the phenomenon, and quantitative predictions that are supported by numerical simulation and laboratory experiments. It is demonstrated how resonant flexural vibrations of the pendulum rod force the disk to slide towards an equilibrium position, against gravity, by a biasing effect of HF excitation. This was hypothesized already by Blekhman and Malakhova (1986), however, without suggesting which mechanism should cause such vibrations in the first place, and consequently with no predictions for the stable states of the system in dependence of the physical parameters. It turns out that the necessary flexural rod vibrations arise due to a slight imperfection, breaking the perfect symmetry of the system. At the same time the

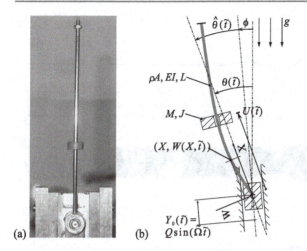

Fig. 7.14 (a) Experimental Chelomei's pendulum (steel rod Ø3 × 150 mm, plastic disk Ø12.2/ Ø3 × 5.3 mm, excitation amplitude 2 mm, frequency 200 Hz); (b) Model with a flexible pendulum rod, and a slight deviation from perfect vertical support oscillation. (Thomsen and Tcherniak 2001)

stiffening effect of HF excitation is responsible for stabilizing the pendulum near the up-pointing position.

With this model, a series of simplifications results in a set of three nonlinearly coupled equations of motion describing, respectively, rigid body motions $\theta(t)$ of the rod, the first-mode flexural vibration amplitude $v(t)$ of the rod, and displacement u (t) of the disk (cf. Fig. 7.14(b)). Using the MDSM, one ends up with considerably more tractable equations describing only the slow or average components of θ, v, and u. It is even possible to set up analytical expressions for the stationary values of these slow components, and to determine their stability.

Fig. 7.15 shows typical time series of numerical solutions (in solid line) of the equations of motion, with parameters corresponding to an experimental pendulum. Dashed lines indicate theoretical predictions for the stationary values of the slow components of motion $\langle \theta \rangle$ and $\langle u \rangle$, and for the amplitude of the flexural rod vibrations v. When the pendulum is released from the inverted vertical position, as appears from Fig. 7.15(a), its rigid body rotation θ performs damped oscillations and then stabilizes at the theoretically predicted value (in dashed line), which is 0.10 radians $\approx 5.7°$ off vertical. The thick appearance of the time history curve reflects that the large and slow oscillations are overlaid by small and rapid ones, as appears from the close-up of a short time interval at the right of the figure. Fig. 7.15 (b) shows how, in response to the off line excitation of the pendulum rod, flexural rod vibrations grow up and stabilize at the predicted amplitude (in dashed line). Finally Fig. 7.15(c) shows how – in response to the average effect of flexural vibrations from the rod and gravity – the disk climbs from the rod hinge to the position predicted theoretically (in dashed line). This position is just below the antinode (max. amplitude) for first-mode vibrations of the hinged-free rod.

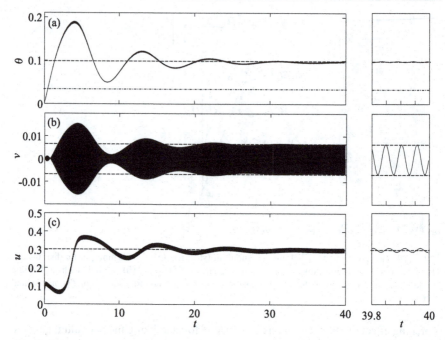

Fig. 7.15 Results of numerical simulation of the equations of motion for Chelomei's pendulum. (a) Rod inclination $\theta(t)$; (b) flexural rod vibrations $v(t)$; and (c) disk position $u(t)$. Dashed lines indicate corresponding theoretical predictions for the average values $\langle\theta\rangle$ and $\langle u\rangle$ in (a, c), and the amplitude of rod vibrations in (b). (Thomsen and Tcherniak 2001)

The theoretical predictions of $\langle\theta\rangle \approx 5.7°$ and $\langle u\rangle = 0.30$ agrees quite closely with the values $5°$ and 0.32 observed with the corresponding experimental pendulum. However, one should not put too much into this agreement of numbers: For the experimental pendulum both $\langle\theta\rangle$ and $\langle u\rangle$ were observed to be weakly modulated by a slow and seemingly chaotic component, e.g. the disk was observed floating at positions from about $u = 0.19$ to 0.32, with θ changing accordingly. And precise results were hard to reproduce due to difficulties with maintaining a sufficiently fine frequency tuning. The explanation for this is provided by theoretical stability diagrams and frequency response curves, showing a strong sensitivity to parameters and very narrow zones in parameter space where the phenomenon can occur. Thus even the slightest change in e.g. excitation frequency has a significant effect on the stationary values of $\langle\theta\rangle$ and $\langle u\rangle$, making consistent experimental results inherently hard to reproduce.

Besides a stunning phenomenon in its own interest, Chelomei's pendulum inspired suggestions for a new principle for vibration damping (cf. Fig. 7.7(d)), as discussed in Sect. 4.7, and in Thomsen (1996a).

7.7.4 Stiffening of a Flexible String

A length of wire may be stabilized upside down by vertical HF excitation, due to an increase in buckling load reflecting increased effective stiffness (Acheson and Mullin 1998; Champneys and Fraser 2000; Fraser and Champneys 2002; Mullin et al. 2003; Shishkina et al. 2008) – a phenomenon that Acheson (1997) has referred to as 'not quite the Indian rope trick'. The implication is that the stiffening effect is to be taken into account under conditions of strong HF excitation, if predictions of derived quantities such as natural frequencies, buckling loads, and equilibrium positions should be meaningful.

Analytical predictions for the stiffening effect of HF excitation for a clamped-free string has been derived (Thomsen 2003b), which compare well with laboratory experiments for a wide range of conditions. We encountered a model for this system in Sect. 7.6.3 (Fig. 7.8 and Eq. (7.123)), and derived analytical predictions for the stiffening effect for the actual case where resonance effect cannot be ignored (GARP theory, Eq. (7.129)). The expressions for the change in effective stiffness can be used to predict how much the string tip will lift in gravity, when the string base is aligned and excited horizontally. This tip lift is a quantity that can easily be observed and measured experimentally, and thus provides a means for checking the quality of the theoretical predictions.

The laboratory experiment involved a clamped-free piano string, horizontally aligned and subjected to longitudinal HF excitation at the clamped end. Gravity causes the string to bend downwards, while the HF excitation generally causes it to stiffen, and thus to straighten somewhat (Fig. 7.16). Measuring the lift of the string tip (using an optical tracker) thus quantifies the stiffening effect.

Fig. 7.17 shows a set of acceleration responses, i.e. the tip lift as a function of base acceleration, with the frequency kept constant at each of four values. As appears the agreement between GARP theory and experiment is generally good.

7.8 Summing Up

Hopefully a beginning understanding of the possible effects that HF excitation may have on mechanical systems, has by now emerged, as well as skills to analyze them systematically. Recall from the introduction to Sect. 7.2 and Sect. 7.2.3 that the MDSM is not the only analytical method available for this analysis; it just seems in many cases to be more convenient than the alternatives. The examples given in this chapter, though quite diverse, are not claimed to be inclusive or representative for every relevant application. They are rather meant to be indicative of typical problems and phenomena, and to provide specific reference for the sections dealing with generalized systems. If you want to expand on a beginning interest in this area, some of the references cited at the end of Sect. 7.1 may be a place to start; in particular Blekhman (2000) is recommended.

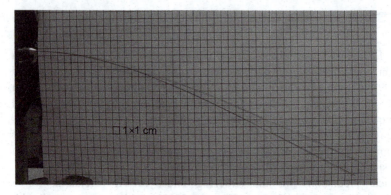

Fig. 7.16 Clamped-free 0.5 × 500 mm piano string excited by horizontal vibrations at the left end. Two images overlaid: One with gravity as the only excitation (lower string image), and one with the base vibrating at 62 Hz and 3.3 mm displacement amplitude (upper image). The string tip is seen to lift about 2 cm. (Thomsen 2003b)

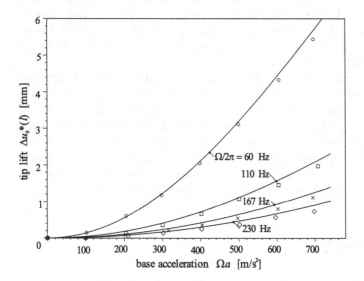

Fig. 7.17 Tip lift of a Ø1 × 550 mm horizontal string with horizontal base excitation, as a function of input acceleration at four values of input frequency. Curves: GARP theory; Symbol markers: experimental measurements. (Thomsen 2003b)

7.9 Problems

Problem 7.1 Fig. P7.1 shows a model of a needle pointing instrument, mounted on a structure that vibrates in direction ψ at small amplitude $qL \ll L$ and high frequency $\Omega \gg \omega = \sqrt{3k/(mL^2)}$. In the absence of imposed vibrations and other external forces the needle pointer has a stable equilibrium at $\theta = 0$. The equation governing needle motions is:

$$\ddot{\theta} + 2\beta\dot{\theta} + \omega^2\theta + \frac{3}{2}q\Omega^2 \sin(\Omega t) \sin(\theta - \psi) = 0. \tag{7.137}$$

(a) Set up an equation governing the slow (i.e. average) motions of the needle, and explain how in principle the imposed vibrations affect the dynamics of the needle pointer.
(b) Derive an expression for the bias error at $\theta = 0$ when ψ is small, and sketch how this error depends on the excitation intensity $q\Omega$.

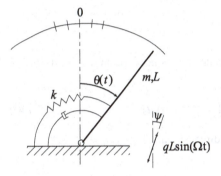

Fig. P7.1

Problem 7.2 Fig. P7.2 shows a flywheel consisting of two pairs of masses connected by rigid and massless bars. The wheel can rotate freely in a horizontal plane, subjected only to viscous damping at the hinge, and kinematic high-frequency oscillation along a fixed line. The equation of motion is:

$$\ddot{\theta} + 2\beta\dot{\theta} + p\Omega^2 \sin(\Omega t)(\sin\theta + \cos\theta) = 0, \tag{7.138}$$

where $\Omega \gg 1$, $p\Omega = O(1)$, and:

$$p = \frac{q}{2}\frac{M-m}{M+m}. \tag{7.139}$$

In the absence of high-frequency excitation the flywheel is (marginally) stable in any position. When $M = m$ this also applies in the presence of high-frequency excitation.

But what happens in the presence of high-frequency excitation when $M \neq m$? First discuss what you would expect. Then examine the existence and stability of possible quasi-equilibriums of the flywheel by setting up the equation governing the slow component of $\theta(t)$.

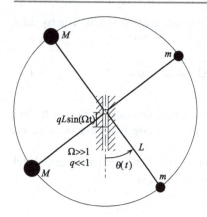

Fig. P7.2

Problem 7.3 A simple model for the instantaneous penetration depth u(t) during vibrational piling (Fig. P7.3) is:

$$\ddot{u} + \gamma u \operatorname{sgn}(\dot{u}) = 1 + q\Omega^2 \sin(\Omega t), \qquad (7.140)$$

where all variables and parameters are nondimensional, $\Omega \gg 1$, $q \ll 1$, $q\Omega = O(1)$, and:

$$\gamma = \frac{\mu E \pi d^2}{Mg}, \quad u = \frac{U}{l}, \quad t = \omega \tilde{t}, \quad \omega^2 = \frac{g}{l}, \quad \Omega = \frac{\tilde{\Omega}}{\omega}, \qquad (7.141)$$

where M is the pile mass and d its diameter, E is the elastic modulus of the surrounding medium, μ is the coefficient of kinetic friction between pile and medium.

(a) Set up an equation governing the slow (i.e. average) component of $u(t)$.
(b) Simplify this equation for the case of a relatively small average piling speed.
(c) Derive and discuss an expression for the vibrational force acting on the pile (i.e. the static force equivalencing the average effect of the fast vibrations).
(d) Modify (7.140) to account for the case where the coefficient of kinetic friction is larger for motions directed inwards, i.e. $\mu = \mu_+$ for $\dot{u} > 0$, $\mu = \mu_-$ for $\dot{u} < 0$, where $\mu_+ > \mu_- > 0$, and repeat Questions a–c for that case.

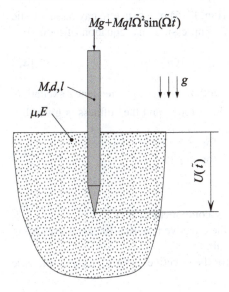

Fig. P7.3

Problem 7.4 For the simplified model of Chelomei's pendulum in Fig. P7.4:

(a) Set up the equations of motion for the case where horizontal oscillations of the support $Z_h(t) = Q_h \sin(\tilde{\Omega} t + \eta)$ appear in addition to the vertical oscillations $Z_v(t) = Q_v \sin(\tilde{\Omega} t)$.

(b) Using the method of direct separation of motions, consider the existence and stability of a state where the rod performs small amplitude vibrations near $\theta = 0$, while the disk performs small amplitude vibrations near a fixed value of $U \in]0;l[$.

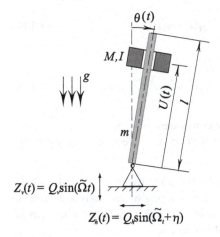

Fig. P7.4

Problem 7.5 Consider a simple model (Fig. P7.5) of a structure having an elastic modulus in tension differing from that in compression, and equation of motion:

$$m\ddot{u} + c\dot{u} + k(u)u = q\Omega^2 \sin(\Omega t) + p(t), \tag{7.142}$$

where $q\Omega^2$ is the force amplitude of a source of strong high frequency excitation, $\Omega \gg \sqrt{\sup(k)/m}$, $p(t)$ is an additional 'slow' force, and the stiffness is given by:

$$k(u) = \begin{cases} k_1 & \text{for } u > 0, \\ k_2 & \text{for } u < 0. \end{cases} \tag{7.143}$$

where k_1 and k_2 are constants.

(a) Set up an equation governing the slow component of $u(t)$.
(b) Use this to provide an expression for the effective restoring force corresponding to the term $k(u)u$, and discuss the result.
(c) Simplify and discuss the expression for the effective restoring force in the case of small but finite $|u| < q/m$.

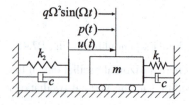

Fig. P7.5

Problem 7.6   Consider a system with nonlinear, position-dependent viscous damping:

$$\ddot{u} + \varepsilon c(u)\dot{u} + \omega_0^2 u = q\Omega^2 \sin(\Omega t), \tag{7.144}$$

where ε is a small parameter, $\Omega \gg \omega_0$, and:

$$c(u) = c_0 + c_1 u + c_2 u^2 + c_3 u^3, \tag{7.145}$$

(a) Set up an equation governing the slow component of $u(t)$.
(b) Use this to state and discuss an expression for the effective damping corresponding to the term $c(u)\dot{u}$.

Problem 7.7 The main mass M in Fig. P7.7 is loaded by gravity g and, and has dry contact with friction coefficient μ to a plane vibrating horizontally at amplitude A and frequency Ω. The main mass has an imbedded linear oscillator with mass m, stiffness k, orientation angle α, and negligible damping.

(a) Set up equations of motion for the position $u(t)$ of the main mass and the position $z(t)$ of the oscillator mass, assuming that the main mass never loses contact with the plane.

(b) Explain why the speed $\dot{u}$ of M could have a non-zero average, so that the main mass moves in one direction along the plane, on the average.

(c) Derive an expression for the average speed of the main mass, and discuss its dependency of the frequency of excitation Ω and the oscillator orientation α.

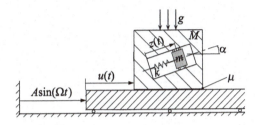

Fig. P7.7

Appendix A: Performing Numerical Simulation

Solving vibration and stability problems typically involves both analytical and computational work (and sometimes also experimental). We here briefly touch upon a few practical aspects of dealing numerically with nonlinear dynamical systems. This concerns solving differential equations, computing chaos-related quantities, programming a helpful user interface, and using relevant Internet resources. The information given here is intended only to give an overview of the kind of tools and skills required for doing *computational dynamics*.

A.1. Solving Differential Equations

Mathematical models of vibrating systems usually come in the form of differential equations having second-order time derivatives. With continuous systems, space derivatives occur as well. Space derivatives are typically eliminated by some discretization method, e.g., Galerkin, finite element, or finite difference – so that in any case a set of second-order ODEs (ordinary differential equations) results. These can almost always be recast into the general first-order form:

$$\dot{\tilde{\mathbf{x}}} = \tilde{\mathbf{f}}(\tilde{\mathbf{x}}, t). \tag{A.1}$$

where $\tilde{\mathbf{x}} \in R^n$ is a vector of state variables (typically generalized displacements and velocities) and $\tilde{\mathbf{f}}$ is a vector of generally nonlinear functions. By introducing a new dependent variable to eliminate the explicit time-dependency, such a system can be further transformed into first-order autonomous form:

$$\dot{\mathbf{x}} = \mathbf{f}(\mathbf{x}), \tag{A.2}$$

where $\mathbf{x} \in R^{n+1}$ and $\mathbf{f} \in R^{n+1}$. For example, the Duffing equation

© The Editor(s) (if applicable) and The Author(s), under exclusive license to
Springer Nature Switzerland AG 2021
J. J. Thomsen *Vibrations and Stability*,
https://doi.org/10.1007/978-3-030-68045-9

$$\ddot{u} + 2\beta\omega\dot{u} + \omega^2 u + \gamma u^3 = q\sin(\Omega t), \tag{A.3}$$

can be recast into the standard form (A.1) by introducing a new variable $v = \dot{u}$, the vectors $\tilde{\mathbf{x}}$ and $\tilde{\mathbf{f}}$ becoming:

$$\tilde{\mathbf{x}} = \begin{Bmatrix} u \\ v \end{Bmatrix}, \quad \tilde{\mathbf{f}}(\mathbf{x}, t) = \begin{Bmatrix} v \\ q\sin(\Omega t) - 2\beta\omega v - \omega^2 u - \gamma u^3 \end{Bmatrix}. \tag{A.4}$$

The form (A.2) is achieved by further introducing the variable $\varphi = \Omega t$, i.e.:

$$\mathbf{x} = \begin{Bmatrix} u \\ v \\ \varphi \end{Bmatrix}, \quad \mathbf{f}(\mathbf{x}) = \begin{Bmatrix} v \\ q\sin\varphi - 2\beta\omega v - \omega^2 u - \gamma u^3 \\ \Omega \end{Bmatrix}. \tag{A.5}$$

Solving ODEs of the forms (A.1)–(A.2) numerically is a standard task. Carefully tested library software exists for doing it. It is recommended to employ for this task the best existing software available, e.g., routines from the commercial software programming libraries IMSL, NAG, or similar – or the built-in functions in numerical application software such as MATLAB (e.g., Biran and Breiner 2002) or GNU OCTAVE. Consult the literature on numerical methods (e.g., Press et al. 2002; Morton and Mayers 1994), should you insist on home-coding an ODE-solver.

Biran and Breiner (2002) provides useful guidance on choosing an ODE-solver suitable for the task, of which we summarize the most important here. The most critical of the decisions is whether to use a *non-stiff* solver or a *stiff solver*, and what *integration order* to use:

Non-stiff solvers For a start, opt for a non-stiff solver, like those based on Runge-Kutta and predictor-corrector methods.

For smooth/continuous systems where polynomial extrapolation provides good predictions, predictor-corrector methods are generally superior to Runge-Kutta.

When polynomial extrapolation fails (as e.g. when predicting a ball bouncing against a table to pass through the table), Runge-Kutta methods perform better and more efficiently (or a stiff solver may be required, see next). Some solvers optionally provides *event handling* procedures for dealing with this (in the bouncing ball example the user then provides an *even-function* for detecting that the ball is about to pass the table).

Stiff solvers If the system is characterized by time constants that differ by at least two order three orders of magnitude, choose a stiff solver. The time constants can be estimated, e.g., by calculating natural frequencies of the system, but other considerations can be necessary. If the system is composed of several subsystems, natural frequencies of the isolated subsystems can provide a cue. If impact processes are involved the impact time(s) provide additional time constants, which may be much shorter than those of the unimpacting system. The same applies when other discontinuous or abruptly changing processes are involved, e.g. as with dry friction or intermittent contact problems.

Stiff solvers operate with large integration steps, at the expense of introducing possibly large errors on the shortest time scales. If this is unacceptable, do not choose a stiff solver, despite the span in time scales.

With stiff solvers the cost of each integration step is usually large. So only choose a stiff solver if the Jacobian of the system does not need to be recomputed by numerical finite differencing at every time step, that is, if the system is linear or weakly linear, or if you can provide the Jacobian in analytical form to the solver.

Go through all options of the solver and choose or experiment with these carefully.

Integration order For *smooth systems* higher-order algorithms (like MATLAB's ode45 Runge-Kutta fourth/fifth order), are usually superior to the corresponding lower-order ones (like MATLAB's ode23 Runge-Kutta second/third order). With non-*smooth/discontinuous systems* or when accuracy requirements are modest, lower-order algorithms can be more efficient.

A.2. Computing Chaos-Related Quantities

For studying chaotic vibrations one need to compute phase plane orbits, Poincaré maps, Lyapunov exponents, and perhaps Lyapunov dimensions and frequency spectra. Transients should be discarded by allowing sufficient simulation time to pass before displaying data.

Phase plane graphs are obtained by displaying two state variables of the system versus each other. For the Duffing equation above, e.g., one displays $v(t)$ versus $u(t)$ where the two quantities are output from the ODE-solver.

Poincaré maps of periodically driven systems are obtained by solving the ODEs with a time-step equal to the period of the driving force. Most ODE-solvers then internally iterate in small time-steps, to output only the solution at the end t_k of the user-specified time interval. To display a Poincaré map for the above Duffing equation, e.g., one displays $v(t_k)$ versus $u(t_k)$ using small dots.

Lyapunov exponents and dimensions are calculated as described in, e.g., Wolf et al. (1985), Parker and Chua (1989), Moon (1987), or Kantz (1994). Freely downloadable software exists for this (see below).

Frequency spectra are obtained by employing an FFT algorithm (Fast Fourier Transform) to the time data output from the ODE-solver. Usually one takes the logarithm and displays the result in dB as a bar graph. Most commercial software libraries contain FFT routines. Consult, e.g., Press et al. (2002), should you consider coding yourself. For a readable treatise on digital signal processing of simulated or experimental vibration data see Brandt (2011).

A.3. Interfacing with the ODE-Solver

There are many ways to design a user-friendly and flexible interface to the ODE-solver. It should be easy to adapt to different mathematical models, allow for easy experimentation with varying system parameters, and aid in verification of analytical and experimental results.

Commercial simulation software exists that allows the user to define a dynamical system in a graphical environment, to integrate the equations, and to display features of the result; examples are MATLAB/SIMULINK and SIMULATIONX (formerly ITI-SIM). However, user-friendliness to general users necessarily implies a lack of flexibility to satisfy individual needs. Some users find that they will have to code their own analysis environment, either in high-level general programming languages such as Python, Fortran, C, or C++, and graphical libraries such as OPENGL – or in dedicated numerically oriented languages such as plane MATLAB and GNU OCTAVE. In many cases the latter will provide an efficient and satisfactory compromise between the conflicting requirements of short program development time and ease of use on the one side, and flexibility and high-speed computing on the other.

If you consider programming your own simulation software, a starting point might be the example menu-skeleton shown in Fig. A.1. The primary menus are as follows.

System/edit Calls up an editor for editing the content of the *parameter file*. This file holds information on 1) the system to be analyzed (physical parameters); 2) the kind

Fig. A.1 Menu-skeleton for a numerical simulation program

☐ **System**
    ☐ Load
    ☐ Edit
    ☐ Save
    ☐ Print
☐ **Compute solution**
☐ **Display solution**
    ☐ Time histories
    ☐ Phase planes
    ☐ Frequency spectra
    ☐ Animation
    ☐ Order analysis
    ☐ Lyapunov exponents
    ☐ Export/Plot/Print
☐ **Simulate in real-time**
☐ **Analytical Predictions**
    ☐ Frequency response
    ☐ Sweep response
    ☐ Export/Plot/Print

of solution aimed at (simple time March, Poincaré map or Lyapunov exponents); 3) Solution parameters (initial conditions, time-span, time-steps, transient cut-off time, error tolerance, etc.); and 4) miscellaneous parameters that are only rarely changed (printer, font size, graphics file format, variable increments for real-time simulation, etc.).

Compute solution Starts the ODE-solver, checks for keyboard interrupts while solving, displays progress in percentage of total task.

Display solution Calls up a sub-menu for choosing a particular feature of the solution to display. Here, 'Order analysis' visualizes the relative magnitude of equation terms (to check possible assumptions regarding these), whereas 'animation' displays an image of the physical system on screen, with its parts moving according to the calculated solution (see below). Time histories, phase planes and spectra are displayed using existing curve plotting software (see below), and may be sent to a printer or saved as graphic files for inclusion in text documents.

Simulate system Loads a 'real-time' simulation/animation module. The differential equations are solved small time-steps ahead, while images of the changing states of the physical system (e.g., a beam or a pendulum) are displayed on screen. The physical parameters of the system can be changed 'on the fly' by pressing pre-defined keys. For example, pressing 'F' may raise the frequency of harmonic excitation in small increments, whereas pressing "f" lowers the frequency. For low-dimensional systems the impression is created that one controls and views the system in real-time, even on a modest PC. A graphical panel displays values of numerical values that could be measured for a similar laboratory experiment, e.g., the instantaneous system time, state, amplitude and frequency.

Animation of the physical system is performed by using the basic graphical routines accompanying most compilers, or – for portability – graphical libraries available for most operating systems and compilers, such as OpenGL with GLUT (freely available, see below). For this one can employ the following procedure: 1) Display the physical system in its instantaneous state while the ODE-solver computes the next state, 2) Redraw the state using background color, and 3) Draw the new state in visible color. Alternatively one can employ two graphical pages, drawing on a background page and swapping pages each time the ODE-solver completes a time-step. To slow down animation, a time-delay is employed while the system is shown in visible color. Delays can be coded via dummy-loops (e.g., adding two numbers an adjustable number of times), or by using the internal timing routines supplied with most compilers.

Analytical predictions Calls up a module for calculating, displaying and exporting results of analytical predictions of system response, e.g., as obtained by theoretical perturbation analysis.

A modular design allows one to rather easily adapt the program to a wide range of different systems. Typically, one should have to change only the definition of differential equations, the set of system parameters, and the graphical image of the system used for animation. As an alternative to a menu-driven user interface, one can of course implement a command-language interface.

A.4. Locating Software on the Internet

A plethora of relevant software is available through the Internet, freely downloadable in many cases. The quality ranges from poor to superior, and one has to employ a critical eye. A few starting points are provided below, valid at the time of writing, just to indicate the kind of resources available. No attempt was made to include all relevant resources, nor to seriously review or recommend any particular piece of software. The information given is prone to change, and may quickly run out of date; still it provides an indication for the novice simulation software programmer on which kind of items there is to look for.

General resources This web pages provide a starting points for searching Internet resources related to computational science, mathematics, and engineering:

- https://gams.nist.gov

Plotting scientific graphs GNUPLOT is a free 2D/3D function- and data-plotting utility available for most platforms. It supports many different image formats (e.g., PostScript, EPS, Latex, WMF, EMF). Executable as a stand-alone or from a user program, it is downloadable via:

- http://www.gnuplot.info

GLE is another free cross-platform 2D/3D function- and data-plotting utility, downloadable from:

- https://glx.sourceforge.io

Numerical libraries IMSL and NAG are well-known libraries with thousands of standard numerical routines (e.g. solvers for differential equations, algebraic equations, and eigenvalue problems) in Fortran, C/C++, which can also be merged with, e.g. MATLAB or Python code:

- https://www.imsl.com
- https://www.nag.com/content/nag-library

It can be difficult to locate, download and install free or shareware libraries of a quality comparable to the commercial ones, but see e.g. the *Netlib Repository*:

- http://www.netlib.org/utk/misc/netlib_query.html (for searching)
- https://netlib.org/master/expanded_liblist.html (for browsing)

or use the *NIST Guide to Available Mathematical Software*

- https://gams.nist.gov

Graphics programming OPENGL is a 2-D and 3-D graphical library that has evolved into some kind of industry standard. It has powerful capabilities to do almost everything graphical you want. OPENGL is part of most common operating systems, and thus your computer probably already has it installed. Start here to learn more about OPENGL:

- https://www.opengl.org

OPENGL is platform independent, i.e. it makes no direct reference or interaction with the operating system. This means your application will run at any platform, which, however, also requires that you deliver some interface between OPENGL and the particular operating system. But for simple applications even this part can be standardized, and that was accomplished by GLUT, the *OpenGL Utility Toolkit*, a system independent toolkit for writing OpenGL programs. GLUT makes it considerably easier to learn about and explore OpenGL programming. The GLUT library has both C, C++, FORTRAN, Python, and Ada programming bindings. The original GLUT is not supported or developed anymore, but has been replaced by the open-source version FreeGLUT:

- http://freeglut.sourceforge.net

A good (though old) reference for learning how to program with GLUT is the freely downloadable document by GLUT's creator, Mark J. Kilgard, in html- or pdf-format, respectively:

- https://www.opengl.org/resources/libraries/glut/spec3/spec3.html
- https://www.opengl.org/resources/libraries/glut/glut-3.spec.pdf

Chaos software To calculate Lyapunov exponents *from a time series*, a MATLAB-version of the algorithm presented in Wolf et al. (1985) (cited about 6000 times) is available at the MathWorks File Exchange:

- https://www.mathworks.com/matlabcentral/fileexchange/48084-wolf-lyapunov-exponent-estimation-from-a-time-series

A MATLAB-version of the algorithm (from the same paper) to calculate Lyapunov exponents *from an ODE-system* is also available:

- https://www.mathworks.com/matlabcentral/fileexchange/4628-calculation-lyapunov-exponents-for-ode

See also the site of the chaos group at the University of Maryland:

- http://www.chaos.umd.edu/chaos.html

Numerical and/or symbolic application software Homepages for some well-known commercial programs for symbolic and numerical computation and simulation are:

- http://www.mathworks.com (MATLAB & SIMULINK)
- http://www.wolfram.com (MATHEMATICA)
- http://www.maplesoft.com (MAPLE)
- http://www.mathcad.com (MATHCAD)
- https://www.simulationx.com (SIMULATIONX; formerly ITI-SIM)

Encyclopedia and databases MATHWORLD is Eric Weisstein's *World of Mathematics*, a comprehensive and interactive mathematics encyclopedia intended for students, teachers, and researchers, freely available at:

- http://mathworld.wolfram.com/

  where there is also a physics encyclopedia:

- http://scienceworld.wolfram.com/physics/

MATWEB holds a database of material properties for a wide range of polymers, metals, fibers, and other engineering materials, freely available at:

- http://www.matweb.com/

Appendix B: Major Exercises

This appendix provides three major exercises, not specific to any particular chapter of the book. To solve them you need to integrate subjects dealt with in several chapters. The exercises can be solved step-by-step while proceeding through the relevant chapters, or as a final check-up or as course work. Most sections of questions form entities, related to specific chapters, which can be solved more or less independently of other questions.

B.1 Tension Control of Rotating Shafts

Background. Lightweight transmission shafts may operate at supercritical operating speeds. Hence, such shafts may have to pass one or more resonance frequencies during startup and shutdown. Accelerations through resonance can cause excessive vibrations, which in turn may cause considerable damage or produce unacceptable noise. Some suggestions for preventing vibrations building up during passage of resonance are based on controlling the stiffness of the shaft system. By this approach the natural frequencies are displaced in a controllable manner during startup and shutdown. The stiffness of the system can be controlled by varying the stiffness of the supports (e.g., by using shape memory alloys). Alternatively the bending stiffness of the shaft can be controlled.

Turkstra and Semercigil (1993) suggested varying the bending stiffness of shafts by applying controlled tension forces. They proposed changing the stiffness instantaneously at pre-determined moments of the startup/shutdown cycle. This is an open-loop control, since no feedback of measured performance is involved in the decision on control action.

Fig. B.1 shows the system as suggested by Turkstra and Semercigil, though, with two extensions: 1) The pneumatic cylinder delivering axial force can provide *compression* as well as *tension*; 2) The control is supplemented by a *closed-loop*

© The Editor(s) (if applicable) and The Author(s), under exclusive license to
Springer Nature Switzerland AG 2021
J. J. Thomsen *Vibrations and Stability*,
https://doi.org/10.1007/978-3-030-68045-9

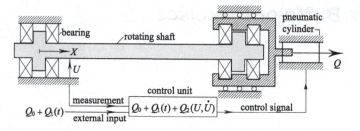

Fig. B.1 Lightweight shaft whose resonance frequencies are controlled by variable tension

element, such that the magnitude of axial force is partly determined by real-time measurements of the shaft vibrations. Your task is to analyze possible dynamic responses of this extended system.

B.1.1 Mathematical Model

Fig. B.2 shows a simplified model of the controlled shaft in a (X, U)-coordinate system. The shaft is considered a Bernoulli-Euler beam of length L, flexural stiffness EI, and mass per unit length ρA. The shaft unbalance is equivalent to a transverse force $P = me\Omega^2\cos(\Omega t)$ at $X = \frac{1}{2}L$, where m is the unbalanced mass and e its eccentricity, and Ω is the speed of shaft rotation. The axial force Q is controllable, $Q = Q_0 + Q_1(t) + Q_2(U, \dot{U})$ where Q_0 is a constant, Q_1 is an explicitly given function of time, and Q_2 is a yet undetermined feedback function of the transverse deformations $U(X, t)$ and velocities $\dot{U}(X, t)$. Axial deformations are neglected and cross-sectional rotations are small, $(U')^2 \ll 1$. The damping forces are assumed to be viscous and mass-proportional, $f_c = -2\beta\rho A\dot{U}(X, t)$ per unit length where β is the damping coefficient.

a) Show that small transverse deformations of the shaft are governed by:

$$\ddot{u} + 2\beta\dot{u} + \frac{EI}{\rho A L^4} u''\!'''' - \frac{1}{\rho A L^2}\left(Q_0 + Q_1(t) + Q_2(u, \dot{u})\right)u''$$
$$= \frac{me\Omega^2}{\rho A L^2}\delta\left(x - \frac{1}{2}\right)\cos\Omega t, \quad u = u(x,t), \quad 0 \le x \le 1, \tag{B.1}$$

where $u \equiv U/L$, $x \equiv X/L$, $\delta(x)$ denote Dirac's delta, $(\ )' \equiv \partial/\partial x$, and the boundary conditions are $u(0, t) = u''(0, t) = u(1, t) = u''(1, t) = 0$.

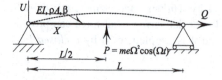

Fig. B.2 Simplified model of the tension-controlled shaft

B.1.2 Eigenvalue Problem, Natural Frequencies and Mode Shapes

a) Define on basis of (B.1) an eigenvalue problem (EVP) for the determination of the undamped linear natural frequencies ω_j and associated mode shapes $\varphi_j(x)$, $j = 1, 2, \ldots$ for the shaft ($Q_1 = Q_2 = me = c = 0$, but $Q_0 \neq 0$).

b) Determine whether the EVP is self-adjoint and completely definite, and supply an interval for Q_0 ensuring the eigenvalues to be real and positive.

c) Compute the Rayleigh quotients $R[u_j]$ for the set of orthogonal test functions $u_j = \sin(j\pi x)$, $j = 1, 2, \ldots$.

d) Show that these Rayleigh quotients and test functions are, respectively, eigenvalues and eigenfunctions for the EVP, and hence that

$$\omega_j^2 = \frac{EI}{\rho A}\left(\frac{j\pi}{L}\right)^4 + \frac{Q_0}{\rho A}\left(\frac{j\pi}{L}\right)^2, \quad j = 1, 2, \cdots, \tag{B.2}$$

$$\varphi_j(x) = \sin j\pi x, \quad 0 \leq x \leq 1,$$

are (squared) natural frequencies and mode shapes for the shaft. Which physical condition corresponds to the case $\omega_1^2 < 0$?

B.1.3 Discretizations, Choice of Control Law

a) Assume that solutions $u(x, t)$ for the nonlinear PDE in (B.1) can be expanded in terms of the n lowest mode shapes φ_j, $j = 1, n$ of the undamped system, that is:

$$u(x, t) = \sum_{j=1}^{n} y_j(t)\varphi_j(x), \quad \varphi_j(x) = \sin j\pi x, \tag{B.3}$$

and show that the modal amplitudes $y_i(t)$ are solutions of the following set of nonlinear ODEs:

$$\ddot{y}_i + 2\beta\dot{y}_i + \left(\omega_i^2 + i^2 q_1(t)\right)y_i + \sum_{j=1}^{n} q_2^{ij}y_j = p_i\Omega^2 \cos\Omega t, \quad i = 1, n, \tag{B.4}$$

where ω_i denote the linear natural frequencies given by (B.2), and:

$$q_1(t) = \frac{Q_1(t)}{\rho A}\left(\frac{\pi}{L}\right)^2, \quad p_i = \frac{2me}{\rho AL^2}\sin\frac{i\pi}{2},$$

$$q_2^{ij} = \frac{-2}{\rho AL^2}\int_0^1 Q_2\left(\sum_{k=1}^{n} y_k\varphi_k, \sum_{k=1}^{n} \dot{y}_k\varphi_k\right)\varphi_i\varphi_j'' \, dx. \tag{B.5}$$

b) It appears from (B.4) that the open-loop and closed-loop elements of the control (Q_1 and Q_2, respectively) affect the system in different ways. When *both* elements are employed, the modal amplitudes are governed by a set of nonlinear, coupled, parametrically and externally excited ODEs.

1. Which kind of equations governs the modal amplitude when only open-loop control is employed? And when only closed-loop is employed?
2. Choosing an open-loop control $Q_1(t) = K \cos(\tilde{\Omega}t)$ may invite problems for certain values of $\tilde{\Omega}$. Which problems, and which $\tilde{\Omega}$?

c) A particular control law is now chosen for closer inspection. We give up the open-loop element but keep the constant element of control, that is, $Q_1(t) = 0$ but $Q_0 \neq 0$. Numerous possibilities then exist for the closed-loop element $Q_2(u,\dot{u})$. Here are some examples (k and $\hat{x}$ are constants, $k \in R$, $\hat{x} \in [0;1]$):

$$Q_2(u,\dot{u}) = \begin{cases} ku(\hat{x},t) & (i) \\ ku^2(\hat{x},t) & (ii) \\ ku(\hat{x},t)\dot{u}(\hat{x},t) & (iii) \\ k\int_0^1 u^2(x,t)\,dx & (iv) \\ k|u(\hat{x},t)| & (v) \\ k\max\limits_{t_n \leq t < t_{n+1}} (u(\hat{x},t)) & (vi) \\ k\dot{u}^2(\hat{x},t) & (vii). \end{cases} \qquad \text{(B.6)}$$

1. Control (*i*) is a bad choice. Why? (Use simple physical reasoning.)
2. Control (*iv*) is difficult to realize. Why?
3. How, in principle, would you handle control (*v*) if a multiple scales analysis of the controlled system was to be performed?
4. We choose, for a closer inspection, a combination of (*ii*) and (*vii*) with $\hat{x} = \frac{1}{2}$ as the measurement point. The closed-loop control then becomes $Q_2(u,\dot{u}) = k_1u^2(\frac{1}{2}, t) + k_2 \dot{u}^2(\frac{1}{2},t)$. Show that in this case the single-mode approximation (Eq. (B.4) with $n = 1$) takes the form:

$$\ddot{y} + 2\beta\dot{y} + \omega^2 y + \kappa_1 y^3 + \kappa_2 \dot{y}^2 y = p\Omega^2 \cos\Omega t, \qquad \text{(B.7)}$$

where $y \equiv y_1$, $\omega \equiv \omega_1$, $p \equiv p_1 = 2me/\rho A$ and $\kappa_j = k_j\pi^2/\rho AL^2$.

d) Eq. (B.7) governs the dynamics of the controlled system when $0 \leq \Omega \ll \omega_2$, and is thus applicable only for angular speeds of the shaft that are well below the second linear natural frequency.

1. Set up the two-mode approximation ((B.4)–(B.5) with $n = 2$) for the case where the system is controlled by the law given in Question c4 above.
2. Can internal resonance occur between the two modes?
3. Do you see any problems in choosing $\hat{x} = \frac{1}{2}$ as the measurement point?

B.1.4 Local Bifurcation Analysis for a Balanced Shaft

a) For the single-mode approximation (B.7) with $p = 0$, introduce a new variable $v = \dot{y}$ and set up the associated system of first-order autonomous ODEs. Determine the singular points, and examine their stability for the case $\beta > 0$, $\omega^2 \in \mathbb{R}$.

b) Classify the singular points (saddle, node, center, focus), and sketch in the (y, v) plane the local flow of orbits near the singular points when $\beta > 0$ and

1. $\omega^2 > 0$, $\kappa_1 > 0$.
2. $\omega^2 > 0$, $\kappa_1 < 0$.
3. $\omega^2 < 0$, $\kappa_1 > 0$.
4. $\omega^2 < 0$, $\kappa_1 < 0$.

c) Show that $(y, v, \omega^2) = (0, 0, 0)$ is a bifurcation point, and sketch the bifurcation in a (ω^2, y) plane when, respectively, $\kappa_1 > 0$ and $\kappa_1 < 0$. Do you recognize this type of bifurcation? Describe (in words) the pre- and post-bifurcation states of the controlled shaft.

B.1.5 Quantitative Analysis of the Controlled System

This section concerns quantitative analysis of the dynamics of the unbalanced and controlled shaft.

a) Assume that $\omega^2 > 0$, so that the static component Q_0 of the axial control force either causes the beam to be tensioned when $u(x, t) = 0$, or causes compressive forces that are less than the critical buckling load. Assume further that the unbalance p is sufficiently small that the shaft performs small but finite oscillations about the statical equilibrium $u(x, t) = 0$. For the single-mode approximation (B.7), determine a first-order multiple scales approximation $y = y_0 + \varepsilon y_1 + O(\varepsilon^2)$ for the response of the controlled shaft near primary external resonance $\Omega \approx \omega$, and β, p, $\kappa_{1,2} = O(\varepsilon)$, $\varepsilon \ll 1$. Also, set up the frequency response equation governing stationary oscillations, and sketch the stationary amplitudes as a function of shaft speed Ω. Briefly discuss the changes in beam response during slow passage of resonance, as compared to the linear case ($\kappa_1 = \kappa_2 = 0$). How does the closed-loop control *qualitatively* affect stationary oscillations of the shaft? (Compare to the uncontrolled case, $\kappa_{1,2} = 0$). How do the effects of the two elements of control (κ_1 and κ_2) differ qualitatively?

b) Assume now that $\omega^2 < 0$, so that when $u(x, t) = 0$ the static control force is compressive and exceeds the critical buckling load. Assume that the shaft performs small but finite oscillations about the buckled state of equilibrium $u(x, t) \neq 0$, and that the control law has no velocity feedback ($\kappa_2 = 0$). For the single-mode approximation (B.7), determine a zero-order multiple scales approximation $y = y_0 + O(\varepsilon)$ for the shaft response near the primary external resonance $\Omega \approx \sqrt{-2\omega^2}$, when terms describing damping, excitation, and nonlinearities are considered $O(\varepsilon)$, $\varepsilon \ll 1$. (Introduce $\tilde{\omega}^2 = -2\omega^2 > 0$ and $\zeta(t) = y\,(t) - \tilde{y}$, where $\tilde{y}$ denotes the buckled state of equilibrium). Set up the frequency response equation governing stationary oscillations.

c) Set up a predictive criterion for chaotic oscillations when $\omega^2 < 0$ and $\kappa_2 = 0$. The criterion can be based on Moon's *Multiwell Potential Criterion* and results from Question b above, but feel free to suggest and examine alternatives if you can. (If Question b was not answered you may assume stationary oscillations of the form $\zeta(t) \equiv y(t) - \tilde{y} = a\cos(\Omega t - \psi) + O(\varepsilon)$ where a is a known function of Ω.)

B.1.6 Using a Dither Signal for Open-Loop Control

Next consider using a rapidly vibrating control input to stabilize the shaft system or reduce its vibrations near the fundamental resonance $\Omega \approx \omega_1$. Such a *dither* signal, which is an open-loop control, could take the form $Q_1(t) = \hat{Q}\hat{\Omega}\cos(\hat{\Omega}\,t)$, with $\hat{Q} = O(1)$, $\hat{\Omega} \gg \Omega$, and $\hat{\Omega}$ away from any resonance of the system. It is easy to show that the single-mode approximation (B.7) governing the amplitude of the lowest vibration mode then changes into:

$$\ddot{y} + 2\beta\dot{y} + \left(\omega^2 + \hat{q}\hat{\Omega}\cos(\hat{\Omega}t)\right)y + \kappa_1 y^3 + \kappa_2\dot{y}^2 y = p\Omega^2\cos(\Omega t), \qquad (B.8)$$

where $\hat{q} \equiv \hat{Q}\pi^2/\rho A L^2$.

Using the method of direct partition of motions, split the modal amplitude y (t) into slow and fast components, $y(t) = z(t) + \hat{\Omega}^{-1}\phi(t,\tau)$, $\tau = \hat{\Omega}\,t$, and set up the differential equation governing the slow (i.e. average) components $z(t)$. What does this equation tell about the effect of the dither signal, in principle? [You don't need to solve the z-equation to answer the latter question, just compare to the y-equation (B.8)].

B.1.7 Numerical Analysis of the Controlled System

This part assumes access to a computer with a program simulating nonlinear systems. The program should be capable of computing solutions for nonlinear ODEs, and to display time-series, phase planes, frequency spectra, Poincaré maps and Lyapunov exponents. For the constants mentioned below, unless otherwise stated, you can assume $\zeta = 0.05$, $\gamma = 0.5$, and $y(0) = \dot{y}(0) = 0.01$.

a) Show that when $\kappa_2 = 0$ the single-mode approximation (B.7) for the closed-loop controlled shaft can be recast into the following nondimensional form:

$$\ddot{y} + 2\zeta\dot{y} + \alpha y + \gamma y^3 = p\tilde{\Omega}^2 \cos\tilde{\Omega}\tau, \tag{B.9}$$

where:

$$\tau = \tilde{\omega}t, \quad \tilde{\omega} = \sqrt{|\omega^2|} > 0, \quad \tilde{\Omega} = \frac{\Omega}{\tilde{\omega}},$$

$$\zeta = \frac{\beta}{\tilde{\omega}}, \quad \gamma = \frac{\kappa_1}{\tilde{\omega}^2}, \quad \alpha = \begin{cases} +1 & \text{for } \omega^2 > 0 \\ -1 & \text{for } \omega^2 < 0, \end{cases} \tag{B.10}$$

where τ is nondimensional time, $\dot{y} \equiv dy/d\tau$, $\zeta \ll 1$ the damping ratio, α denotes whether the static equilibrium of the shaft is straight ($\alpha = 1$) or buckled ($\alpha = -1$), γ is the nondimensional control gain, and $\tilde{\Omega}$ the nondimensional shaft speed.

b) For $p = 0$ and initial conditions $y(0) = 0$, $\dot{y}(0) = 0.10$, examine the simulated transient response of the unloaded system when, respectively, $\alpha = 1$ and $\alpha = -1$. Is it as you would expect? Compare with the answers of Questions B.1.4a–c.

c) Check the quantitative predictions obtained in Question B.1.5a when $\alpha = 1$, $\tilde{\Omega} \approx 1$, and $p \ll 1$. Do the stationary amplitudes and dominating frequencies of the numerical solution agree with the theoretical predictions? For example, what are the deviations between predicted and simulated amplitudes when, respectively, $p = 0.05$ and $p = 0.2$?

d) Simulate the system for $\alpha = 1$, $p = 0.3$ and increasing/decreasing values of $\tilde{\Omega}$. This corresponds to tracing an amplitude curve of the frequency response obtained in Question B.1.5a. Do the simulation results agree with theoretical predictions?

e) When $\alpha = -1$ chaotic solutions may exist. For $\alpha = -1$, choose a frequency of excitation $\tilde{\Omega} < 2$ and an amplitude of excitation $p < 0.5$ for which the phase plane orbits appear to be chaotic. Make sure that the corresponding solutions really are chaotic by examining frequency spectrum, Poincaré map, and largest Lyapunov exponent (remember to cut off transients). Study then the routes to chaos involved when $\tilde{\Omega}$ is varied while p is held constant. Attempt on this basis to sketch a response-versus-frequency bifurcation diagram. Do you observe period-doublings, intermittent chaos, transient chaos, or quasiperiodic responses?

f) Test the predictive chaos criterion obtained in Question B.1.5c, and discuss its applicability for the present system

B.1.8 Conclusions

Summarize your findings in brief form, discussing implications in terms of the physical system, and bring up a few suggestions for a possible extended analysis.

B.2 Vibrations of a Spring-Tensioned Beam

The spring-tensioned beam in Fig. B.3 supports an unbalanced engine. Your task is to analyze possible responses of the system.

B.2.1 Mathematical Model

The beam in Fig. B.3 is plane and simply supported at the ends. It is made of linearly elastic material, and has length L, flexural stiffness EI and mass per unit length ρA. Transverse deformations are described by $U(X, t)$. Axial deformations, shear deformations and rotatory inertia can be neglected. Rotations of cross sections are sufficiently small for the approximation $\kappa = U''(X, t)$ to be valid, where κ is the curvature and $()' \equiv \partial/\partial X$. The dissipation of energy can be approximated by viscous mass-proportional damping forces, $f_c = -2\beta\rho A\,\dot{U}(X, t)$ per unit length, where β is the coefficient of viscous damping.

The engine is mounted halfway along the beam. Due to unbalanced rotating components it exerts a time-varying transverse load $P(t) = me\Omega^2\cos(\Omega t)$ on the beam, where Ω is the angular speed of the unbalanced mass m with eccentricity e.

The right beam-support is movable, and is connected to a fixed point in space through a linear spring having stiffness K. The elongation of the spring is Z_0 when the beam is undeformed (Fig. B.3(a), (b)), where Z_0 can be positive, negative or zero. When the beam deforms transversally the two supports come closer, causing a further elongation ΔZ of the spring (Fig. B.3(c)).

a) Show that the elongation ΔZ accompanying transverse beam deformations $U(X, t)$ is, approximately:

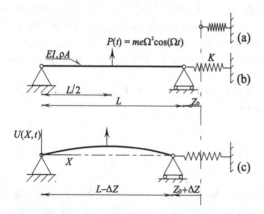

Fig. B.3 Beam subjected to adjustable pre-tension and transverse time-varying load. **a** Un-deformed spring. **b** Beam pre-tensioned by stretched spring. **c** Beam pre-tensioned and transversally deformed

$$\Delta Z \approx \frac{1}{2} \int_0^L (U')^2 \, dX \quad \text{for} \, (U')^2 << 1. \tag{B.11}$$

b) Assuming $(U')^2 \ll 1$, show that small but finite transverse vibrations of the beam are governed by:

$$\ddot{u} + 2\beta \dot{u} + \frac{EI}{\rho A L^4} u'''' - \frac{K}{\rho A L} \left(\frac{Z_0}{L} + \frac{1}{2} \int_0^1 (u')^2 \, dx \right) u'',$$

$$= \frac{m e \Omega^2}{\rho A L^2} \delta \left(x - \frac{1}{2} \right) \cos \Omega t; \quad u = u(x, t), \quad 0 \le x \le 1, \tag{B.12}$$

$$u(0, t) = u''(0, t) = u(1, t) = u''(1, t) = 0,$$

where $u \equiv U/L$, $x \equiv X/L$, $\delta(x)$ denote Dirac's delta, $(u')^2 \ll 1$, and $( )' \equiv \partial/\partial x$.

B.2.2 Eigenvalue Problem, Natural Frequencies and Mode Shapes

a) Define on the basis of (B.12) an eigenvalue problem (EVP) for the determination of undamped linear natural frequencies ω_j and associated mode shapes $\varphi_j(x)$, $j = 1, 2, \dots$ for the beam.
b) Determine whether the EVP is self-adjoint and completely definite, and supply an interval for Z_0 ensuring the eigenvalues to be real and positive.
c) Compute the Rayleigh quotients $R[u_j]$ for the set of orthogonal test functions $u_j = \sin(j\pi x)$, $j = 1, 2, \dots$.
d) Show that these Rayleigh quotients and test functions are, respectively, eigenvalues and eigenfunctions for the EVP, and hence that

$$\omega_j^2 = \frac{EI}{\rho A} \left(\frac{j\pi}{L} \right)^4 + \frac{K Z_0}{\rho A} \left(\frac{j\pi}{L} \right)^2, \quad j = 1, 2, \dots, \tag{B.13}$$

$$\varphi_j(x) = \sin(j\pi x), \quad 0 \le x \le 1,$$

are (squared) natural frequencies and mode shapes for the beam. Which physical condition corresponds to the case $\omega_1^2 < 0$?

B.2.3 Discrete Models

a) Assume that solutions $u(x, t)$ for the nonlinear PDE (B.12) can be expanded in terms of the n lowest mode shapes φ_j, $j = 1, n$ of the undamped system, that is:

$$u(x, t) = \sum_{j=1}^n y_j(t) \varphi_j(x), \quad \varphi_j(x) = \sin j\pi x, \tag{B.14}$$

and show that the modal amplitudes $y_i(t)$ are solutions of the following set of nonlinear ODEs:

$$\ddot{y}_i + 2\beta\dot{y}_i + \left(\omega_i^2 + \gamma_i \sum_{k=1}^{n} k^2 y_k^2\right) y_i = p_i \Omega^2 \cos \Omega t, \quad i = 1, n \tag{B.15}$$

where ω_i denote the linear natural frequencies given by (B.13), and:

$$\gamma_i \equiv \frac{i^2 K \pi^4}{4\rho AL} > 0, \quad p_i \equiv \frac{2me}{\rho AL^2} \sin \frac{i\pi}{2}. \tag{B.16}$$

b) With $n = 1$ (B.15) take the form

$$\ddot{y} + 2\beta\dot{y} + \omega^2 y + \gamma y^3 = p\Omega^2 \cos \Omega t \tag{B.17}$$

where $\omega \equiv \omega_1$, $\gamma \equiv \gamma_1 = K\pi^4/(4\rho AL)$ and $p \equiv p_1 = 2me/(\rho AL^2)$. At certain conditions this single-mode approximation adequately describes the dynamic response of the beam. Which conditions?

c) Write down the two-mode approximation corresponding to (B.15) with $n = 2$. What are the conditions for *external* resonance? Can *internal* resonance occur between the two modes? If yes, for which values of the tension force KZ_0?

B.2.4 Local Bifurcation Analysis for the Unloaded System

In this section the engine mounted on the beam is assumed to be perfectly balanced, that is, $p = me = 0$.

a) For the single-mode approximation (B.17) with $p = 0$, introduce a new variable $v = \dot{y}$ and set up the associated system of first-order autonomous ODEs. Determine the singular point of this system, and examine their stability for the case $\beta > 0$, $\gamma > 0$ and $\omega^2 \in \mathbb{R}$.

b) Classify the singular points (saddle, node, center, focus), and sketch in the (y, v) plane the local flow of orbits near the singular points when $\gamma > 0$ and

1. $\omega^2 > 0, 0 < \beta \ll \omega$
2. $\omega^2 < 0, 0 < \beta \ll \sqrt{-2\omega^2}$

c) Show that $(y, v, \omega^2) = (0, 0, 0)$ is a bifurcation point, and sketch the bifurcation in a (ω^2, y) plane. Do you recognize this type of bifurcation? Describe the pre- and post-bifurcation states in terms of the physical system.

B.2.5 Quantitative Analysis of the Loaded System

a) Assume that $\omega^2 > 0$, so that the pre-tensioned spring either causes the beam to be tensioned when $u(x, t) = 0$, or causes compressive forces that are less than the critical buckling load. Assume further that the load p from the unbalanced engine is sufficiently small that the beam performs small (but finite) oscillations about the statical equilibrium position $u(x, t) = 0$. For the single-mode approximation (B.17), determine a first-order multiple scales approximation $y = y_0 + \varepsilon y_1 + O(\varepsilon^2)$ for the beam response near primary external resonance $\Omega \approx \omega$, when $\beta, \gamma, p = O(\varepsilon)$, $\varepsilon \ll 1$. Compute the frequency response equation governing stationary oscillations, and sketch stationary amplitudes as a function of excitation frequency Ω. Discuss briefly the changes in beam response that occur when Ω is varied slowly across ω, as compared to the linear case ($\gamma = 0$).

b) Assume now that $\omega^2 < 0$, so that when $u(x, t) = 0$ the supporting spring loads the beam by a compressive force exceeding the critical buckling load. Assume further that the load p is sufficiently small for the beam to perform small but finite oscillations about a buckled state of equilibrium $u(x, t) \neq 0$. For the single-mode approximation (B.17), determine a first-order multiple scales approximation $y = y_0 + \varepsilon y_1 + O(\varepsilon^2)$ for the beam response near primary external resonance $\Omega \approx \tilde{\omega}$ when $\beta, \gamma, p = O(\varepsilon)$, $\varepsilon \ll 1$. (Introduce $\tilde{\omega}^2 \equiv -2\omega^2 > 0$ and $\eta(t) \equiv y(t) - \tilde{y}$, where $\tilde{y}$ denotes the buckled state of statical equilibrium). Compute the frequency response equation governing stationary oscillations. Discuss briefly the dynamic response as compared to the unbuckled case treated in Question B.2.5a above.

c) Set up a predictive criterion for chaotic oscillations to appear when $\omega^2 < 0$. The criterion should be based on Moon's *Multiwell Potential Criterion* and results from Question B.2.5b above, but feel free to examine alternatives. (If this question was not answered you may assume stationary oscillations having the form $\eta(t) = y(t) - \tilde{y} = a\cos(\Omega t - \psi) + O(\varepsilon)$, where a is a known function of Ω.)

B.2.6 Numerical Analysis

This part assumes you have access to a computer with a program simulating nonlinear systems. The program should be capable of computing solutions for the single-mode approximation (B.17), that is, for:

$$\ddot{y} + 2\beta\dot{y} + \omega^2 y + \gamma y^3 = p\Omega^2 \cos(\Omega t). \tag{B.18}$$

Further, the program should provide plots of time-series, phase planes, frequency spectra, Poincaré maps and Lyapunov exponents. Unless otherwise stated you can assume $\beta = 0.05$ and $\gamma = 0.5$ for the damping and the nonlinearity, respectively, and $y(0) = \dot{y}(0) = 0.01$ for the initial conditions.

a) For $p = 0$ and initial conditions $y(0) = \dot{y}(0) = 0.10$, examine the numerically computed transient response of the unloaded system when, respectively, $\omega^2 = 1$ and $\omega^2 = -\frac{1}{2}$. Is it as could be expected? Compare it with the answers of Questions B.2.4a–c.

b) Check the quantitative predictions of Question B.2.5a when $\omega^2 = 1$, $\Omega \approx \omega$ and $p \ll 1$. Do the stationary amplitudes and dominating frequencies of the numerical solution agree with the theoretical predictions? What are the deviations between predicted and simulated amplitudes when, respectively, $p = 0.01$ and $p = 0.1$?

c) Check, as in B.2.6(b) above, the quantitative predictions of Question B.2.5b when $\omega^2 = -\frac{1}{2}$, $\Omega \approx 1$ and $p \ll 1$.

d) Simulate the system for $\omega^2 = 1$, $\Omega = 1.2$ and increasing values of p. What is observed?

e) Simulate the system for $\omega^2 = 1$, $p = 0.2$ and increasing values of Ω. This corresponds to tracing an amplitude curve of the frequency response obtained in Question B.2.5a. Do the simulation results agree with the theoretical predictions?

f) When $\omega^2 < 0$ many of the solutions will be chaotic. For $\omega^2 = -\frac{1}{2}$, choose a frequency of excitation $\Omega < 2$ and an amplitude of excitation $p < 0.5$ for which the phase plane orbits appear chaotic. Make sure that the corresponding solutions are chaotic by examining frequency spectrum, Poincaré map and largest Lyapunov exponent (remember to cut off transients). Investigate then the routes to chaos involved when, respectively, Ω is varied while p is held constant, and vice versa. Sketch the associated bifurcation diagrams for Ω and p. Do you observe period-doublings, intermittent chaos, transient chaos, or quasiperiodic responses?

g) Test the predictive chaos criterion obtained in Question B.2.5c, and discuss its applicability for the present system.

B.2.7 Conclusions

Summarize your findings in brief form, discussing implications in terms of the physical system, and bring up a few suggestions for a possible extended analysis.

B.3 Dynamics of a Microbeam

Background. The simply supported microbeam system in Fig. B.4 has been proposed as an alternative to the typical cantilever design for microbeams used in atomic force microscopes (AFM). Your task is to analyze possible responses of the new proposal, and to provide theoretical results that can aid in designing a first experimental model. When reporting you should have two readerships in mind: The

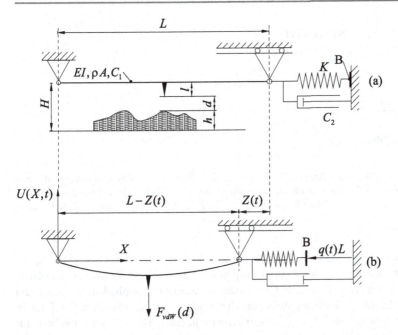

Fig. B.4 System definition: **a** undeformed and **b** deformed microbeam for a non-contact atomic force microscope

managers of your department, who are uninterested in equations and only wants to know what you think they mean, – and skilled senior scientists, whom the managers might call up in case your calculations and conclusions needs revision.

AFMs provide pictures of atoms on surfaces. They work by scanning a fine ceramic or semiconductor tip over a surface. The tip is usually positioned at the end of a cantilever beam, with typical dimensions length × width × thickness = 200 × 20 × 1 μm, a fundamental natural frequency of the order 100 kHz, and outer tip radius about 100 Å. As the tip is repelled by or attracted to the surface by interatomic Van der Waals forces, the cantilever beam deflects. The magnitude of the deflection is captured using optical methods; e.g., by tracing a laser beam reflected from the surface of the beam. A plot of the laser deflection versus tip position over the sample surface provides a picture of the hills and valleys that constitute the topography of the surface. The AFM can work with the tip touching the sample (contact mode), or the beam can be made to vibrate in resonance without touching the surface (non-contact mode) or intermittently touching it (tapping mode). These operating modes correspond to using, respectively, strong (contact), weak (non-contact), or mixed (intermittent contact) interatomic van der Waals forces (cf. Fig. B.5).

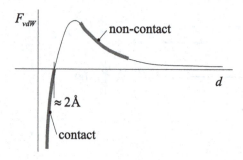

Fig. B.5 Van der Waal interatomic force F_{vdw} as a function of AFM tip-to-sample distance d. Thick parts of the curve mark the operating ranges of contact mode and non-contact mode AFMs, respectively. The function describing F_{vdW} is: $F_{vdW}(d) = k_1 d^{-2} - k_2 d^{-8}$, where k_1 and k_2 are known constants

The new design concept that you are asked to examine concerns replacing the cantilever microbeam with a simply supported beam, whose dynamical properties to some extent can be adjusted by means of externally applied axial thrust and damping. In this preliminary study, the dynamics of the microbeam in itself, i.e. in the absence of interatomic forces, is of primary interest. This will form the basis for understanding and predicting dynamical responses in the presence of a sample to be scanned, i.e. when the atoms of the AFM tip are interacting with atoms on a sample surface.

B.3.1 System Description

The beam in Fig. B.4(a) is simply supported at the ends, is made of linearly elastic material, and has length L, flexural stiffness EI, and mass per unit length ρA. Transverse deformations are described by $U(X, t)$, whereas stretching of the beam axis, shear deformations, longitudinal inertia, and rotary inertia can be neglected. Rotations of beam segments are sufficiently small that $\kappa = U''(X, t)$ is a good approximation, with κ being curvature and $()' \equiv \partial/\partial X$. Dissipation of beam energy can be modeled as internal viscous damping, i.e. the internal bending moment is $EI\kappa + C_1 EI\,\dot{\kappa}$, where $(\cdot) \equiv \partial/\partial t$ and C_1 is the coefficient of internal damping.

One beam support is movable, and is connected to a point B through a linear spring having stiffness K. Fig. B.4(a) shows the beam and the spring undeformed, whereas Fig. B.4(b) shows a deformed state. The position $q(t)L$ of point B with respect to a fixed boundary is externally controlled by piezoelectric ceramics, so that $q(t) = q_0 + q_1\cos(\Omega t)$, where q_0, q_1, and Ω are the control parameters (q_0 can be assigned positive as well as negative values). The compression of the spring is $q(t) L - Z(t)$, where $q(t)$ is known (controlled), whereas $Z(t)$ is the shortening in the distance between beam supports that occurs when the beam deforms, $U(X, t) \neq 0$. In Fig. B.4(b) Z is shown greatly exaggerated. In addition, to supply more control possibilities, the proposal involves coupling the moveable beam support to the fixed boundary through a linear viscous damper with coefficient C_2.

The AFM tip at the middle of the beam has negligible mass ($\ll \rho AL$) and length l, and has instantaneous distance $d(t)$ above the actual scan point of the sample (this corresponds to the distance the outer atoms of the tip can be moved before experiencing strong "contacting" interaction forces with the atoms of sample, cf. Fig. B.5. The sample to be scanned is positioned on a base a distance H beneath the beam base. Using piezoelectric linear drives the AFM adjusts this distance all the time, however, at a rate that can be considered quasi-static as compared to the rapid vibrations of the beam. Thus, ignoring small horizontal movements of the tip due to beam bending, the tip-to-sample distance is $d(t) = U(L/2, t) + H - h - l$.

B.3.2 Mathematical Model

a) Show that the end-shortening $Z(t)$ accompanying transverse beam deformations $U(X, t)$ is:

$$Z \approx \frac{1}{2} \int_0^L (U')^2 \, dX \quad \text{for } (U')^2 \ll 1, \quad (U' = \partial U / \partial X). \tag{B.19}$$

b) Show that small but finite transverse vibrations of the beam are governed by:

$$\ddot{u} + \frac{C_1 EI}{\rho AL^4} u'''' + \frac{EI}{\rho AL^4} u'''' + \frac{F_{vdW}\left(u(\frac{1}{2}, t)L + H - h - l\right)}{\rho AL^2} \delta\left(x - \frac{1}{2}\right)$$

$$+ \left[\frac{K}{\rho AL} \left(q_0 + q_1 \cos(\Omega t) - \frac{1}{2} \int_0^1 (u')^2 \, dx \right) - \frac{C_2}{\rho AL} \int_0^1 u' \dot{u}' \, dx \right] u'' = 0, \tag{B.20}$$

with boundary conditions $u(0, t) = u''(0, t) = u(1, t) = u''(1, t) = 0$, where $u = u(x, t)$, $u \equiv U/L$, $x \equiv X/L \in [0;1]$, $(u')^2 \ll 1$, $\delta(x)$ is Dirac's delta function, $(\cdot) \equiv \partial/\partial t$, and now $()' \equiv \partial/\partial x$. Describe the physical meaning of each term or group of terms, and identify all nonlinearities. How would you characterize this type of equation?

B.3.3 Eigenvalue Problem, Natural Frequencies and Mode Shapes

a) Define on basis of (B.20) an eigenvalue problem (EVP) for determining the beams undamped linear natural frequencies ω_j and associated mode shapes $\varphi_j(x)$, $j = 1, 2, \ldots$, in the presence of static axial loading ($q_0 \neq 0$, $q_1 = 0$) and absence of interatomic forcing ($F_{vdW} = 0$).

b) Examine whether the EVP is self-adjoint and completely definite, and supply an interval for q_0 ensuring the eigenvalues to be real and positive.

c) Compute the Rayleigh quotients $R[u_j]$ for the test functions $u_j = \sin(j\pi x)$, $j = 1$, 2, ...

d) Show that these Rayleigh quotients and test functions are, respectively, eigenvalues and eigenfunctions for the EVP, so that

$$\omega_j^2 = \frac{EI}{\rho A}\left(\frac{j\pi}{L}\right)^4 - \frac{Kq_0 L}{\rho A}\left(\frac{j\pi}{L}\right)^2, \quad j = 1, 2, \ldots,$$

$$\varphi_j(x) = \sin(j\pi x), \quad 0 \le x \le 1,$$

(B.21)

are squared natural frequencies and mode shapes for the beam. Which physical condition corresponds to the case $\omega_1^2 < 0$?

B.3.4 Discrete Models, Mode Shape Expansion

a) Assume that solutions $u(x, t)$ to (B.20) can be expanded in terms of the n lowest mode shapes φ_j, $j = 1, n$ determined above, that is:

$$u(x,t) = \sum_{j=1}^{n} y_j(t)\varphi_j(x), \quad \varphi_j(x) = \sin(j\pi x).$$

(B.22)

Show that the modal amplitudes $y_i(t)$ are solutions to

$$\ddot{y}_i + 2\beta i^4 \dot{y}_i + \left(\omega_i^2 + qi^2\cos(\Omega t) + \gamma i^2 \sum_{r=1}^{n} r^2 y_r^2 + \eta i^2 \sum_{r=1}^{n} r^2 y_r \dot{y}_r\right) y_i$$

$$= f_{vdW,i}(y_i), \quad i = 1, n,$$

(B.23)

where ω_i denote the linear natural frequencies given by (B.21), and

$$\gamma \equiv \frac{K\pi^4}{4\rho A L} > 0, \quad q \equiv -\frac{q_1 K\pi^2}{\rho A L}, \quad \beta \equiv \frac{C_1 EI\pi^4}{2\rho A L^4}, \quad \eta \equiv \frac{C_2\pi^4}{2\rho A L},$$

$$f_{vdW,i}(y_i) = \frac{-2\sin(i\pi/2)}{\rho A L^2} F_{vdW}\left(H - h - l + L\sum_{r=1}^{n} y_r \sin(r\pi/2)\right).$$

(B.24)

b) For the n-mode approximation (B.23), at which values of excitation frequency Ω do you expect primary resonances to occur? Can you say anything about the possible existence of internal resonances? (for both questions: assume $\omega_i^2 > 0$ and base your answers on simple reasoning, known examples, and rough calculations; you are not requested to perform a full perturbation analysis).

c) With $n = 1$ the system (B.23) take the form

$$\ddot{y} + 2\beta\dot{y} + (\omega^2 + q\cos(\Omega t))y + \gamma y^3 + \eta y^2 \dot{y} = f_{vdW}(y), \qquad \text{(B.25)}$$

where $\omega \equiv \omega_1$ $y \equiv y_1$, and $f_{vdW}(y) = -2F_{vdW}(H - h - l + Ly)/\rho AL^2$. Under which conditions does this single-mode approximation adequately describe the dynamic response of the beam? What does each term represent physically?

B.3.5 Local Bifurcation Analysis for the Statically Loaded System

This section concerns possible behaviors of the beam when there is no time-varying component of the axial loading, and no sample to be scanned, i.e. $q = 0$ and $f_{vdW} = 0$.

a) For the single-mode approximation (B.25) with $q = f_{vdW} = 0$, introduce a new variable $v = \dot{y}$, and set up the associated system of first-order autonomous ordinary differential equations. Determine the singular point of this system, and examine their stability when $\beta > 0$, $\gamma > 0$, $\eta > 0$, $\omega^2 \in \mathbb{R}$, $\beta^2 \ll |\omega^2|$, and $\eta/\gamma \ll |1|$.

b) Classify the singular points (saddle, node, center, focus), and sketch in the (y, v) plane the local flow of orbits near the singular points for these two cases: I) $\omega^2 > 0$; and II) $\omega^2 < 0$.

c) Show that $(y, v, \omega^2) = (0, 0, 0)$ is a bifurcation point. Sketch the bifurcation in a (ω^2, y) plane, indicating the stability of branches. Do you recognize this type of bifurcation? Is it subcritical or supercritical? Describe the pre- and post-bifurcation states in terms of the physical system.

B.3.6 Quantitative Analysis of the Loaded System

This section considers quantitative analysis of the beam dynamics when there is harmonic axial loading, but no sample to be scanned, i.e. $q \neq 0$ but $f_{vdW} = 0$.

a) Assume that $\omega^2 > 0$, so that the pre-load of the spring (Kq_0L) either causes the beam to be tensioned when $u(x, t) = 0$, or causes compressive forces that are less than the critical buckling load. Assume further that the dynamic part of load is sufficiently small that the beam performs small but finite oscillations about the statical equilibrium $u(x, t) = 0$. For the single-mode approximation (B.25), determine a first-order multiple scales approximation $y = y_0 + \varepsilon y_1 + O(\varepsilon^2)$ for the beam response near primary resonance when $(\beta, q, \gamma, \eta) = O(\varepsilon)$, $\varepsilon \ll 1$. Also, set up the frequency response equation governing stationary oscillations, and sketch the stationary amplitudes as a function of excitation frequency Ω. Briefly discuss the changes in beam response during slow passage of resonance, as compared to the linear case ($\gamma = \eta = 0$).

b) Assume now that $\omega^2 < 0$, so that when $u(x, t) = 0$ the spring loads the beam with a compressive force exceeding the critical buckling load. Assume further that q is sufficiently small that the beam performs small but finite oscillations about a buckled state of equilibrium $u(x, t) \neq 0$, and that the nonlinear damping is negligible ($\eta = 0$). For the single-mode approximation (B.25), determine a first-order multiple scales approximation $y = y_0 + O(\varepsilon)$ for the beam response near the primary external resonance $\Omega \approx \sqrt{-2\omega^2}$ when $(\beta, q, \gamma) = O(\varepsilon)$, $\varepsilon \ll 1$. (Introduce $\tilde{\omega}^2 = -2\omega^2 > 0$ and $\zeta(t) = y\,(t) - \tilde{y}$, where $\tilde{y}$ denotes the buckled state of equilibrium). Set up the frequency response equation governing stationary oscillations.

c) Set up a predictive criterion for chaotic oscillations when $\omega^2 < 0$ and $\eta = 0$. The criterion can be based on Moon's *Multiwell Potential Criterion* and results from Question b above, but feel free to suggest and examine alternatives. (If Question b was not answered you may assume stationary oscillations of the form $\zeta(t) \equiv y(t) - \tilde{y} = a\cos(\Omega t - \psi) + O(\varepsilon)$ where a is a known function of Ω.)

B.3.7 Numerical Analysis

This part assumes you to use a computer program capable of simulating nonlinear systems of the form (B.25). In addition, the program should provide plots of time series, phase planes, frequency spectra, Poincaré maps, and Lyapunov exponents. Unless otherwise stated you can assume $\beta = 0.02$, $\gamma = 0.9$, $\eta = 0.1$, $f_{vdw} = 0$, and use $y(0) = \dot{y}(0) = 0.01$ for the initial conditions.

a) For $q = 0$ and initial conditions $y(0) = \dot{y}(0) = 0.1$, examine the numerically computed transient response for the purely statically loaded system when, $\omega^2 = 1$ and $\omega^2 = -\frac{1}{2}$, respectively. Is it as could be expected? Compare to the predictions obtained with Question 4.

b) Check the quantitative predictions of Question 5a for $\omega^2 = 1$, $\Omega \approx 2\omega$, and small q. Do the stationary amplitudes and dominating frequencies of the numerical solution agree with the theoretical predictions? For example, what are the deviations between predicted and simulated amplitudes when, respectively, $q = 0.1$ and $q = 0.5$?

c) Check, as in Question b above, the quantitative predictions obtained in Question 5b for $\omega^2 = -\frac{1}{2}$, $\Omega \approx \tilde{\omega}$, $\eta = 0$, small q, and initial conditions $y(0) = 1.01$, $\dot{y}(0) = 0.01$.

d) Simulate the system for $\omega^2 = 1$, $q = 0.3$ and increasing/decreasing values of Ω. This corresponds to tracing an amplitude curve of the frequency response obtained in Question 5a. Do the simulation results agree with the theoretical predictions?

e) When $\omega^2 < 0$ chaotic solutions may exist. For $\omega^2 = -\frac{1}{2}$, choose a frequency of excitation $\Omega < 4$ and an amplitude of excitation $q < 1$ for which the phase plane orbits appear to be chaotic. Make sure that the corresponding solutions really are chaotic by examining frequency spectrum, Poincaré map, and largest Lyapunov exponent (remember to cut off transients).

Then study the routes to chaos involved when Ω is varied while q is held constant. Attempt on this basis to sketch a response-versus-frequency bifurcation diagram. Do you observe period-doublings, intermittent chaos, transient chaos, or quasiperiodic responses?

f) Test the predictive chaos criterion obtained in Question 5c, and discuss its applicability for the present system

B.3.8 Conclusions

Summaries the essences of your findings. Write for a readership with a knowledge on vibrations and stability corresponding to basic courses in strength of materials and vibrations, that is: for a reader who does not know about the advanced analysis techniques you are using (and does not want to know), but still needs to understand what your analysis implies. Express yourself as briefly and clearly as possible, and avoid mathematical symbols and unnecessary jargon. Bring up at least two suggestions for possible future work on the microbeam system.

Appendix C: Mathematical Formulas

C.1 Formulas Typically Used in Perturbation Analysis

In particular perturbation analysis involves using a lot of mathematical standard operations involving complex variables, powers of two-term sums, and averaging integrals. Here are formulas for most of them. Unless otherwise stated all variables are real-valued, i denotes the imaginary unit, overbars complex conjugation, and cc complex conjugates of preceding terms.

C.1.1 Complex Numbers

In the below formulas A, B, and i are complex-valued, while all other variables are real-valued.

$$\bar{i} = -i,$$
$$\overline{a+ib} = a - ib,$$
$$\overline{AB} = \bar{A}\bar{B}, \; A, B \in C,$$

$$\overline{A^n} = \bar{A}^n, A \in C,$$
$$A = ae^{i\varphi}, \text{ where } a = |A| = \sqrt{\text{Re}(A)^2 + \text{Im}(A)^2},$$
$$\text{and } \varphi = \angle A = \arctan(\text{Im}(A)/\text{Re}(A)), \qquad \text{(C.1)}$$
$$\overline{e^{ia}} = e^{-ia},$$
$$e^{ia} = \cos a + i \sin a,$$
$$\cos\varphi = \tfrac{1}{2}(e^{i\varphi} + e^{-i\varphi}), \sin\varphi = \tfrac{1}{2i}(e^{i\varphi} - e^{-i\varphi}),$$
$$\sqrt{a+ib} = \alpha + i\beta, \text{ where, } \alpha\beta = \tfrac{1}{2}b, \alpha^2 - \beta^2 = a,$$
$$\text{and } \alpha = \pm\sqrt{\tfrac{1}{2}\left(a + \sqrt{a^2 + b^2}\right)}, \beta = \pm\sqrt{\tfrac{1}{2}\left(-a + \sqrt{a^2 + b^2}\right)}.$$

© The Editor(s) (if applicable) and The Author(s), under exclusive license to
Springer Nature Switzerland AG 2021
J. J. Thomsen *Vibrations and Stability*,
https://doi.org/10.1007/978-3-030-68045-9

C.1.2 Powers of Two-Term Sums

$$(a+b)^3 = a^3 + b^3 + 3a^2b + 3ab^2,$$
$$(a+b)^4 = a^4 + b^4 + 4a^3b + 4ab^3 + 6a^2b^2, \tag{C.2}$$
$$(a+b)^5 = a^5 + b^5 + 5a^4b + 5ab^4 + 10a^3b^2 + 10a^2b^3.$$

C.1.3 Averaging Integrals

These integrals often occur with averaging or 'slow/fast' analysis with harmonic excitation. Let $T = 2\pi/\Omega$ be the averaging period and Ω the frequency, then:

$$\frac{1}{T}\int_0^T (\sin \Omega t)^{2i-1}(\cos \Omega t)^{j-1} dt$$

$$= \frac{1}{T}\int_0^T (\cos \Omega t)^{2i-1}(\sin \Omega t)^{j-1} dt = 0, \quad \text{for } i,j = 1,2,\cdots,$$

$$\frac{1}{T}\int_0^T \sin^2(\Omega t)dt = \frac{1}{T}\int_0^T \cos^2(\Omega t)dt = \frac{1}{2},$$

$$\frac{1}{T}\int_0^T \sin^2(\Omega t)\cos^2(\Omega t)dt = \frac{1}{8}, \tag{C.3}$$

$$\frac{1}{T}\int_0^T \sin^4(\Omega t)dt = \frac{1}{T}\int_0^T \cos^4(\Omega t)dt = \frac{3}{8},$$

$$\frac{1}{T}\int_0^T \cos(\Omega t)\cos(\Omega t - \varphi)dt = \frac{1}{T}\int_0^T \sin(\Omega t)\sin(\Omega t - \varphi)dt = \frac{1}{2}\cos\varphi,$$

$$\frac{1}{T}\int_0^T \sin(\Omega t)\cos(\Omega t - \varphi)dt = -\frac{1}{T}\int_0^T \cos(\Omega t)\sin(\Omega t - \varphi)dt = \frac{1}{2}\sin\varphi.$$

In terms of a phase variable $\psi = \Omega t$ the same integrals becomes:

$$\frac{1}{2\pi} \int_0^{2\pi} (\sin\psi)^{2i-1}(\cos\psi)^{j-1}\, d\psi$$

$$= \frac{1}{2\pi} \int_0^{2\pi} (\cos\psi)^{2i-1}(\sin\psi)^{j-1} d\psi = 0, \quad \text{for } i,j = 1,2,\cdots,$$

$$\frac{1}{2\pi} \int_0^{2\pi} \sin^2\psi\, d\psi = \frac{1}{2\pi} \int_0^{2\pi} \cos^2\psi\, d\psi = \frac{1}{2},$$

$$\frac{1}{2\pi} \int_0^{2\pi} \sin^2\psi\cos^2\psi\, d\psi = \frac{1}{8}, \tag{C.4}$$

$$\frac{1}{2\pi} \int_0^{2\pi} \sin^4\psi\, d\psi = \frac{1}{2\pi} \int_0^{2\pi} \cos^4\psi\, d\psi = \frac{3}{8},$$

$$\frac{1}{2\pi} \int_0^{2\pi} \cos\psi\cos(\psi-\varphi)d\psi = \frac{1}{2\pi} \int_0^{2\pi} \sin\psi\sin(\psi-\varphi)d\psi = \frac{1}{2}\cos\varphi,$$

$$\frac{1}{2\pi} \int_0^{2\pi} \sin\psi\cos(\psi-\varphi)d\psi = -\frac{1}{2\pi} \int_0^{2\pi} \cos\psi\sin(\psi-\varphi)d\psi = \frac{1}{2}\sin\varphi.$$

C.1.4 Dirac's Delta Function

If $x_0 \in [x_1; x_2]$ then $\int_{x_1}^{x_2} \delta(x-x_0)\, dx = 1$. If further $f(x)$ is a function on $[x_1;x_2]$, then $\int_{x_1}^{x_2} \delta(x-x_0)f(x)\, dx = f(x_0)$. Also, $\delta(x-x_0) = 0$ for $x \neq x_0$, $\delta(x-x_0) \to \infty$ for $x \to x_0$.

C.1.5 Fourier Series of a Periodic Function

The Fourier series of a T-periodic function $f(t)$ is, in complex-valued form:

$$f(t) = \frac{1}{2}c_0 + \sum_{k=1}^{\infty} c_k e^{ik\omega t} + cc, \tag{C.5}$$

where $\omega = 2\pi/T$, and the complex-valued Fourier coefficients c_k are given by:

$$c_k = \frac{1}{T} \int_0^T f(t)e^{-ik\omega t}\, dt. \tag{C.6}$$

The corresponding real-valued form is:

$$f(t) = \frac{1}{2}a_0 + \sum_{k=1}^{\infty} (a_k \cos(k\omega t) + b_k \sin(k\omega t)), \tag{C.7}$$

where the real-valued Fourier coefficients a_k and b_k are given:

$$a_k = \frac{2}{T} \int_0^T f(t) \cos(k\omega t)dt, \quad b_k = \frac{2}{T} \int_0^T f(t) \sin(k\omega t)dt. \qquad (C.8)$$

If f and df/dt are piecewise continuous, then the series in (C.5) or (C.7) converges to $f(t)$ if t is a point of continuity, and to $\frac{1}{2}(f(t_+) + f(t_-))$ if t is a point of discontinuity.

C.2 Formulas Used in Stability Analysis

C.2.1 Sylvester's Criterion

Sylvester's criterion provides an often convenient way to determine if an $n \times n$ Hermitian/self-adjoint matrix $\mathbf{A} = \bar{\mathbf{A}}^T$ is positive-definite. It says that $\mathbf{A}$ is positive definite if and only if $|\mathbf{A}_j| > 0$ for all $j = 1, n$, where $\mathbf{A}_j$ is $\mathbf{A}$'s leading principal minor of order n (i.e. the $j \times j$ upper left corner of $\mathbf{A}$).

C.2.2 the Routh-Hurwitz Criterion

The Routh-Hurwitz criterion can be used to determine whether all the roots of a polynomial have negative real parts. This is relevant in stability analysis, which often boils down to checking the sign of the real parts of eigenvalues, which in turn are roots of a characteristic polynomial.

Let $f(x) = a_0 x^n + a_1 x^{n-1} + \cdots + a_{n-1}x + a_n$ be the polynomial, where $a_0 > 0$ and $a_j \in R, j = 0, n$. Then a *necessary* condition that all roots x of $f = 0$ have negative real parts is that $a_j > 0, j = 0, n$. The *Routh-Hurwitz criterion* states a necessary *and* sufficient condition (e.g., Flügge 1962; Pearson 1974; Pedersen 1986; Ziegler 1968): All roots of $f = 0$ have negative real parts if and only if $D_j > 0, j = 1, n$, where D_n is the *Hurwitz determinant* of order n, and $D_j, j = 1, n-1$ is D_n's *principal minor* of order j. The Hurwitz determinant is formed as follows:

$$D_n = \begin{vmatrix} a_1 & a_0 & 0 & \cdots & & & & & 0 \\ a_3 & a_2 & a_1 & a_0 & 0 & & & & 0 \\ a_5 & a_4 & a_3 & a_2 & a_1 & a_0 & 0 & \cdots & 0 \\ \vdots & \vdots & & & & & & & \vdots \\ a_{2n-1} & a_{2n-2} & \cdots & & & & & & a_n \end{vmatrix}, \qquad (C.9)$$

where the system of changing subscripts should be obvious, and elements with subscripts less than 0 or larger than n are replaced by zeroes. The principal minor D_1 is then the determinant of the upper 1×1 sub-matrix of D_n (equal to a_1), D_2 is the determinant of the upper 2×2 matrix of D_n, and so on. For the lowest few values of n we provide the following derived conditions, necessary and sufficient, for all roots to have negative real parts:

$$
\begin{aligned}
&n = 1: \ a_j > 0, j = 0, 1, \\
&n = 2: \ a_j > 0, j = 0, 2, \\
&n = 3: \ a_j > 0, j = 0, 3, \ \text{and} \ D_2 = a_1 a_2 - a_0 a_3 > 0, \\
&n = 4: \ a_j > 0, j = 0, 4, \ \text{and} \ D_3 = (a_1 a_2 - a_0 a_3)a_3 - a_1^2 a_4 > 0 .
\end{aligned}
\tag{C.10}
$$

Pedersen (1986) provides derived conditions as those above for n up to eight, and for n up to four for polynomials with complex-valued coefficients.

C.2.3 Mathieu's Equation: Stability of the Zero-Solution

A typical form of *Mathieu's equation* is:

$$
\ddot{u} + (\delta + 2\varepsilon \cos(2\tau))u = 0,
\tag{C.11}
$$

where $u = u(\tau)$ and $(\cdot) = d/d\tau$. Mathieu's equation belongs to a more general class of *Mathieu-Hill equations*, which are linear differential equations with periodically varying coefficients. They occur very often when the stability of equilibriums for structures with parametric excitation is to be examined. Fig. C.1 is a *Strutt diagram* (van der Pol and Strutt 1928), depicting regions in the (δ, ε)-parameter plane where the solution $u = 0$ to (C.11) is unstable. The transition curves separating stable from unstable regions emanate from the $\varepsilon = 0$ axis at $\delta = j^2$, $j = 0, 1, 2, \ldots$ On these curves the solutions to (C.11) are periodic. Exact analytical expressions for the transition curves in the form $\varepsilon(\delta)$ or $\delta(\varepsilon)$ do not exist; they have to be determined numerically, e.g. by solving eigenvalue problems set up by using *Floquet theory* (Nayfeh and Mook 1979). However, for cases of weak damping and excitation, perturbation methods can be used.

A *damped* form of Mathieu's equation is:

$$
\ddot{u} + 2\mu\dot{u} + (\delta + 2\varepsilon \cos(2\tau))u = 0,
\tag{C.12}
$$

where μ is the damping parameter. For this equation Nayfeh and Mook (1979) provides the following approximate perturbation expressions, defining the three first regions of instability, corresponding to the transition curves emanating from $\delta = 0$, 1, and 4, respectively, in Fig. C.1 (where $\mu = 0$):

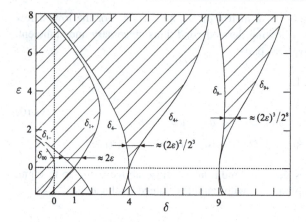

Fig. C.1 Strutt diagram, depicting instability regions (hatched) of the zero-solution to Mathieu's equation, $\ddot{u} + (\delta + 2\varepsilon \cos(2t))u = 0$

$$\delta_{00} = -\frac{1}{2}\varepsilon^2 + O(\varepsilon^3),$$

$$\delta_{1-} = 1 - \sqrt{\varepsilon^2 - 4\mu^2} - \frac{1}{8}\varepsilon^2 + O(\varepsilon^3),$$

$$\delta_{1+} = 1 + \sqrt{\varepsilon^2 - 4\mu^2} - \frac{1}{8}\varepsilon^2 + O(\varepsilon^3),$$

$$\delta_{4-} = 4 + \frac{1}{6}\varepsilon^2 - \sqrt{\frac{1}{16}\varepsilon^4 - 16\mu^2} + O(\varepsilon^3),$$

$$\delta_{4+} = 4 + \frac{1}{6}\varepsilon^2 + \sqrt{\frac{1}{16}\varepsilon^4 - 16\mu^2} + O(\varepsilon^3),$$

(C.13)

which are valid for $\mu \ll 1$ and $\varepsilon \ll 1$, to the order of magnitude indicated.

A damped Mathieu equation can be transformed into the undamped standard form (C.11). For example, the variable shift $u(\tau) \rightarrow u(\tau)\exp(-\mu\tau)$ transforms (C.12) into the form (C.11) (where δ then changes to $\delta - \mu^2$).

Similarly, the equation of motion for a linear single-degree-of-freedom oscillator with parametric excitation typically takes this form:

$$\ddot{x} + 2\beta\dot{x} + (\omega^2 + q\cos(\Omega t))x = 0, \qquad (C.14)$$

where $x = x(t)$ and $(\cdot) = d/dt$. To transform this into the undamped standard form (C.11), we first introduce a new independent variable $\tau = \frac{1}{2}\Omega t$ (to get the argument of the cosine right), and then introduce a new dependent variable u by letting $x(\tau) = u(\tau)\exp(\gamma\tau)$. Inserting this into (C.14), it is found that the damping term vanishes by choosing $\gamma = -2\beta/\Omega$, and that the undamped form (C.11) then is obtained with the following values of δ and ε:

$$\delta = \frac{4}{\Omega^2}(\omega^2 - \beta^2); \quad 2\varepsilon = \frac{4q}{\Omega^2}. \tag{C.15}$$

Without damping, $\beta = 0$, the transition curves emanates at the $\varepsilon = 0$ axis from $\delta = j^2, j = 0, 1, \ldots$, corresponding to $\Omega = 2\omega / j$. The primary parametric resonance corresponding to $j = 1$ then occurs for $\Omega = 2\omega$, i.e. when the frequency of excitation is twice the natural frequency of the oscillator. Increased damping β tends to lift the instability regions in the Strutt diagram off from the $\varepsilon = 0$ ($q = 0$) axis, such that instability occurs at higher values of ε (or q).

Appendix D: Natural Frequencies and Mode Shapes for Structural Elements

Natural frequencies and mode shapes are provided here for some basic structural members: strings, rods, beams, rings, membranes, and plates. The main sources are Harris (1996), Den Hartog (1985), Inman (2014), Flügge (1962). Table D.1 shows the notation, which, unless otherwise stated, is used throughout the appendix.

Table D.1 Notation

A	Cross-sectional area
c_0	Longitudinal wave speed = $\sqrt{E/\rho}$
E	Young's modulus
G	Shear modulus = $\frac{1}{2} E/(1 + v)$
h	Thickness
I	Planar moment of inertia
J	Mass moment of inertia = ρI
k	Shear-deflection coefficient
K	Plate bending stiffness = $Eh^3/12(1 - v^2)$
l	Length
L	Wavelength
N	Axial/longitudinal tension force
r	Radius of gyration = $\sqrt{I/A}$
s	Slenderness ratio = $\tilde{L}/r, \tilde{L} \in \{l, L\}$
u, v	Longitudinal or in-plane deformation
w	Lateral deformation
v	Poisson's ratio
ρ	Density
φ	Mode shape, eigenfunction
ω	Natural frequency (angular)
$(\dot{\ })$	Time derivative
$(\ )'$	Spatial derivative, direction understood
$(\ )_x'$	Spatial derivative, direction x
∇^2	Laplacian operator, $\partial^2/\partial x^2 + \partial^2/\partial y^2$

J. J. Thomsen *Vibrations and Stability*,
https://doi.org/10.1007/978-3-030-68045-9

D.1 Strings

For a uniform elastic string or cable with negligible bending stiffness, and subjected to constant tension N at the fixed ends, the free and undamped transverse motions $w(x, t)$ are governed by:

$$\ddot{w} = \bar{c}_0^2 w'', \tag{D.1}$$

where $\bar{c}_0 = \sqrt{N/\rho A}$ is the transverse wave speed. The jth natural frequency is:

$$\omega_j = j\pi \frac{\bar{c}_0}{l}, \quad j = 1, 2, \cdots, \tag{D.2}$$

with corresponding mode shapes

$$\varphi_j(x) = \sin\left(j\pi \frac{x}{l}\right). \tag{D.3}$$

D.2 Rods

D.2.1 Longitudinal Vibrations

For a uniform rod the free and undamped longitudinal motions $u(x, t)$ are governed by:

$$\ddot{u} = c_0^2 u''. \tag{D.4}$$

Table D.2. lists natural frequencies and mode shapes for some different boundary conditions.

Table D.2 Natural frequencies and mode shapes for longitudinally vibrating uniform rods

Boundary supports	Natural frequency ω_j	Mode shape $\varphi_j(x)$
Fixed-fixed	$j\pi\frac{c_0}{l}$, $j = 1, 2, \ldots$	$\sin\left(j\pi\frac{x}{l}\right)$
Fixed-free	$(j - \frac{1}{2})\pi\frac{c_0}{l}$, $j = 1, 2, \ldots$	$\sin\left((j - \frac{1}{2})\pi\frac{x}{l}\right)$
Free-free	$j\pi\frac{c_0}{l}$, $j = 0, 1, \ldots$	$\cos\left(j\pi\frac{x}{l}\right)$

D.2.2 Torsional Vibrations

For a rod of uniform circular cross-section the free and undamped torsional motions are governed by:

$$\ddot{\theta} = \tilde{c}_0^2 \theta'', \tag{D.5}$$

where $\tilde{c}_0 = \sqrt{G/\rho}$ is the torsional wave speed. This equation of motion is the same as Eq. (D.4) for longitudinal vibrations, so the natural frequencies and mode shapes are obtained by substituting $\tilde{c}_0$ for c_0 in Table D.2.

D.3 Beams

D.3.1 Bernoulli-Euler Beam Theory

The classical *Bernoulli-Euler theory* holds for vibrations where the spatial wavelength of the vibration is much larger than the height of the beam, i.e. $s \gg 1$; this means that rotary inertia and shear deflections can be ignored. Slender beams vibrating in lower modes satisfy this assumption. For thick beams or higher vibration modes, use Timoshenko theory (Sect. D.3.2).

For a uniform *Bernoulli-Euler beam* the free and undamped transverse motions $w(x, t)$ are governed by:

$$\rho A \ddot{w} + E I w'''' = 0. \tag{D.6}$$

The jth natural frequency ω_j is:

$$\omega_j = \frac{\lambda_j^2}{l^2} \sqrt{\frac{EI}{\rho A}} = \lambda_j^2 \frac{c_0}{sl}, \quad j = 1, 2, \cdots, \tag{D.7}$$

where λ_j is the jth root of the frequency equation, which is given in Table D.3 along with the corresponding mode shapes $\varphi(x)$ for various boundary conditions. Boundary conditions defining the different beam types are listed in Table D.4.

Table D.3 Natural frequencies and mode shapes for flexurally vibrating uniform Bernoulli-Euler beams[i,ii]

Boundary supports	Frequency equation	Roots λ_j of the frequency equation[iii]	Mode shape[iv] $\varphi_j(x)$
a) Clamped-clamped	$\cos\lambda\cosh\lambda = 1$	4.7300 7.8532 10.9956 14.1372 $\rightarrow (2j+1)\pi/2$	$J\left(\lambda_j \frac{x}{l}\right) - \frac{J(\lambda_j)}{H(\lambda_j)} H\left(\lambda_j \frac{x}{l}\right)$
b) Clamped-hinged	$\tan\lambda = \tanh\lambda$	3.9266 7.0686 10.2102 13.3518 $\rightarrow (4j+1)\pi/4$	$J\left(\lambda_j \frac{x}{l}\right) - \frac{J(\lambda_j)}{H(\lambda_j)} H\left(\lambda_j \frac{x}{l}\right)$
c) Clamped-free (= cantilever)	$\cos\lambda\cosh\lambda = -1$	1.8751 4.6941 7.8548 10.9955 $\rightarrow (2j-1)\pi/2$	$J\left(\lambda_j \frac{x}{l}\right) - \frac{G(\lambda_j)}{F(\lambda_j)} H\left(\lambda_j \frac{x}{l}\right)$
d) Clamped-guided	$\tan\lambda = -\tanh\lambda$	2.3650 5.4978 8.6394 11.7810 $\rightarrow (4j-1)\pi/4$	$J\left(\lambda_j \frac{x}{l}\right) - \frac{F(\lambda_j)}{J(\lambda_j)} H\left(\lambda_j \frac{x}{l}\right)$
e) Hinged-hinged (= simply supported)	$\sin\lambda = 0$	$j\pi$	$\sin\left(j\pi \frac{x}{l}\right)$
f) Hinged-guided	$\cos\lambda = 0$	$(2j-1)\pi/2$	$\sin\left((j-\frac{1}{2})\pi \frac{x}{l}\right)$
g) Guided-guided	$\sin\lambda = 0$	$j\pi$	$\cos\left(j\pi \frac{x}{l}\right)$
h) Free-free	$\cos\lambda\cosh\lambda = 1$	Same as for (a)	$G\left(\lambda_j \frac{x}{l}\right) - \frac{J(\lambda_j)}{H(\lambda_j)} F\left(\lambda_j \frac{x}{l}\right)$
i) Free-hinged	$\tan\lambda = \tanh\lambda$	Same as for (b)	$G\left(\lambda_j \frac{x}{l}\right) - \frac{G(\lambda_j)}{F(\lambda_j)} F\left(\lambda_j \frac{x}{l}\right)$
j) Free-guided	$\tan\lambda = -\tanh\lambda$	Same as for (d)	$G\left(\lambda_j \frac{x}{l}\right) - \frac{H(\lambda_j)}{F(\lambda_j)} F\left(\lambda_j \frac{x}{l}\right)$

[i]The natural frequencies are $\omega_j = \lambda_j^2 (EI/\rho A)^{1/2}/l^2$, where λ_j is given in the table
[ii]Adapted from Flügge (1962) with permission from the publisher
[iii]Lowest roots numerically, or all roots analytically, or limits for large j (indicated by $\rightarrow$)
[iv]$F(u) = \sinh u + \sin u$; $G(u) = \cosh u + \cos u$; $H(u) = \sinh u - \sin u$; $J(u) = \cosh u - \cos u$

The data in Table D.3 assumes there is no axial loading. Generally the presence of axial tension increases the effective transverse stiffness, and thus also increases the natural frequencies. Conversely, compressive loading reduces natural frequencies (the lowest becoming zero with the approach to buckling loading). The effect of axial loading on natural frequencies increases with beam slenderness, i.e. as a beam become more 'string-like' the effect of tension increases (because relatively more of the transverse stiffness comes from tension). Bokaian (1988, 1990) provides formulas for natural frequencies and mode shapes of axially tensioned or compressed beams for different standard boundary conditions (see also Brøns et al. 2021, Sah et al. 2018, Hermansen and Thomsen 2018).

Table D.4 Boundary conditions for Bernoulli-Euler beams in flexure

Boundary support	Boundary conditions	Quantity vanishing at the boundary
Clamped	$\varphi = \varphi' = 0$	Translation and slope
Hinged	$\varphi = \varphi'' = 0$	Translation and bending moment
Guided	$\varphi' = \varphi''' = 0$	Slope and transverse force
Free	$\varphi'' = \varphi''' = 0$	Bending moment and transverse force

D.3.2 Timoshenko Beam Theory

Timoshenko theory is used when the spatial wavelength of the vibration is *not* large compared to the beam height. This means that effects of rotary inertia and shear deflections need to be accounted for. Use Timoshenko theory for thick beams or higher vibration modes. For slender beams vibrating in lower modes, the simpler Bernoulli-Euler theory can be used instead (Sect. D.3.1).

The free vibrations of a uniform *Timoshenko beam* are governed by a pair of coupled equations (Flügge 1962; Marguerre and Wölfel 1979; Elishakoff et al. 2015; Brøns and Thomsen 2019):

$$\rho A \ddot{w} = kGA(w' - \theta)',$$
$$J\ddot{\theta} = EI\theta'' + kGA(w' - \theta), \tag{D.8}$$

where $w(x,t)$ is the total transverse deflection of the beam (due to both bending and shear) and $\theta(x,t)$ is the slope due to bending. The general expression for the eigenfunctions $\varphi(x)$ and $\psi(x)$ corresponding, respectively, to the w- and the θ-motions, are[1]:

$$\varphi(x) = C_1 \cosh\left(\frac{\alpha x}{l}\right) + C_2 \sinh\left(\frac{\alpha x}{l}\right) + C_3 \cos\left(\frac{\beta x}{l}\right) + C_4 \sin\left(\frac{\beta x}{l}\right),$$

$$\psi(x) = \frac{p^2 E/kG + \alpha^2}{\alpha l}\left(C_2 \cosh\left(\frac{\alpha x}{l}\right) + C_1 \sinh\left(\frac{\alpha x}{l}\right)\right) \tag{D.9}$$
$$+ \frac{p^2 E/kG - \beta^2}{\beta l}\left(-C_4 \cos\left(\frac{\beta x}{l}\right) + C_3 \sin\left(\frac{\beta x}{l}\right)\right),$$

where:

$$\left.\begin{matrix}\alpha^2\\\beta^2\end{matrix}\right\} = \frac{p^2}{2}\left[\mp\left(1 + \frac{E}{kG}\right) + \sqrt{\left(1 - \frac{E}{kG}\right)^2 + \left(\frac{2l}{pr}\right)^2}\right], \quad p = \frac{\omega l}{c_0}. \tag{D.10}$$

---

[1]Typo(s) in the source (Flügge 1962) corrected here.

The frequency equation for the determination of natural frequencies ω is calculated by substituting a set four boundary conditions into (D.9), and requiring the determinant of the coefficient matrix for the unknown constants C_j, $j = 1, 4$ to vanish identically. Common types of boundary conditions are: clamped ($\varphi = \psi = 0$), hinged ($\varphi = \psi' = 0$), guided ($\varphi' = \psi = 0$), and free ($\varphi' - \psi = \psi' = 0$).

Formulas for a few common set of boundary conditions are given below. Only hinged-hinged boundary conditions allow for exact solutions, while for other cases the frequency equation is transcendent and needs numerical solution.

Hinged-Hinged Beam For a beam hinged at $x = 0$ and $x = l$ the boundary conditions are $\varphi(0) = \psi'(0) = \varphi(l) = \psi'(l) = 0$, and the natural frequencies are:

$$\omega_{j1} = \frac{j\pi c_0}{l} \lambda_{j1}, \omega_{j2} = \frac{j\pi c_0}{l} \lambda_{j2}, j = 1, 2, \cdots, \tag{D.11}$$

where the frequency parameter λ_{jk} is given by[2]:

$$\left.\begin{array}{c}\lambda_{j1}^2 \\ \lambda_{j2}^2\end{array}\right\} = \frac{1}{2}\left(\eta_j \mp \sqrt{\eta_j^2 - 4\frac{kG}{E}}\right) \quad \text{where} \quad \eta_j = 1 + \frac{kG}{E}\left(1 + \left(\frac{l}{j\pi r}\right)^2\right). \tag{D.12}$$

The corresponding eigenfunctions are $\varphi_j(x) = A_j\sin(j\pi x/l)$ and $\psi_j(x) = B_j\cos(j\pi x/l)$, with amplitude ratios given by:

$$\left(\frac{B_j}{A_j}\right)_k = \frac{j\pi}{l}\left(1 - \frac{E}{kG}\lambda_{jk}^2\right), k = 1, 2, \cdots. \tag{D.13}$$

Clamped-Free Beam (=Cantilever) For a beam clamped at $x = 0$ and free at $x = l$ the boundary conditions are $\varphi(0) = \psi(0) = \varphi'(l) - \psi(l) = \psi'(l) = 0$, and the frequency equation is (Flügge 1962):

$$2 + \frac{\alpha^2 - \beta^2}{\alpha\beta}\sinh\alpha\sin\beta + \left(\frac{\beta^2 - p^2}{\alpha^2 + p^2} + \frac{\alpha^2 + p^2}{\beta^2 - p^2}\right)\cosh\alpha\cos\beta = 0. \tag{D.14}$$

Free-Free Beam Taking here the origin at the beam mid-point, with the free ends at $x = \pm l/2$, the boundary conditions for the *symmetric* (even) modes are $\psi(0) = \varphi'(0) - \psi(0) = \psi'(l/2) = \varphi'(l/2) - \psi(l/2) = 0$, and the frequency equation is[2]:

$$\tan\frac{\beta}{2} + \frac{\beta}{\alpha}\frac{\alpha^2 - p^2 E/kG}{\beta^2 + p^2 E/kG}\tanh\frac{\alpha}{2} = 0, \tag{D.15}$$

---

[2]Typo(s) in the source (Flügge 1962) corrected here.

while for the *antisymmetric* (odd) modes the boundary conditions at $x = 0$ changes to $\varphi(0) = \psi'(0) = 0$, leading to the frequency equation[2]:

$$\tan\frac{\beta}{2} - \frac{\alpha}{\beta}\frac{\beta^2 + p^2 E/kG}{a^2 - p^2 E/kG}\tanh\frac{\alpha}{2} = 0. \tag{D.16}$$

Axial loading As for Bernoulli-Euler beams axial tension, if present, increases the natural frequencies while compressive loads reduce them. Abramovich (1992) provides formulas for natural frequencies of Timoshenko beams under axial loads for various standard boundary conditions.

D.4 Rings

D.4.1 In-plane Bending

Of the many possible modes of vibration of a full circular ring of radius R, the most important are often bending vibrations in the plane of the ring. Their natural frequencies are given by:

$$\omega_j = \frac{j(j^2 - 1)}{\sqrt{j^2 + 1}}\sqrt{\frac{EI}{\rho AR^4}} = \frac{j(j^2 - 1)}{\sqrt{j^2 + 1}}\frac{c_0 r}{R^2}, \, j = 1, 2, \cdots \tag{D.17}$$

where j is the number of full sinusoidal waves on the developed circumference of the ring. The corresponding mode shapes describe purely radial displacement, $u = \cos(j\theta)$, where θ is the angle from a reference radius.

D.4.2 Out-of-Plane Bending

For a full circular ring of circular cross section, the natural frequencies for flexural vibrations perpendicular to the plane of the ring is close to those in (D.17):

$$\omega_j = \frac{j(j^2 - 1)}{\sqrt{j^2 + 1 + \nu}}\sqrt{\frac{EI}{\rho AR^4}} = \frac{j(j^2 - 1)}{\sqrt{j^2 + 1 + \nu}}\frac{c_0 r}{R^2}, \, j = 1, 2, \cdots. \tag{D.18}$$

The mode shapes are complicated, flexural motions being coupled with torsion.

D.4.3 Extension

For a full circular ring, the natural frequencies for extensional vibrations are:

$$\omega_j = \sqrt{j^2 + 1}\frac{c_0}{R}, \, j = 0, 1, \cdots. \tag{D.19}$$

The mode shapes are given by radial and tangential displacements, respectively, $u = \cos(j\theta)$ and $v = -j\sin(j\theta)$; thus for $j = 0$ the mode shape is purely radial.

D.5 Membranes

We consider a membrane (i.e. a thin, stretchable structure having negligible bending stiffness), being stretched in all directions with a constant force N per unit boundary length. The equation governing transverse motions w in rectangular coordinates (x, y), or polar coordinates (r, θ), is:

$$\begin{aligned}
\ddot{w} &= \hat{c}_0^2\left(w''_{xx} + w''_{yy}\right) \\
&= \hat{c}_0^2\left(r^{-1}\left(rw'_r\right)'_r + r^{-2}w''_{\theta\theta}\right),
\end{aligned} \tag{D.20}$$

where $\hat{c}_0^2 = N/\rho h$.

D.5.1 Rectangular Membranes

With the four edges fixed along $x = 0$, $x = a$, $y = 0$, and $y = b$, the natural frequencies and mode shapes of a rectangular membrane are, respectively:

$$\begin{aligned}
\omega_{ij} &= \pi\hat{c}_0\sqrt{\left(\frac{i}{a}\right)^2 + \left(\frac{j}{b}\right)^2}, \, i, j = 1, 2, \cdots, \\
\varphi_{ij}(x, y) &= \sin\left(i\pi\frac{x}{a}\right)\sin\left(j\pi\frac{y}{b}\right).
\end{aligned} \tag{D.21}$$

D.5.2 Circular Membranes

The frequency equation for a circular membrane fixed at $r = R$ is:

$$J_n(\omega R/\hat{c}_0) = 0, n = 0, 1, \cdots, \tag{D.22}$$

where J_n is the Bessel function of the first kind and nth integral order. The solutions of this equation has the form:

Table D.5 Values of α_{nj} to be used with Eq. (D.23)[a]

	$n = 0$	1	2	3
$j = 1$	2.404	3.832	5.135	6.379
2	5.520	7.016	8.417	9.760
3	8.654	10.173	11.620	13.017
4	11.792	13.323	14.796	16.224

[a]Adapted from Flügge (1962) with permission from the publisher

$$\omega_{nj} = \frac{\alpha_{nj}\hat{c}_0}{R}, j = 1, 2, \cdots, \tag{D.23}$$

where numerical values for the constant α_{nj} are listed in Table D.5. The corresponding mode shapes are:

$$\varphi_{nj} = J_n\left(\omega_{nj}r/\hat{c}_0\right)\cos(n\theta). \tag{D.24}$$

D.6 Plates

The equation governing free transverse motions w of a thin plate is:

$$\frac{\rho h}{K}\ddot{w} + \nabla^2\nabla^2 w = 0, \tag{D.25}$$

where K is the plate bending stiffness:

$$K = \frac{Eh^3}{12(1 - v^2)}. \tag{D.26}$$

Using $(\bar{n}, \bar{t})$ to denote coordinates normal and tangential, respectively, to the boundary, the common types of boundary conditions are: clamped ($w = w'_{\bar{n}} = 0$), hinged ($w = w''_{\bar{n}\bar{n}} + vw''_{\bar{t}\bar{t}} = 0$), and free ($w''_{\bar{n}\bar{n}} + vw''_{\bar{t}\bar{t}} = [w''_{\bar{n}\bar{n}} + (2 - v)w''_{\bar{t}\bar{t}}]'_{\bar{n}} = 0$).

D.6.1 Rectangular Plates

With a rectangular plate having side lengths a and b, we assume it to be supported along the lines $x = 0$, $x = a$, $y = 0$, and $y = b$.

When the plate is *hinged*, i.e. *simply supported*, at all four edges, the natural frequencies and mode shapes are, respectively:

$$\omega_{ij} = \left(\left(\frac{i\pi}{a} \right)^2 + \left(\frac{j\pi}{b} \right)^2 \right) \sqrt{\frac{K}{\rho h}}, \quad i,j = 1,2,\cdots,$$

$$\varphi_{ij}(x,y) = \sin\left(i\pi\frac{x}{a} \right) \sin\left(j\pi\frac{y}{b} \right).$$

(D.27)

It is also possible to obtain analytical solutions for rectangular plates that are *simply supported at two opposite edges*, e.g. at $x = 0$ and $x = a$, with any boundary conditions at the other edges. To do this insert $w(x,y,t) = \varphi(y)\sin(j\pi x/a)\sin(\omega t)$, $j = 1, 2, \ldots$ into (D.25), and determine the unknown eigenfunction component $\varphi(y)$ just as for one-dimensional beam problems (cf. Sect. D.3.1).

In all other cases approximate methods are needed, e.g. as described in Chap. 2 for general eigenvalue problems, and detailed specifically for plate vibration problems in Leissa (1993a). Flügge (1962) provides nondimensional numerical constants for the calculation of the lowest few natural frequencies for various combination of clamped, free, and hinged/simple supports for a few side length ratios.

D.6.2 Circular Plates

For a circular plate of radius R the natural frequencies are:

$$\omega_{ij} = \frac{\lambda_{ij}}{R^2} \sqrt{\frac{K}{\rho h}}, i,j = 1,2,\cdots,$$

(D.28)

where λ_{ij} is a root of a frequency equation involving Bessel functions. Numerical values of λ_{ij} for the lowest values of i and j are given in Table D.6 for circular plates with clamped or free edges. The mode shapes are defined by the values of i and j, so that i is the number of nodal diameters, and j the number of nodal circles.

D.7 Other Structures

Harris (1996) and Flügge (1962) also give brief treatments of other structures, such as rotating disks and thin shells. For in depth treatments regarding vibration of plates, disks, and shells see Leissa (1993a, b), Niordson (1985).

Table D.6 Values of λ_{ij} to be used with Eq. D.28[a]

	Clamped boundary			Free boundary			
	$i = 0$	1	2	$i = 0$	1	2	3
$j = 0$	10.2	21.2	34.8	–	–	5.25	12.2
1	39.8	60.8	88.4	9.08	20.5	35.2	52.9
2	88.9	120.2	157.9	38.5	59.9	88	120

[a]Adapted from Flügge (1962) with permission from the publisher

Appendix E: Properties of Engineering Materials

E.1 Friction and Thermal Expansion Coefficients

Friction coefficients may vary for as much as 100% for seemingly identical conditions. The values given are thus only indicative of the magnitude to expect (Tables E.1, E.2 and E.3).

E.2 Density and Elasticity Constants

Table E.1 Coefficients of friction[1]

Contacting surface	Static, μ_s	Kinetic, μ_k
Brake lining on cast iron	0.4	0.3
Brass on ice (0 °C)	–	0.02
Brass on steel (dry)	0.5	0.4
Glass on glass	0.9–1.0	–
Metal on ice	–	0.02
Steel on steel (dry)	0.6	0.4
Steel on steel (greasy)	0.1	0.05
Teflon on steel	0.04	0.04
Wood on wood	0.25–0.5	–
Ceramics on ceramics	0.05–0.5	
Metal or ceramic on polymer	0.04–0.5	
Polymer on polymer	0.05–1.0	
Steel on ceramics	0.1–0.5	

[1]Ashby and Jones (1991), Meriam and Kraige (1998), Weast (1981)

J. J. Thomsen *Vibrations and Stability*, https://doi.org/10.1007/978-3-030-68045-9

Table E.2 Coefficients of thermal expansion[2]

Material	α $(10^{-6}/°C)$	Material	α $(10^{-6}/°C)$
Aluminum alloys	23	Nylon	70–140
Brass	19.1–21.2	Polyethylene	140–290
Bronze	18–21	Rock	5–9
Cast iron	9.9–12	Rubber	130–200
Concrete	7–14	Steel	10–18
Copper & alloys	16.6–17.6	Titanium alloys	8.1–11
Glass	5–11	Tungsten	4.3

[2]Ashby and Jones (1991)

Table E.3 Density and elasticity data for common engineering materials[3]

Material	Density ρ (kg/m^3)	Young's modulus[4, 5] E (GPa)	Yield strength[5] σ_y (MPa)	Poisson's ratio ν (1)
Aluminum alloys	2600–2800	70–79	35–500	0.33
Brass	8400–8600	96–100	70–550	0.34
Bronze	8200–8800	96–120	82–690	0.34
Cast iron	7000–7400	83–170	120–290	0.2–0.3
Concrete, plain	2300	17–31[6]	–	0.1–0.2
Copper	2300	110–120	55–760	0.33–0.36
Epoxies	1100–1400	3	–	–
Foamed polymers	10–200	0.001–0.01	–	–
Glass	2400–2800	48–83	–	0.17–0.27
Glass-fiber/epoxy	1800	35–45	–	–
Gold	19,300	82	205	–
Ice	920	9.1	–	–
Iron	7900	196	276–621	–
Nylon	880–1100	2.1–3.4	44–59	0.4
PMMA	1200	3.4	53.8–73.1	–
Polycarbonate	1200–1300	2.6	62.1	–
Polyesters	1100–1500	1–5	–	–
Polyethylene	960–1400	0.7–1.4	9.0–27.6	0.4
Polypropylene	880–910	0.9	31.0–37.2	–
Polystyrene	1000–1100	3–3.4	–	–
PVC	1300–1600	0.003–0.01	40.7–44.8	–
Quartz, SiO_2	2600	94	–	–
Rock, granite	2600–2900	40–100[6]	–	0.2–0.3
Rubber	960–1300	0.0007–0.004	1–7	0.45–0.50
Silicon	2500–3200	107	–	–
Silver	10,500	76	–	–
Steel	7850	190–210	200–1600	0.27–0.30
Tin and alloys	7300–8000	41–53	11	–

(continued)

Table E.3 (continued)

Material	Density ρ (kg/m^3)	Young's modulus[4, 5] E (GPa)	Yield strength[5]σ_y (MPa)	Poisson's ratio v (1)
Titanium	4500	100–120	760–1000	0.33
Tungsten	1900	340–380	760	0.2
Wood, $\perp$ to grain	480–720	0.6–1.0	30–50	–
Wood, $\parallel$ to grain	480–720	9–16	30–50	–

[3]Ashby and Jones (1991), Gere and Timoshenko (1997), Callister (2003)
[4]The shear modulus G is given by the other constants, $G = \frac{1}{2}E/(1 + v)$
[5]In tension
[6]In compression

References

Abramovich H (1992) Natural frequencies of Timoshenko beams under compressive axial loads. J Sound Vib 157(1):183–189

Abusoua A, Daqaq MF (2017) On using a strong high-frequency excitation for parametric identification of nonlinear systems. ASME Journal of Vibration and Acoustics 139:051012 (1–7)

Abusoua A, Daqaq MF (2018) Changing the nonlinear resonant response of an asymmetric mono-stable oscillator by injecting a hard high-frequency harmonic excitation. J Sound Vib 436:262–272

Acheson DJ (1993) A pendulum theorem. Proc Roy Soc Lond A 443:239–245

Acheson DJ (1995) Multiple-nodding oscillations of a driven inverted pendulum. Proc Roy Soc Lond A 448:89–95

Acheson D (1997) From calculus to chaos: an introduction to dynamics. Oxford University Press, Oxford

Acheson D, Mullin T (1993) Upside-down pendulums. Nature 366:215–216

Acheson D, Mullin T (1998) Ropy magic. New Sci 157(2122):32–35

Adhikari S (2013) Structural dynamic analysis with generalized damping models: analysis. Wiley

Adhikari S (2014) Structural dynamic analysis with generalized damping models: identification. Wiley

Amabili M (2008) Nonlinear vibrations and stability of shells and plates. Cambridge University Press, Cambridge

Amabili M (2018) Nonlinear damping in large-amplitude vibrations: modelling and experiments. Nonlinear Dyn 93:5–18

Andersen SB, Thomsen JJ (2002) Post-critical behavior of Beck's column with a tip mass. Int J Non-linear Mech 37(1):135–151

Anderson TJ, Nayfeh AH, Balachandran B (1996) Experimental verification of the importance of the nonlinear curvature in the response of a cantilever beam. ASME J Vib Acoust 118(1):1–27

Anderson GL, Tadjbakhsh IG (1989) Stabilization of Ziegler's pendulum by means of the method of vibrational control. J Math Anal Appl 143:198–223

Antman SS (1995) Nonlinear problems of elasticity. Appl Math Sci 107 (Springer, New York)

Apffel B, Novkoski F, Eddi A, Fort E (2020) Floating under a levitating liquid. Nature 585 (7823):48–52

Arczewski K, Pietrucha J, Leech CM (1993) Mathematical modeling of complex mechanical systems, vol 1. Discrete models. Ellis Horwood, New York

© The Editor(s) (if applicable) and The Author(s), under exclusive license to
Springer Nature Switzerland AG 2021
J. J. Thomsen *Vibrations and Stability*,
https://doi.org/10.1007/978-3-030-68045-9

Armstrong-Hélouvry B, Dupont P, Canudas de Wit C (1994) A survey of models, analysis tools and compensation methods for the control of machines with friction. Automatica 30(7):1083–1138

Ashby MF, Jones DRH (1991) Engineering materials—an introduction to their properties and applications. Pergamon Press, Oxford

Atluri S (1973) Nonlinear vibrations of a hinged beam including nonlinear inertia effects. ASME J Appl Mech 40(1):121–126

Avramov KV, Mikhlin YV (2013) Review of applications of nonlinear normal modes for vibrating mechanical systems. Appl Mech Rev 65:020801(1–20)

Awrejcewicz J, Krysko VA, Papkova IV, Krysko AV (2016) Deterministic Chaos in one-dimensional continuous systems. World Scientific, Singapore

Babitsky VI (1998) Theory of vibro-impact systems and applications. Springer, Berlin

Babitsky VI, Veprik AM (1993) Damping of beam forced vibration by a moving washer. J Sound Vib 166(1):77–85

Balachandran B, Nayfeh AH (1991) Observations of modal interactions in resonantly forced beam-mass structures. Nonlinear Dyn 2:77–117

Barton DAW (2017) Control-based continuation: Bifurcation and stability analysis for physi-cal experiments. Mech Syst Signal Process 84:54–64

Barton DAW, Mann BP, Burrow SG (2012) Control-based continuation for investigating nonlinear experiments. J Vib Control 18(4):509–520

Bathe K, Wilson EL (1976) Numerical methods in finite element analysis. Prentice-Hall, Englewood Cliffs, N.J.

Belhaq M, Fahsi A (2008) 2:1 and 1:1 frequency-locking in fast excited van der Pol-Mathieu-Duffing oscillator. Nonlinear Dyn 53(1–2):139–152

Belhaq M, Sah SM (2008) Fast parametrically excited van der Pol oscillator with time delay state feedback. Int J Non-linear Mech 43(2):124–130

Bellman RE, Bentsman J, Meerkov SM (1986) Vibrational control of nonlinear systems: vibrational stabilizability. IEEE Trans Autom Control AC-31(8):710–716

Bert CW (1973) Material damping: an introductory review of mathematic measures and experimental technique. J Sound Vib 29(2):129–153

Bikdash M, Balachandran B, Nayfeh AH (1994) Melnikov analysis for a ship with a general roll-damping model. Nonlinear Dyn 6:101–124

Billah Y, Scanlan R (1991) Resonance, Tacoma Narrows bridge failure, and undergraduate physics textbooks. Am J Phys 59(2):118–123

Biran A, Breiner M (2002) MATLAB 6 for Engineers. Prentice Hall, London

Bishop SR, Clifford MJ (1996) Zones of chaotic behaviour in the parametrically excited pendulum. J Sound Vib 189(1):142–147

Bishop SR, Xu D (1996) Stabilizing the parametrically excited pendulum onto high order periodic orbits. J Sound Vib 194(2):287–293

Blackburn JA, Zhou-jing Y, Vik S, Smith HJT, Nerenberg MAH (1987) Experimental study of chaos in a driven pendulum. Physica 26D:385–395

Blaquiére A (1966) Nonlinear system analysis. Academic, New York

Blekhman II (1976) Method of direct motion separation in problems of vibration acting on nonlinear mechanical systems. Mech Solids 11(4):7–19

Blekhman II (1994) Vibrational mechanics. Fizmatlit Publishing Company, Moscow (In Russian)

Blekhman II (2000) Vibrational mechanics—nonlinear dynamic effects, general approach, applications. World Scientific, Singapore. (English translation of Blekhman (1994))

Blekhman II (ed) (2004) Selected topics in vibrational mechanics. World Scientific, Singapore

Blekhman II (2008) Vibrational dynamic materials and composites. J Sound Vib 317:657–663

Blekhman II, Dzhanelidze GY (1964) Vibrational displacement. Nauka Press, Moscow (in Russian)

Blekhman II, Malakhova OZ (1986) Quasi-equilibrium positions of the Chelomei pendulum. Sov Phys Dokl 31(3):229–231

Blekhman II, Sorokin VS (2010) On the separation of fast and slow motions in mechanical systems with high-frequency modulation of the dissipation coefficient. J Sound Vib 329:4936–4949

Blekhman II, Sorokin VS (2016) Effects produced by oscillations applied to nonlinear dynamic systems: a general approach and examples. Nonlinear Dyn 83(4):2125–2141

Bogdanoff JL, Citron SJ (1965) Experiments with an inverted pendulum subject to random parametric excitation. J Acoust Soc Am 38:447–452

Bogoliubov NN, Mitropolskii YA (1961) Asymptotic methods in the theory of nonlinear oscillations. Gordon and Breach, New York

Bokaian A (1988) Natural frequencies of beams under compressive axial loads. J Sound Vib 126 (1):49–65

Bokaian A (1990) Natural frequencies of beams under tensile axial loads. J Sound Vib 142 (3):481–498

Bolotin VV (1963) Nonconservative problems of the theory of elastic stability. Pergamon Press, Oxford

Bolotin VV (1964) The dynamic stability of elastic systems. Holden-Day, San Francisco

Brandt A (2011) Noise and vibration analysis: signal analysis and experimental procedures. Wiley, Chichester, UK

Brillouin L (1946) Wave propagation in periodic structures. Dover Publications, USA

Broch JT (1984) Mechanical vibration and shock measurements. Brüel & Kjær Sound and Vibration, Copenhagen

Brøns M, Thomsen JJ (2019) Experimental testing of Timoshenko predictions of supercritical natural frequencies and mode shapes for free-free beams. J Sound Vib 459:114856(1–14)

Brøns M, Thomsen JJ, Sah SM, Tcherniak D, Fidlin A (2021) Estimating bolt tension from vibrations: transient features, nonlinearity, and signal processing. Mech Syst Signal Process 150:107224(1–17)

Brumberg RM (1970) On the motion of a solid body along a vibrating tube without breaking away from it. Izv. AN SSSR. Mekhanika Tverdogo Tela 5:46–51 (in Russian)

Bukhari M, Malla A, Kim H, Barry O, Zuoa L (2020) On a self-tuning sliding-mass electromagnetic energy harvester. AIP Adv 10:095227

Bureau E, Schilder F, Elmegård M, Santos IF, Thomsen JJ, Starke J (2014) Experimental bifurcation analysis of an impact oscillator—determining stability. J Sound Vib 333(21):5464–5474

Bureau E, Schilder F, Santos IF, Thomsen JJ, Starke J (2013) Experimental bifurcation analysis of an impact oscillator—tuning a non-invasive control scheme. J Sound Vib 332(22):5883–5897

Burton R (1968) Vibration and impact. Dover Publications, New York

Burton TD, Rahman Z (1986) On the multi-scale analysis of strongly non-linear forced oscillators. Int J Non-linear Mech 21:135–146

Cabboi A, Putelat T, Woodhouse J (2016) The frequency response of dynamic friction: Enhanced rate-and-state models. J Mech Phys Solids 92:210–236

Cabboi A, Segeren M, Hendrikse H, Metrikine A (2020) Vibration-assisted installation and decommissioning of a slip-joint. Eng Struct 209:109949(1–16)

Callister WD (2003) Fundamentals of materials science and engineering—an interactive E·text. http://ecal-admin.mme.tcd.ie/MSEInteractive/tb04.pdf

Carr J (1981) Applications of centre manifold theory. Springer, New York

Cartmell MP (1990) Introduction to linear, parametric and nonlinear vibrations. Chapman and Hall, London

Cartmell MP, Lawson J (1994) Performance enhancement of an autoparametric vibration absorber. J Sound Vib 177(2):173–195

Champneys AR, Fraser WB (2000) The 'Indian rope trick' for a parametrically excited flexible rod: linearized analysis. Proc Roy Soc Lond A 456(1995):553–570

Chatterjee S, Chatterjee S, Singha TK (2005) On the generation of steady motion using fast-vibration. J Sound Vib 283:1187–1204

Chatterjee S, Singha TK, Karmakar SK (2003) Non-trivial effect of fast vibration on the dynamics of a class of non-linearly damped mechanical systems. J Sound Vib 260(4):711–730

Chatterjee S, Singha TK, Karmakar SK (2004) Effects of high-frequency excitation on a class of mechanical systems with dynamic friction. J Sound Vib 269(1–2):61–89

Chelomei SV (1981) Dynamic stability upon high-frequency parametric excitation. Sov Phys Dokl 26(4):390–392

Chelomei VN (1956) On the possibility of increasing the stability of elastic systems by using vibration. Dokl Akad Nauk SSSR 110(3):345–347 (in Russian)

Chelomei VN (1983) Mechanical paradoxes caused by vibrations. Sov Phys Dokl 28(5):387–390

Chen Y (1966) Vibrations: theoretical methods. Addison-Wesley, Reading, Massachusetts

Chen G (ed) (2000) Controlling chaos and bifurcations in engineering systems. CRC Press, Boca Raton, Florida

Cicconofri F, DeSimone A (2016) Motility of a model bristle-bot: a theoretical analysis. Int J Non-Linear Mech 76:233–239

Cochard A, Bureau L, Baumberger T (2003) Stabilization of frictional sliding by normal load modulation. ASME J Appl Mech 70(March):220–226

Collatz L (1963) Eigenwertaufgaben mit Technischen Anwendungen. Akademische Verlagsgesellschaft Geest & Portig K.-G, Leipzig (in German)

Cook GR, Simiu E (1991) Periodic and chaotic oscillations of modified stoker column. J Eng Mech 117(9):2049–2064

Cook RD, Malkus DS, Plesha ME (1989) Concepts and applications of finite element analysis. Wiley, New York

Crandall SH (1970) The role of damping in vibration theory. J Sound Vib 11(1):3–18

Cross MC, Hohenberg PC (1994) Spatiotemporal chaos. Science 264:1569–1570

Dahl PR (1976) Solid friction damping of mechanical vibrations. AIAA J 14(12):1675–1682

Dankowicz H, Schilder F (2013) Recipes for continuation. SIAM—Society for Industrial and Applied Mathematics, Philadelphia, USA

Den Hartog JP (1985) Mechanical vibrations. Dover Publications, New York

Denis V, Jossic M, Renault A, Giraud-Audine C, Thomas O (2017) Robust measurement of backbone curves of a nonlinear piezoelectric beam. In: Stépán G, Csernák G (eds) (2017) Proceedings of the 9th EUROMECH nonlinear dynamics conference (ENOC 2017), June 25–30, 2017, Budapest, Hungary, 2 pp

Dewdney AK (1985, August) A computer microscope zooms in for a look at the most complex objects in mathematics. Sci Am 8–14

Dimarogonas AD, Haddad S (1992) Vibration for engineers. Prentice-Hall, London

Dimentberg MF, Iourtchenko DV (2004) Random vibrations with impacts: a review. Non-linear Dyn 36:229–254

Ditto WL, Pecora LM (1993, August) Mastering chaos. Sci Am 62–68

Ditto WL, Rauseo SN, Spano ML (1990) Experimental control of chaos. Phys Rev Lett 65 (26):3211–3214

Dohnal F (2008) Damping by parametric stiffness excitation: resonance and anti-resonance. J Vib Control 14(5):669–688

Dowell EH (1982) Flutter of a buckled plate as an example of chaotic motion of a deterministic autonomous system. J Sound Vib 85(3):333–344

Dowell EH, Pezeshki C (1986) On the understanding of chaos in Duffing's equation including a comparison with experiment. ASME J Appl Mech 53(1):5–9

Dowell EH, Pezeshki C (1988) On necessary and sufficient conditions for chaos to occur in Duffing's equation: a heuristic approach. J Sound Vib 121(2):195–200

Elishakoff I (2005) Controversy associated with the so-called "Follower Forces": Critical overview. Appl Mech Rev 58(2):117–142

Elishakoff I, Kaplunov J, Nolde E (2015) Celebrating the centenary of Timoshenko's study of effects of shear deformation and rotary inertia. Appl Mech Rev 67(6):060802(1–11)

Elmegård M, Krauskopf B, Osinga HM, Starke J, Thomsen JJ (2014) Bifurcation analysis of a smoothed model of a forced impacting beam and comparison with an experiment. Nonlinear Dyn 77(3):951–966

El Naschie MS (1990a) Stress, stability and Chaos. McGraw-Hill, London

El Naschie MS (1990b) On the susceptibility of local elastic buckling to chaos. Zeitschrift für Angew Math Mechanik 70(12):535–542

El Naschie MS, Al Athel S (1989 On the connection between statical and dynamical chaos. Zeitschrift für Naturforschung 44a:645–650

Enz S, Thomsen JJ (2011a) Predicting phase shift effects for vibrating fluid-conveying pipes due to Coriolis forces and fluid pulsation. J Sound Vib 330(21):5096–5113

Enz S, Thomsen JJ, Neumeyer S (2011b) Experimental investigation of zero phase shift effects for Coriolis flowmeters due to pipe imperfections. Flow Meas Instrum 22(1):1–9

Euler L (1744) Methodus Inveniendi Lineas Curvas Maximi Minimive Properietate Gau-dentes (Appendix, de Curvis Elasticis). Marcum Michaelem Bousquet, Lausanne and Geneva (in Latin)

Evan-Iwanowski RM (1976) Resonance oscillations in mechanical systems. Elsevier, Amsterdam

Ewins DJ (2000) Modal testing: theory, practice and application, 2nd edn. Wiley, New York

Farmer JD, Ott E, Yorke JA (1983) The dimension of chaotic attractors. Physica 7D:153–180

Feeny B, Guran A, Hinrichs N, Popp K (1998) A historical review on dry friction and stick-slip phenomena. ASME Appl Mech Rev 51(5):321–341

Feeny BF, Moon BF (2000) Quenching stick-slip chaos with dither. J Sound Vib 237(1):173–180

Feigenbaum MJ (1978) Qualitative universality for a class of nonlinear transformations. J Stat Phys 19(1):25–52

Feigenbaum MJ (1980, Summer) Universal behavior in nonlinear systems. Los Alamos Sci 4–27

Fenn JG, Bayne DA, Sinclair BD (1998) Experimental investigation of the "effective potential" of an inverted pendulum. Am J Phys 66(11):981–984

Fidlin AY (1991) On averaging in systems with a variable number of degrees of freedom. J Appl Math Mech 55:507–510

Fidlin A (1999) On the separation of motions in systems with a large fast excitation of general form. Eur J Mech A/Solids 18:527–538

Fidlin A (2000) On asymptotic properties of systems with strong and very strong high-frequency excitation. J Sound Vib 235(2):219–233

Fidlin A (2001) On the asymptotic analysis of discontinuous systems. ZAMM 82(2):75–88

Fidlin A (2002) On the oscillations in discontinuous and unconventionally strong excited systems: asymptotic approaches and dynamic effects. Doctoral dissertation, Department of Mechanical Engineering, University of Karlsruhe

Fidlin A (2004a) On asymptotic analysis of systems with fast excitation. In: Blekhman II (ed) Selected topics in vibrational mechanics. World Scientific, Singapore

Fidlin A (2004b) On the averaging of discontinuous systems. In: Blekhman II (ed) Selected topics in vibrational mechanics. World Scientific, Singapore

Fidlin A (2005) On the strongly nonlinear behavior of an oscillator in a clearance. In: van Campen DH, Lazurko MD, van der Oever WPJM (eds) (2005) Proceedings of ENOC-2005, Eindhoven, Netherlands, 7–12 August 2005, Technical University of Eindhoven, pp 389–398

Fidlin A (2006) Nonlinear oscillations in mechanical engineering. Springer, Berlin Heidelberg

Fidlin A, Thomsen JJ (2001) Predicting vibration-induced displacement for a resonant friction slider. Eur J Mech A/Solids 20:155–166

Fidlin A, Thomsen JJ (2007) Nontrivial effects of high-frequency excitation for strongly damped mechanical systems. Int J Non-Linear Mech 43:569–578

Flügge W (1962) Handbook of engineering mechanics. McGraw-Hill, New York

Fraser WB, Champneys AR (2002) The 'Indian rope trick' for a parametrically excited flexible rod; nonlinear and subharmonic analysis. Proc Roy Soc Lond A 458(2022):1353–1373

Galileo G (1638) Discorsi e Dimostrazioni Matematiche, Intorno a Due Nuove Scienze (Eng.: Two New Sciences). Elsevier, Leiden

Gendelman OV (2014) Nonlinear normal modes in damped forced systems. In: Kerschen G (ed) Modal Analysis of Nonlinear Mechanical Systems. CISM International Centre for Mechanical Sciences, vol 555, Springer, Vienna

Gendelman O, Manevitch LI, Vakakis AF, M'Closkey R (2001) Energy pumping in nonlinear mechanical oscillators: part I—dynamics of the underlying Hamiltonian systems. ASME J Appl Mech 68(1):34–41

Gere JM, Timoshenko SP (1997) Mechanics of materials. PWS Publishing, Boston

Ginsberg JH (2001) Mechanical and structural vibrations—theory and applications. Wiley, New York

Givois A, Tan J-J, Touzé C, Thomas O (2020) Backbone curves of coupled cubic oscillators in one-to-one internal resonance: bifurcation scenario, measurements and parameter identification. Meccanica 55:481–503

Gleick J (1987) Chaos: making a new science. Sphere Books, London

Goldstein G (1990) Francis Moon: Coming to terms with chaos. Interview Mech Eng 112(1): 40–47

Gottwald GA, Melbourne I (2004) A new test for chaos in deterministic systems. Proc Roy Soc Lond A 460:603–611

Grebogi C, Ott E, Yorke JA (1985) Superpersistent chaotic transients. Ergodic Theor Dyn Syst 5:341–372

Greenwood DT (2003) Advanced dynamics. Cambridge University Press, Cambridge

Gross D, Hauger W, Schröder J, Wall WA, Bonet J (2011) Engineering mechanics 2—mechanics of materials. Springer, Heidelberg

Guckenheimer J, Holmes P (1983) nonlinear oscillations, dynamical systems, and bifurcations of vector fields. Springer, New York

Guran A, Plaut RH (1993) Stability of Ziegler's pendulum with eccentric load and load-dependent stiffness. Archive Appl Mech 63:170–175

Hale JK, Kocak H (1991) Dynamics and bifurcations. Springer, New York

Hansen MH (1999) Aeroelasticity and dynamics of spinning disks. Ph.D. thesis, DCAMM Report S85, Department of Solid Mechanics, Technical University of Denmark

Hansen MH (2000) Effect of high-frequency excitation on natural frequencies of spinning disks. J Sound Vib 234(4):577–589

Harris CM (ed) (1996) Shock and vibration handbook. McGraw-Hill, New York

Hatwal H, Mallik AK, Ghosh A (1983a) Forced nonlinear oscillations of an autoparametric system —part 1: periodic responses. ASME J Appl Mech 50(3):657–662

Hatwal H, Mallik AK, Ghosh A (1983b) Forced nonlinear oscillations of an autoparametric system —part 2: chaotic responses. ASME J Appl Mech 50(3):663–668

Haxton RS, Barr ADS (1972) The autoparametric vibration absorber. ASME J Eng Ind 94(1):119– 125

Hendriks F (1983) Bounce and chaotic motion in print hammers. IBM J Res Dev 27(3):273–280

Hermansen MH, Thomsen JJ (2018) Vibration-based estimation of beam boundary parameters. J Sound Vib 429:287–304

Hill TL, Cammarano A, Neild SA, Barton DAW (2017) Identifying the significance of nonlinear normal modes. Proc Roy Society A 473(2199):20160789(1–25)

Hirsch P (1930) Das pendel mit oszillierendem aufhängepunkt. Z Angew Math Mechanik 10 (1):41–52

Hodges DH (1984) Proper definition of curvature in nonlinear beam kinematics. AIAA J 22:1825– 1827

Holman JP (1994) Experimental methods for engineers. McGraw-Hill, Singapore

Holmes PJ (1977) Bifurcations to divergence and flutter in flow-induced oscillations: a finite dimensional analysis. J Sound Vib 53(4):471–503

Holmes PJ, Moon FC (1983) Strange attractors and chaos in nonlinear mechanics. ASME J Appl Mech 50(4b):1021–1032

Hsu L (1983a) Analysis of critical and post-critical behaviour of nonlinear dynamical systems by the normal form method, part i: normalization formulae. J Sound Vib 89(2):169–181

Hsu L (1983b) Analysis of critical and post-critical behaviour of nonlinear dynamical systems by the normal form method, part ii: divergence and flutter. J Sound Vib 89(2):183–194

Hunnekens BGB, Fey RHB, Shukla A, Nijmeijer H (2011) Vibrational self-alignment of a rigid object exploiting friction. Nonlinear Dyn 65:109–129

Hunt JB (1979) Dynamic vibration absorbers. Mechanical Engineering Publications Ltd., London

Høgsberg J, Krenk S (2012) Balanced calibration of resonant shunt circuits for piezoelectric vibration control. J Intell Mater Syst Struct 23(17):1937–1948

Ibrahim RA (1992a) Friction-induced vibration, chatter, squeal, and chaos: part I—mechanics of friction. In: Friction-induced vibration, chatter, squeal, and chaos, ASME DE-vol 49, pp 107–121

Ibrahim RA (1992b) Friction-induced vibration, chatter, squeal, and chaos: part II—dynamics and modeling. In: Friction-induced vibration, Chatter, Squeal, and Chaos, ASME DE-vol 49, pp 123–138

Inman DJ (1989) Vibration, with control, measurement and stability. Prentice-Hall, London

Inman DJ (2014) Engineering vibration, 4th international edn. Pearson International Edition, Essex, England

Ivanov AP (1997) Dynamics of systems with mechanical collisions. International Program of Education, Moscow (in Russian)

Jackson EA (1991) Perspectives of nonlinear dynamics, vol 1 & 2. Cambridge University Press, Cambridge

Jang SK, Bert CW (1989) Free vibration of stepped beams: exact and numerical solutions. J Sound Vib 130(2):342–346

Jensen JS (1996) Transport of continuous material in vibrating pipes. In Proceedings of the EUROMECH 2nd European nonlinear oscillation conference, Prague, vol 1, 9–13 Sept 1996, pp 211–214

Jensen JS (1997) Fluid transport due to nonlinear fluid-structure interaction. J Fluids Struct 11:327–344

Jensen JS (1998) Non-linear dynamics of the follower-loaded double pendulum with added support-excitation. J Sound Vib 215(1):125–142

Jensen JS (1999a) Articulated pipes conveying fluid pulsating with high frequency. Nonlinear Dyn 19(2):173–193

Jensen JS (1999b) Non-trivial effects of fast harmonic excitation. Ph.D. thesis, DCAMM Report S83, Department of Solid Mechanics, Technical University of Denmark

Jensen JS (2000a) Buckling of an elastic beam with added high-frequency excitation. Int J Non-Linear Mech 35(2):217–227

Jensen JS (2000b) Effects of high-frequency bi-directional support-excitation of the follower-loaded double pendulum. In: Lavendelis E, Zakrzhevsky M (eds) Klüwer series: solid mechanics and its applications, vol 37. Klüwer, Dordrecht, pp 169–178

Jensen JS (2003) Phononic band gaps and vibrations in one- and two-dimensional mass-spring structures. J Sound Vib 266(5):1053–1078

Jensen JS, Tcherniak DM, Thomsen JJ (2000) Stiffening effects of high-frequency excitation: experiments for an axially loaded beam. ASME J Appl Mech 67(2):397–402

Jensen JS, Thomsen JJ, Tcherniak DM (2004) Non-trivial effects of high-frequency excitation for pendulum type systems. In: Blekhman II (ed) Selected topics in vibrational mechanics. World Scientific, Singapore

Jiang X, McFarland DM, Bergman LA, Vakakis AF (2003) Steady state passive nonlinear energy pumping in coupled oscillators: theoretical and experimental results. Nonlinear Dyn 33(1):87–102

Kaas-Petersen C (1989) PATH—user's guide. Department of Mathematical Studies, University of Leeds

Kalmus HP (1970) The inverted pendulum. Am J Phys 38(7):874–878

Kantz H (1994) A robust method to estimate the maximal Lyapunov exponent of a time series. Phys Lett A 185:77–87

Kapelke S, Seemann W, Hetzler H (2017) The effect of longitudinal high-frequency in-plane vibrations on a 1-dof friction oscillator with compliant contact. Nonlinear Dyn 88(4):3003–3015

Kapitaniak T (1993) Analytical method of controlling chaos in Duffing's oscillator. J Sound Vib 163(1):182–187

Kapitaniak T (1996) Controlling chaos: theoretical and practical methods in non-linear dynamics. Academic, London

Kapitza PL (1951) Dynamic stability of a pendulum with an oscillating point of suspension. Z Eksperimental'noj i Teoreticeskoj Fiziki 21(5):588–597 (in Russian)

Kapitza PL (1965) Collected papers by P. L. Kapitza, vol 2. In: Ter Haar D (ed) Pergamon Press, London, pp 714–726

Kerschen G (Ed.) (2014) Modal analysis of nonlinear mechanical systems. CISM International Centre for Mechanical Sciences, vol 555. Springer, Vienna

Kerschen G, Peeters M, Golinval J-C, Vakakis AF (2009) Nonlinear normal modes, Part I: a useful framework for the structural dynamicist. Mech Syst Signal Process 23(1):170–194

Keller HB (1986) Lectures on numerical methods in bifurcation problems. Springer, Heidelberg

Kelly SG (1993) Fundamentals of mechanical vibrations. McGraw-Hill, New York

Kelly SG (2007) Advanced vibration analysis. CRC Press Inc., Boca Raton, Florida

Khalily F, Golnaraghi MF, Heppler GR (1994) On the dynamic behaviour of a flexible beam carrying a moving mass. Nonlinear Dyn 5:493–513

Kobayashi AS (ed) (1993) Handbook on experimental mechanics. VCH Publishers, New York

Kobrinskii AE (1969) Dynamics of mechanisms with elastic connections and impact systems. Iliffe Books, London

Korenev BG, Reznikov LM (1993) dynamic vibration absorbers, theory and technical applications. Wiley, Chichester

Koughan J (1996) The collapse of the Tacoma Narrows Bridge, evaluation of competing theories of its demise, and the effects of the disaster of succeeding bridge designs. Undergraduate Engineering Review, The University of Texas at Austin, Department of Mechanical Engineering, http://www.me.utexas.edu/~uer/papers/paper_jk.html

Kounadis AN (1991) Chaos-like phenomena in the non-linear dynamic stability of discrete damped or undamped systems under step loading. Int J Non-Linear Mech 26(3/4):301–311

Kovacic I (2020) Nonlinear oscillations—exact solutions and their approximations. Springer, Berlin

Kovacic I, Brennan M (eds) (2011) The Duffing equation: nonlinear oscillators and their behaviour. Wiley

Kozol JE, Brach RM (1991) Two-dimensional vibratory impact with chaos. J Sound Vib 148 (2):319–327

Krack M, Aboulfotoh N, Twiefel J, Wallaschek J, Bergman LA, Vakakis AF (2017) Toward understanding the self-adaptive dynamics of a harmonically forced beam with a sliding mass. Arch Appl Mech 87:699–720

Krenk S, Høgsberg J (2014) Tuned mass absorber on a flexible structure. J Sound Vib 333 (6):1577–1595

Krauskopf B, Osinga HM, Galán-Vioque J (2007) Numerical continuation methods for dynamical systems—path following and boundary value problems. Springer, Dordrecht

Kremer E (2016) Slow motions in systems with fast modulated excitation. J Sound Vib 383(295–308)

Kremer E (2018) Low-frequency dynamics of systems with modulated high-frequency stochastic excitation. J Sound Vib 437(422–436)

Krishnan A (1998) Use of finite difference method in the study of stepped beams. Int J Mech Eng Educ 26(1):11–24

Krylov V, Sorokin SV (1997) Dynamics of elastic beams with controlled distributed stiffness parameters. Smart Mater Struct 6:573–582

Kuo C-C, Morino L, Dugundji J (1972) Perturbation and harmonic balance methods for nonlinear panel flutter. Am Inst Aeronaut Astronaut J 10(11):1479–1484

Lacarbonara W (1999) Direct treatment and discretizations of non-linear spatially continuous systems. J Sound Vib 221(5):849–866

Lacarbonara W, Carboni B, Quaranta G (2016) Nonlinear normal modes for damage detection. Meccanica 51:1–17

Lacarbonara W, Rega G, Nayfeh AH (2003) Resonant non-linear normal modes. Part I: analytical treatment for structural one-dimensional systems. Int J Non-Linear Mech 38:851–872

Lacarbonara W, Yabuno H (2006) Refined models of elastic beams undergoing large in-plane motions: theory and experiment. Int J Solids Struct 43(17):5066–5084

Lakrad F, Belhaq M (2002) Periodic solutions of strongly non-linear oscillators by the multiple scales method. J Sound Vib 258(4):677–700

Lancioni G, Lenci S, Galvanetto U (2016) Dynamics of windscreen wiper blades: squeal noise, reversal noise and chattering. Int J Non-Linear Mech 80:132–143

Lanczos C (1962) The variational principles of mechanics. University of Toronto Press, Toronto

Landau LD, Lifshitz EM (1976) Course of theoretical physics, vol 1. Mechanics. Pergamon Press, Oxford

Langthjem M (1995a) Finite element analysis and optimization of a fluid-conveying pipe. Mech Struct Mach 23(5):343–376

Langthjem MA (1995b) On dynamic stability of an immersed fluid-conveying tube. DCAMM Report 512, Technical University of Denmark

Langthjem MA, Sugiyama Y (1999) Vibration and stability analysis of cantilevered two-pipe systems conveying different fluids. J Fluids Struct 13(2):251–268

Langthjem MA, Sugiyama Y (2000) Dynamic stability of columns subjected to follower loads: a survey. J Sound Vib 238(5):809–851

Lazan BJ (1968) Damping of materials and members in structural mechanics. Pergamon Press, London

Lazarov B, Snaeland SO, Thomsen JJ (2010) High-frequency effects in 1D spring-mass systems with strongly non-linear inclusions. In: Leonov G, Nijmeijer H, Pogromsky A, Fradkov A (eds) Dynamics and control of hybrid dynamical systems, World Scientific, 20 pp

Lazarov B, Snaeland, SO, Thomsen JJ (2008) Using strong nonlinearity and high-frequency vibrations to control effective properties of discrete elastic waveguides. In Andrievsky BR, Fradkov AL (eds),Proceedings of the sixth EUROMECH nonlinear dynamics conference (ENOC 2008), June 30–July 4, 2008, St. Petersburg, Russia, 6 pp

Lazarov BS, Thomsen JJ (2009) Using high-frequency vibrations and non-linear inclusions to create metamaterials with adjustable effective properties. Int J Non-linear Mech 44:90–97

Lehman B, Graef JR, Sahay D (1996) Vibrational control of one- and two-species harvested population models with a delay. Autom Remote Control 57(2):177–186

Leipholz H (1977) Direct variational methods and eigenvalue problems in engineering. Publishing, Leyden, Noordhoff Int

Leissa AW (1993a) Vibration of plates. Acoustical Society of America, New York

Leissa AW (1993b) Vibration of Shells. Acoustical Society of America, New York

Lenci S, Rega G (2003) Optimal control of nonregular dynamics in a duffing oscillator. Nonlinear Dyn 33:71–86

Leven RW, Koch BP (1981) Chaotic behaviour of a parametrically excited damped pendulum. Phys Lett 86A(2):71–74

Leven RW, Pompe B, Wilke C, Koch BP (1985) Experiments on periodic and chaotic motions of a parametrically forced pendulum. Physica 16D:371–384

Levi M (1999) Geometry and physics of averaging with applications. Phys D 132:150–164

Li GX, Païdoussis MP (1994) Stability, double degeneracy and chaos in cantilevered pipes conveying fluid. Int J Non-linear Mech 29(1):83–107

Li GX, Rand RH, Moon FC (1990) Bifurcations and chaos in a forced zero-stiffness impact oscillator. Int J Non-linear Mech 25(4):417–432

Lian J-W (2005) Identifying Coulomb and viscous damping from free-vibration acceleration decrements. J Sound Vib 282(3–5):1208–1220

Liang JW, Feeny BF (1998) Identifying coulomb and viscous friction from free-vibration decrements. Nonlinear Dyn 16:337–347

Londoño JM, Neild SA, Cooper JE (2015) Identification of backbone curves of nonlinear systems from resonance decay responses. J Sound Vib 348:224–238

Lorenz H (1924) Lehrbuch der Technischen Physik. Band 1: Technische Mechanik Starrer Gebilde, Verlag von Julius Springer, Berlin

Lowenstern ER (1932) The stabilizing effect of imposed oscillations of high frequency on a dynamical system. Philos Mag 13:458–486

Lurie KA (2007) An Introduction to the mathematical theory of dynamic materials. Springer, New York

Mailybaev A, Seyranian AP (2009) Stabilization of statically unstable columns by axial vibration of arbitrary frequency. J Sound Vib 328:203–212

Makrides GA, Edelstein WS (1992) Some numerical studies of chaotic motions in tubes conveying fluid. J Sound Vib 152(3):517–530

Mandelbrot BB (1982) The fractal geometry of nature. WH Freeman, San Francisco

Marguerre K, Wölfel H (1979) mechanics of vibration. Sijthoff & Noordhoff, The Netherlands

Marsden JE, McCracken M (1976) The Hopf Bifurcation and its applications. Springer, New York

Meerkov SM (1980) Principle of vibrational control: theory and applications. IEEE Trans Autom Control AC-25(4):755–762

Meirovitch L (1967) Analytical methods in vibrations. Macmillan, New York

Meirovitch L (2001) Fundamentals of vibrations. McGraw-Hill, New York

Meriam JL, Kraige LG (1998) Engineering mechanics, vol 2. Dynamics. Wiley, New York

Michaelis MM (1985) Stroboscopic analysis of the inverted pendulum. Am J Phys 53(11):1079–1083

Michaux MA, Ferri AA, Cunefare KA (2008) Effect of waveform on the effectiveness of tangential dither forces to cancel friction-induced oscillations. J Sound Vib 311:802–823

Mickens RE (1984) Comments on the method of harmonic balance. J Sound Vib 94(3):456–460

Mikhlin YV, Avramov KV (2010) Nonlinear normal modes for vibrating mechanical systems. Review of Theoretical developments. Appl Mech Rev 63(6):060802(1–21)

Miranda EC, Thomsen JJ (1998) Vibration induced sliding: theory and experiment for a beam with a spring-loaded mass. Nonlinear Dyn 16(2):167–186

Mitropolsky YA (1965) Problems of the asymptotic theory of nonstationary vibrations. Israel Program for Scientific Translations Ltd., Jerusalem

Mitropolsky YA, Nguyen VD (1997) Applied asymptotic methods in nonlinear oscillations. Kluwer, Dordrecht

Moon FC (1980) Experiments on chaotic motions of a forced nonlinear oscillator: strange attractors. ASME J Appl Mech 47(3):638–644

Moon FC (1987) Chaotic vibrations. Wiley, New York

Moon FC, Broschart T (1991) Chaotic source of noise in machine acoustics. Arc Appl Mech 61:438–448

Moon FC, Cusumano J, Holmes PJ (1987) Evidence for homoclinic orbits as a precursor to chaos in a magnetic pendulum. Physica 24D:383–390

Moon FC, Li GX (1990) Experimental study of chaotic vibrations in a pin-jointed space truss structure. AIAA J 28(5):915–921

Moon FC, Shaw SW (1983) Chaotic vibrations of a beam with non-linear boundary conditions. Int J Non-Linear Mech 18(6):465–477

Moore DB, Shaw SW (1990) The experimental response of an impacting pendulum system. Int J Non-Linear Mech 25(1):1–16

Morton KW, Mayers DF (1994) Numerical Solution of Partial Differential equations. Cambridge University Press, Cambridge

Müller F, Krack M (2020a) Explanation of the self-adaptive dynamics of a harmonically forced beam with a sliding mass. Archive Appl Mech. (Published online 6 Mar 2020, 10.1007/s00419-020-01787-z)

Müller F, Krack M (2020b) Correction to: explanation of the self-adaptive dynamics of a harmonically forced beam with a sliding mass. Arc Appl Mech. (Published online 14 Oct 2020, 10.1007/s00419-020-01787-z)

Mullin T, Champneys A, Fraser WB, Galan J, Acheson D (2003) The 'Indian wire trick' via parametric excitation: a comparison between theory and experiment. Proc Roy Soc Lond A 459 (2031):539–546

Myneni K, Barr TA, Corron NJ, Pethel SD (1999) New method for the control of fast chaotic oscillations. Phys Rev Lett 83(11):2175–2178

Narayanan S, Jayaraman K (1991) Chaotic vibration in a non-linear oscillator with coulomb damping. J Sound Vib 146(1):17–31

Nath J, Chatterjee S (2016) Tangential acceleration feedback control of friction induced vibration. J Sound Vib 377:22–37

Nayfeh AH (1973) Perturbation methods. Wiley, New York

Nayfeh AH (1989) Application of the method of multiple scales to nonlinearly coupled oscillators. In: Hirschfelder et al. (eds) Lasers molecules and methods. Wiley, New York, pp 137–196

Nayfeh AH (1993) Method of normal forms. Wiley, New York

Nayfeh AH (1997) Nonlinear interactions. Wiley, New York

Nayfeh AH (1998) Reduced-order models of weakly nonlinear spatially continuous systems. Nonlinear Dyn 16:105–125

Nayfeh AH, Raouf RA (1987) Nonlinear forced response of infinitely long circular cylindrical shells. ASME J Appl Mech 54(3):571–577

Nayfeh AH, Balachandran B (1989) Modal interactions in dynamical and structural systems. Appl Mech Rev 42(11):175–201

Nayfeh AH, Balachandran B (1995) Applied nonlinear dynamics. Wiley, New York

Nayfeh AH, Balachandran B, Colbert MA, Nayfeh MA (1989) An experimental investigation of complicated responses of a two-degree-of freedom structure. ASME J Appl Mech 56(4):960–967

Nayfeh AH, Khdeir AA (1986) Nonlinear rolling of ships in regular seas. Int Shipbuilding Prog 33:40–49

Nayfeh AH, Mook DT (1979) Nonlinear oscillations. Wiley, New York

Nayfeh AH, Zavodney LD (1988) Experimental observation of amplitude- and phase-modulated responses of two internally coupled oscillators to a harmonic excitation. ASME J Appl Mech 55(3):706–710

Neumeyer S, Sorokin VS, van Gastel MHM, Thomsen JJ (2019) Frequency detuning effects for a parametric amplifier. J Sound Vib 445:77–87

Neumeyer S, Sorokin VS, Thomsen JJ (2017) Effects of quadratic and cubic nonlinearities on a perfectly tuned parametric amplifier. J Sound Vib 386:327–335

Newhouse S, Ruelle D, Takens F (1978) Occurrence of strange axiom-A attractors near quasiperiodic flows on Tm, $m \geq 3$. Commun Math Phys 64:35–40

Newton I (1686) Philosophiae Naturalis Principia Mathematica (English translation by A. Motte, 1729). University of California Press, Berkeley and Los Angeles, 1962

Niordson FI (1985) Shell theory. Elsevier, Amsterdam

Noël JP, Kerschen G (2017) Nonlinear system identification in structural dynamics: 10 more years of progress. Mech Syst Signal Process 83:2–35

Ott E (1993) Chaos in dynamical systems. Cambridge University Press, Cambridge

Ott E, Grebogi C, Yorke JA (1990) Controlling chaos. Phys Rev Lett 64(11):1196–1199

Païdoussis MP, Li GX, Moon FC (1989) Chaotic oscillations of the autonomous system of a constrained pipe conveying fluid. J Sound Vib 135(1):1–19

Païdoussis MP, Li GX, Rand RH (1991) Chaotic motions of a constrained pipe conveying fluid: comparison between simulation, analysis, and experiment. ASME J Appl Mech 58(2):559–565

Païdoussis MP, Semler C (1993) Nonlinear dynamics of a fluid-conveying cantilevered pipe with an intermediate spring support. J Fluids Struct 7:269–298

Panovko YG, Gubanova II (1965) Stability and oscillations of elastic systems; paradoxes, fallacies and new concepts. Consultants Bureau, New York

Parker TS, Chua LO (1989) Practical numerical algorithms for chaotic systems. Springer-, New York

Pearson CE (1974) Handbook of applied mathematics. Van Nostrand Reinhold Co., New York

Pecora LM, Caroll TL (1990) Synchronization in chaotic systems. Phys Rev Lett 64(8):821–824

Pedersen P (1986) Quantitative dynamic stability. Doctoral dissertation, Technical University of Denmark

Peeters M, Kerschen G, Golinval JC (2011) Dynamic testing of nonlinear vibrating structures using nonlinear normal modes. J Sound Vib 330(3):486–509

Peter S, Riethmüller R, Leine RI (2016) Tracking of backbone curves of nonlinear systems using phase-locked-loops. In: Kerschen G (ed) Nonlinear dynamics, vol 1. Conference proceedings of the society for experimental mechanics series. Springer, Cham

Peterson I (1990) Rock and roll bridge. Sci News 137(November):344–346

Pilipchuk VN (1988) Transformation of the vibratory-systems by means of a pair of non-smooth periodic-functions (In Ukrainian), Dopovidi Akademii Nauk Ukrainskoi Rsr Seriya A-Fiziko-Matematichni Ta Technichni Nauki 4:36–38

Pilipchuk VN (2002) Some remarks on non-smooth transformations of space and time for vibrating systems with rigid barriers. PMM J Appl Math Mech 66:31–37

Poddar B, Moon FC, Murkherjee S (1988) Chaotic motion of an elastic-plastic beam. ASME J Appl Mech 55(1):185–189

Poincaré H (1899) Les Méthods Nouvelles de la Mécanique Céleste, vol 1–3. Gauthier-Villars, Paris (in French)

Polyanin AD, Zaitsev VF (2003) Handbook of exact solutions for ordinary differential equations, 2nd edn. Chapman & Hall/CRC, Boca Raton, Florida

Popp K (1992) Some model problems showing stick-slip motion and chaos. In: Friction-induced vibration, chatter, squeal, and chaos, ASME DE-vol 49, pp 1–12

Press WH, Teukolsky SA, Vetterling WT, Flannery BP (2002) Numerical recipes. Cambridge University Press, Cambridge

Quinn D, Rand R, Bridge J (1995) The dynamics of resonant capture. Nonlinear Dyn 8(1):1–20

Rakaric Z, Kovacic I (2016) Mechanical manifestations of bursting oscillations in slowly rotating systems. Mech Syst Signal Process 81:35–42

Rayleigh JWS (1877) The theory of sound, vol 1–2, 2nd edn, 1945 re-issue. Dover Publications, New York

Rebouças GFS, Santos IF, Thomsen JJ (2017) Validation of vibro-impact force models by numerical simulation, perturbation methods and experiments. J Sound Vib 413:291–307

Rebouças GFS, Santos IF, Thomsen JJ (2019) Unilateral vibro-impact systems—experimental observations against theoretical predictions based on the coefficient of restitution. J Sound Vib 440:346–371

Rega G, Settimi V, Lenci S (2020) Chaos in one-dimensional structural mechanics. Nonlinear Dyn 102:785–834

Renson L, Gonzalez-Buelga A, Barton DAW, Neild SA (2016) Robust identification of backbone curves using control-based continuation. J Sound Vib 367:145–158

Renson L, Kerschen G, Cochelin B (2016) Numerical computation of nonlinear normal modes in mechanical engineering. J Sound Vib 364:177–206

Renson L, Shaw AD, Barton DAW, Neild SA (2019) Application of control-based continuation to a nonlinear structure with harmonically coupled modes. Mech Syst Signal Process 120:449–464

Rhoads JF, Jeffrey F, Miller NJ, Nicholas J, Shaw SW, Feeny BF (2008) Mechanical domain parametric amplification. ASME J Vib Acoust 130(6):061006(1–7)

Rhoads JF, Shaw SW (2010) The impact of nonlinearity on degenerate parametric amplifiers. Appl Phys Lett 96(23):234101(1–3)

Romeiras FJ, Bondeson A, Ott E, Antonsen TM, Grebogi C (1987) Quasiperiodically forced dynamical systems with strange nonchaotic attractors. Physica 26D:277–294

Rosenberg RM (1960) Normal modes of nonlinear dual-mode systems. ASME J Appl Mech 27 (2):263–268

Rosenstein MT, Collins JJ, De Luca CJ (1993) A practical method for calculating largest Lyapunov exponents from small data sets. Physica D 65:117–134

Ross SS (1984) Construction disasters. McGraw-Hill, New York

Rossikhin YA, Shitikova MV (1995) Analysis of nonlinear free vibrations of suspension bridges. J Sound Vib 186(3):369–393

Rudowski J, Szemplinska-Stupnicka W (1987) On an approximate criterion for chaotic motion in a model of a buckled beam. Ing Archiv 57:243–255

Sah SM, McGehee CC, Mann BP (2013) Dynamics of a horizontal pendulum driven by high-frequency rocking. J Sound Vib 332:6505–6518

Sah SM, Thomsen JJ, Brøns M, Fidlin A, Tcherniak D (2018) Estimating bolt tightness using transverse natural frequencies. J Sound Vib 431:137–149

Sanders JA, Verhulst F (1985) Averaging methods in nonlinear dynamical systems. Appl Math Sci 59 (Springer, New York)

Sayag MR, Dowell EH (2016) Linear versus nonlinear response of a cantilevered beam under harmonic base excitation: theory and experiment. ASME J Appl Mech 83(10):101002(1–8)

Schmidt G, Tondl A (2009) Non-Linear vibrations. Cambridge University Press, Cambridge

Schmitt JM, Bayly PV (1998) Bifurcations in the mean angle of a horizontally shaken pendulum: analysis and experiment. Nonlinear Dynamics 15:1–14

Schuster HG (1989) Deterministic chaos. VCH Verlag, Weinheim

Sethna PR (1967) An extension of the method of averaging. Q Appl Math 25:205–211

Seydel R (2010) Practical bifurcation and stability analysis. Springer, New York

Shabana AA (1996) Theory of vibration I: an introduction. Springer, New York

Shabana AA (1997) Theory of vibration II: discrete and continuous systems. Springer, New York

Shapiro B, Zinn B (1997) High-frequency nonlinear vibrational control. IEEE Trans Autom Control 42(1):83–90

Shaw SW and Holmes PJ (1983) A periodically forced piecewise linear oscillator. J Sound Vib 90 (1):129–155

Shaw SW, Pierre C (1993) Normal modes for non-linear vibratory systems. J Sound Vib 164 (1):85–124

Sheheitli H, Rand RH (2011) Dynamics of three coupled limit cycle oscillators with vastly different frequencies. Nonlinear Dyn 64:131–145

Sheheitli H, Rand RH (2012) Dynamics of a mass-spring-pendulum system with vastly different frequencies. Nonlinear Dyn 70(1):25–41

Shin Y-H, Choi J, Kim SJ, Kim S, Maurya D, Sung T-H, Priya S, Kang C-Y, Song H-C (2020) Automatic resonance tuning mechanism for ultra-wide bandwidth mechanical energy harvesting. Nano Energy 77:104986

Shinbrot T, Ott E, Grebogi C, Yorke JA (1990) Using chaos to direct trajectories to targets. Phys Rev Lett 65(26):3215–3219

Shishkina EV, Blekhman II, Cartmell MP, Gavrilov SN (2008) Application of the method of direct separation of motions to the parametric stabilization of an elastic wire. Nonlinear Dyn 54:313–331

Siddiqui SAQ, Golnaraghi MF, Heppler GR (2003) Large free vibrations of a beam carrying a moving mass. Int J Non-Linear Mech 38:1481–1493

Sieber J, Krauskopf B (2008) Control based bifurcation analysis for experiments. Nonlinear Dyn 51(3):365–377

Sieber J, Krauskopf B, Wagg D, Neild S, Gonzalez-Buelga A (2011) Control-based continuation of unstable periodic orbits. J Comput Nonlinear Dyn 6(1):011005(1–9)

Skiadas CH (ed) (2016) The foundations of chaos revisited: from poincaré to recent advancements. Springer, Switzerland

Skiadas CH, Skiadas C (eds) (2016) Handbook of applications of chaos theory. Chapman and Hall/CRC

Smale S (1967) Differential dynamical systems. Bull Am Math Soc 747–817

Sorokin SV, Terentév AV (1998) On Modal interaction, stability and nonlinear dynamics of a model two d.o.f. mechanical system performing snap-through motion. Nonlinear Dyn 16 (3):239–257

Sorokin VS, Blekhman II, Thomsen JJ (2010) Motions of elastic solids in fluids under vibration. Nonlinear Dyn 60:639–650

Sorokin VS, Blekhman II (2020) Vibration overcomes gravity on a levitating fluid. Nature 585 (7823):31–32

Sorokin VS, Thomsen JJ (2015a) Eigenfrequencies and eigenmodes of a beam with periodically continuously varying spatial properties. J Sound Vib 347:14–26

Sorokin VS, Thomsen JJ (2015b) Vibration suppression for strings with distributed loading using spatial cross-section modulation. J Sound Vib 335:66–77

Sorokin VS, Thomsen JJ (2016) Effects of weak nonlinearity on the dispersion relation and frequency band-gaps of a periodic Bernoulli-Euler beam. Proc Roy Soc A 472 (2186):20150751(1–22)

Stephenson A (1908a) On a new type of dynamic stability. Mem Proc Manchester Literary Philos Soc 52(8):1–10

Stephenson A (1908b) On induced stability. Philos Mag 15:233–236

Stephenson A (1909) On induced stability. Philos Mag 17:765–766

Stoker JJ (1950) Nonlinear vibrations. Interscience, New York

Sudor DJ, Bishop SR (1999) Inverted dynamics of a tilted pendulum. Eur J Mech A/Solids 18:517–526

Sugiyama Y, Langthjem MA (2007) Physical mechanism of the destabilizing effect of damping in continuous non-conservative dissipative systems. Non-Linear Mech 42(1):132–145

Sugiyama Y, Langthjem MA, Katayama K (2019) Dynamic stability of columns under nonconservative forces—theory and experiment. Springer Nature, Cham, Switzerland

Sugiyama Y, Langthjem MA, Ryo B-J (1999) Realistic follower forces. J Sound Vib 225(4):770–782

Sung CK, Yu WS (1992) Dynamics of a harmonically excited impact damper: bifurcations and chaotic motion. J Sound Vib 158(2):317–329

Szemplinska-Stupnicka W (1992) A discussion on necessary and sufficient conditions for steady state chaos. J Sound Vib 152(2):369–372

Szemplinska-Stupnicka W, Plaut RH, Hsieh J-C (1989) Period doubling and chaos in unsymmetric structures under parametric excitation. ASME J Appl Mech 56(4):947–952

Szemplinska-Stupnicka W, Rudowski J (1992) Local methods in predicting the occurrence of chaos in two-well potential systems: superharmonic frequency region. J Sound Vib 152(1):57–72

Tabarrok B, Rimrott FPJ (1994) Variational methods and complementary formulations in dynamics. Kluwer, Dordrecht

Tang DM, Dowell EH (1988) On the threshold force for chaotic motions for a forced buckled beam. ASME J Appl Mech 55(1):190–196

Tcherniak D, Thomsen JJ (1998) Slow effects of fast harmonic excitation for elastic structures. Nonlinear Dyn 17(3):227–246

Tcherniak DM (1999) The influence of fast excitation on a continuous system. J Sound Vib 227 (2):343–360

Tcherniak DM (2000) Using fast vibration to change the nonlinear properties of mechanical systems. In: Lavendelis E, Zakrzhevsky M (eds) Klüwer series: solid Mechanics and its Applications, vol 37. Klüwer, Dordrecht, pp 227–236

Temple G, Bickley WG (1933) Rayleigh's principle and its applications to engineering. Oxford University Press, London

Thomas O, Mathieu F, Mansfield W, Huang C, Trolier-McKinstry S, Nicu L (2013) Efficient parametric amplification in micro-resonators with integrated piezoelectric actuation and sensing capabilities. Appl Phys Lett 102(16):163504(1–5)

Thompson JMT, Bishop SR (eds) (1994) Nonlinearity and chaos in engineering dynamics. Wiley, Chichester

Thompson JMT, Stewart HB (1986) Nonlinear dynamics and chaos; Geometrical methods for engineers and scientists. Wiley, Chichester

Thomsen JJ (1992) Chaotic vibrations of non-shallow arches. J Sound Vib 153(2):239–258

Thomsen JJ (1995) Chaotic dynamics of the partially follower-loaded elastic double pendulum. J Sound Vib 188(3):385–405

Thomsen JJ (1996a) Vibration suppression by using self-arranging mass: effects of adding restoring force. J Sound Vib 197(4):403–425

Thomsen JJ (1996b) Vibration induced sliding of mass: non-trivial effects of rotatory inertia. In: Proceedings of the EUROMECH 2nd European nonlinear oscillation conference, vol 1, Prague, September 9–13, 1996, pp 455–458

Thomsen JJ (1999) Using fast vibrations to quench friction-induced oscillations. J Sound Vib 228 (5):1079–1102

Thomsen JJ (2000) Vibration-induced displacement using high-frequency resonators and friction layers. In: Lavendelis E, Zakrzhevsky M (eds) Klüwer series: solid Mechanics and its Applications, vol 37. Klüwer, Dordrecht, pp 237–246

Thomsen JJ (2002) Some general effects of strong high-frequency excitation: stiffening, biasing, and smoothening. J Sound Vib 253(4):807–831

Thomsen JJ (2003a) Dynamic effects of nonlinearity and fast vibrations: stiffening, biasing, smoothening, chaos. Doctoral dissertation, Technical University of Denmark

Thomsen JJ (2003b) Theories and experiments on the stiffening effect of high-frequency excitation for continuous elastic systems. J Sound Vib 260(1):117–139

Thomsen JJ (2005) Slow high-frequency effects in mechanics: problems, solutions, potentials. Int J Bifurcat Chaos 15:2799–2818

Thomsen JJ (2008a) Effective properties of mechanical systems under high-frequency excitation at multiple frequencies. J Sound Vib 311:1249–1270

Thomsen JJ (2008b) Using strong nonlinearity and high-frequency vibrations to control effective mechanical stiffness. In: Andrievsky BR, Fradkov AL (eds) Proceedings of the sixth EUROMECH nonlinear dynamics conference (ENOC 2008), June 30–July 4, 2008, St. Petersburg, Russia, 9 pp

Thomsen JJ, Blekhman II (2007) Using nonlinearity and spatiotemporal property modulation to control effective structural properties: dynamic rods. In: M. Papadrakakis et al. (eds) Proceedings of COMPDYN2007 (ECCOMAS thematic conference on computational methods in structural dynamics and earthquake engineering), 13–16 June 2007, National Technical University of Athens, Greece, 12 pp

Thomsen JJ, Fidlin A (2003) Analytical approximations for stick-slip vibration amplitudes. Int J Non-Linear Mech 38(3):389–403

Thomsen JJ, Fidlin A (2008) Near-elastic vibro-impact analysis by discontinuous transformations and averaging. J Sound Vib 311(1–2):386–407

Thomsen JJ, Dahl J (2010) Analytical predictions for vibration phase shifts along fluid-conveying pipes due to Coriolis forces and imperfections. J Sound Vib 329(15):3065–3081

Thomsen JJ, Fuglede N (2020) Perturbation-based prediction of vibration phase shift along fluid-conveying pipes due to Coriolis forces, nonuniformity, and nonlinearity. Nonlinear Dyn 99(1):173–199

Thomsen JJ, Tcherniak DM (2001) Chelomei's pendulum explained. Proc Roy Soc London A 457 (2012):1889–1913

Thomson WT, Dahleh MD (1998) Theory of vibrations with applications. Prentice-Hall, Upper Saddle River, NJ

Timoshenko S, Young DH, Weaver W (1974) vibration problems in engineering. Wiley, New York

Tondl A (1991) Quenching of self-excited vibrations. Elsevier, Amsterdam

Tondl A, Nabergoj R, Ruijgrok M, Verhulst F (2000) Autoparametric resonance in mechanical systems. Cambridge University Press, Cambridge

Troger H, Steindl A (1991) Nonlinear stability and bifurcation theory. Springer, New York

Tsonis AA (1992) Chaos—from theory to applications. Plenum, New York

Tufillaro NB, Ramshankar R, Gollub JP (1989) Order-disorder transition in capillary ripples. Phys Rev Lett 62(4):422–425

Tufillaro NB, Abott T, Reilly J (1992) An experimental approach to nonlinear dynamics and chaos. Addison-Wesley, Redwood City, California

Turkstra TP, Semercigil SE (1993) Elimination of resonance with a switching tensile support. J Sound Vib 163(2):359–362

Ueha S, Tomikawa Y (1993) Ultrasonic motors: theory and applications. Clarendon Press, Oxford

Vakakis AF (1997) Non-linear normal modes (NNMs) and their applications in vibration theory: an overview. Mech Syst Signal Process 11(1):3–22

Vakakis AF, Gendelman O (2001) Energy pumping in nonlinear mechanical oscillators: Part II—resonance capture. Asme J Appl Mech 68(1):42–48

van der Pol B, Strutt MJO (1928) On the stability of the solutions to Mathieu's equation. Philos Mag 5:18–38

Verhulst F (1996a) Nonlinear differential equations and dynamical systems. Springer, Berlin

Verhulst F (1996b) Autoparametric resonance, survey and new results. In: Proceedings of the EUROMECH 2nd European Nonlinear Oscillation Conference, Prague, September 9–13, 1996, vol 1, pp 483–488

Veskos P, Demiris Y (2006) Experimental comparison of the van der Pol and Rayleigh nonlinear oscillators for a robotic swinging task. In: Proceedings of the AISB 2006 conference, adaptation in artificial and biological systems, Bristol, pp 197–202

Weast RC (ed) (1981) Handbook of chemistry and physics. CRC Press, Boca Raton

Weibel S, Kaper TJ, Baillieul J (1997) Global dynamics of a rapidly forced cart and pendulum. Nonlinear Dyn 13:131–170

Wiercigroch M (2000) Modelling of dynamical systems with motion dependent discontinuities. Chaos, Solitons Fractals 11:2429–2442

Wolf A, Swift JB, Swinney HL, Vastano JA (1985) Determining Lyapunov exponents from a time series. Physica 16D:285–317

Woodhouse J (1998) Linear damping models for structural vibration. J Sound Vib 215(3):547–569

Wu B, Zhou Y, Lim CW, Sun W (2018) Analytical approximations to resonance response of harmonically forced strongly odd nonlinear oscillators. Arch Appl Mech 88:2123–2134

Yabuno H, Tsumoto K (2007) Experimental investigation of a buckled beam under high-frequency excitation. Arch Appl Mech 77:339–351

Yagasaki K, Sakata M, Kimura K (1990) Dynamics of a weakly nonlinear system subjected to combined parametric and external excitation. ASME J Appl Mech 57(1):209–217

Yim SCS, Lin H (1991) Nonlinear impact and chaotic response of slender rocking objects. J Eng Mech 117(9):2079–2100

Zak M (1984) Elastic continua in high frequency excitation field. Int J Non-Linear Mech 19 (5):479–487

Zames G, Shneydor NA (1976) Dither in nonlinear systems. IEEE Trans Autom Control 21 (5):660–667

Zhou Y, Wu B, Lim CW, Sun W (2020) Analytical approximations to primary resonance response of harmonically forced oscillators with strongly general nonlinearity. Appl Math Model 87:534–545

Zhuravlev VF (1976) A method for analyzing vibration-impact systems by means of special functions. Mechanics of Solids (English translation of Izv. AN SSSR. Mekhanika Tverdogo Tela) 11:23–27

Zhuravlev VF, Klimov DM (1988) Applied methods in the theory of nonlinear oscillations. Nauka, Moscow (in Russian)

Ziegler H (1968) Principles of structural stability. Blaisdell, Waltham, Massachusetts

Zienkiewicz OC (1982) The finite element method. McGraw-Hill, London

Index

Printed in the United States
by Baker & Taylor Publisher Services